FLORA OF TROPICAL EAST AFRICA

COMPOSITAE (Part 3)

H. BEENTJE, C. JEFFREY & D.J.N. HIND*

SENECIONEAE**

Cass. in J. Phys. Chim. Hist. Nat. 88: 196 (1819); C. Jeffrey in K.B. 41: 873–943
(1986); K.Bremer, Asteraceae Clad. & Class.: 479–520 (1994)

Herbs, shrubs or trees; sometimes dioecious (not in ours), sometimes succulent. Leaves alternate, sometimes rosulate, rarely opposite (not in ours). Capitula solitary, cymose or paniculate, radiate, disciform or discoid. Involucre of phyllaries in one row, with or without a outer row of small bracts (calyculus), occasionally in two (in ours only in *Dendrosenecio*) or more rows, sometimes connate; receptacle epaleate. Ray florets female, fertile; outer florets in disciform heads female; disc florets usually perfect and 5-lobed. Anthers ecalcarate, with or without tails, the apical appendage flat, ovate to oblong; style branches oblong, apically usually truncate and penicillate, sometimes with an appendage of fused hairs. Achenes with pappus of bristles, rarely absent or reduced.

120 genera and ± 3000 species, worldwide.

1. Petiole not prehensile; anther bases without
   sterile tailed auricles 2
   Climbers with the petiole prehensile and
   thickened at its base, often remaining after fall
   of leaves; anther bases with sterile tailed auricles 11
2. Involucre calyculate with at least one, but usually
   more than one, smaller outer bract 3
   Involucre ecalyculate, consisting of a single row
   completely devoid of outer bracts 7
3. Trees or shrubs; if woody herbs 0.6–2 m high
   then with terminal rosettes of leaves over 28 ×
   8 cm; phyllaries 11–16 mm long; rays 6–20,
   9–29 × 2–5 mm . . . . **88. Dendrosenecio** (p. 548)
   Leaves smaller or (in a few species of
   Crassocephalum and *Senecio*) without terminal
   rosettes, with fewer ray florets or smaller
   phyllaries 4

* Senecioneae by C. Jeffrey & H. Beentje, with *Mikaniopsis & Austrosynotis* by I. Malombe and *Dendrosenecio* by E.B. Knox
  Heliantheae (p. 702) by H. Beentje & D.J.N. Hind, with *Bidens* by Mesfin Tadesse
  Eupatorieae (p. 819) by S.A.L. Smith

HB would like to thank Charles Jeffrey and Quentin Luke for their continuing interest and their careful checking of all the text; and Charlie Jarvis, as always, for his help with Linnean types.

** by C. Jeffrey & H.J. Beentje, c/o Royal Botanic Gardens, Kew

NOTE: the differences between the last three genera are technical rather than immediately obvious, and especially the character of the last key lead is difficult to observe.

Roldana petasitis (Sims) H.Rob. & Brettell has been cultivated in Nairobi gardens. It is a shrubby herb with large lobed leaves to 20 cm wide, purple petioles, and large inflorescences of discoid yellow capitula. *Gillett* 21028!, 21757!

88. DENDROSENECIO*

Hauman in Rev. Zool. Bot. Afr. 28: 31 (1935); A.V.P.: 226 (1957); Mabb. in K.B. 28: 80 (1973); B. Nord. in Opera Bot. 44: 40 (1978); C. Jeffrey in K.B. 41: 887 (1986); E. B. Knox in Contr. Univ. Michigan Herb. 19: 243 (1993)

Perennial giant-rosette plants to 10 m tall with woody trunks; typically upright and polycarpic, with branching initiated below each terminal inflorescence, sometimes monocarpic or creeping; primary stem 2–12 cm wide, comprised mainly of pith, which provides internal water capacitance; secondary growth produces abundant white wood and a grey, furrowed bark. Seedlings and young plants with leaves

* by Eric B. Knox

commonly constricted toward the base, forming a pseudo-petiole, particularly when growing in partial shade. Developing leaves of mature plants revolutely folded, the upper lamina glabrescent; commonly bathed in mucilage, an ice-nucleating pectinaceous material that in some species is secreted among the leaf bases and may collect in the leaf-rosette. Leaves alternate (although appearing whorled in densely packed leaf-rosettes), to 120 cm long and 35 cm wide, simple, in rosettes of 10–100, with developing leaves often reaching almost mature length in a massive 'apical bud' before unfurling to join the leaf-rosette; leaf base wide, sheathing, often with a dense cushion of white trichomes on the upper surface; leaf margin occasionally entire, but usually serrate or dentate, tipped with hydathodes; indumentum of lower lamina comprising multicellular, uniseriate trichomes, variable among species, but generally villose along the lower midvein. Inflorescence to 2.5 m tall, erect, pyramidal-paniculate or cylindrical, densely pubescent to glabrescent; bracteate, with a range of transitional shapes from the mature vegetative leaves subtending the peduncle to the small bracts subtending each capitulum; capitula numerous, with biseriate involucral bracts; outer bracts 5–23, filiform, 5–18 mm long; inner bracts 10–25, narrow to broadly ovate, 11–16 × 2–6 mm, with apical papillae; receptacle nude. Ray florets absent or 6–20, yellow, female, with staminodes sometimes present, ray 2–23 × 2–5 mm; tube 4–8 mm long. Disc florets 30–380, yellow, hermaphrodite, 8–10 mm long; style with a continuous stigmatic surface; anthers non-auriculate, filament collars balusteriform. Achenes glabrous, 8–10-ribbed; pappus of many bristles.

Eleven species in tropical Africa.

NOTE. All herbarium specimens are inherently fragmentary because of the size of these plants, and the exact nature of the collected material is rarely indicated in the collection notes. Most fertile specimens include bracts of various shapes and sizes, not mature vegetative leaves. Sterile specimens generally comprise vegetative leaves, but these are frequently developing leaves or leaves from young plants, and only rarely include the leaf base because of the effort and damage done to the plant in order to remove a complete leaf from a densely packed leaf-rosette. The giant senecios also present a mosaic of morphological variation due to divergence as they spread from mountain to mountain and convergence as they occupied similar habitats on different mountains. The subspecies recognized below were all previously described as distinct species, and in all cases are inferred to represent incipient speciation. Although compelling arguments can be made to reinstate these taxa at specific rank, they are maintained as subspecies because in each case populations can be found with mixed or intergrading morphological features, suggesting that limited interbreeding still occurs. As a result, some specimens do not fall neatly into one or another subspecies, and these uncommon individuals are, of course, over-represented in herbaria.

1. Plants from Tanzania (**T** 2) 2
   Plants from Kenya or Uganda 4
2. Plants from Kilimanjaro 3
   Plants from Mt Meru . . . .   3. *D. meruensis*
3. Primary stem > 4 cm wide; leaf-rosette densely packed;
   foliage marcescent; capitula with > 75 florets . . . . .   1. *D. kilimanjari*
   Primary stem < 4 cm wide; leaf-rosette lax because of
   stem growth between leaf nodes; foliage deciduous;
   capitula with < 75 florets . . . .   2. *D. johnstonii*
4. Plants from Kenya (**K** 3, 4, 5) or eastern Uganda (**U** 3) 5
   Plants from western Uganda (**U** 2) 10
5. Plants from Mt Kenya or Nyandarua/Aberdares 6
   Plants from the Cherangani Hills or Mt Elgon 9
6. Plants growing on drained soils; 3–8 m tall in flower;
   with prominent, aerial branches produced after
   flowering 7
   Plants growing on saturated soils; 1.5–5 m tall in
   flower; monocarpic or with branches produced
   near the soil surface after flowering 8

7. Ray florets prominent; upper leaf-base with a well-
  developed hair cushion ....   4. *D. battiscombei*
  Ray florets absent; upper leaf-base with long, yellowish
  trichomes not forming a well-developed hair cushion   7. *D. keniodendron*
8. Plants monocarpic; 1.5–5 m tall in flower; secreting
  copious quantities of mucilaginous fluid that
  collects in the vegetative leaf-rosette; lower lamina
  surface with sparse to dense, greyish indumentum;
  inflorescence paniculate; endemic to Nyandarua/
  Aberdares ....   5. *D. brassiciformis*
  Plants polycarpic; to 1.5 m tall in flower; lateral
  branches produced near the soil surface and
  capable of adventitious rooting; mucilaginous fluid
  present among the leaf bases, but not accumulating
  in an open leaf-rosette; lower lamina surface with
  dense, cream-coloured indumentum that ends
  abruptly toward thebase; inflorescence cylindrical;
  endemic to Mt Kenya ....   6. *D. keniensis*
9. Plants from the Cherangani Hills ....   8. *D. cheranganiensis*
  Plants from Mt Elgon ....   9. *D. elgonensis*
10. Plants from the Ruwenzori Mts 11
  Plants from the Virunga Mts ....   11. *D. erici-rosenii*
11. Primary stem > 4 cm wide; foliage marcescent, or with
  retained leaf bases; ray florets absent or if present,
  then shorter or slightly longer than the bracts of
  the capitulum ....   10. *D. adnivalis*
  Primary stem < 4 cm wide; leaf bases retained, but
  lamina decomposing after death of the leaf; ray
  florets prominent ....   11. *D. erici-rosenii*

1. **Dendrosenecio kilimanjari** (*Mildbr.*) *E.B.Knox* in Contr. Univ. Michigan Herb.
19: 244 (1993). Type: Tanzania, Kilimanjaro, above Moshi, *Uhlig* 72 (B†, syn., K!,
lecto., chosen by C. Jeffrey)

Upright, polycarpic plant to 10 m tall, with trunk to 40 cm in diameter; pith 4–5 cm
in diameter; stem bears densely packed leaf-rosettes of 25–80 leaves, without
internode elongation; cloaked with marcescent foliage. Infrequent reproduction
yields sparsely branched, columnar plants that rarely exceed three reproductive
cycles. Leaf lamina lanceolate or cordate, to 65 cm long and 22 cm wide, with lower
portion of leaf constricted to form a pseudo-petiole; hair cushion present on upper
leaf-base, lanate along upper midvein, lower surface glabrous or tomentose, villose
along lower midvein. Inflorescence broadly paniculate to 180 cm tall, 80 cm wide;
capitula pendulous. Ray florets prominent or inconspicuous to absent; disc florets
100–380.

Ray florets prominent; leaf lamina extending along the
  midvein toward the base, not forming a pseudo-petiole;
  lower lamina surface not densely tomentose, though
  villose along the midvein........................   a. subsp. *kilimanjari*
Ray florets absent or if present, then not significantly longer
  than the bracts of the capitulum; leaf lamina
  constricted toward the base, forming a pseudo-petiole;
  lower lamina surface densely tomentose ....   b. subsp. *cottonii*

a. subsp. **kilimanjari**

Plant to 7 m tall, with trunk to 30 cm in diameter. Leaf-rosettes of 25–45 leaves. Leaf lamina lanceolate, to 69 cm long and 27 cm wide, constricted toward the base, but lamina generally extending along the midvein and flared at the base; lower surface glabrous or sparsely pubescent. Inflorescence to 150 cm tall, 80 cm in diameter. Ray florets 12–18, to 29 mm long; disc florets 100–200.

TANZANIA. Kilimanjaro, S slope between Umbwe and Weru Weru Rivers, Aug. 1932, *Greenway* 3151! & Between Mandara [Bismark] Hut and Horombo [Peter's] Hut, Feb. 1934, *Greenway* 3793! & Above Marangu, June 1948, *Hedberg* 1388!
DISTR. **T** 2; endemic to Kilimanjaro
HAB. Moist sites in ericaceous moorland; able to withstand low-intensity fire; most abundant on the wet south flank of the mountain where the fire-dominated moorland meets the evergreen montane forest; 3000–3800 m
USES. None recorded on specimens from our area
CONSERVATION NOTES. Protected in the National Park; least concern (LC)

SYN. *Senecio johnstonii* Oliv. in Trans. Linn. Soc. Bot. ser. 2, 2: 340, t. 60 (1887) pro parte et quoad figs. 1, 3, *non* Oliv. sensu stricto
    S. kilimanjari Mildbr. in F.R. 18: 229 (1922); Cotton in Hook. Ic. Pl. 33: t. 3289 (1935); T.T.C.L.: 158 (1949); A.V.P.: 227 (1957)
    S. johnstonii Oliv. subsp. *johnstonii*: Mabb. in K.B. 28: 87 (1973) pro parte
    Dendrosenecio johnstonii (Oliv.) B.Nord. subsp. *johnstonii*: B. Nord. in Opera Bot. 44: 42 (1978) pro parte
    Senecio johnstonii Oliv. subsp. *johnstonii* var. *kilimanjari* (Mildbr.) C.Jeffrey in K.B. 41: 888 (1986)

b. subsp. **cottonii** (*Hutch. & G.Taylor*) *E.B.Knox* in Contr. Univ. Michigan Herb. 19: 244 (1993). Type: Tanzania, Kilimanjaro, *Cotton & Hitchcock* 34 (K!, holo.)

Plant to 10 m tall, with trunk to 40 cm in diameter. Leaf-rosettes of 40–80 leaves. Leaf lamina cordate, to 46 cm long and 22 cm wide, lower portion of leaf constricted to form a pseudo-petiole, broadly flared at the base, to 19 cm long; leaf tomentose on lower surface. Inflorescence to 180 cm tall, 80 cm in diameter. Ray florets absent or 12–16; when present, to 17 mm long; disc florets 150–380.

TANZANIA. Kilimanjaro, N of Horombo [Peter's] Hut, Feb. 1934, *Greenway* 3772! & Shira Plateau, Nov. 1948, *Salt* 21! & Trail from Barranco Hut to Shira Plateau, Jan. 1989, *Knox* 827!
DISTR. **T** 2; endemic to Kilimanjaro
HAB. Moist sites in the afro-alpine zone, particularly along streams and larger ravines; most abundant on the wet, south flank below Kibo Peak; 3600–4275 m
USES. None recorded on specimens from our area
CONSERVATION NOTES. Protected in the National Park; least concern (LC)

SYN. *Senecio cottonii* Hutch. & G.Taylor in K.B. 1930: 15 (1930); Cotton in Hook. Ic. Pl. 33: t. 3290 (1935); T.T.C.L.: 158 (1949); A.V.P.: 227 (1957)
    S. johnstonii Oliv. subsp. *cottonii* (Hutch. & G.Taylor) Mabb. in K.B. 28: 87 (1973)
    Dendrosenecio johnstonii (Oliv.) B.Nord. subsp. *cottonii* (Hutch. & G.Taylor) B.Nord. in Opera Bot. 44: 43 (1978)
    Senecio johnstonii Oliv. subsp. *johnstonii* var. *cottonii* (Hutch. & G.Taylor) C.Jeffrey in K.B. 41: 889 (1986)

2. **Dendrosenecio johnstonii** (*Oliv.*) *B.Nord.* in Opera Bot. 44: 42 (1978) pro parte. Type: Tanzania, Kilimanjaro, *Johnston* 15 (K!, lecto., BM!, isolecto., chosen by C. Jeffrey)

Upright, polycarpic plant to 10 m tall, with trunk 40 cm or more in diameter; pith 1–2 cm in diameter; stem with up to 1 cm between successive leaf nodes; reproducing and branching repeatedly to form a broad, dense canopy with 50–80 aerial meristems in mature trees. Leaf lamina cordate, to 53 cm long and 40 cm wide, lower portion of leaf constricted to form a sheathing pseudo-petiole to 27 cm long; leaf

Fig. 116. *DENDROSENECIO JOHNSTONII* — **1**, habit; **2**, leaf; **3**, inflorescence; **4**, phyllary; **5**, ray floret; **6**, disc floret; **7**, pappus bristle; **8**, stamens; **9**, style branches. Drawn by M. Smith. From Transactions of the Linnean Society ser. 2, Botany, vol. 2, plate 60 (1887).

generally glabrous except sparsely pubescent along lower venation. Inflorescence broadly pyramidal to 60 cm tall, 40 cm in diameter; capitula presented horizontally. Ray florets 11–15; to 25 mm long; disc florets 30–50. Fig. 116 (page 552).

TANZANIA. Kilimanjaro, above Marangu, Oct. 1893, *Volkens* 1131! & S slope between Umbwe and Weru Weru Rivers, Aug. 1932, *Greenway* 3150! & SE side, Jan. 1934, *Schlieben* 4575!
DISTR. **T** 2; endemic to Kilimanjaro
HAB. In the upper afro-montane evergreen forest and forest patches protected from fire in the lower ericaceous moorland; the largest known plants occur along the Mweka Route; 2750–3350 m
USES. None recorded on specimens from our area
CONSERVATION NOTES. Protected in the National Park; least concern (LC)

SYN. *Senecio johnstonii* Oliv. in Trans. Linn. Soc. Bot. ser. 2, 2: 340, t. 60, fig. 2 (1887); Cotton in Hook. Ic. Pl. 33: t. 3288 (1935); T.T.C.L.: 158 (1949)
    S. johnstonii Oliv. subsp. *johnstonii*: Mabb. in K.B. 28: 87 (1973) pro parte, autonym
    Dendrosenecio johnstonii (Oliv.) B. Nord. subsp. *johnstonii*: B. Nord. in Opera Bot. 44: 42 (1978) pro parte, autonym
    Senecio johnstonii Oliv. subsp. *johnstonii* var. *johnstonii*: C. Jeffrey in K.B. 41: 888 (1986), autonym

NOTE. The entry under this name in U.K.W.F. ed. 2: 223 (1994) refers to several taxa of *Dendrosenecio*.

3. **Dendrosenecio meruensis** (*Cotton & Blakelock*) *E.B.Knox* in Contr. Univ. Michigan Herb. 19: 245 (1993). Type: Tanzania, Mt Meru, *B.D. Burtt* 4050 (K!, holo., BM!, BR!, EA!, FT!, G!, P!, S!, iso.)

Upright, polycarpic plant to 7 m tall, with trunk to 35 cm in diameter; pith to 2 cm in diameter; stem with lax leaf-rosettes of 15–25 leaves, with little internode elongation, covered with retained leaf-bases for 1–2 m below the leaf-rosettes. Repeated reproduction and branching yields a sprawling, open canopy with up to 10–30 aerial meristems in mature trees. Leaf lamina elliptic, to 113 cm long and 34 cm wide, hair cushion weakly to strongly developed on upper leaf-base, lower surface glabrous or sparsely pubescent, villose along lower midvein. Inflorescence broadly paniculate to 100 cm tall, 50 cm in diameter; capitula presented horizontally. Ray florets yellow, 12–17, to 23 mm long; disc florets 40–65.

TANZANIA. Mt Meru, NE wall of crater, Oct. 1932, *B.D. Burtt* 4051! & S slope, June 1946, *Greenway* 7802! & W slopes above Olkakola estate, Oct. 1948, *Hedberg* 2398!
DISTR. **T** 2; endemic to Mt Meru
HAB. Moist sites in upper afro-montane evergreen forest and ericaceous moorland; largest and most abundant plants grow along small ravines with abundant moisture and light; 2850–3350 m
USES. None recorded on specimens from our area
CONSERVATION NOTES. Protected in the National Park; least concern (LC)

SYN. *Senecio meruensis* Cotton & Blakelock in K.B. 2: 135 (1948)
    S. johnstonii Oliv. subsp. *johnstonii*; Mabb. in K.B. 28: 87 (1973) pro parte
    Dendrosenecio johnstonii (Oliv.) B.Nord. subsp. *johnstonii*; B. Nord. in Opera Bot. 44: 42 (1978) pro parte
    Senecio johnstonii Oliv. subsp. *johnstonii* var. *meruensis* (Cotton & Blakelock) C.Jeffrey in K.B. 41: 888 (1986)

4. **Dendrosenecio battiscombei** (*R.E.Fr. & T.C.E.Fr.*) *E.B.Knox* in Contr. Univ. Michigan Herb. 19: 243 (1993). Type: Kenya, Mt Kenya, *Fries & Fries* 1304 (UPS!, holo., BM!, K!, S!, iso.)

Upright, polycarpic plant to 7 m tall, with trunk to 35 cm in diameter; pith to 4 cm in diameter; stem with densely packed leaf-rosettes of 25–75 leaves, with no internode elongation, cloaked with marcescent foliage or at least covered with retained leaf-bases for 1–3 m below the leaf-rosettes. Infrequent reproduction yields sparsely branched, spreading plants that rarely exceed four reproductive cycles. Leaf lamina oblanceolate, to 78 cm long and 21 cm wide, hair cushion present on upper leaf-base, sometimes extending along upper midvein, lower surface glabrous or with a sparse to dense pubescence, villose along lower midvein. Inflorescence broadly pyramidal to 120 cm tall, 80 cm in diameter; capitula presented horizontally. Ray florets 6–17; to 20 mm long; disc florets 20–90.

KENYA. Nyandarua/Aberdares, Mt Satima, June 1932, *Dale* K 2848!; Mt Kenya, Dec. 1931, *Rammell* 2676! & Dec. 1943, *Bally* 3358!
DISTR. **K** 3, 4; endemic to Nyandarua/Aberdares and Mt Kenya
HAB. Along streams in the upper afro-montane evergreen forest and ericaceous moorland; more abundant and larger in the afro-alpine zone, reaching the summit of Sattima Peak; less common on Mt. Kenya where it is typically restricted to rocky outcrops and other sites in the moorlands that are protected from fire; 2950–4000 m
USES. None recorded on specimens from our area
CONSERVATION NOTES. Protected in the National Park; least concern (LC)

SYN. *Senecio battiscombei* R.E.Fr. & T.C.E.Fr. in Svensk Bot. Tidskr 16: 334 (1922); A.V.P.: 229, 357 (1957); K.T.S.: 159, pl. 11 (1961)
    S. aberdaricus R.E.Fr. & T.C.E.Fr. in Svensk Bot. Tidskr 16: 331 (1922). Type: Kenya, Nyandarua/Aberdares, Sattima, *Fries & Fries* 2436a (UPS!, lecto., BM!, BR!, S!, isolecto., chosen by Hedberg)
    S. johnstonii Oliv. subsp. *battiscombei* (R.E.Fr. & T.C.E.Fr.) Mabb. in K.B. 28: 86 (1973)
    Dendrosenecio johnstonii (Oliv.) B.Nord. subsp. *battiscombei* (R.E.Fr. & T.C.E.Fr.) B.Nord. in Opera Bot. 44: 43 (1978)
    Senecio johnstonii Oliv. subsp. *battiscombei* (R.E.Fr. & T.C.E.Fr.) Mabb. var. *battiscombei*; C. Jeffrey in K.B. 41: 891 (1986); K.T.S.L.: 561 (1994)

NOTE. There is considerable variation in leaf morphology throughout the elevation range on Nyandarua/Aberdares, with plants from the lower elevations having smaller, less pubescent leaves. Additionally, the leaf of *Schmitt* 692 has a cordate lamina and is constricted below to form a pseudo-petiole, but otherwise conforms to *D. battiscombei*.

5. **Dendrosenecio brassiciformis** (*R.E.Fr. & T.C.E.Fr.*) *Mabb.* in Vuilleumier and Monasterio, High Altitude Tropical Biogeography: 100 (1986). Type: Kenya, Nyandarua/Aberdares, Sattima, *Fries & Fries* 2400 (UPS!, lecto., BM!, BR!, K!, S! isolecto., chosen by Hedberg)

Upright, monocarpic plant to 5 m tall, with trunk to 20 cm in diameter, pith to 6 cm in diameter; stem with a single, densely packed leaf-rosette of 40–110 leaves, with no internode elongation, cloaked with marcescent foliage. The leaf-rosette secretes copious quantities of a mucilaginous fluid, and this dried mucilage is generally apparent on the base of the upper surface of the leaf. Lateral branching is initiated below the developing inflorescence, but the plants almost never survive the first reproductive cycle. Leaf lamina oblanceolate, to 58 by 20 cm. Inflorescence narrowly paniculate to 220 by 100 cm; capitula horizontal. Ray florets 16–17, to 23 mm long; disc florets 50–80.

KENYA. Nyandarua/Aberdares, Kinangop, March 1910, *Galpin* 7905! & Nyeri–Naivasha road, Feb. 1966, *Greenway* 12,335! & Eland Hill, Oct. 1970, *Mabberley* 362!
DISTR. **K** 3, 4; endemic to Nyandarua/Aberdares
HAB. In saturated soils along valley bottoms in the afro-alpine zone; scattered plants sometimes occur in wet sedge-meadows on the plateau; 2950–3950 m
USES. None recorded on specimens from our area
CONSERVATION NOTES. Protected in the National Park; least concern (LC)

Syn. *Senecio brassiciformis* R.E.Fr. & T.C.E.Fr. in Svensk Bot. Tidskr 16: 338 (1922); A.V.P.: 227, 358 (1957); K.T.S.: 159 (1961)
   S. brassica R.E.Fr. & T.C.E.Fr. subsp. *brassiciformis* (R.E.Fr. & T.C.E.Fr.) Mabb. in K.B. 28: 89 (1973), *nom. incorr.*
   Dendrosenecio brassica B.Nord. subsp. *brassiciformis* (R.E.Fr. & T.C.E.Fr.) B.Nord. in Opera Bot. 44: 43 (1978), *nom. incorr.*
   Senecio keniensis Baker f. subsp. *brassiciformis* (R.E.Fr. & T.C.E.Fr.) C.Jeffrey in K.B. 41: 892 (1986); K.T.S.L.: 562 (1994)

Note. The plants that grow on the plateau are significantly smaller than those up in the afro-alpine zone. Because these lower plants initiate flowering near ground-level, a lateral branch may occasionally survive the first reproductive cycle, but these plants do not normally have the creeping growth-form seen in *D. keniensis.*

6. **Dendrosenecio keniensis** (*Baker f.*) *Mabb.* in Vuilleumier and Monasterio, High Altitude Tropical Biogeography: 100 (1986). Type: Kenya, Mt Kenya, Höhnel Valley, *Gregory* s.n. quoad inflorescentiam et flores (BM!, lecto., chosen by C. Jeffrey)

Polycarpic plant to 1.5 m tall (in flower), initially upright, with trunk to 5 cm in diameter, pith to 2 cm in diameter; stem with densely packed leaf-rosettes of 30–40 leaves, with no internode elongation, cloaked with marcescent foliage. Lateral branches produced near ground-level readily capable of adventitious rooting to support a 'creeping' horizontal growth-form. Leaf lamina oblanceolate, to 56 cm long and 18 cm wide, basal portion wide, non-photosynthetic, and capable of secreting limited quantities of a mucilaginous fluid containing ice-nucleating polysaccharides, hair cushion present on upper leaf-base, which is also often coated with dried mucilage, lower surface with a dense, felty indumentum on the photosynthetic portion only. Inflorescence narrowly conical to 110 cm tall, 20 cm in diameter; capitula pendulous. Ray florets 12–16, to 25 mm long; disc florets 60–80.

Kenya. Mt Kenya, Jan. 1932, *Rammell* 2674! & Sirimon Track, Jan. 1963, *Verdcourt* 3539! & Lower Hinde Valley, ENE of Urumandi Hut, Jan. 1985, *Townsend* 2265!
Distr. **K** 4; endemic to Mt Kenya
Hab. In saturated soils along valley bottoms and in other wet sites of the afro-alpine zone; most abundant in the Teleki Valley; 3300–4275 m
Uses. None recorded on specimens from our area
Conservation notes. Protected in the National Park; least concern (LC)

Syn. *Senecio keniensis* Baker f. in J.B. 32: 140 (1894); U.K.W.F. ed. 2: 223 (1994)
   Lobelia gregoriana sensu Baker f. in J. B. 32: 66 (1894) pro parte, quoad folia
   Senecio brassica R.E.Fr. & T.C.E.Fr. in Svensk Bot. Tidskr 16: 336 (1922), *nom. illegit. superfl.*; A.V.P.: 226 (1957); K.T.S.: 159 (1961). Type as for *S. keniensis* Baker f.
   Dendrosenecio brassica B. Nord. in Opera Bot. 44: 43 (1978), *nom. illegit.*
   Senecio keniensis Baker f. subsp. *keniensis*: C. Jeffrey in K.B. 41: 892 (1986); K.T.S.L.: 562 (1994), autonym

Note. A mix-up with the *Gregory* material resulted in descriptions that united the leaf of *Lobelia gregoriana* with the inflorescence of this species and vice versa. *Senecio keniensis* was rejected as a *nomen confusum*, but this practice is no longer permitted by the Code and the replacement name, *S. brassica*, is now superfluous and other names based on this basionym are correspondingly illegitimate. Although Fries and Fries (1922) cited the *Gregory* material for *S. brassica*, Hedberg (1957) subsequently lectotypified this name with *Fries & Fries* 1305 (UPS!).

7. **Dendrosenecio keniodendron** (*R.E.Fr. & T.C.E.Fr.*) *B.Nord.* in Opera Bot. 44: 43 (1978).Type: Kenya, Mt Kenya, *Fries & Fries* 1356 (UPS!, lecto., BM!, K!, S!, isolecto., chosen by Hedberg)

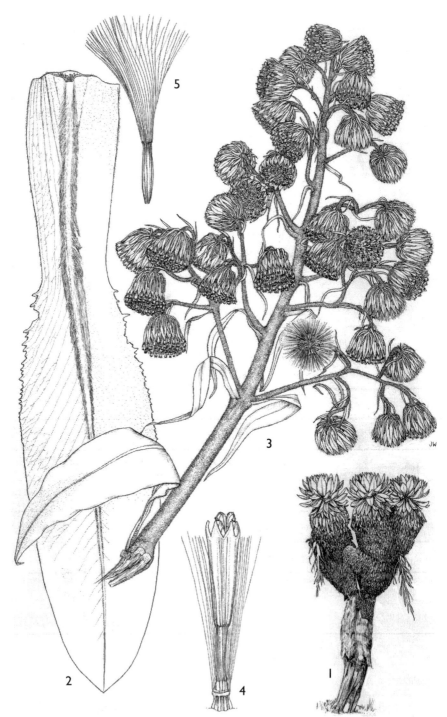

FIG. 117. *DENDROSENECIO KENIODENDRON* — **1**, habit; **2**, young leaf, × ¹/₂; **3**, part of inflorescence, × ¹/₂; **4**, floret, × 5; **5**, achene, × 3. 1 from photos of *Gilbert Rogers* s.n., 14.vi.1933; 2 from *Fries & Fries* 1356; 3–5 from *Gilbert Rogers* s.n., 14.vi.1933; all with reference to photos supplied by Eric Knox and accompanying *Hedberg* 1726. Drawn by J.Williamson.

Upright, polycarpic plant to 7 m tall, with trunk to 50 cm in diameter, pith to 8 cm in diameter; stem with densely packed leaf-rosettes of 50–100 leaves, with no internode elongation; cloaked with marcescent foliage. Infrequent reproduction yields sparsely branched, columnar plants that rarely exceed three reproductive cycles. Leaf lamina oblanceolate, to 82 cm long and 22 cm wide, upper leaf-base with long, yellowish trichomes, sometimes extending along upper midvein, lower surface glabrous or sparsely pubescent, villose along lower midvein. Inflorescence paniculate to 250 cm tall, 120 cm in diameter; capitula pendulous. Ray florets absent; disc florets 80–140. Fig. 117 (page 556).

KENYA. Nyandarua/Aberdares, Jan. 1933, *Rammell* 3063!; Mt Kenya, Teleki Tarn, July 1948, *Hedberg* 4350! & Naro Moru Track, Dec. 1988, *Knox* 763!
DISTR. **K** 3, 4; endemic to Mt Kenya and Nyandarua/Aberdares
HAB. Moist, drained soils in the afro-alpine zone; largest and most abundant plants grow on the west flank where afternoon cloud ameliorates evapo-transpiration; on Nyandarua/Aberdares known only from the very summit of Sattima Peak; 3650–4350 m
USES. None recorded on specimens from our area
CONSERVATION NOTES. Protected in the National Park; least concern (LC)

SYN. *Senecio keniodendron* R.E.Fr. & T.C.E.Fr. in Svensk Bot. Tidskr 16: 328 (1922); A.V.P.: 231 91957); K.T.S.: 160 (1961); U.K.W.F. ed. 2: 223 (1994); K.T.S.L.: 562 (1994)

NOTE. F1 hybrids of *D. keniensis* x *D. keniodendron* are occasionally found on Mt Kenya (named as *Senecio* × *saundersii* W. Sauer & E. Beck), but with no evidence of back-crossing or the maintenance of hybrid swarms there is little reason to recognize formally such hybrids.

8. **Dendrosenecio cheranganiensis** (*Cotton & Blakelock*) *E.B.Knox* in Contr. Univ. Michigan Herb. 19: 243 (1993). Type: Kenya, Cherangani Hills, *Dale* 3392 (K!, holo., EA!, FT!, iso.)

Upright, polycarpic plant to 6 m tall or 'dwarfed' to 1.5 m tall, pith 2–3 cm in diameter; stem with densely packed leaf-rosettes of 40–70 leaves, with no internode elongation, cloaked with marcescent foliage or at least covered with retained leaf-bases for 1–2 m below the leaf-rosettes. Leaf lamina oblanceolate, to 94 cm long and 25 cm wide, hair cushion weakly to strongly developed on upper leaf-base, lower surface glabrous or sparsely to densely pubescent, villose along lower midvein. Inflorescence broadly paniculate; capitula presented horizontally. Ray florets 10–15, to 25 mm long; disc florets 35–70.

Plant 2–6 m in flower, growing on drained soils . . . . . .   a. subsp. *cheranganiensis*
Plant to 1.5 m in flower, growing on saturated soils . . .   b. subsp. *dalei*

a. subsp. **cheranganiensis**

Plant to 6 m tall, with trunk to 25 cm in diameter, pith to 2 cm in diameter. Leaf-rosettes of 40–70 leaves. Infrequent reproduction yields sparsely branched, columnar plants that rarely exceed three reproductive cycles. Leaf to 94 cm long and 25 cm wide. Inflorescence to 100 cm tall, 70 cm in diameter. Ray florets 10–13; disc florets 35–50.

KENYA. Cherangani Hills, Marakwet Hills, Sep. 1934, *Dale* K 3240! & Kamalagon [Kameligon], Dec. 1970, *Mabberley* 512! & Kamalagon [Kamelogon], Dec. 1988, *Knox* 719!
DISTR. **K** 3; endemic to the Cherangani Hills
HAB. Moist, drained soils along streams or margins of wet sedge-meadows in the upper afro-montane evergreen forest and ericaceous moorlands; occasionally growing in seasonally wet sites at the lower elevations; 2600–3400 m
USES. None recorded on specimens from our area
CONSERVATION NOTES. No Park protection; at least vulnerable (VU-B1ab)

SYN. *Senecio cheranganiensis* Cotton & Blakelock in K.B. 1937: 364 (1937); K.T.S.: 159 (1961)
     S. johnstonii Oliv. subsp. *cheranganiensis* (Cotton & Blakelock) Mabb. in K.B. 28: 85 (1973)
     Dendrosenecio johnstonii (Oliv.) B.Nord. subsp. *cheranganiensis* (Cotton & Blakelock) B.Nord. in Opera Bot. 44: 42 (1978)

Senecio johnstonii Oliv. subsp. *battiscombei* (R.E.Fr. & T.C.E.Fr.) Mabb. var. *cheranganiensis* (Cotton & Blakelock) C.Jeffrey in K.B. 41: 891 (1986); K.T.S.L.: 561 (1994)

b. subsp. **dalei** (*Cotton & Blakelock*) *E.B.Knox* in Contr. Univ. Michigan Herb. 19: 244 (1993). Type: Kenya, Cherangani Hills, *Dale* 3393 (K!, holo., EA!, S!, iso.)

Plant to 1.5 m tall (in flower), with trunk to 10 cm in diameter, pith to 3 cm in diameter. Leaf-rosettes of 40–60 leaves. Frequent reproduction combined with limited stem growth yields compact, densely branched plants that rarely exceed three reproductive cycles. Leaf to 53 cm long and 15 cm wide. Inflorescence to 75 cm tall, 30 cm in diameter. Ray florets 12–15; disc florets 50–70.

KENYA. Cherangani Hills, Marakwet Hills, Sep. 1934, *Dale* 3239! & Chepkotet, Aug. 1968, *Thulin & Tidigs* 229! & Kamalagon [Kameligon], Dec. 1970, *Mabberley* 511!
DISTR. **K** 3; endemic to the Cherangani Hills
HAB. In saturated soils of wet sedge-meadows in the ericaceous moorlands; 3050–3500 m
USES. None recorded on specimens from our area
CONSERVATION NOTES. No Park protection; at least vulnerable (VU-B1ab)

SYN. *Senecio dalei* Cotton & Blakelock in K.B. 1937: 365 (1937): K.T.S.: 160 (1961)
   S. johnstonii Oliv. subsp. *dalei* (Cotton & Blakelock) Mabb. in K.B. 28: 85 (1973)
   Dendrosenecio johnstonii (Oliv.) B Nord. subsp. *dalei* (Cotton & Blakelock) B.Nord. in Opera Bot. 44: 42 (1978)
   Senecio johnstonii Oliv. subsp. *battiscombei* (R.E.Fr. & T.C.E.Fr.) Mabb. var. *dalei* (Cotton & Blakelock) C.Jeffrey in K.B. 41: 891 (1986); K.T.S.L.: 561 (1994)

9. **Dendrosenecio elgonensis** (*T.C.E.Fr.*) *E.B.Knox* in Contr. Univ. Michigan Herb. 19: 244 (1993). Type: Uganda, Mt Elgon, *Dümmer* 3382 (B†, holo., K!, UPS!, US!, iso.)

Upright, polycarpic plant to 8 m tall, with trunk to 50 cm in diameter; pith 2.5–5 cm in diameter; stem with densely packed leaf-rosettes of 30–50 leaves, with no internode elongation; cloaked with marcescent foliage or at least covered with retained leaf-bases for 1–3 m below the leaf-rosettes. Branching sparse to dense; spreading. Leaf lamina elliptic or cordate, to 97 cm long and 32 cm wide, with lower portion of leaf constricted, hair cushion present on upper leaf-base, sometimes extending along upper midvein, lower surface glabrous or with a dense indumentum, villose along lower midvein. Inflorescence broadly paniculate to 120 cm tall, 80 cm in diameter; capitula pendulous or presented horizontally. Ray florets 11–13, to 24 mm long, or absent; disc florets 40–130.

Ray florets prominent; leaf lamina extending along the midvein toward the base, not forming a pseudo-petiole, although sometimes constricted to the midvein just below a cordate base of the main portion of the lamina, but flaring again toward the base; lower lamina surface not covered with a dense, felty indumentum, though villose along the midvein . . . .   a. subsp. *elgonensis*
Ray florets absent; leaf lamina constricted toward the base, forming a pseudo-petiole; lower lamina surface covered with a dense, felty indumentum . . . .   b. subsp. *barbatipes*

a. subsp. **elgonensis**

Plant to 7 m tall, with trunk to 30 cm in diameter; pith 2.5–3 cm in diameter; stem normally cloaked with marcescent foliage (or with retained leaf-bases after fire) but eventually lost as bark develops. Periodic reproduction yields sparsely branched, spreading plants that rarely exceed five reproductive cycles. Leaf lamina elliptic or cordate, to 97 × 32 cm, with lower portion of leaf constricted with the lamina extending along the midvein and flared at the base, lower leaf surface glabrous except along midvein. Capitula presented horizontally. Ray florets 11–13, to 24 mm long; disc florets 40–70.

UGANDA. Mt Elgon, Sep. 1932, *A.S. Thomas* 628! & First Baguishi Hut, Jan. 1934, *Dale* K 3211! & above Bulambuli, Aug. 1934, *Synge* 888!
KENYA. Mt Elgon, Feb. 1932, *Lugard & Lugard* 699! & near Kassowai River, 1932, *Porter* 2732! & Telel, Jan. 1935, *Dale* 3378!
DISTR. U 3; K 3, 5; endemic to Mt Elgon
HAB. Moist sites in the upper afro-montane evergreen forest, ericaceous moorlands, and afro-alpine zone; found along streams and small ravines at the lower elevations; large populations extend along the west, south, and east flanks of the mountain and within the crater; infrequent at the higher elevations; 2750–4200 m
USES. None recorded on specimens from our area
CONSERVATION NOTES. Protected in the National Park; least concern (LC)

SYN. *Senecio elgonensis* T.C.E.Fr. in Svensk Bot. Tidskr 17: 229 (1923); I.T.U.: 97, photo 13 (1952); A.V.P.: 228 (1957); K.T.S.: 160 (1961); Hamilton, Field Guide Uganda For. Trees: 79 (1981)
    S. amblyphyllus Cotton in K.B. 1932: 473 (as *amblyophyllus*) (1932), corr. Cotton in K.B. 1937: 368 (1937); I.T.U.: 97 (1952); K.T.S.: 159 (1961); Hamilton, Field Guide Uganda For. Trees: 79 (1981). Type: Kenya, Mt Elgon, small stream on Kassowai River, *Fairbairn* 2678 (K!, holo., EA!, iso.)
    S. johnstonii Oliv. subsp. *elgonensis* (T.C.E.Fr.) Mabb. in K.B. 28: 85 (1973)
    Dendrosenecio johnstonii (Oliv.) B.Nord. subsp. *elgonensis* (T.C.E.Fr.) B.Nord. in Opera Bot. 44: 42 (1978)
    Senecio johnstonii Oliv. subsp. *elgonensis* (T.C.E.Fr.) Mabb. var. *elgonensis*; C. Jeffrey in K.B. 41: 889 (1986); K.T.S.L.: 561 (1994)

NOTE. Young plants in partly shaded situations at lower elevations may display some internode elongation.

b. subsp. **barbatipes** (*Hedberg*) *E.B.Knox* in Contr. Univ. Michigan Herb. 19: 244 (1993). Type: Kenya, Mt Elgon, *Honoré* 2520 (K!, holo., BM!, EA!, iso.)

Plant to 8 m tall, with trunk to 50 cm in diameter; pith 4–5 cm in diameter; dead lamina quickly shed, only leaf-bases retained until eventually shed by developing bark. Periodic reproduction yields densely branched, spreading plants that regularly attain five reproductive cycles. Leaf lamina cordate, to 51 cm long and 23 cm wide, lower portion of leaf constricted to form a pseudo-petiole to 32 cm long, lower leaf surface with a sparse to dense, felty indumentum. Capitula pendulous. Ray florets absent; disc florets 90–130.

UGANDA. Mt Elgon, 1939, *Dale* A2/41 (U.73)! & Jackson's Summit, Jan. 1948, *Eggeling* 5753!
KENYA. Mt Elgon, Feb. 1930, *Gardner* 2269! & W slope of Koitobos, May 1948, *Hedberg* 872! & trail to Koitobos, no date, *Ekkens* 622
DISTR. U 3; K 3, 5; endemic to Mt Elgon
HAB. Drained soils in the upper afro-alpine zone; abundant on the outer and inner crater rim and surrounding slopes; 3750–4225 m
USES. None recorded on specimens from our area
CONSERVATION NOTES. Protected in the National Park; least concern (LC)

SYN. *Senecio gardneri* Cotton in K.B. 1932: 471 (1932), *nom. illegit.*, *non* C. B. Clarke (1876); I.T.U.: 98 (1952). Type as for *D. elgonensis* subsp. *barbatipes*
    S. gardneri Cotton var. *ligulatus* Cotton & Blakelock in K.B. 1937: 370 (1937). Type: Kenya, Mt Elgon, *Taylor* 3719 (BM!, holo., S!, iso.)
    S. barbatipes Hedberg, A.V.P.: 230, 358 (1957); K.T.S.: 159 (1961); Hamilton, Field Guide Uganda For. Trees: 79 (1981)
    S. johnstonii Oliv. subsp. *barbatipes* (Hedberg) Mabb. in K.B. 28: 86 (1973)
    Dendrosenecio johnstonii (Oliv.) B.Nord. subsp. *barbatipes* (Hedberg) B.Nord. in Opera Bot. 44: 42 (1978)
    Senecio johnstonii Oliv. subsp. *elgonensis* (T.C.E.Fr.) Mabb. var. *ligulatus* (Cotton & Blakelock) C.Jeffrey in K.B. 41: 889 (1986); K.T.S.L.: 561 (1994)

NOTE. Var. *ligulatus* likely results from residual gene-flow along the zone of contact of subsp. *barbatipes* and subsp. *elgonensis*. However, such gene-flow must be uncommon because the two subspecies do not synchronize their mass flowering. Although *Taylor* 3719 has ray florets present, the preponderance of morphological attributes are those of subsp. *barbatipes*.

10. **Dendrosenecio adnivalis** (*Stapf*) *E.B.Knox* in Contr. Univ. Michigan Herb. 19: 243 (1993). Type: Uganda, Ruwenzori, *Dawe* 663 (K!, lecto., chosen by C. Jeffrey)

Upright, polycarpic plant to 10 m tall, with trunk to 40 cm in diameter; pith to 3–4 cm in diameter; stem with densely packed leaf-rosettes of 25–60 leaves, with no internode elongation; cloaked with marcescent foliage or at least covered with retained leaf-bases for 1–3 m below the leaf-rosettes. Periodic reproduction yields sparsely to densely branched, spreading plants that regularly attain five reproductive cycles. Leaf lamina oblanceolate or elliptic to cordate, to 96 cm long and 26 cm wide, with lower portion of leaf constricted to form a pseudo-petiole, hair cushion present on upper leaf-base, sometimes extending along the midvein, lower surface glabrous or sparsely to densely pubescent, or with a felty indumentum, villose along lower midvein. Inflorescence narrowly paniculate to 160 cm tall, 60 cm in diameter; capitula pendulous. Ray florets 9–20, to 16 mm long, or absent; disc florets 90–250.

SYN. [*Senecio johnstonii* sensu Stuhlmann in Mit Emin Pascha ins Herz von Afrika: 292, 295, 300 (1894), *non* Oliv.]
    S. adnivalis Stapf subsp. *adnivalis*: Hauman in Rev. Bot. Zool. Afr. 28: 48 (1935); A.V.P.: 232, 359 (1957), autonym
    S. adnivalis Stapf subsp. *refractisquamatus* (De Wild.) Hauman in Rev. Bot. Zool. Afr. 28: 47 (1935) pro parte
    S. johnstonii Oliv. subsp. *refractisquamatus* (De Wild.) Mabb. in K.B. 28: 82 (1973) pro parte
    Dendrosenecio johnstonii (Oliv.) B.Nord. subsp. *refractisquamatus* (De Wild.) B.Nord. in Opera Bot. 44: 42 (1978) pro parte

Mature leaves normally > 60 cm long; lower lamina surface
    glabrous or pubescent . . . .   a. subsp. *adnivalis*
Mature leaves normally < 60 cm long; lower lamina surface
    covered with a dense, felty indumentum . . . .   b. subsp. *friesiorum*

a. subsp. **adnivalis**

Leaves 60–96 cm long, 15–26 cm wide, margin entire or serrulate, lower surface glabrous or with sparse to dense pubescence.

var. **adnivalis**

Leaf lamina extending along the midvein toward the base, not forming a pseudo-petiole.

UGANDA. Ruwenzori Mts, Bujuku Valley, Aug. 1933, *Eggeling* 1324! & Kabambe, Mobuku Valley, July 1953, *Osmaston* 3219! & Lower Kitandara Lake, June 1968, *Hamilton* 762!
DISTR. U 2; Congo (Kinshasa), endemic to Ruwenzori Mts
HAB. Afro-alpine zone; on small hummocks in wet sedge-meadows (bogs) at the lowest elevations; growing to the edge of the glaciers; 3250–4500 m
USES. None recorded on specimens from our area
CONSERVATION NOTES. Protected in the National Park; least concern (LC)

SYN. *Senecio adnivalis* Stapf in J.L.S. 37: 521 (1906); I.T.U.: 96 (1952); Hamilton, Field Guide Uganda For. Trees: 79 (1981)
    S. lanuriensis De Wild., Pl. Bequaert. 5:143 (1929), *non* De Wild., tom. cit.: 107 (1929). Type: Congo (Kinshasa), Ruwenzori, Lanuri Valley, *Bequaert* 4533 (BR!, holo., K!, iso.)
    S. refractisqamatus De Wild., Pl. Bequaert. 5: 148 (1929). Type: Congo (Kinshasa), Ruwenzori, Butagu Valley, *Bequaert* 3850 (BR!, holo.)
    S. erioneuron Cotton in K.B. 1932: 438 (1932). Type: Congo (Kinshasa), Ruwenzori, *Humbert* 8929 (K!, holo., BR!, P!, iso.); I.T.U.: 97 (1952)
    S. adnivalis Stapf var. *adnivalis*: Hauman in Rev. Bot. Zool. Afr. 28: 48 (1935), autonym
    S. adnivalis Stapf var. *intermedia* Hauman in Rev. Bot. Zool. Afr. 28: 48 (1935). Syntypes: Uganda/Congo (Kinshasa), Ruwenzori, *Hauman* 471bis (K!, syn.) & *Humbert* 8930 (B!, BR!, GH!, K!, P!, US!, syn.); Uganda/Rwanda/Congo (Kinshasa), Virunga Mts, *Burtt* 2797 (BR!, EA!, syn.) & *Burtt* 2811 (K!, P!, syn.) & *Burtt* 3050 (P!, syn.) & *Burtt* 3098 (K!, P!, syn.) pro parte

S. alticola T.C.E.Fr. var. *subcalvescens* Hauman in Rev. Bot. Zool. Afr. 28: 50 (1935).
Syntypes: Uganda/Rwanda/Congo (Kinshasa), Virunga Mts, *Burtt* 3012 (K!, syn.) &
Thomas 1122 (K!, syn.); Uganda/Congo (Kinshasa), Ruwenzori, *Chapin* 89 (BM!, BR!,
NY!, syn.) pro parte
S. erioneuron Cotton var. *erioneuron*; Hauman in Rev. Bot. Zool. Afr. 28: 52 (1935); A.V.P.:
232, 359 (1957), autonym
S. erioneuron Cotton var. *oligochaeta* Hauman in Rev. Bot. Zool. Afr. 28: 52 (1935). Type:
Congo (Kinshasa), Ruwenzori, Kérésé Valley, *Hauman* 470 (BR!, holo., K!, iso.)
S. refractisquamatus De Wild. var. *intermedia* (Hauman) Robyns in Fl. Parc Nat. Albert 2: 578
(1947) pro parte
S. adnivalis Stapf var. *erioneuron* (Cotton) Hedberg, A.V.P.: 232 (1957)
S. johnstonii Oliv. subsp. *adnivalis* (Stapf) C.Jeffrey var. *adnivalis* (Stapf) Lisowski, Asterac.
Fl. Afr. Centr.: 323 (1991)

var. **petiolatus** (*Hedberg*) *E.B.Knox* in Contr. Univ. Michigan Herb. 19: 243 (1993). Type:
Uganda, Ruwenzori, *Fishlock & Hancock* 84 (K!, holo.)

Leaf lamina constricted toward the base, forming a pseudo-petiole; margin serrate.

Uganda. Ruwenzori Mts, Namwamba Valley, April 1932, *Humphreys* 1370! & above upper
Kitandara Lake, Aug. 1953, *Osmaston* 3275! & near Bujuku Hut, June 1968, *Hamilton* 711!
Distr. U 2; Congo (Kinshasa), endemic to Ruwenzori Mts
Hab. Moist slopes in upper afro-alpine zone; most abundant in gullies and other sites with
ample water and good drainage; 3600–4250 m
Uses. None recorded on specimens from our area
Conservation notes. Protected in the National Park; least concern (LC)

Syn. *Senecio petiolatus* Hauman in Rev. Bot. Zool. Afr. 28: 54 (1935); I.T.U.: 98 (1952), *nom. non*
rite publ.
S. stanleyi Hauman in Rev. Bot. Zool. Afr. 28: 52 (1935). Type: Congo (Kinshasa),
Ruwenzori, *Hauman* 472 (BR!, holo., K!, iso.)
S. adnivalis Stapf var. *petiolatus* Hedberg, A.V.P.: 233 (1957). Type as for *D. adnivalis* subsp.
adnivalis var. *petiolatus*
S. adnivalis Stapf var. *stanleyi* (Hauman) Hedberg, A.V.P.: 233 (1957)

Note. Much of the variation in leaf morphology of *D. adnivalis* subsp. *adnivalis* was shown by
Mabberley (1973) to intergrade, and the early profusion of names was a result of insufficient
collecting. Although intermediates between var. *adnivalis* and var. *petiolatus* can be found, var.
adnivalis is much more common and var. *petiolatus*, where it does grow, forms relatively pure
stands or intermingles with var. *adnivalis* with limited intergradation. Hence, var. *petiolatus* is
retained for those plants that clearly possess a pseudo-petiole.

b. subsp. **friesiorum** (*Mildbr.*) *E.B.Knox* in Contr. Univ. Michigan Herb. 19: 243 (1993). Type:
Congo (Kinshasa), Ruwenzori, *Mildbraed* 2599 (B†, holo.) & Ruwenzori, west slope, *Humbert*
8934 (K!, neo., B!, BR!, GH!, NY!, P!, US!, isoneo., chosen by Mabb.)

Plant to 7 m tall, with trunk to 35 cm in diameter. Leaf lamina oblanceolate; to 45 cm long
and 13 cm wide, lower leaf surface with a dense, felty indumentum. Ray florets 11–15, to 10 mm
long; disc florets 150–250.

Uganda. Ruwenzori Mts, SW ridge of Mt Stanley, Aug. 1953, *Osmaston* 3260! & below edge of
Batoda Plateau, July 1988, *Knox* 270!
Distr. U 2; Congo (Kinshasa), endemic to Ruwenzori Mts
Hab. Upper afro-alpine zone; endemic to the western flank of Mt. Stanley (down to Lac Vert
and Kiondo) and the Batoda Plateau (down to Lake Batoda); 3900–4200 m
Uses. None recorded on specimens from our area
Conservation notes. Protected in the National Park; least concern (LC)

Syn. *Senecio hypoleucus* Muschl. in Z.A.E.: 665 (1914), *nom. nud.*
S. friesiorum Mildbr. in F.R. 18: 231 (1922); A.V.P.: 231, 358 (1957)
S. albescens De Wild., Pl. Bequaert. 5: 140 (1929). Type: Congo (Kinshasa), Ruwenzori,
Butagu Valley, *Bequaert* 3847 (BR!, holo.)
S. johnstonii Oliv. subsp. *refractisquamatus* (De Wild.) Mabb. var. *friesiorum* (Mildbr.) Mabb.
in K.B. 28: 84 (1973)

Dendrosenecio johnstonii (Oliv.) B.Nord. subsp. *refractisquamatus* (De Wild.) B.Nord. var.
friesiorum (Mildbr.) B.Nord. in Opera Bot. 44: 42 (1978)
Senecio johnstonii Oliv. subsp. *adnivalis* (Stapf) C.Jeffrey var. *friesiorum* (Mildbr.) Mabb. in
K.B. 41: 891 (1986); Lisowski, Asterac. Fl. Afr. Centr.: 327 (1991)

11. **Dendrosenecio erici-rosenii** (*R.E.Fr.* & *T.C.E.Fr.*) *E.B.Knox* in Contr. Univ.
Michigan Herb. 19: 244 (1993). Type: Congo (Kinshasa), Mt Nyiragongo, *Fries* 1716
(UPS!, holo.)

Upright, polycarpic plant to 9 m tall, with trunk to 50 cm in diameter; pith to 2 cm
in diameter; stem with densely packed leaf-rosettes of 10–30 leaves, with no internode
elongation; covered with retained leaf-bases for 1–3 m below the leaf-rosettes.
Frequent reproduction yields densely branched, spreading plants that regularly attain
five or more reproductive cycles. Leaf lamina oblanceolate, to 67 cm long and 26 cm
wide, hair cushion present or absent on upper leaf-base, lower surface glabrous or
sparsely pubescent to densely lanate, villose along lower midvein. Inflorescence
narrowly pyramidal to 100 cm tall, 40 cm in diameter; capitula pendulous or
presented horizontally. Ray florets 9–14, to 22 mm long; disc florets 35–120.

SYN. *Senecio johnstonii* Oliv. subsp. *refractisquamatus* (De Wild.) Mabb. in K.B. 28: 82 (1973)
　　　pro parte
　　Dendrosenecio johnstonii (Oliv.) B.Nord. subsp. *refractisquamatus* (De Wild.) B.Nord. in
　　　Opera Bot. 44: 42 (1978) pro parte
　　Senecio johnstonii Oliv. subsp. *adnivalis* (Stapf) C.Jeffrey in K.B. 41: 889 (1986) pro parte

Ray florets prominent; hair cushion absent or weakly
　　developed; lower lamina surface glabrous or weakly
　　pubescent (although may be villose along lower
　　midvein) . . . . a. subsp. *erici-rosenii*
Ray florets absent or inconspicuous; hair cushion strongly
　　developed; lower lamina surface floccose, lanate, or
　　tomentose . . . . b. subsp. *alticola*

a. subsp. **erici-rosenii**

Plant to 9 m tall. Upper leaf-base glabrous or with a weakly formed hair cushion; lower leaf
surface glabrous or sparsely pubescent. Capitula presented horizontally. Ray florets to 22 mm
long; disc florets 35–80.

UGANDA. Virunga Mts, Mgahinga Crater, *Eggeling* 1076! & Mt Muhavura, Oct. 1948, *Hedberg*
　　2056!; Ruwenzori Mts, Aug. 1938, *Purseglove* 285!
DISTR. U 2; Congo (Kinshasa), Rwanda; endemic to Mt Muhi, Mt Kahuzi, Virunga Mts and
　　Ruwenzori Mts
HAB. Moist, drained sites in the upper afro-montane evergreen forest, ericaceous forest, and the
　　lower afro-alpine zone; restricted to protected situations at the higher elevations; 2750–4200 m
USES. Stem pith eaten by gorilla (*Knox*)
CONSERVATION NOTES. Protected in the National Park; least concern (LC)

SYN. *Senecio erici-rosenii* R.E.Fr. & T.C.E.Fr. in Svensk Bot. Tidskr. 16: 330 (1922); I.T.U.: 97
　　　(1952); A.V.P.: 228, 359 (1957); Hamilton, Field Guide Uganda For. Trees: 79 (1981)
　　S. longiligulatus De Wild., Pl. Bequaert. 5: 145 (1929), as *longeligulatus*. Type: Congo
　　　(Kinshasa), Ruwenzori, Lanuri Valley, *Bequaert* 4657 (BR!, holo., K!, iso.)
　　S. kahuzicus Humb. in Bull. Soc. Bot. Fr. 81:842 (1935). Type: Congo (Kinshasa), Mt
　　　Kahuzi, *Humbert* 7713 (P!, holo., B!, BR!, K!, iso.)
　　S. johnstonii Oliv. subsp. *adnivalis* (Stapf) C.Jeffrey var. *erici-rosenii* (R.E.Fr. & T.C.E.Fr.)
　　　C.Jeffrey in K.B. 41: 850 (1986); Lisowski, Asterac. Fl. Afr. Centr.: 324, t. 71 (1991)

b. subsp. **alticola** (T.C.E.Fr.) *E.B.Knox* in Contr. Univ. Michigan Herb. 19: 244 (1993). Type:
Congo (Kinshasa), Mt Karisimbi, *Mildbraed* 1609 (B†, holo.); Uganda, Mt Muhavura, *Hedberg*
2086 (UPS!, neo., EA!, K!, S!, isoneo., chosen by Hedberg)

Plant to 7 m tall, with trunk to 40 cm in diameter. Leaf to 50 cm long and 23 cm wide, hair cushion present on upper leaf-base; lower surface densely floccose, lanate or tomentose. Capitula pendulous. Ray florets absent or inconspicuous; if present to 12 mm long; disc florets 100–120.

UGANDA. Virunga Mts, Mt Muhavura, NE face, no date, *Eggeling* 1001! & Mt Muhavura, Oct. 1929, *Snowden* 1555! & Mt Muhavura, Nov. 1954, *Stauffer* 743!
DISTR. U 2; Congo (Kinshasa), Rwanda; endemic to Virunga Mts, on Mts Muhavura, Karisimbi and Mikeno
HAB. Upper afro-alpine zone; 3400–4475 m
USES. Stem pith eaten by gorilla (*Knox*)
CONSERVATION NOTES. Protected in the National Park; least concern (LC)

SYN. *S. erici-rosenii* R.E.Fr. & T.C.E.Fr. var. *alticola* Mildbr. in F.R. 18: 230 (1922), *nom. illegit.*
   S. alticola T.C.E.Fr. in Svensk Bot. Tidskr 17: 229 (1923); I.T.U.: 97 (1952)
   S. adnivalis Stapf var. *intermedia* Hauman in Rev. Bot. Zool. Afr. 28: 48 (1935). Syntypes: Uganda/Congo (Kinshasa), Ruwenzori, *Hauman* 471*bis* (K!, syn.) & *Humbert* 8930 (B!, BR!, GH!, K!, P!, US!, syn.); Uganda/Rwanda/Congo (Kinshasa), Virunga Mts, *Burtt* 2797 (BR!, EA!, syn.) & *Burtt* 2811 (K!, P!, syn.) & *Burtt* 3050 (P!, syn.) & *Burtt* 3098 (K!, P!, syn.) pro parte
   S. alticola T.C.E.Fr. var. *alticola*; Hauman in Rev. Bot. Zool. Afr. 28: 49 (1935), autonym
   S. alticola T.C.E.Fr. var. *subcalvescens* Hauman in Rev. Bot. Zool. Afr. 28: 50 (1935). Syntypes: Uganda/Rwanda/Congo (Kinshasa), Virunga Mts, *Burtt* 3012 (K!, syn.) & *Thomas* 1122 (K!, syn.); Uganda/Congo (Kinshasa), Ruwenzori, *Chapin* 89 (BM!, BR!, NY!, syn.) pro parte
   S. refractisquamatus De Wild. var. *intermedia* (Hauman) Robyns in Fl. Parc Nat. Albert 2: 578 (1947) pro parte
   S. adnivalis Stapf var. *alticola* (T.C.E.Fr.) Hedberg, A.V.P.: 233, 359 (1957)
   S. johnstonii Oliv. subsp. *adnivalis* (Stapf) C.Jeffrey var. *alticola* (T.C.E.Fr.) C.Jeffrey in K.B. 41: 890 (1986); Lisowski, Asterac. Fl. Afr. Centr.: 327 (1991)

NOTE. Mildbraed in F.R. 18: 230 (15 Oct. 1922) published *Senecio erici-rosenii* R.E.Fr. & T.C.E.Fr. var.? *alticola*; he indicated that the species name was forthcoming by Fries & Fries in Svensk Bot. Tidskr. (which it was, later that year). Mildbraed's 'description' of the species (rather than the variety) is so scanty that the consensus is that it is invalid, which also invalidates the variety. Thore Fries subsequently elevated *alticola* to *S. alticola*, citing it as (Mildbr.) T.C.E.Fr. in Svensk Bot. Tidskr 17: 229 (1923), but as Mildbraed's name was invalidly published, this constitutes the first valid publication of the name.

89. **CINERARIA**

L., Sp. Pl. ed. 2: 1242 (1763) & Gen. Pl., ed. 6: 426 (1764)

Perennial herbs or sub-shrubs. Leaves alternate, mostly pinnatisect-lobed, auriculate, palmately veined. Capitula in lax corymbose cymes, radiate or rarely discoid. Involucre 1-seriate, with calyculus; receptacle epaleate. Florets yellow, the outer female, radiate, in 1 series; the inner hermaphrodite, infundibuliform or narrowly campanulate. Anthers obtuse or slightly sagittate at base, with ovate apical appendage; style branches truncate to obtuse. Achenes compressed with thickened or winged margins; pappus of many fine bristles.

15–20 species, mostly South African but with a few in Madagascar and tropical Africa and SW Arabia.

Cineraria deltoidea *Sond.* in Linnaea 23: 68 (1850); F.P.S. 3: 3: 17 (1956); Hilliard, Comp. Natal: 379 (1977); Maquet in Fl. Rwanda 3: 670, fig. 209/3 (1985); C. Jeffrey in K.B. 41: 930 (1986); Blundell, Wild Flow. E. Afr.: fig. 357 (1987); Lisowski, Aster. Fl. Afr. Centr. 2: 434 (1991); U.K.W.F. ed. 2: 225 (1994). Type: South Africa, Natal, *Gueinzius* 343 (MEL, holo., W!, iso.)

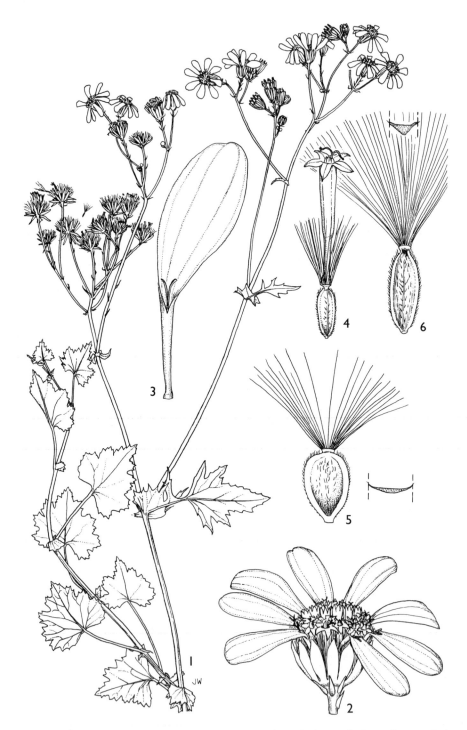

FIG. 118. *CINERARIA DELTOIDEA* — **1**, habit, × ²/₅; **2**, capitulum, × 3; **3**, ray floret, × 8; **4**, disc floret, × 8; **5**, marginal achene (with cross-section), × 8; **6**, central achene (with diagrammatic cross-section), × 8. 1, 5–6 from *Smith et al.* 226; 2–4 from *Burtt* 4400. Drawn by Juliet Williamson.

Perennial herb, erect or more usually scandent or trailing, 15–300 cm high or long; stems branched, often purplish, floccose-tomentose, thinly arachnoid or sparsely pubescent, glabrescent. Leaves petiolate, except the uppermost, very broadly ovate to narrowly triangular in outline, 1–7 cm long, 1–9.3 cm wide, base subtruncate to deeply cordate or emarginate, obscurely to moderately deeply 5–11-lobed, dentate on the distal margins of the lobes, apex acute to rounded and apiculate or sometimes attenuate, green and glabrescent except for main veins or glabrous above, floccose-tomentose to sparsely pubescent beneath, especially on veins, often glabrescent; petiole often narrowly winged, rarely with 1–4 small oblong lateral lobes, expanded and usually auriculate at the base, 0.8–6.5 cm long; uppermost stem leaves sessile, smaller. Capitula rarely solitary, normally in erect to pendulous, often copious, terminal cymes. Involucre cylindrical, 4–7.5 mm high; bracts of calyculus 2–8, lanceolate, 1–3 mm long; phyllaries 8–14, 3.5–7 mm long, pale green with brownish or reddish-brown tips, glabrous or sometimes thinly arachnoid. Ray florets 4–14, rays golden yellow to lemon yellow, 4–14 mm long, 1.5–3.5 mm wide, tube 2–3 mm long, sparsely hairy or rarely glabrous; disc florets yellow, corolla 3–6.2 mm long, tube usually glabrous, lobes 0.5–1 mm long. Achenes dark-coloured, 2–3.5 mm long, the outer compressed and slightly winged, the inner 3-angled, shortly ciliate on the margins and sometimes also on the faces, or completely glabrous; pappus 3–6 mm long. Fig. 118 (page 564).

UGANDA. Kigezi District: Muhavura–Mgahinga Saddle, Sep. 1946, *Purseglove* 2186! & Mt Mgahinga, Dec. 1959, *Miller* 30!; Mbale District: Elgon Caldera, Dec. 1967, *Hedberg* 4484!
KENYA. Laikipia District, Sep. 1884, *Thomson* s.n.!; Mt Kenya, Sirimon Track, Sep. 1970, *Kokwaro* 2396!; Masai District: Ngong Hills, Nov. 1971, *Kibue* 159!
TANZANIA. Arusha District: Mt Meru, Oct. 1959, *Carmichael* 720!; Ufipa District: Mbizi Forest Reserve, Oct. 1987, *Ruffo & Kisena* 2651!; Mbeya District: Mbeya Peak Forest Reserve, Aug. 1958, *Myembe* 55!
DISTR. U 2, 3; K 1, 3–6; T 2, 4, 6, 7; Congo (Kinshasa), Rwanda, Sudan, Ethiopia, Zambia, Malawi, Mozambique, Zimbabwe, South Africa
HAB. Grassland, montane forest and forest margins, glades in bamboo and *Hypericum* zone, moorland, especially in moist places, in lava crevices; may be temporarily co-dominant in secondary vegetation; (1100–)2100–4050 m
USES. None recorded on specimens from our area, apart from 'readily eaten by horses' (*Thorold*)
CONSERVATION NOTES. Widespread; least concern, LC

SYN. *C. grandiflora* Vatke in Linnaea 39: 503 (1875); F.P.S. 3: 17 (1956); A.V.P.: 222 & 349 (1957); Cribb & Leedal, Mount. Flow. E. Africa: 153, t. 41a (1982). Type: Ethiopia, Dschan Meda, *Schimper* 1517 (B†, holo., BM, K!, S, iso.)
    C. kilimandscharica Engl. in Abh. K. Preuss. Akad. Wiss. Berlin 1891 (2): 439 (1892). Type: Tanzania, Kilimanjaro, *Johnston* 129 or 4 (K!, lecto. chosen by Jeffrey)
    C. bracteosa Engl. in Götzen, Afrika Ost West: 377 & 383 (1895). Type: Congo (Kinshasa), *von Götzen* 64, 106 (B†, syn.)
    C. prittwitzii Engl. in Götzen, Afrika Ost West: 375 & 383 (1895). Type: Congo (Kinshasa), *von Götzen* 29 (B†, holo.)
    C. buchananii S.Moore in J.L.S. 35: 352 (1902). Type: Malawi, without locality, *Buchanan* 10 (BM!, holo.)
    C. foliosa O.Hoffm. in E.J. 30: 434 (1902). Type: Tanzania, Njombe District, Ukinga Mts, Kipengere Ridge, *Goetze* 973 (B†, holo., BM!, K!, iso.), **syn. nov.**
    Senecio schubotzianus Muschl. in Z.A.E. 2: 405 (1911). Type: Congo (Kinshasa), *Mildbraed* 1416 (B†, holo., BR, fragm., not found)
    S. kirschsteinianus Muschl. in Z.A.E. 2: 405 (1911). Type: Congo (Kinshasa), SE of Karisimbi, *Mildbraed* 1647 (B†, holo., BR, fragm., not found)
    Cineraria monticola Hutch. in K.B. 1931: 251 (1931). Type: South Africa, Zoutpansberg, *Hutchinson & Gillett* 3201 (K!, holo.)
    C. bequaertii De Wild., Pl. Bequaert. 5: 441 (1932). Type: Congo (Kinshasa), between Tongo and Mokule, *Bequaert* 5863 (BR!, holo.)

NOTE. A very variable species. Distinctive forms with tomentose lower leaf surfaces (and sometimes also somewhat lyrato-pinnately lobed leaves) occur on some mountains, e.g. Moroto in Uganda, Suji-Mamba, Kisarawe and near Mpwapwa in Tanzania as well as further south.

90. **EURYOPS**

(Cass.) Cass. in Dict. Sci. Nat. 16: 49 (1820); C. Jeffrey in K.B. 41: 931 (1986)

Othonna L. subgen. *Euryops* Cass. in Bull. Sc. Soc. Philom. Paris: 140 (1818)

Shrubs, rarely herbs. Leaves alternate or rarely rosulate. Capitula solitary, usually radiate; involucre without a calyculus; phyllaries usually connate. Ray and disc florets yellow; style branches truncate or obtuse. Achenes ribbed or smooth; pappus of many bristles or occasionally absent.

A genus of about 100 species, mostly southern African, extending through tropical Africa to Socotra and the SW Arabian peninsula.

1. Leaves entire, up to 3.2 mm wide; pappus present 2
   Leaves pinnately divided, 1–3 cm wide; pappus absent .. 5. *E. chrysanthemoides*
2. Leaves ericoid, appressed, 5.5–10 mm long .... 2. *E. dacrydioides*
   Leaves linear, ± spreading, 10–60 mm long 3
3. Rays 12–21 mm long; leaves shortly pubescent or hyaline-ciliate in a strip along the margins, especially in lower two-thirds .... 3. *E. elgonensis*
   Rays 9.5–15 mm long; leaves with short hairs on the margins near the base, otherwise glabrous, or with some long woolly hairs at the base and in the leaf-axils 4
4. Leaves minutely serrulate or entire, glabrous except for the basal marginal hairs .... 4. *E. brownei*
   Leaves entire; with long woolly hairs at the base and in the leaf-axils .... 1. *E. jacksonii*

1. **Euryops jacksonii** *S.Moore* in J.L.S. 35: 362 (1902); B. Nord. in Op. Bot. 20: 52 (1968); U.K.W.F. ed. 2: 226 (1994). Type: Kenya, Kiambu District: Kikuyu, *Jackson* s.n. (BM!, holo.)

Woody herb 15–45 cm tall; stems much branched, densely leafy. Leaves sessile, linear or slightly oblanceolate, entire, 1.2–6 cm long, 0.1–0.15 cm wide, with a slightly expanded membraneous base, apex obtuse to subtruncate and obscurely acuminate, glabrous except for marginal hairs towards the base and for some long woolly hairs at the base and in the leaf axils. Capitula solitary on long axillary peduncles, radiate; stalks of individual capitula ebracteolate, 5–13 cm long, glabrous; phyllaries 13, broadly ovate or ovate-oblong, 5–6 mm long, united in lower $^1/_3$ or $^1/_2$, acute, ciliate on upper margins, otherwise glabrous, 3–5-veined. Ray florets (12–)13(–14), yellow, rays 9.5–12 × 2.5–3.5 mm, usually 4-veined; disc florets many, yellow, 2.8–4 mm long, the tube cylindrical but basally inflated. Achenes narrowly oblong, 2.5 mm long, 5–8-ribbed, pilose; pappus of many white bristles 1–2 mm long.

KENYA. Nyandarua/Aberdare Mts, Sasamua Dam, Jan. 1953, *Verdcourt* 877!; Nakuru/Kericho District: SW Mau Forest, Farm 1645, *Whitall* 88!; Masai District: Olokurto, May 1961, *Glover et al.* 943!
DISTR. **K** 3/4, 5, 6; known only from the Mau and the Nyandarua/Aberdares
HAB. Upland grassland or bushed grassland, clearings in bamboo zone; (2000–)2400–3300 m
USES. Minor medicinal for childbirth (*Glover*); grazed by all stock (*Glover*)
CONSERVATION NOTES. Though the distribution range is small, the altitude range is considerable and the habitat common; least concern, LC

SYN. *E. fruticulosa* R.E. Fr. in Acta Hort. Berg. 9 (6): 157 (1928). Type: Kenya, Nyandarua/Aberdares, Kinangop Plateau, *Fries & Fries* 2502 (UPS!, lecto., BM!, S!, isolecto., chosen by Nordenstam)

2. **Euryops dacrydioides** *Oliv.* in Hook., Ic. Pl. 16, t. 1508 (1886); T.T.C.L.: 150 (1949); A.V.P.: 245 & 366 (1957); B. Nord. in Op. Bot. 20: 120 (1968). Type: Tanzania, Kilimanjaro, *Johnston* 153 (K!, holo., BM!, iso.)

Evergreen shrub 50–180 cm tall, much branched, ericoid; stems densely leafy, brittle, at first grey with woolly hairs, glabrescent and showing leaf-scars after fall of the rather persistent bases of the dead leaves. Leaves densely crowded, sessile, appressed, firm, acicular or subulate, 0.55–1 cm long, 0.1 cm wide, keeled, subacute, minutely apiculate, bright green, laxly woolly in the leaf axils, otherwise glabrous. Capitula solitary on short upper axillary peduncles, radiate; stalks of individual capitula 0.7–2.3 cm long; phyllaries (11–12–)13, ovate or ovate-lanceolate, 3–4.5 mm long, united in lower third, briefly acuminate, shortly bearded at apex, otherwise glabrous, 3–5-veined. Ray florets 13–16, yellow, rays 9–10 mm long, 2.5 mm wide, 4-veined; disc florets many, 3.5–4.5 mm long. Achenes oblong, 2.5 mm long, 5–8-ribbed, glabrous; pappus of several bristles 1.5–2 mm long.

TANZANIA. Kilimanjaro, near Horombo [Peters] Hut, Feb. 1934, *Greenway* 3763! & idem, June 1948, *Hedberg* 1167! & Shira Plateau, Feb. 1982, *Sigara* 240!
DISTR. **T** 2; only known from Kilimanjaro
HAB. Rocky ground from the giant heath zone to the perpetual snow; 3300–4500(–4700) m
USES. None recorded on specimens from our area
CONSERVATION NOTES. Though the distribution range is small, the altitude range is considerable and the habitat common; probably least concern, LC

NOTE. Abundant in the Horombo Hut area.

3. **Euryops elgonensis** *Mattf.* in E.J. 59, Beibl. 133: 42 (1925); A.V.P.: 246 & 366 (1957); B. Nord. in Op. Bot. 20: 122, fig. 13a (1968); U.K.W.F. ed. 2: 226 (1994). Type: Uganda, Mt Elgon, *Dummer* 3386 (B†, holo., K!, lecto., chosen by Nordenstam)

Shrub 60–120 cm tall; stems blackish, showing leaf-scars after fall of the rather persistent bases of the dead leaves. Leaves crowded, sessile, lush green and succulent, lanceolate-linear, 1–2.5 cm long, 0.15–0.32 cm wide, slightly narrowed above the expanded membraneous base, obtuse, apiculate, shortly pubescent or hyaline-serrulate on margins especially in proximal two-thirds, glabrescent with age. Capitula solitary on upper axillary peduncles, radiate; stalks of individual capitula about equalling the leaves, 1.7–2.2 cm long, sparsely pubescent; phyllaries 16–18, ovate or ovate-lanceolate, almost free, acuminate, glabrous, 4–4.5 mm long. Ray florets 15–16, rays yellow, 12–21 mm long, 2.5–3.5 mm wide; disc florets many, 4–5 mm long. Achenes oblong, 2–2.5 mm long, 5–8-ribbed, glabrous; pappus of bristles 1–2 mm long.

UGANDA. Elgon, above Bulambuli, Nov. 1933, *Tothill* 2369! & Elgon, Dec. 1938, *A.S. Thomas* 2738! & W part of caldera, Dec. 1967, *Hedberg* 4525!
KENYA. Elgon, Dec. 1930, *Lugard & Lugard* 368! & Jan. 1944, *Tweedie* 18 & E slope of Koitoboss, May 1948, *Hedberg* 863!
DISTR. **U** 3; **K** 3/5; only known from Elgon
HAB. Upper bamboo zone and moorlands, may be locally dominant; 3000–4200 m
USES. None recorded on specimens from our area
CONSERVATION NOTES. Though the distribution range is small, the altitude range is considerable and the habitat common; probably least concern, LC

4. **Euryops brownei** *S.Moore* in J.B. 54: 284 (1916); A.V.P.: 246 & 366 (1957); B. Nord. in Op. Bot. 20: 125, t. 13/b–e (1968); Blundell, Wild Flow. E. Afr.: fig. 363 (1987); K.T.S.L.: 557, fig. (1994); U.K.W.F. ed. 2: 225, t. 95 (1994). Type: Kenya, Mt Kenya, Feb. 1914, *Orde Browne* s.n. (BM!, holo.)

FIG. 119. *EURYOPS BROWNEI* — **1**, habit, × ²/₃; **2**, capitulum, × 3; **3**, disc floret, × 10. 1 from *Townsend* 2230, 2–3 from *Tweedie* 1975. Drawn by Juliet Williamson.

Woody herb or shrub 60–300 cm tall; stem blackish, showing leaf-scars after fall of the rather persistent bases of the dead leaves. Leaves crowded, sessile, linear, light green, 1–3.2 cm long, 0.1–0.25 cm wide, slightly narrowed above the expanded membraneous base, minutely serrulate or entire, bluntly apiculate, shortly ciliate on the margins near the base, otherwise glabrous. Capitula radiate, numerous, solitary on upper axillary peduncles; stalks of individual capitula 0.8–4.6 cm long, very sparsely hairy to densely tomentose; involucres woolly at the base; phyllaries 13–16, ovate or ovate-lanceolate, 4.5–6.5 mm long, united in lower half or third, obtuse to acuminate, shortly hairy at the apex. Ray florets (11–)13(–21), bright yellow, rays 10–15 mm long, 2.7–4 mm wide; disc florets many, yellow, 2.3–4 mm long. Achenes oblong, 2.5–4.5 mm long, 5–8-ribbed, glabrous or sparsely pilose; pappus of bristles 1–2.5 mm long. Fig. 119 (page 568).

KENYA. Elgeyo District: Cherangani, E of Kamelogon, Jan. 1971, *Tweedie* 3910!; Nyandarua/Aberdares, N Kinangop, July 1948, *Hedberg* 1550!; Mt Kenya, 34 km NW of Nanyuki, Mar. 1968, *Mwangangi & Fosberg* 570!
TANZANIA. Arusha District: Mt Meru, *Uhlig* 534; & W slopes above Olkakola, Nov. 1948, *Hedberg* 2433!
DISTR. **K** 2–4; **T** 2; endemic to Elgon, Cherangani, Nyandarua/Aberdare Mts, Mt Kenya and Mt Meru
HAB. Giant heath zone, where it may be locally dominant; (1550–)2700–3750 m
USES. None recorded on specimens from our area
CONSERVATION NOTES. Fairly widespread with a large altitude range; probably least concern, LC

SYN. *E. agrianthoides* Mattf. in E.J. 33: 43 (1924); T.T.C.L.: 150 (1949). Type: Tanzania, Arusha District, Mt Meru, *Uhlig* 534 (B†, holo., EA, lecto., chosen by Nordenstam)
  E. brownei S.Moore subsp. *aberdarica* R.E.Fr. in Act. Hort. Berg. 9: 158 (1928); Chiov., Racc. Bot. Miss. Consol. Kenya: 72 (1935). Type: Kenya, Nyandarua/Aberdares, Sattima, *Fries & Fries* 2466 (UPS!, holo., BM!, S!, iso.)

5. **Euryops chrysanthemoides** (*DC.*) *B.Nord.* in Op. Bot. 20: 365, t. 62/c–g (1968). Type: South Africa, Cape Province, Uitenhage, Olifantshoek, *Ecklon & Zeyher* 10.9 (G DC, lecto., K!, isolecto.)

Shrub 0.5–2 m high, nearly completely glabrous. Leaves pinnatilobate, 3–10 cm long, 1–3 cm wide. Capitula solitary on stalks 5–20 cm long. Ray and disc florets yellow, glabrous. Achenes black, pappus absent.

KENYA. Nandi District: 9 km from Kaimosi to Kapsabet, Oct. 1981, *Gilbert & Mesfin* 6704! & Nairobi District: Chiromo campus, behind Entomology laboratory, near Computer Science, Jan. 1983, *Gachathi* 272
DISTR. **K** 4, 5; **T** 3 (see note); cultivated plant, native of South Africa (Cape Province and Natal)
HAB. Roadside; 1500–2100 m
USES. An ornamental plant, often used as a live hedge
CONSERVATION NOTES. Least concern, LC

SYN. *Gamolepis chrysanthemoides* DC., Prodr. 6: 40 (1838)

NOTE. Reported to be establishing itself near Kaimosi. Planted as a hedge in Pare District: S Pare Mts, Chabaru R.C. Church, Aug. 1987, *Ruffo* 2481!

91. **LOPHOLAENA**

DC., Prodr. 6: 335 (1838)

Shrubs or perennial herbs with thick rootstock. Leaves alternate, fleshy. Capitula solitary or in corymbose panicles, discoid. Involucre 1-seriate, without calyculus; phyllaries connate at base; receptacle epaleate. Ray florets absent; disc florets white, pink or purplish; anthers obtuse at base, with ovate or oblong apical appendage; style branches with filiform pappillose appendages. Achenes with pappus of many fine bristles; pappus in 2–3 series.

A genus of about 15 species, mostly from southern Africa.

1. Capitula solitary 2
   Capitula in compound corymbs; involucre 2–3 mm in
   diameter .... 3. *L. trianthema*
2. Involucre 13–15 mm long, 7–10 mm in diameter .... 1. *L. ussanguensis*
   Involucre 9.5–15 mm long, 2.5–3 mm in diameter .... 2. *L. dolichopappa*

1. **Lopholaena ussanguensis** (*O.Hoffm.*) *S.Moore* in J.B. 67: 275 (1929); T.T.C.L.: 155 (1949). Type: Tanzania, Njombe District, Ussangu, Kinga Mts, *Goetze* 1257 (B†, holo., K!, iso.)

Shrub to 1 m. Leaves sessile, fleshy, slightly obovate, 2–7 cm long, 1–5 cm wide, broadly cuneate towards the subamplexicaul slightly decurrent base, margins entire and slightly thickened, apex obtuse or rounded, glabrous, faintly 3-veined. Capitula solitary, terminal; stalks of the individual capitula 13–15 mm long, slightly expanded below the capitulum. Involucres broadly cylindrical, 13–15 mm long, 7–10 mm wide; phyllaries 5–6, oblong, broadly to narrowly triangular, ciliate near apex, with curious double margins, the inner wide and membranous. Florets 25–30, ± 18 mm long, white. Achenes to 6 mm long and 3 mm in diameter, with faint ribs, villous; pappus white, ± 18 mm long.

TANZANIA. Njombe District: Ussangu, Kinga Mts, 1899, *Goetze* 1257!
DISTR. **T** 7; known from only the type
HAB. Rocky slope; 2900 m
USES. None recorded on specimens from our area
CONSERVATION NOTES. At least vulnerable (VU-D2) and possibly extinct, as it has not been collected for over a hundred years

SYN. *Senecio ussanguensis* O.Hoffm. in E.J. 30: 438 (1901)

NOTE. The protologue says the leaves are decurrent to 3 cm, but this not visible in the specimen – maybe mm?

2. **Lopholaena dolichopappa** (*O.Hoffm.*) *S.Moore* in Bull. Herb. Boiss. ser. 2, 4: 1021 (1904); T.T.C.L.: 154 (1949). Type: Tanzania, Mbeya/Chunya Districts, Unyiha, *Goetze* 1462 (B†, lecto., BM!, isolecto.)

Shrubby herb 30–60 cm tall from a perennial woody rootstock, semi-succulent; stems trailing or erect, becoming somewhat woody with age. Leaves sessile, narrowly oblanceolate to spatulate, 1.5–6 cm long, 0.3–1.5 cm wide, attenuate into a petioloid base, margins entire, apex obtuse or minutely apiculate, glabrous except for a few hairs in the leaf axils. Capitula discoid, solitary, axillary and terminal on lateral branches, erect; stalks of the individual capitula 1.2–7 cm long, gradually slightly expanded below the capitula. Involucres cylindrical, 9.5–15 mm long, 2.5–3 mm wide; phyllaries 5, green or brown, oblong, elongating to up to 2 cm in fruit, glabrous, finely shortly ciliate on the margins, shortly acuminate. Florets white, sometimes tinged lilac, 3–6, exserted and exceeding the pappus at anthesis; anthers purple. Achenes obovoid-cylindrical, 7.5–8.5 mm long, 2.8–3.5 mm wide, swollen, broadly shallowly ribbed, densely pubescent, glabrescent with age; pappus much elongated in fruit, white or pale brown, up to 2.8–4 cm long. Fig. 120 (page 571).

TANZANIA. Mbeya District: Mporoto Mts, Galijembe Forest Reserve, Nov. 1992, *Mwasumbi* 16544!; Njombe District: km 3, Njombe–Mbeya road, Sep. 1956, *Semsei* 2459! & Yakobi, 20 km S of Njombe, Nov. 1987, *Mwasumbi et al.* 13449!

Fig. 120. *LOPHOLAENA DOLICHOPAPPA* — **1**, habit, × ²/₃; **2**, capitulum, × 2; **3**, capitula at mature seed stage, × ²/₃; **4**, mature achenes, × 1. 1 from *Gillett* 17839, 2 from *van Rensburg* 522, 3–4 from *Richards* 18499. Drawn by Juliet Williamson.

DISTR. **T** 7; Zambia, Malawi
HAB. Upland rocky grassland, where it may be common after burning, and bushland with
    Protea and *Tarchonanthus*; 1600–2750 m
USES. None recorded on specimens from our area
CONSERVATION NOTES. Widespread in a common habitat, probably least concern (LC)

SYN. *Senecio dolichopappus* O.Hoffm. in E.J. 30: 438, t. 18 (1901)
    Lopholaena pauciflora Thell. in Vierteljahrsschr. Nat. Ges. Zurich 66: 242 (1921); T.T.C.L.:
        155 (1949). Type: Tanzania, Rungwe District, Kyimbila, *Stolz* 2267 (Z!, holo.)
    L. brevipes Phillips & C.A.Sm. in Trans. Roy. Soc. S. Afr. 21: 236 (1933); T.T.C.L.: 154 (1949).
        Type: Tanzania, Mbeya/Chunya Districts, Usafwa, *Goetze* 1075 (B†, BM!, syn., K!, isosyn.)
    L. whyteana sensu T.T.C.L.: 155 (1949); Cribb & Leedal, Mount. Flow. E. Africa: 161 (1982),
        non (Britten) Phillips & C.A.Sm.

3. **Lopholaena trianthema** (*O.Hoffm.*) *B.L.Burtt* in Notes Roy. Bot. Gard. Edinb. 34:
85 (1975); Cribb & Leedal, Mount. Flow. S. Tanz.: 161, t. 44b (1982); C. Jeffrey in
K.B. 41: 932 (1986); Lisowski, Aster. Fl. Afr. Centr. 2: 436 (1991). Type: Tanzania,
Mbeya/Chunya Districts, Unyika, *Goetze* 1352 (B†, holo., K!, iso.)

Woody herb 100–150 cm tall, erect; rootstock woody; stems winged. Leaves
somewhat succulent, sessile, elliptic or slightly obovate, 4–10 cm long, 1–4 cm wide,
cuneate-attenuate into a petioloid base, margins entire, apex obtuse, minutely
apiculate, glabrous, somewhat 3-veined. Capitula discoid, numerous in dense
compound corymbs, the ultimate units ± umbelliform, bracteate, stalked, the bract-
axils lanate; stalks of the individual capitula 7–17 mm long. Involucre cylindrical,
green, 8–14 mm long, 2–2.5 mm wide, glabrous; phyllaries 3(–4), green, spreading,
obtuse and slightly bearded on the margins at the apex. Florets 3–4, white, cream to
orangish, exserted and exceeding the pappus at anthesis. Achenes ellipsoid-
cylindrical, 6–6.5 mm long, 3 mm wide, obtusely ribbed, swollen, shortly pubescent,
glabrescent; pappus white, 8 mm long in flower, 12–20 mm long in fruit.

TANZANIA. Mbeya District: Chunya Escarpment, Aug. 1970, *Richards* 25790! & near Mbeya, Nov.
    1963, *Procter* 2444! & Chunya–Mbeya road, Sep. 1954, *Smith* 1248!
DISTR. **T** 7; Congo (Kinshasa), Zambia, Malawi
HAB. Pyrophyte in upland grassland, *Protea* bushland and open sites in light woodland, old
    cultivations; 1400–2150 m
USES. None recorded on specimens from our area
CONSERVATION NOTES. Widespread in a common habitat, probably least concern (LC)

SYN. *Senecio trianthemos* O.Hoffm. in E.J. 30: 437 (1901)
    Lopholaena sp. of T.T.C.L.: 155 (1949)

92. EMILIA

(Cass.) Cass. in Dict. Sc. Nat. 34: 393 (1825); C. Jeffrey in K.B. 41: 908 (1986)

Emilia Cass. in Bull. Soc. Philom. Paris 1817: 68 (1817) sine dignitate definita
Senecio L. subgen. *Emilia* (Cass.) O.Hoffm. in E.& P. Pf. 4, 5 (54): 297 (1890)

Annual or perennial herbs. Leaves alternate, simple, sometimes rosulate. Capitula
often corymbose, sometimes solitary, discoid or radiate. Involucre without calyculus,
phyllaries free or connate at base; receptacle epaleate. Florets of various colours;
anthers obtuse or slightly sagittate at base, appendaged at apex; style branches
truncate to obtuse, sometimes with appendages of fused papillae. Achenes with
pappus of many fine bristles.

100 species, paleotropical.

1. Capitula radiate 2
   Capitula discoid, the ray florets absent 7
2. Involucres narrow, up to 2.5 (–?4 in *abyssinica*) mm in
   diameter; capitula usually several; rays 4–8 3
   Involucres > 2 mm in diameter; capitula solitary; rays
   8–26 4
3. Rays 3.5–6 mm long . . . . 1. *E. helianthella* (p. 575)
   Rays 1.5–2.5 mm long . . . . 2. *E. abyssinica* (p. 575)
4. Phyllaries pubescent . . . . 6. *E. tricholepis* (p. 579)
   Phyllaries glabrous 5
5. Leaves linear to oblanceolate-linear, up to 0.4 cm wide 5. *E. ukambensis* (p. 578)
   Leaves spatulate or obovate-spatulate 6
6. Plants subscapose; disc florets 6–8 mm long . . . . 4. *E. somalensis* (p. 578)
   Plants with leafy stems; disc florets 3.5–6.5 mm long . . . 3. *E. discifolia* (p. 576)
7. Leaves (at least the upper) with auriculate or sagittate
   base and semi-amplexicaul 8
   All leaves with attenuate, cuneate or rounded base,
   neither sagittate nor auriculate nor semi-amplexicaul 20
   (leaves mostly absent in *E. sp. aff. rigida*, see Note for that taxon)
8. Florets white, yellow, orange or red 9
   Florets purple or mauve, violet, pink or magenta 14
9. Capitula solitary, terminal; peduncles markedly
   inflated below the capitula . . . . 17. *E. cenioides* (p. 585)
   Capitula not solitary or if so, then peduncles not
   markedly inflated below the capitula and/or plant
   scapigerous, the flowering stems without leaves
   except at the base 10
10. Achenes glabrous or with only few hairs near apex 11
    Achenes hairy all over 12
11. Cauline leaves elliptic to linear-lanceolate, 3–4 × as
    long as wide . . . . 11. *E. caespitosa* (p. 581)
    Cauline leaves (at least the lower ones) elliptic to
    obovate, ± 1.5 × as long as wide . . . . 13. *E. vanmeelii* (p. 582)
12. Style arms apically truncate, not particularly appendaged 13
    Style arms terminating in a subulate appendage of
    fused papillae 42
13. Leaves sparsely to densely pubescent; florets 4–5.5 mm
    long . . . . 10. *E. kivuensis* (p. 580)
    Leaves glabrous to puberulous beneath; florets
    7.5–10 mm long . . . . 12. *E. jeffreyana* (p. 582)
14. Stem and/or leaves glabrous; phyllaries 8–14 15
    Stem and/or leaves hairy, at least partly so 16
15. Limb of the corolla tube completely exserted from the
    involucre at anthesis; pappus 2–4 mm long . . . . 25. *E. violacea* (p. 589)
    Limb of corolla tube included in the involucre at
    anthesis; pappus 3.5–4 mm long . . . . 26. *E. aurita* (p. 589)
16. Involucre 7.5–10 mm long; pappus 5.5–8 mm long . . 40. *E. sonchifolia* (p. 596)
    Involucre 2.5–5.5 mm long; pappus 1.5–3 mm long 17
17. Phyllaries pubescent; involucre 5–5.5 mm long . . . . . . 27. *E. kilwensis* (p. 589)
    Phyllaries glabrous; involucre 2.5–4.5 mm long 18
18. Corollas 5–5.5 mm long . . . . 39. *E. decipiens* (p. 594)
    Corollas 2.5–3.5 mm long 19
19. Involucres 2.5 mm long; median stem leaves petiolate . . 38. *E. myriocephala* (p. 594)
    Involucres 2.5–4.5 mm long; median stem leaves sessile 37. *E. pammicrocephala*
    (p. 593)

41. Basal leaves obovate, oblanceolate or elliptic; phyllaries
    pubescent . . . . 35. *E. tenera* (p. 592)
    Basal leaves oblanceolate linear to narrowly elliptic or
    linear-lanceolate; phyllaries glabrous . . . . 36. *E. limosa* (p. 593)
42. Lower leaves with base cuneate or attenuate into a
    winged exauriculate petioloid base; phyllaries 8–21,
    4–9 mm long; corolla orange or red . . . . 14. *E. coccinea* (p. 583)
    Lower leaves almost sessile; phyllaries 5–6, 4–5 mm long;
    corolla white . . . . 41. *E. mbagoi* (p. 596)

Lisowski, Aster. Fl. Afr. Centr. 2: 402 (1991) states *E. robynsiana* Lisowski occurs in Tanzania; we have seen no specimens from this country. He also states that *E. lisowskiana* occurs in Uganda; again, we have seen no specimens from there.

1. **Emilia helianthella** *C.Jeffrey* in K.B. 41: 911 (1986). Type: Tanzania, Iringa District, Ruaha National Park, 5 km W of Magangwe Ranger Post, *Bjørnstad* 1726 (K!, holo.)

Annual herb 30–80 cm tall; stems erect, sparsely pubescent towards the base. Leaves sessile, pale green, elliptic, obovate or the lowermost sometimes spatulate, 2–6 cm long, 0.3–1.8 cm wide, upper smaller and narrower, base cuneate, obscurely to coarsely sinuate-dentate or serrate, apex rounded or obtuse, sparsely pubescent, glabrescent. Capitula numerous in lax terminal corymbs, radiate; involucre cylindrical, 5–5.5 mm long, 1.5 mm across; phyllaries 8, pale glaucous green, 4.5–5 mm long, glabrous. Ray florets 4–6, yellow, tube 3 mm long, glabrous except at apex, rays 3.5–6 mm long, 1.5–2.5 mm wide, 4-veined; disc florets few, yellow, corolla 4 mm long, tube glabrous, lobes 0.5–0.7 mm long. Achenes 1.2–1.5 mm long, glabrous; pappus 3.5 mm long.

TANZANIA. Kigoma District: Sabaga, July 1975, *Kahurananga et al.* 2741!; Dodoma District: 36 km
    S of Itigi Station on Chunya road, Apr. 1964, *Greenway & Polhill* 11607!; Mbeya District: Lupa
    Forest Reserve, Aug. 1962, *Boaler* 642!
DISTR. **T** 4, 5, 7; not known elsewhere
HAB. *Brachystegia* woodland on sandy soils, may be common on roadsides; 1200–1500 m
USES. None recorded on specimens from our area
CONSERVATION NOTES. Fairly widespread in a common habitat, probably least concern (LC)

NOTE. *Humbert* 7210b from **T** 4, Kigoma area, has a few hairs on the achenes and shorter rays,
    and is possibly a hybrid with *E. abyssinica.*

2. **Emilia abyssinica** (*A.Rich.*) *C.Jeffrey* in K.B. 41: 911 (1986); Lisowski, Aster. Fl. Afr. Centr. 2: 361 (1991); U.K.W.F. ed. 2: 223, t. 94 (1994). Type: Ethiopia, Adowa, *Schimper* 67 (P!, holo., K!, M!, P!, iso.)

Annual herb 10–120 cm tall; stems erect, usually ± pubescent, especially towards the base, sometimes glaucous. Leaves sessile, pale green, sometimes purple underneath or glaucous, spatulate to oblanceolate, 0.8–7.5 cm long, 0.1–4 cm wide, base subtruncate, cuneate or attenuate into a slightly semi-amplexicaul petioloid base, sinuate-dentate to coarsely sinuate-serrate, apex rounded to obtuse, varyingly pubescent. Capitula radiate, 1–12 in lax to dense terminal corymbs; stalks of the individual capitula slender, glabrous; involucre 4–6.5 mm long, 1–2.5(–4?) mm wide, glabrous; phyllaries green, (5–)8(–11), 4–6.5 mm long, glabrous. Ray florets 4–8, yellow, tube 2–3.5 mm long, glabrous or with a few hairs at the apex, rays 1.5–2.5 mm long, 0.5–1.3 mm wide, 4-veined; disc florets yellow or orange, 3–4 mm long, tube glabrous, lobes 0.2–0.5 mm long. Achenes 1–2.5 mm long, shortly hairy; pappus 3–4 mm long.

var. **abyssinica**

Ray florets 1.5–2 mm long, 0.5–0.8 mm wide.

UGANDA. W Nile District: Maracha Rest Camp, Aug. 1953, *Chancellor* 121!; Ankole District: Bunyanguru, July 1939, *Purseglove* 885!; Masaka District: Lake Victoria, Kasokero, May 1969, *Lye et al.* 2908!

KENYA. Turkana District: Naitamajong Hill, June 1934, *Champion* T328!; W Suk District: Suk Escarpment, Nov. 1959, *Tweedie* 1922!

TANZANIA. Mwanza District: Buhumbi, May 1952, *Tanner* 789!; Ngara District: Mabawe, Bugofi, Dec. 1959, *Tanner* 4652!; Bukoba, Aug. 1931, *Haarer* 2087!

DISTR. U 1–4; **K** 2, 3 (fide UKWF), 6; **T** 1, 2, 4–8; Nigeria, Cameroon, Central African Republic, Congo (Kinshasa), Rwanda, Burundi, Sudan, Ethiopia, Djibouti, Zambia, Malawi, Mozambique, Zimbabwe

HAB. Weed in cultivation (rice, sorghum), ruderal sites, grassland, open woodland; 650–1900 m

USES. Minor medicinal for cattle (*Tanner*)

CONSERVATION NOTES. Widespread in a common habitat, least concern (LC)

SYN. *Senecio abyssinicus* A.Rich., Tent. Fl. Abyss. 1: 438 (1848); Oliv. & Hiern, F.T.A. 3: 410 (1877); F.P.S. 3: 48 (1956); Verdcourt & Trump, Common Poison. Pl. E. Afr.: 154 (1969); Maquet in Fl. Rwanda 3: 668, fig. 208/1 (1985)

   S. bellidifolius A.Rich., Tent. Fl. Abyss. 1: 439 (1848), *nom. illegit.*, *non* Kunth in H.B.K. (1818). Type: Ethiopia, Shire, *Dillon* s.n. (P!, holo.)

   S. quartinianus Aschers. in Schweinf., Beitr. Fl. Aethiop.: 157 (1867). Type as for *S. bellidifolius* A. Rich.

NOTE. Poisonous to livestock.

   var. **macroglossa** *C.Jeffrey* in K.B. 41: 911 (1986). Type: Tanzania, Mwanza District, Beda, *Rounce* 256 (K!, holo.)

   Ray florets 2.2–2.5 mm long, 1–1.3 mm wide.

TANZANIA. Mwanza District: Igongwa, Mbaruka, May 1952, *Tanner* 730!; Shinyanga, Mar. 1932, *Burtt* 3733! & Feb. 1933, *Bax* 228!

DISTR. **T** 1; not known elsewhere

HAB. Open disturbed places; 1100–1200 m

USES. None recorded on specimens from our area

CONSERVATION NOTES. Restricted in area, but in a common habitat; data deficient (DD) but probably Vulnerable

   3. **Emilia discifolia** (*Oliv.*) *C.Jeffrey* in K.B. 41: 912 (1986); Lisowski, Aster. Fl. Afr. Centr. 2: 363, t. 79 (1991); U.K.W.F. ed. 2: 223, t. 94 (1994). Type: Uganda, *Grant* s.n. (K!, holo.)

   Annual or perhaps rarely short-lived perennial herb 2–90 cm tall, much branched or not branched at all; stems erect or decumbent, pale green, sometimes bluish, sparsely to densely white pubescent especially when young and in axils in lower part, glabrous or almost so in upper part. Leaves pale to bluish green, sometimes glaucous, subsucculent, sessile, spatulate or rotundate-spatulate, the uppermost sometimes oblanceolate, 0.8–10.5 cm long, 0.2–5.6 cm wide, base rounded or cuneate into a winged petioloid base, margins sinuate-dentate especially or at least 3–5-toothed on the upper margins, apex subtruncate or rounded, sparsely to densely pubescent, sometimes glabrescent. Capitula terminal, solitary, long-stalked, radiate; capitulum stalk glabrous; involucre cylindrical, 3.5–8 mm long, 4.5–8 mm in diameter; phyllaries 6–13, bright green, sometimes tinged reddish, 3–7.5 mm long, glabrous. Florets yellow or orange, the disc florets often slightly darker in colour. Ray florets 8–26, most often 13 or 18, tube 2.7–5.5 mm long, hairy at the apex, ray 3.5–7 × 1.5–3 mm, 4-veined; disc florets bright yellow to orange, corolla 3.5–6.5 mm long, tube glabrous, lobes 0.5–0.7 mm long. Achenes 1.8–3 mm long, shortly hairy; pappus white, 3.5–7 mm long. Fig. 121 (page 577).

UGANDA. W Nile District: below Paida Rest Camp, Aug. 1953, *Chancellor* 174!; Kigezi District: Kachwekano Farm, May 1949, *Purseglove* 2882!; Mengo District: Mawanda, Feb. 1936, *Hazel* 391!

FIG. 121. *EMILIA DISCIFOLIA* — **1**, habit, × ²/₃; **2**, habit of much smaller plant, × ²/₃; **3**, capitulum, × 3; **4**, ray floret, × 9; **5**, style branches of disc floret, × 24; **6**, achene, × 6. 1, 4 from *Chancellor* 174, 2 from *Eggeling* 2937, 3 & 5 from *Hancock* 22, 6 from *Miringu & Beentje* 24. Drawn by Juliet Williamson.

KENYA. Northern Frontier District: Lerogi Forest, 28 km N of Maralal, Nov. 1977, *Carter & Stannard* 404!; Embu District: Kijegge, 2 km SE of Chakariga, Feb. 1996, *Miringu & Beentje* 24!; Masai District: Migori Bridge, June 1961, *Glover et al.* 1795!
TANZANIA. Moshi District: W Kilimanjaro Forest Reserve, Apr. 1960, *Kanywa* 7!; Lushoto District: W Usambara Mts, E of Mlalo, Feb. 1985, *Borhidi et al.* 85/545!; Kilosa District: Mt Luemba summit, Feb. 1933, *B.D. Burtt* 4528!
DISTR. U 1–4; **K** 1–7; **T** 1–3, 5, 6; Congo (Kinshasa), Rwanda, Burundi, Sudan, Ethiopia, Zambia, Zimbabwe
HAB. Grassland, especially where disturbed, roadsides, in cultivated areas; may be locally common; 450–2500 m
USES. Used for salt (*Chancellor*); minor medicinal against hiccoughs, against syphilis (*Tanner*)
CONSERVATION NOTES. Widespread in a common habitat, least concern (LC)

SYN. *Senecio discifolius* Oliv. in Trans. Linn. Soc. 29: 100 (1873); Oliv. & Hiern, F.T.A. 3: 410 (1877); F.P.S. 3: 48 (1956); F.P.U.: 178, fig. 119 (1962); Verdcourt & Trump, Common Poison. Pl. E.Afr.: 153 (1969)
  S. hoffmanianus Muschl. in E.J. 43: 62 (1909). Type: Rwanda, Niansa Mt, *Kandt* 38 (B†, holo., EA!, iso.)

4. **Emilia somalensis** (*S.Moore*) *C.Jeffrey* in K.B. 41: 912 (1986); U.K.W.F. ed. 2: 223 (1994). Type: Kenya, Northern Frontier District: Marsabit, *Delamere* s.n. (BM!, holo.)

Short-lived perennial subscapigerous herb 5–30(–60) cm tall; stems glabrous or sparsely floccose in upper parts. Leaves sessile, subsucculent, spatulate to oblanceolate-spatulate, 2–8.5 cm long, 0.3–2 cm wide, long-attenuate and decurrent onto an entire petioloid base, broadly sinuate-dentate or at least with 3 teeth at the apex, obtuse, glabrous or scattered-pubescent. Capitula terminal, solitary, long-stalked, radiate; involucre cylindrical, 7.5–9.5 mm long, 4.5–5 mm in diameter; phyllaries 12–14, green with reddish tips and orange stripes, 7–9.5 mm long, glabrous. Florets bright yellow to orange-yellow. Ray florets 18–20, tube 4.5–5.5 mm long, hairy at the apex, rays 6–10 mm long, 2–2.7 mm wide, 4-veined; disc florets 6–8 mm long, tube glabrous, lobes 0.7–1 mm long. Achenes 3 mm long, hairy; pappus 5–8 mm long.

KENYA. Northern Frontier District: Mt Nyiru, Mar. 1995, *Bytebier et al.* 80!; Naivasha District: Mt Longonot, Feb. 1963, *Verdcourt* 3586!; Masai District: Maparasha Hills 12 km ESE of Bissel, Sep. 1981, *Gilbert* 6366!
DISTR. **K** 1, 3, 6, ?7 (see note); Ethiopia
HAB. Short grassland on rocky slopes and mountain summits; 1400–2750 m
USES. None recorded on specimens from our area
CONSERVATION NOTES. Three specimens seen from Ethiopia, and fifteen from Kenya. Occurring over a large area, but in few localities; probably least concern (LC)

SYN. *Senecio discifolius* Oliv. var. *scaposus* O. Hoffm. in P.O.A. C: 417 (1895). Type: Kenya, Teita, *Hildebrandt* 2557 (B†, holo.)
  Euryops somalensis S.Moore in J.B. 38: 459 (1900)
  S. megamontanus Cufod. in Nuov. Giorn. Bot. Ital. n.s. 50: 113 (1943). Type: Ethiopia, Mega, *Corradi* 2077, 2079 (FT!, syn.)
  Senecio sp. B of U.K.W.F.: 478 (1974)

NOTE. The specimen on which Jeffrey (in K.B. 41: 912 (1986)) bases the **K** 7 record is *Wakefield* s.n., comm. May 1880. This bears the note 'Ribe to Galla country'. It is most likely from **K** 7 but then the altitude would be much lower then elsewhere. Despite the name, there is no material from Somalia.

5. **Emilia ukambensis** (*O.Hoffm.*) *C.Jeffrey* in K.B. 41: 912 (1986); U.K.W.F. ed. 2: 223 (1994). Type: Kenya, 'Central Province', *Hildebrandt* 2704, 2711 (both B†, syn.)

Erect perennial herb 30–60 cm tall; stems virgate, glabrous or sparsely pubescent. Leaves sessile, ± crowded, linear to oblanceolate, 2.5–7.5 cm long, 0.1–0.4 cm wide, base attenuate, remotely sinuate-denticulate or sinuate-dentate, revolute at least when dry, apex obtuse to acute, apiculate, glabrous or pubescent. Capitula solitary, radiate, long-stalked; stalks of the individual capitula glabrous; involucre cylindrical, 5–7 mm long, 3–4 mm in diameter; phyllaries 10–13, glabrous, 4.5–6.5 mm long. Florets bright yellow. Ray florets 13–14, tube 4.3–5 mm long, hairy at the apex, rays yellow, 4–4.7 × 1.5–1.8 mm, 4-veined; disc florets 5.2–8.5 mm long, tube glabrous, lobes 0.5–0.7 mm long. Achenes 3.2–3.7 mm long, hairy; pappus 5–7 mm long.

KENYA. Machakos/Masai Districts: Chyulu Hills, Aug. 1947, *North* in *Bally* 5254! & Chyulu Hills N, May 1938, *Bally* 8179! & idem, May 1981, *Gibert* 6183!
TANZANIA. Moshi District: Ran Rambo, Jan. 1929, *Haarer* 1768!; Mbulu District: near Babati, Nov. 1930, *Haarer* 1829! & Mbulumbulu, July 1943, *Greenway* 6780!
DISTR. **K** 4, 6; **T** 2; not known elsewhere
HAB. Rocky grassland; 900–2000 m
USES. None recorded on specimens from our area
CONSERVATION NOTES. Known from five sites, possibly vulnerable (VU-D2)

SYN. *Senecio ukambensis* O.Hoffm. in E.J. 20: 235 (1894)

NOTE. *Gibert* 5375 from **K** 1, Lolokwe, is an annual which looks very similar to this taxon. It has fewer and wider phyllaries. It was growing on a sedge tussock on a wet flush at 1650 m altitude.

6. **Emilia tricholepis** *C.Jeffrey* in K.B. 41: 912 (1986); U.K.W.F. ed. 2: 224 (1994). Type: Kenya, Turkana District, 3 km from Kacheliba, *Leippert* 5058 (K!, holo., EA, iso.)

Short-lived perennial tufted herb 30–60 cm tall; stems glabrous or almost so. Leaves sessile, subsucculent, glaucous, spatulate, 3–5.6 cm long, 1–2 cm wide, cuneate or attenuate into a petioloid base, coarsely sinuate-dentate on distal margins and sometimes also near base, apex rounded in outline, scattered hairy to glabrous. Capitula solitary, radiate, long-stalked; stalks of the individual capitula glabrous; involucre cylindrical, 7.5–9 mm long, 3–5 mm in diameter; phyllaries (12–)13(–14), 7–8.5 mm long, pubescent. Florets yellow to orange-yellow. Ray florets 13, tube 3.8–6 mm long, hairy at apex, rays 6.7–9 × 2.8–4 mm; disc florets 6.5–7 mm long, tube with a few hairs about the middle, lobes 0.5–0.7 mm long. Achenes 3–3.5 mm long, shortly hairy; pappus 5–6.5 mm long.

KENYA. Northern Frontier District: near Mugurr, June 1970, *Mathew* 6672!; Turkana District: Lorengipe, Mar. 1965, *Newbould* 7335!; W Suk District: 5 km S of Orrtum, July 1969, *Mabberley & McCall* 74!
DISTR. **K** 1–4, 7; not known elsewhere
HAB. Sparse vegetation on rocky slopes, in rock crevices, as a weed on cleared fields; 700–2000 m
USES. None recorded on specimens from our area
CONSERVATION NOTES. Known from 15 specimens, probably least concern (LC)

7. **Emilia hockii** (*De Wild. & Muschl.*) *C.Jeffrey* in K.B. 41: 912 (1986); Lisowski, Aster. Fl. Afr. Centr. 2: 365 (1991). Type: Congo (Kinshasa), Lubumbashi, *Hock* s.n. (BR!, holo.)

Scapigerous perennial herb 6–30 cm tall with creeping rhizome. Stems pubescent. Leaves slightly fleshy, grey-green above, sessile, narrowly obovate or elliptic, 1.1–5 cm long, 0.2–2.8 cm wide, base attenuate, margins entire, apex obtuse to rounded, scattered-pubescent or finely puberulous, glabrescent. Scapes 1–4, glabrous, 1-headed; capitula terminal, solitary, discoid; involucre cylindrical, 11.5–13.5 mm long, ± 5.5 mm in diameter; phyllaries 10–13, glabrous, dark green, 10–13 mm long. Ray florets absent; disc florets yellow or orange, 7–11 mm long, tube glabrous, lobes 0.5–0.7 mm long. Achenes 4–5.5 mm long, hairy; pappus 7.5–9 mm long.

TANZANIA. Ufipa District: Mbisi, Nov. 1949, *Bullock* 1861!; Mbeya District: Mbozi, Aug. 1933, *Greenway* 3622!; Iringa District: Iringa–Mbeya road due N of Lake Ngwazi, Sep. 1971, *Perdue & Kibuwa* 11431!

Distr. **T** 4, 7; Congo (Kinshasa), Zambia, Malawi
Hab. Grassland, regularly burned grassland, roadsides, in cultivation; 1500–2250 m
Uses. None recorded on specimens from our area
Conservation notes. Widespread; least concern (LC)

Syn. *Senecio hockii* De Wild. & Muschl. in B.S.B.B. 49: 231 (March 1913)
      S. rogersii S.Moore in J.B. 51: 185 (Apr. 1913). Type: Congo (Kinshasa), Lubumbashi [Elisabethville], *Rogers* 10176 (BM!, holo.)

8. **Emilia bellioides** (*Chiov.*) *C.Jeffrey* in K.B. 41: 913 (1986). Type: Somalia, Mogadishu, *Paoli* 9 (FT!, syn.), *Senni* 564 (FT!, syn.)

Annual herb 5–25 cm tall; stems scattered-pubescent. Leaves sessile, light green or grey-green, slightly succulent, broadly obovate-spatulate or the median oblanceolate, 1–5.5 cm long, 0.3–2.8 cm wide, base cuneate into a narrow, exauriculate petioloid base, sinuate-dentate to slightly 3-lobed on distal margins, rounded to subtruncate, with scattered to dense pubescence. Capitula solitary and terminal, discoid, on slender pubescent stalk 2.5–10 cm long; involucre cylindrical, 6.5–8.5 mm long, 2–3 mm in diameter; phyllaries 8–12, 5.5–8 mm long, shortly pubescent or hispid. Ray florets absent; disc florets yellow, 5–6.4 mm long, tube glabrous, lobes 1–1.3 mm long. Achenes 2.5 mm long, shortly hairy; pappus 4–6 mm long.

Kenya. Lamu District: between Mkokoni and Aswe, Sep. 1956, *Rawlins* 105! & Simambaya Is., E side, Oct. 1957, *Greenway & Rawlins* 9421! & Kiunga Point, July 1961, *Gillespie* 19!
Distr. **K** 7; Somalia
Hab. Sand dunes, in crevices in coral; near sea level–10 m
Uses. None recorded on specimens from our area
Conservation notes. Restricted area and habitat; 5 Kenyan specimens; data deficient (DD) but probably Vulnerable due to the delicate nature of the habitat

Syn. *Senecio bellioides* Chiov., Fl. Somal. 2: 272 (1932)

9. **Emilia basifolia** *Baker* in K.B. 1898: 154 (1898); C. Jeffrey in K.B. 41: 913 (1986); Lisowski, Aster. Fl. Afr. Centr. 2: 368 (1991). Type: Malawi, Zomba, *Whyte* s.n. (K!, holo.)

Annual herb 20–50 cm tall, sometimes slightly scrambling; stems pubescent in lower part. Leaves often dense near base of plant, sessile, narrowly obovate or elliptic, 1.5–4.5 cm long, 0.5–2.4 cm wide, base cuneate, margins ± sinuate, obscurely denticulate, apex obtuse, pubescent. Capitula solitary, terminal on long stalk, discoid, stalks of the individual capitula glabrous; involucre cylindrical, 5.5–6.5 mm long, 3.5 mm in diameter; phyllaries 11, 5–7 mm long, glabrous. Ray florets absent; disc florets orange, 4.5–5.2 mm long, lobes 1.5–1.7 mm long. Achenes 1.5–2.2 mm long, glabrous; pappus 3.5–4 mm long.

Tanzania. Ufipa District: between Lake Tanganyika and Lake Rukwa, 1896, *Nutt* s.n.! & 15 km on Sumbawanga–Chala road, Mar. 1994, *Bidgood et al.* 2556! & Tatanda Mission, Feb. 1994, *Bidgood et al.* 2454!
Distr. **T** 4; Congo (Kinshasa), Zambia, Malawi
Hab. Moist grassland; 1750–1800 m
Uses. None recorded on specimens from our area
Conservation notes. 3 specimens from Tanzania, but probably least concern (LC)

10. **Emilia kivuensis** (*Muschl.*) *C.Jeffrey* in K.B. 41: 913 (1986); U.K.W.F. ed. 2: 223 (1994). Type: Rwanda/Uganda, Mt Muhavura, *Mildbraed* 1853 (B†, holo., BR!, iso.)

Erect annual herb 20–50 cm high, sometimes described as perennial; stems sparsely to densely pubescent or villous in lower part. Leaves sessile, lanceolate, oblanceolate or spatulate, 2.3–8 cm long, 0.5–1.8 cm wide, the basal cuneate-attenuate into exauriculate base, the median and upper auriculate and semi-amplexicaul, margins shortly remotely sinuate-dentate, apex acute to rounded or obtuse, apiculate, very sparsely to densely shortly pubescent. Capitula discoid, erect, long-stalked, 2–10 in lax terminal corymbs; stalks of the individual capitula glabrous. Involucre cylindrical 5.5–7 mm long, 2–3.5 mm in diameter; phyllaries 7–9(–13), 5–6.5 mm long, glabrous. Ray florets absent; disc florets yellow or orange-yellow; corolla 4–5.5 mm long, tube glabrous, lobes 0.7–1 mm long. Achenes 2.2–2.5 mm long, hairy; pappus 4–5 mm long.

UGANDA. Kigezi District: Buambara, Nov. 1950, *Purseglove* 3503!; Mengo District: Kyadondo, Naguru Hill, May 1969, *Rwaburindore* 29!; Masaka District: Masaka city S, Apr. 1972, *Lye* 6728!
KENYA. Uasin Gishu District: Kipkarren, Mar. 1932, *Brodhurst-Hill* 748!
TANZANIA. Bukoba District: Bunazi, Oct. 1931, *Haarer* 2329!; Ngara District: Bushubi, Keza, Nov. 1959, *Tanner* 4485!; Biharamulo District: Nyabugombe, Mwibumba, Nov. 1960, *Tanner* 5584a!
DISTR. U 2, 4; **K** 3; **T** 1; Rwanda
HAB. Grassland, often with *Loudetia kagerensis*; 1050–1650 m
USES. Minor medicine agains yaws (*Tanner*)
CONSERVATION NOTES. Due to its distribution and habitat, probably least concern (LC)

SYN. *Senecio kivuensis* Muschl. in Z.A.E. 2: 390 (1911)
     S. sp. A of U.K.W.F.: 478 (1974)

NOTE. *Tallantire* 661 from **U** 2, mile 40 Masaka–Mbarara road, might be a hybrid between *E. kivuensis* and *E. caespitosa.* Phyllaries 6 mm long, sparsely crispate; florets orange; outer florets female, subradiate with 4-lobed outer lip and 1-lobed inner lip; outer lip 1.5 mm long; tube 5 mm long; inner florets tubular; ovary 1.5 mm long, hairy; pappus 4.5 mm long; corolla 6 mm long, the tube with a few hairs, lobes 1 mm long.
     Lisowski, Aster. Fl. Afr. Centr. 2: 261 (1991) has *E. kivuensis* as a synonym of *E. debilis.*

11. **Emilia caespitosa** *Oliv.* in Trans. Linn. Soc. 29: 100 (1873); Oliv. & Hiern, F.T.A. 3: 405 (1877); Maquet in Fl. Rwanda 3: 646, fig. 200/1 (1985); C. Jeffrey in K.B. 41: 913 (1986); Lisowski, Aster. Fl. Afr. Centr. 2: 383 (1991); C. Jeffrey in K.B. 52: 210 (1997). Type: Tanzania, Bukoba District, Karagwe, *Grant* 464 (K!, holo.)

Annual herb 4–90 cm high; stem erect, sparsely pubescent at least towards the base, or glabrous. Leaves subsucculent, light green, sometimes purplish beneath or near margins, sessile, the lower circular, spatulate or obovate, upper obovate, elliptic to linear-lanceolate, 0.8–15 cm long, 0.4–4 cm wide, base cuneate or attenuate, in upper semi-amplexicaul, margins remotely sinuate-denticulate to -serrate, obtuse to acute, sparsely pubescent or glabrous. Capitula discoid, 2–4 in terminal subumbelliform to lax corymbs; involucre broadly cylindrical, 4–8 mm long, 2.5–3 mm in diameter; phyllaries 8–13, 4–8 mm long, sparsely pubescent or glabrous. Disc florets bright orange, occasionally red, 5.5–12.5 mm long, shortly hairy; lobes 1.5–2.8 mm long. Achenes 2.2–4 mm long, glabrous or rarely short-hairy; pappus 3.5–7 mm long.

UGANDA. Kigezi District: Kisoro, Nov. 1948, *Eggeling* 5865!; Ankole District: Ruizi R., Nov. 1950, *Jarrett* 173! & 6 km NE of Rubaare, Dec. 1968, *Lye & Lester* 600!
TANZANIA. Mwanza District: Rubondo I., Oct. 1985, *FitzGibbon & Barcock* 48!; Kondoa District: mountain near Kondoa-Irangi, Aug. 1932, *Geilinger* 1705!; Mbeya District: Tukuyu, Musekera stream, Mar. 1986, *Bidgood et al.* 110!
DISTR. U 2; **T** 1, 5–7; Congo (Kinshasa), Rwanda, Burundi, Sudan, Ethiopia, Angola, Zambia, Malawi, Mozambique, Zimbabwe
HAB. Grassland, abandoned cultivations, woodland; may be locally common; 300–1900 m
USES. Minor medicine against syphilis (*Roscoe*)
CONSERVATION NOTES. Widespread, so least concern (LC)

SYN. *E. macaulayae* Garab. in K.B. 1924: 140 (1924). Type: Zambia, near Mumbwa, *Macaulay* 659 (K!, holo.)

E. humbertii Robyns in B.J.B.B. 17: 102 (1943). Type: Congo (Kinshasa), mountains to SW of Lake Edward, *Humbert* 8276 (BR!, holo.)

E. humbertii Robyns var. *angustifolia* Robyns in B.J.B.B. 17: 102 (1943). Type: Congo (Kinshasa), near Nzulu, Gahogo, *de Witte* 1399 (BR!, holo.)

12. **Emilia jeffreyana** *Lisowski* in Polish Bot. Stud. 1: 91, t. 4c (1991) & Aster. Fl. Afr. Centr. 2: 379 (1991): C. Jeffrey in K.B. 52: 208 (1997). Type: Congo (Kinshasa), Mt Kahuzi, *Lisowski* 85628 (POZG, holo.)

Annual herb 0.4–1.5 m high; stems erect, branched, sparsely pilose near base. Leaves sessile, the basal oblanceolate-spatulate, 6–11 cm long, 1.5–3 cm wide, base attenuate into a pseudopetiole, apex acute; stem leaves lanceolate to oblong, base auriculate with dentate margins and semi-amplexicaul, apex acute to acuminate, glabrous above, puberulous to glabrous beneath. Capitula 2–4, eventually long-stalked, discoid, many-flowered; involucre 6–10 mm long, 2.5 mm in diameter, glabrous; phyllaries 8–13, 6–10 mm long, with black papillose apex. Disc florets yellow to bright orange, corolla 7.5–10 mm long, lobes 1–2 mm long. Achenes 1.5–3 mm long, hairy; pappus 3.5–6 mm long.

KENYA. Northern Frontier District: Sololo, Aug. 1952, *Gillett* 13691!
DISTR. **K** 1; Congo (Kinshasa), Rwanda, Burundi, Ethiopia
HAB. Swamp edge, on black cotton soil; ± 740 m
USES. None recorded on specimens from our area
CONSERVATION NOTES. Widespread, so least concern (LC)

SYN. *E. coccinea* auct., *non* (Sims) G. Don, pro parte

NOTE. Lisowski reports that this species occurs in Uganda; we have seen no specimens from Uganda.
　　Sterile specimens resemble *Sonchus*, because of the long leaves that are sagittate or hastate at base.

13. **Emilia vanmeelii** *Lawalrée* in B.J.B.B. 19: 225 (1949); Lisowski, Aster. Fl. Afr. Centr. 2: 381 (1991); C. Jeffrey in K.B. 52: 210 (1997). Type: Tanzania, (**T** 4; district: not found), Utinta, *van Meel* 918 (BR!, holo.)

Annual herb 30–100 cm high; stems erect, pubescent in lower part. Leaves slightly fleshy, often purplish beneath and on the margins, the lower broadly reniform-spatulate to obovate, 2–13 cm long, 1–7 cm wide, decurrent onto a narrow exauriculate petioloid base, margins subentire, apex rounded; the median and upper obovate, oblong, ovate or lanceolate, 2–9 cm long, 0.5–6.5 cm wide, base auriculate and semi-amplexicaul, margins shallowly sinuate-dentate, apex rounded to acute, sparsely pubescent, especially beneath. Capitula 1–4 in terminal corymbs, dense to lax, discoid; stalks glabrous or shortly pubescent; involucre broadly cylindrical, 5–7.5 mm long, 3–3.5 mm in diameter; phyllaries 13–21, 4.5–7 mm long, glabrous or pubescent. Disc florets red, sometimes (but rarely in East Africa) orange or yellow, corolla 6–10.5 mm long, lobes ± 1.8 mm long. Achenes 2.5–3.5 mm long, sparsely hairy or glabrous; pappus 3–6 mm long.

TANZANIA. Ufipa District: Kipili, Feb. 1950, *Bullock* 2392!; Rungwe District: Kyela, Lema Forest, May 1983, *R. Abdallah* 1363! & Chivanje, Tukuyu Tea Estates, Mar. 1987, *Lovett & Congdon* 1780!
DISTR. **T** 4, 5, 7; Congo (Kinshasa), Zambia, Malawi, Mozambique
HAB. Woodland, more common in disturbed situations; 750–1700 m
USES. None recorded on specimens from our area
CONSERVATION NOTES. Widespread, so least concern (LC)

NOTE. Specimens with hairy achenes may represent the result of introgression from *E. coccinea*.

14. **Emilia coccinea** (*Sims*) *G.Don* in Sw., Hort. Bot. ed. 3: 382 (1839); Oliv. & Hiern, F.T.A. 3: 407 (1877); F.P.S. 3: 28 (1956); C. Jeffrey in K.B. 41: 913 (1986); Blundell, Wild Flow. E. Afr.: fig. 424 (1987); U.K.W.F. ed. 2: 224, t. 94 (1994); C. Jeffrey, Fl. Masc. 109: 160 (1993) & in K.B. 52: 209 (1997). Type: from a plant cultivated at Vauxhall, London (†); lectotype: t. 564 in Sims, Bot. Mag. 16 (1803), chosen by Jeffrey

Annual herb, 15–120 cm high, erect; stems pubescent or glabrous. Leaves ± subsucculent, often glaucous, sometimes purplish beneath and on margins, spatulate, obovate or elliptic, the upper also ovate or oblong, 1.2–19.5 cm long, 0.4–6 cm wide, cuneate or attenuate into a winged exauriculate petioloid base in lower leaves, upper leaves also cordate, hastate or sagittate and semi-amplexicaul, margins remotely sinuate-dentate or subentire, apex rounded to obtuse, rarely acute, apiculate, varyingly pubescent beneath at least on midrib, glabrescent. Capitula in terminal corymbs of 1–6, first sub-umbelliform, later lax, discoid; capitulum stalk glabrous or with a few hairs; involucre broadly cylindrical, 4.5–9.5 mm long, 2–5 mm in diameter; phyllaries light green, 8–21, 4–9 mm long, glabrous to densely setiferous or hirsute at least towards the apex. Disc florets bright orange, less often orange-red, on the coast occasionally with a tinge of yellow, corolla 5.3–9.5 mm long, lobes 1–2.2 mm long, glabrous or shortly hairy in upper part. Achenes 2–4.7 mm long, shortly hairy at least in upper part; pappus 3–6 mm long. Fig. 122 (page 584).

UGANDA. Karamoja District: Kabong, May 1959, *J. Wilson* 731!; Busoga District: Busoga, July 1905, *E. Brown* 288!; Mengo District: Bombo–Masindi road, Apr. 1959, *Lind* 2393!
KENYA. W Suk District: near Mt Marobus, Oct. 1977, *Carter & Stannard* 15!; Masai District: Masai Mara Game Reserve, Aug. 1971, *Kokwaro & Mathenge* 2681!; Kwale District: Mukaka, Aug. 1982, *Robertson* 3344!
TANZANIA. Shinyanga, May 1931, *B.D. Burtt* 2445!; Pare District: Kisiwani, Jan. 1960, *Williams Sangai* 726!; Kilosa District: Ilonga, Matarawe R., May 1968, *Renvoize & Abdallah* 2359!; Zanzibar, Dunga, Jan. 1929, *Greenway* 1160!
DISTR. **U** 1, 3, 4; **K** 1–7; **T** 1–7; **Z**, **P**; Congo (Kinshasa), Burundi, Sudan, Angola, Zambia, Mozambique, Zimbabwe
HAB. Grassland, bushed grassland; one of the commonest weeds of cultivation on the coast and to a lesser degree elsewhere; 0–1800(–2400) m
USES. Leaves widely used as vegetable; minor medicinal against syphilis (*Koritschoner, Tanner*), nose disease (*Koritschoner*), swollen legs (*Tanner*) and swollen throat (*Tanner*), as chest medicine (*Jeffery*) and as burn treatment (*Gaetan*)
CONSERVATION NOTES. Widespread, so least concern (LC)

SYN. *Cacalia coccinea* Sims in Bot. Mag. 16, t. 564 (1803)
    Emilia flammea sensu auct., *non* Cass.
    E. sagittata sensu auct. e.g. Oliv. & Hiern, F.T.A. 3: 405 (1877); U.O.P.Z.: 241 (1949), *non* DC.
    E. javanica sensu Agnew, Upl. Kenya Wild Fl.: 480 cum ic. (1974), *non* (Burm.f.) C.B. Rob.

NOTE. The name *E. coccinea* has been applied in a broad sense so as to include most of the large-headed African *Emilia*. Lisowski (1991) formalized the informal groups of Jeffrey (1986), and Jeffrey (1997) discussed correct application of names for these entities.

15. **Emilia tenuipes** *C.Jeffrey* in K.B. 41: 913 (1986). Type: Tanzania, Dodoma District, 11 km S of Dodoma, *Polhill & Paulo* 1262 (K!, holo.)

Erect annual herb to 60 cm tall; stems almost glabrous, slightly mauve-tinged. Leaves petiolate; blade broadly reniform, 2–6.5 cm long, 3–7 cm wide, exauriculate, base cordate to cordate-subtruncate or broadly cuneate in upper leaves, margins shallowly sinuate-dentate, apex rounded, pilose beneath near base, otherwise glabrous; petiole slender, slightly mauve-tinged to purple, glabrous, 3–8.5 cm long. Capitula solitary, discoid, long-stalked; involucre cylindrical, 10–11 mm long, ± 7 mm in diameter; phyllaries 8, green, glabrous, 10–11 mm long. Ray florets absent; disc

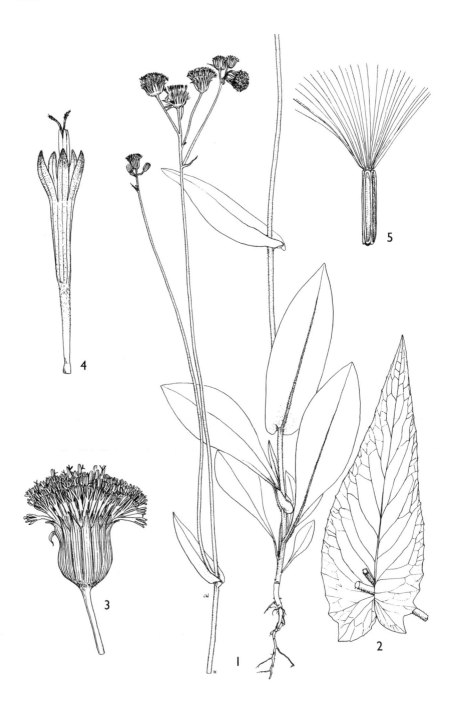

Fig. 122. *EMILIA COCCINEA* — **1**, habit, × ¹/₂; **2**, leaf, × ¹/₂; **3**, capitulum, × 3; **4**, disc floret, × 8; **5**, achene, × 8. 1 from *Tanner* 3611, 2 from *Drummond & Hemsley* 3492, 3 & 5 from *Renvoize & Abdallah* 1524, 4 from *Sandford* 2. Drawn by Juliet Williamson.

florets bright orange to orange-yellow, 9–11 mm long, tube glabrous, lobes 3–3.5 mm long, those of the outer florets the longer. Achenes 6.5 mm long, slightly rostrate, shortly hairy; pappus 6–8 mm long.

TANZANIA. Dodoma District: 11 km S of Dodoma, Jan. 1962, *Polhill & Paulo* 1262!; Iringa District: Udzungwa Mts National Park, Ikula, June 2002, *Luke et al.* 8701!
DISTR. **T** 5, 7; not known elsewhere
HAB. Dry woodland, edge of dry bushland; 950–1250 m
USES. None recorded on specimens from our area
CONSERVATION NOTES. Known from only two specimens; probably vulnerable (VU-D2)

16. **Emilia juncea** *Robyns* in B.J.B.B. 17: 104 (1943); Maquet in Fl. Rwanda 3: 646, fig. 200/4 (1985); C. Jeffrey in K.B. 41: 914 (1986); Lisowski, Aster. Fl. Afr. Centr. 2: 395, t. 83 (1991). Type: Congo (Kinshasa), Rutshuru Plain, May-ya-Moto, *de Witte* 2033 (BR!, holo.)

Erect annual herb 30–90(–120) cm tall; stems glabrous. Leaves sessile, oblanceolate, narrowly elliptic or almost linear, 2.4–13 cm long, 0.2–1 cm wide, base attenuate, margins subentire or minutely sinuate-dentate, apex obtuse, glabrous or almost so. Capitula 2 or more in lax corymbs, discoid; stalks of the individual capitula long, glabrous; involucre cylindrical, 5–7.5 mm long, 1.5–2.5 mm in diameter; phyllaries 7–13, usually 8, 4.5–7 mm long, glabrous. Ray florets absent; disc florets yellow or orange; corolla 5.7–8 mm long; tube glabrous; lobes 0.5–0.8 mm long. Achenes 2.2–3.4 mm long, angular, pubescent; pappus 4–6 mm long.

var. **juncea**; C. Jeffrey in K.B. 41: 914 (1986)

Corolla 5.7–6.5 mm long. Achenes 2.2–2.5 mm long; pappus ± 4.5 mm long.

UGANDA. Ankole District: Igara, Lubare Ridge, May 1970, *Lye & Katende* 5479!; Kigezi District: Bufumbira, Bukimbiri, Oct. 1947, *Purseglove* 2499!; Mubende District: Singo, SW of Niko Hill, Mar. 1970, *Lye et al.* 5132!
TANZANIA. Bukoba District: Minziro Forest Reserve, Nov. 1999, *Kayombo & Mwiga* 3145! & idem, same date, *Festo et al.* 471!
DISTR. **U** 2, 4; **T** 1; Congo (Kinshasa), Burundi
HAB. Grassland on rocky hill-sides; 1150–1950 m
USES. None recorded on specimens from our area
CONSERVATION NOTES. 5 specimens from Uganda, 2 from Tanzania; because of its distribution and habitat probably least concern (LC)

NOTE. The material from Tanzania seems to have a rootstock, so possibly this is occasionally a short-lived perennial.

var. **iringensis** *C.Jeffrey* in K.B. 41: 914 (1986). Type: Tanzania, Iringa District, Ruaha National Park, Magangwe Hill, *Bjørnstad* 2686 (K!, holo.)

Corolla ± 8 mm long. Achenes ± 3.4 mm long; pappus ± 6 mm long.

TANZANIA. Iringa District: Ruaha National Park, Magangwe Hill, Mar. 1973, *Bjørnstad* 2686!
DISTR. **T** 7; known from only the type
HAB. Mixed miombo woodland; ± 1540 m
USES. None recorded on specimens from our area
CONSERVATION NOTES. Known from only the type; at least vulnerable (VU-D2)

17. **Emilia cenioides** *C.Jeffrey* in K.B. 41: 914 (1986). Type: Tanzania, Mbulu District, Tarangire National Park, *Richards* 25360 (K!, holo.)

Erect annual herb 45–80 cm tall; stems almost glabrous. Stem leaves sessile, lanceolate to oblanceolate (in the lower) or oblong, 3–10 cm long, 0.4–2 cm wide, base auriculate and semi-amplexicaul, margins subentire, apex obtuse to subacute, apiculate, almost glabrous. Capitula solitary, long-stalked, discoid; stalk glabrous, swollen beneath the capitulum; involucre broadly cylindrical, 6.5–8.5 mm long, 4–12 mm in diameter; phyllaries 8–9, 6–8.5 mm long, glabrous. Ray florets absent; disc florets bright orange or yellow; corolla 8–8.2 mm long, lobes 2 mm long. Achenes 1.6–2.4 mm long, shortly hairy; pappus 3.5–4.5 mm long.

TANZANIA. Mbulu District: Tarangire National Park, Tarangire Swamp, Feb. 1970, *Richards* 25360 (K!, holo.); Singida District: Mgori Dam, Aug. 1999, *Kuchar* 22341! & Apr. 2000, *Kuchar* 23373!
DISTR. **T** 2, 5; not known elsewhere
HAB. Sedge or grass swamp, or grassland on black clay; 1050–1300 m
USES. None recorded on specimens from our area
CONSERVATION NOTES. Known from only the cited specimens, possibly vulnerable (VU-D2)

18. **Emilia leucantha** *C.Jeffrey* in K.B. 41: 914 (1986). Type: Tanzania, Ufipa District, Malonje Plateau, *Richards* 8426 (K!, holo.)

Erect annual herb 15–50 cm tall; stems scattered-pubescent. Leaves sessile, narrowly elliptic or lanceolate to obovate, 1.5–11.5 cm long, 0.3–1.3 cm wide, base cuneate, margins remotely obscurely sinuate-denticulate, apex acute, scattered-pubescent to almost glabrous. Capitula 1–3 in lax terminal cymes, discoid, long-stalked; stalk shortly hairy; involucre cylindrical, 6–6.5 mm long, 2 mm in diameter; phyllaries 5.5–6 mm long, shortly setulose, green. Florets white (Peter says yellow); corolla 4.8–5.7 mm long, tube glabrous, lobes 1.5–2 mm long, those of the outer florets the longer. Achenes 2.2 mm long, sparsely hairy; pappus 3–4 mm long.

TANZANIA. Buha District: Mkigo to Lussimbi, Mar. 1926, *Peter* 38779!; Ufipa District: Malonje Plateau, Nsanga Mt., Mar. 1959, *Richards* 12141! & 15 km on Sumbawanga–Chala road, Mar. 1994, *Bidgood et al.* 2559!
DISTR. **T** 4; Zambia
HAB. Damp sites in grassland, seepage areas on sand; 1550–2100 m
USES. None recorded on specimens from our area
CONSERVATION NOTES. 4 specimens from Tanzania; data deficient (DD)

19. **Emilia debilis** S.*Moore* in J.L.S. 37: 172 (1905); C. Jeffrey in K.B. 41: 913 (1986); Lisowski, Aster. Fl. Afr. Centr. 2: 394 (1991); U.K.W.F. ed. 2: 224, t. 95 (1994). Type: Uganda, Kigezi District, Rukiga, *Bagshawe* 444 (BM!, holo.)

Annual herb 10–45 cm tall; stems erect, light green, scattered-pubescent. Leaves light green, sometimes purple beneath, sessile, elliptic, the upper also obovate or narrowly obovate, elliptic or linear, 1.8–7.5 cm long, 0.3–1.4 cm wide, cuneate into a petioloid base or in upper leaves often attenuate, margins subentire, remotely shallowly sinuate-serrate, apex obtuse, apiculate, shortly pubescent. Capitula 2–5 in lax terminal corymbs; stalks of the individual capitula shortly pubescent; involucre cylindrical, 5.5–8 mm long, 1.5–2 mm wide; phyllaries (7–)8(–9), green with yellow to brown tips, 5–8 mm long, pilose at least in upper half. Florets white, cream or pale yellow, corolla 4.5–6 mm long, tube glabrous, lobes 1–1.5 mm long. Achenes 1.7–2.2 mm long, glabrous; pappus 3.5–7 mm long.

UGANDA. Kigezi District, Rukiga, no date, *Bagshawe* 444!; Mbale District: Sebei, Kabururon, Aug. 1955, *Norman* 273!
KENYA. Trans Nzoia District: 13 km from Kitale on Cherangani road, Aug. 1961, *Symes* 751!; Nandi District: 12 km from Nandi Hills on Lessos road, Oct. 1981, *Gilbert & Mesfin* 6739!; N Nyeri District: Liki R., no date, *Battiscombe* 1313!
DISTR. **U** 2, 3; **K** 3–5; Congo (Kinshasa), Rwanda, Burundi

HAB. Moist sites in grassland and woody grassland, drier spots in swamps; 1600–2750 m
USES. None recorded on specimens from our area
CONSERVATION NOTES. 16 East African specimens; probably least concern (LC)

SYN. *E. kikuyorum* R.E.Fr. in Act. Hort. Berg. 9: 147 (1928); Maquet in Fl. Rwanda 3: 646, fig.
200/2 (1985). Type: Kenya, Mt Kenya, Forest Station, *Fries & Fries* 302 (UPS!, holo.,
BM!, BR, iso.)

NOTE. Two specimens from Tanzania key out to this taxon: **T** 4, Buha District: Keza Mission, May
1994, *Bidgood & Vollesen* 3291! and **T** 7, Iringa District: Mufindi, Lake Ngwazi, Mar. 1991, *Bidgood
& Vollesen* 2152! Both are from *Brachystegia* woodland and from shallow pockets of soil on rocks,
or from heavy lateritic loam; they are erect annual herbs with leaves that are purple underneath,
and the flowers are yellow or pale yellow. They differ from *E. debilis* in having two basal leaves
that are broadly elliptic to almost round and 1–2.5 cm long, while the very few cauline leaves are
narrower and 1.5–3 cm long and 0.4–1.1 cm wide. Inflorescence, floral and fruit characters are
otherwise the same as for *E. debilis* but it does not have the feel, or the habitat of, that species.

20. **Emilia ukingensis** (*O.Hoffm.*) *C.Jeffrey* in K.B. 41: 915 (1986). Type: Tanzania,
Njombe District, Ukinga Mts, *Goetze* 971 (B†, holo., K!, iso.)

Tufted subscapose perennial herb 5–40 cm tall; stems sparsely pubescent to almost
glabrous. Leaves sessile, oblanceolate, oblanceolate-linear or linear, 1.5–5.3 cm long,
0.1–0.7 cm wide, base cuneate-attenuate, margins sinuate-denticulate or remotely
sinuate-dentate, apex obtuse, shortly scattered-pubescent, especially beneath.
Capitula solitary, long-stalked, discoid; stalk almost glabrous except for fine hairs just
below capitulum, or entirely glabrous; involucre cylindrical, 5.5–6 mm long, 2–3 mm
in diameter; phyllaries 8, green, very sparsely shortly crispate or glabrous, 4.5–5.5 mm
long. Disc florets yellow; corolla 4–5 mm long, tube glabrous, lobes 1.2–2 mm long.
Achenes 2 mm long, shortly hairy; pappus 2–3.5 mm long.

TANZANIA. Iringa District: main road between Mafinga [Sao Hill] and Lugoda turn-off, Mar.
1986, *Bidgood & Keeley* 341!; Njombe District: Mlangali–Njombe road, 2500 m, Nov. 1978,
Richards 2662! & Kipengere Mts, Jan. 1957, *Richards* 7723!
DISTR. **T** 7; not known elsewhere
HAB. Moist sites in grassland, rock crevices; 1800–2950 m
USES. None recorded on specimens from our area
CONSERVATION NOTES. 8 specimens seen, ?near threatened (?NT)

SYN. *Senecio ukingensis* O.Hoffm. in E.J. 30: 435 (1901)

NOTE. A polymorphic species, varying considerably in leaf-width and degree of development of
a caespitose habit.

21. **Emilia tenuis** *C.Jeffrey* in K.B. 41: 915 (1986). Type: Tanzania, Ufipa District,
Chapota, *Richards* 8491 (K!, holo.)

Weakly erect herb ± 40 cm high; stems glabrous. Leaves sessile, narrowly oblong to
linear, 1–5.2 cm long, 0.1–0.5 cm wide, base attenuate, margins entire, apex
attenuate, glabrous. Capitula 1–2 in lax terminal corymbs, long-stalked, discoid;
stalks of the individual capitula glabrous or minutely puberulous below the
capitulum; involucre cylindrical, 6.5 mm long, 1.5–2 mm in diameter; phyllaries 8,
green, glabrous or sparsely puberulous, 6 mm long. Disc florets orange; corolla
7–7.3 mm long, tube glabrous, lobes 1.2–2.2 mm long, those of the outer florets the
larger. Achenes 2 mm long, glabrous; pappus 4.5–5 mm long.

TANZANIA. Ufipa District: Chapota, Mar. 1957, *Richards* 8491!
DISTR. **T** 4; known from only the type
HAB. Marsh; 1650 m
USES. None recorded on specimens from our area
CONSERVATION NOTES. Known from only the type; at least vulnerable (VU-D2)

22. **Emilia longifolia** *C.Jeffrey* in K.B. 41: 915 (1986). Type: Uganda, Teso, Serere, *Chandler* 83 (K!, holo.)

Slender erect herb 45–60 cm tall; stems glabrous. Leaves linear or linear-lanceolate, 4.5–10 cm long, 0.3–0.6 cm wide, base attenuate, margins subentire, apex acute, shortly pubescent and glabrescent or glabrous. Capitula 1–4, lax, discoid; stalks of the individual capitula slender, glabrous or almost so; involucre cylindrical, 4.5–6.5 mm long, 2 mm in diameter; phyllaries 8–9, 4–6 mm long, glabrous. Disc florets white or very pale yellow; corolla 4.7–7.2 mm long, lobes 1.1–1.7 mm long. Achenes 1.8–2.3 mm long, glabrous; pappus 2.3–3.5 mm long.

UGANDA. Teso District: Serere, Nov.–Dec. 1931, *Chandler* 83! & Omunyal Swamp, Sep. 1954, *Lind* 371!
TANZANIA. Ngara District: Nyakisasa, Kibirizi, Mar. 1961, *Tanner* 5882!
DISTR. **U** 3; **T** 1; not known elsewhere
HAB. Swamp; 1050–1650 m
USES. None recorded on specimens from our area
CONSERVATION NOTES. Two specimens from Uganda, one from Tanzania, in a vulnerable habitat; at least vulnerable (VU-D2)

23. **Emilia cryptantha** *C.Jeffrey* in K.B. 41: 915 (1986). Type: Tanzania, Bukoba District, Bugandika, *Haarer* 2457 (K!, holo.)

Slender erect annual herb 30–55 cm tall; stems varyingly pubescent, especially in lower part, to glabrous or almost so. Leaves sessile, the basal obovate, the median and upper oblanceolate, narrowly elliptic or lanceolate, 2.4–10.5 cm long, 0.2–1.5 cm wide, attenuate with petioloid base, margins remotely obscurely sinuate-serrate or sinuate-denticulate, apex obtuse, minutely apiculate, pilose to almost glabrous. Capitula 2–5 in lax terminal corymbs, discoid, long-stalked; stalks of the individual capitula slender, shortly sparsely pubescent, especially just below the capitulum, to glabrous or almost so; involucre cylindrical, 6–8.5 mm long, 1–2 mm in diameter; phyllaries (4–)8, 5.5–8 mm long, glabrous or sparsely pubescent. Disc florets deep magenta to purple, corollas 6.3–6.8 mm long, lobes 1–1.2 mm long. Achenes 2.2–2.4 mm long, glabrous; pappus 5–6 mm long.

UGANDA. Masaka District: 4–5 km N of Lake Nabugabo, Sep. 1969, *Lye et al.* 4344! & Kalungu county, 0.5 km S of W Mengo–Masaka border, Dec. 1970, *Lye* 5824!; Mengo District: Kampala, Kings Lake, July 1935, *Chandler-Hancock* 5!
TANZANIA. Bukoba District: near Kitwe, Oct. 1931, *Haarer* 2200! & Bugandika, Jan. 1932, *Haarer* 2457!
DISTR. **U** 4; **T** 1; not known elsewhere
HAB. Swamp grassland; 1100–1250 m
USES. None recorded on specimens from our area
CONSERVATION NOTES. Five specimens from Uganda, two from Tanzania, in a vulnerable habitat. Said to be frequent in one swamp. At least vulnerable (VU-D2)

24. **Emilia longipes** *C.Jeffrey* in K.B. 41: 915 (1986). Type: Tanzania, Manyoni District, Kazikazi, *B.D. Burtt* 3717 (K!, holo.)

Annual herb, caespitose, erect, much branched from the base, 15–40 cm tall; stems sparingly pubescent in lower part, glabrescent, or glabrous. Leaves sessile, crowded, oblanceolate, elliptic or linear-lanceolate, 2.5–10.8 cm long, 0.3–1.8 cm wide, base sometimes somewhat attenuate, margins remotely obscurely to distinctly sinuate-dentate or sinuate-serrate, apex obtuse to acute, sparsely pubescent with long hairs to almost glabrous, glabrescent. Capitula solitary, discoid, long-stalked, stalks of the individual capitula glabrous; involucre cylindrical, 6–7.5 mm long, 2.5–3 mm in diameter; phyllaries 7–13, green, 5.5–7 mm long, glabrous. Disc florets pale lilac to purple, corolla 5.2–7.5 mm long, lobes 1.3–2 mm long. Achenes 2.5–3.4 mm long, shortly hairy, with tufts of longer hairs at base and apex; pappus 2.5–5.3 mm long.

TANZANIA. Mwanza, no date, *Davis* 193!; Shinyanga District: near Seke, May 1931, *B.D. Burtt* 2459!; Tabora District: Itobo, Mar. 1979, *Acres* 112!
DISTR. **T** 1, 4, 5; not known elsewhere
HAB. Marshy grassland, seasonally wet grassland, moist grassland; 1100–1300 m
USES. None recorded on specimens from our area
CONSERVATION NOTES. Known from ten specimens, but from a fairly common habitat; probably least concern (LC)

25. **Emilia violacea** *Cronquist* in B.J.B.B. 22: 309 (1952); C. Jeffrey in K.B. 41: 915 (1986); Lisowski, Aster. Fl. Afr. Centr. 2: 371 (1993). Type: Burundi, Muhovuozi Plains, *Michel & Reed* 982 (BR!, holo.)

Erect or decumbent annual herb 14–80 cm tall; stems light green, smooth, glabrous. Leaves sessile, lanceolate, 2.2–13 cm long, 0.4–2 cm wide, base sagittate, auriculate and semi-amplexicaul, margins minutely obscurely to clearly sinuate-denticulate, especially on the auricles, apex obtuse, apiculate, glabrous. Capitula 2–4 in lax, divaricate terminal cymes, discoid; stalks of the individual capitula slender, glabrous; involucre cylindrical, 5–7 mm long, 2 mm in diameter; phyllaries 8 or 13, 4.5–7 mm long, glabrous. Disc florets pale mauve or violet, corolla 5.5–6.5 mm long, tube glabrous, lobes 1.2–2 mm long, those of the outer florets the larger. Achenes 2.8–4 mm long, glabrous; pappus 2–4 mm long.

TANZANIA. Mwanza, no date, *Davis* 273!; Kigoma District: Uvinza, Feb. 1958, *Friend* 35!; Mpanda District: 15 km on Karema road from Mpanda–Uvinza road, Mar. 1994, *Bidgood et al.* 2713!
DISTR. **T** 1, 4; Congo (Kinshasa), Burundi, Zambia
HAB. Moist grassland or swamp grassland; 950–1150 m
USES. None recorded on specimens from our area
CONSERVATION NOTES. Known from six specimens altogether, data deficient (DD)

26. **Emilia aurita** *C.Jeffrey* in K.B. 41: 915 (1986). Type: Tanzania, Rungwe District, Kyimbila, *Stolz* 1267 (K!, holo., M!, iso.)

Presumably annual herb ± 50 cm tall; stems smooth, glabrous. Leaves sessile, lanceolate, 5–11 cm long, 0.4–1.6 cm wide, base sagittate, auriculate and semi-amplexicaul, margins obscurely sinuate-denticulate, especially on the auricles, apex obtuse, apiculate, glabrous. Capitula 4–6 in lax, divaricate terminal cymes, discoid; stalks of the individual capitula slender, glabrous; involucre cylindrical, 5–6.5 mm long, 2 mm in diameter; phyllaries 12–14, 4.5–5.5 mm long, glabrous. Disc florets mauve, corolla 5.5 mm long, tube glabrous, lobes 1–1.2 mm long. Achenes 1.4 mm long (very immature), glabrous; pappus 3.5–4 mm long.

TANZANIA. Rungwe District: Kyimbila area, May 1912, *Stolz* 1267!
DISTR. **T** 7; known from only the type
HAB. Not recorded
USES. None recorded on specimens from our area
CONSERVATION NOTES. Known from only the type; at least vulnerable (VU-D2) but probably extinct (Ex)

27. **Emilia kilwensis** *C.Jeffrey* in K.B. 41: 916 (1986). Type: Tanzania, Kilwa District, Selous Game Reserve, Mtwatawa, *Ludanga* 980 (K!, holo.)

Erect annual herb ± 40 cm tall; stems scattered-pubescent. Leaves sessile, the lower broadly spatulate, median oblanceolate or elliptic, upper lanceolate, 1–7 cm long, 0.3–2 cm wide, base cuneate in lower leaves, shortly sagittate in upper ones, margins obscurely sinuate, apex obtuse, apiculate, scattered-pubescent, at least on midrib beneath. Capitula 3–4 in lax terminal corymbs, discoid; stalks of the individual capitula pilose near apex; involucre cylindrical, 5–5.5 mm long, 1.5 mm in diameter; phyllaries 8, 5.5–6 mm long, pubescent. Disc florets presumably purple; corolla 5–5.7 mm long, tube glabrous, lobes 1.3–1.5 mm long, those of the outer florets the larger; ovary 0.8 mm long, hairy; pappus 3 mm long.

TANZANIA. Kilwa District: Selous Game Reserve, Mtwatawa, Mar. 1970, *Ludanga* 980!
DISTR. **T** 8; known from only the type
HAB. Not recorded; ± 125 m
USES. None recorded on specimens from our area
CONSERVATION NOTES. Known from only the type; at least vulnerable (VU-D2)

28. **Emilia micrura** *C.Jeffrey* in K.B. 41: 916 (1986). Type: Tanzania, Tunduru District, W of Pucha Pucha, *Milne-Redhead & Taylor* 7722 (K!, holo.)

Annual herb 22–30 cm tall; stems erect or ascending, pale green, glabrous. Leaves pale green, linear or linear-elliptic, 2.8–6 cm long, 0.2–0.4 cm wide, base attenuate, apex attenuate and obtuse, glabrous. Capitula 3–4, lax, discoid, on slender peduncles; involucre cylindrical, 4.5–5 mm long, ± 2.5 mm in diameter; phyllaries 10–14, very pale green, 3–4 mm long, glabrous. Disc florets magenta, style arms white, anthers purple, corolla 4.6–5.5 mm long, lobes 1.2–1.5 mm long. Achenes 1.4 mm long, sparsely hairy; pappus 2 mm long.

TANZANIA. Tunduru District: W of Pucha Pucha and 1.5 km E of Mawese R., Dec. 1956, *Milne-Redhead & Taylor* 7722!
DISTR. **T** 8; known from only the type
HAB. Boggy grassland on sand; ± 450 m
USES. None recorded on specimens from our area
CONSERVATION NOTES. Known from only the type; at least vulnerable (VU-D2)

29. **Emilia crispata** *C.Jeffrey* in K.B. 41: 916 (1986); Lisowski, Aster. Fl. Afr. Centr. 2: 389 (1991). Type: Tanzania, Iringa District, Kilolo, *Polhill & Paulo* 1422 (K!, holo., BR!, iso.)

Erect annual herb 24–60 cm tall; stems pubescent. Leaves sessile, oblanceolate or elliptic, 2–6 cm long, 0.3–0.7 cm wide, base obtuse, margins entire, apex acute, densely shortly pubescent. Capitula solitary, rarely in pairs, discoid; stalk slender, dull purple-brown, almost glabrous; involucre cylindrical, 5–6 mm long, ± 4 mm in diameter; phyllaries 8–9, green at base, purplish distally, almost glabrous, 5–6 mm long. Disc florets purple-mauve, anthers purple, style mauve with white arms, corolla 5.5–6 mm long, lobes 1.5 mm long. Achenes 1.7–2 mm long, glabrous; pappus 2.5–4 mm long.

TANZANIA. Tanzania, Iringa District, Dabaga Highlands, Kilolo, Feb. 1962, *Polhill & Paulo* 1422!
DISTR. **T** 7; Congo (Kinshasa)
HAB. Grassland with scattered trees; ± 1830 m
USES. None recorded on specimens from our area
CONSERVATION NOTES. Known from only the type and a single specimen from Congo (Kinshasa); at least vulnerable (VU-D2)

30. **Emilia simulans** *C.Jeffrey* in K.B. 41: 916 (1986). Type: Tanzania, Tabora District, SE of Goweko, *Peter* 35099 (K!, holo.)

Erect annual herb 30–45 cm high; stems pubescent. Leaves sessile, narrowly elliptic to lanceolate, 2.5–6.3 cm long, 0.5–0.7 cm wide, base obtuse, margins obscurely sinuate-dentate, apex obtuse, apiculate, pilose. Capitula 2–3, lax, discoid; stalks of the individual capitula slender, glabrous; involucre 5.3–7 mm long, 2–4 mm in diameter; phyllaries 8, 5 mm long, glabrous. Disc florets white, corolla 6 mm long, lobes 1–1.5 mm long. Achenes 2 mm long, shortly sparsely hairy; pappus 3–4.5 mm long.

TANZANIA. Tabora District, SE of Goweko, Jan. 1926, *Peter* 35099!
DISTR. **T** 4; known from only the type
HAB. 'Pori', possibly thorn-bushland; ± 1180 m
USES. None recorded on specimens from our area
CONSERVATION NOTES. Known from only the type; at least vulnerable (VU-D2)

31. **Emilia fugax** *C.Jeffrey* in K.B. 41: 916 (1986). Type: Tanzania, Kigoma District, Lake Tanganyika, Kibwesa Point, *Newbould & Jefford* 1667 (K!, holo.)

Annual herb 40–50 cm high; stems smooth, glabrous. Leaves sessile, narrowly elliptic, 1–5 cm long, 0.1–0.4 cm wide, base obtuse, margins entire, apex attenuate and obtuse, glabrous. Capitula solitary or several in lax long-stalked groups, discoid; involucre 5–6.5 mm long, 2.5–3 mm in diameter; phyllaries 7–8, ± 5 mm long, glabrous but for the penicillate apex. Disc florets pale yellow, corolla 4–4.6 mm long, lobes 0.8–1.4 mm long. Achenes 2.5–3.2 mm long, very shortly hairy; pappus 2–3 mm long.

TANZANIA. Kigoma District: Lake Tanganyika, Kibwesa Point, Aug. 1956, *Newbould & Jefford* 1667!
DISTR. **T** 4; Zambia
HAB. Small patch of *Cyperus* in bushland; 900 m
USES. None recorded on specimens from our area
CONSERVATION NOTES. One specimen from Tanzania; data deficient (DD)

NOTE. 'Rapid ephemeral'; the Tanzanian specimen, in full flower, was collected ten days after rain.

32. **Emilia flaccida** *C.Jeffrey* in K.B. 41: 916 (1986). Type: Tanzania, Kigoma District, 22 km N of Nguruka, Kitolo, *Windisch-Graetz* in *Bally* 7625 (K!, holo.)

Slender herb at least 50 cm high; stems glabrous. Leaves narrowly lanceolate, 3–8.5 cm long, 0.2–0.3 cm wide, attenuate at base and apex, glabrous. Capitula 2–3, lax, discoid; stalks of the individual capitula glabrous; involucre cylindrical, 5 mm long, 4 mm in diameter; phyllaries 10, 4.5 mm long, glabrous but for the penicillate apex. Disc florets mauve; corolla 5–6 mm long, lobes 1–1.2 mm long. Achenes 2 mm long, glabrous; pappus 4 mm long.

TANZANIA. Kigoma District: 22 km N of Nguruka, Kitolo, Oct. 1949, *Windisch-Graetz* in *Bally* 7625!
DISTR. **T** 4; known from only the type
HAB. Lakeside grassland; ± 1150 m
USES. None recorded on specimens from our area
CONSERVATION NOTES. Known from only the type; at least vulnerable (VU-D2)

33. **Emilia rigida** *C.Jeffrey* in K.B. 41: 916 (1986). Type: Tanzania, Kigoma District, 58 km S of Uvinza [Uvinsa], *Bullock* 3277 (K!, holo.)

Erect annual herb 30–40 cm tall; stems slender, glabrous. Leaves sessile, lanceolate, 1–2 cm long, 0.1–0.2 cm wide, base cuneate, margins sinuate-denticulate, apex acute, glabrous. Capitula 2 in lax terminal corymbs, discoid; stalks of the individual capitula very slender, glabrous; involucre cylindrical, 4.5 mm long, 1.5 mm in diameter; phyllaries 8, 4 mm long, glabrous. Disc florets purple; outer corollas 4.5 mm long, unequally lobed, with 2 large outer lobes 2.5 mm long and 3 smaller inner lobes 1.8 mm long; inner corollas 3.7 mm long, with lobes 1.5 mm long. Achenes 2 mm long, glabrous; pappus 1.5 mm long, scant.

TANZANIA. Kigoma District, 58 km S of Uvinza [Uvinsa], Aug. 1950, *Bullock* 3277!
DISTR. **T** 4; known from only the type
HAB. Peaty bog overlying rock; ± 1700 m
USES. None recorded on specimens from our area
CONSERVATION NOTES. Known from only the type; at least vulnerable (VU-D2)

NOTE. *Bidgood, Leliyo & Vollesen* 4512, collected in May 2000 from **T** 4, Mpanda District: Uzondo
Plateau, 05°29'S 30°3 32'E at 1675 m, is a small annual from seasonally inundated grassland.
It is virtually leafless but the few lower leaves visible are sub-orbicular and stalked, and ± 6 ×
4 mm. The few upper leaves visible are semi-amplexicaul at base and up to 20 × 3 mm. This
would key out near *E. rigida* but differs in the shape of both upper and lower leaves, and is
smaller in size (7–25 cm high) as well as in all of its parts. Lisowski in his Aster. Fl. Afr. Centr.
2 (1991) also has no taxon resembling this collection. It is placed here provisionally as *E. sp.
aff. rigida.*

34. **Emilia integrifolia** *Baker* in K.B. 1895: 69 (1895); C. Jeffrey in K.B. 41: 917
(1986); Lisowski, Aster. Fl. Afr. Centr. 2: 370 (1991); U.K.W.F. ed. 2: 224, t. 94 (1994).
Type: Zambia, Fwambo, *Carson* 102 (K!, lecto., chosen by Jeffrey)

Annual or short-lived perennial herb 10–125 cm tall; stems slender, erect or basally
decumbent and then erect, pubescent to glabrous, sometimes purplish distally.
Leaves sessile, somewhat subsucculent, the basal spatulate to oblanceolate or
narrowly elliptic, 1.1–6.5 cm long, 0.2–1.3 cm wide, attenuate into petioloid base,
obscurely sinuate-denticulate, apex obtuse to acute; median and upper sessile,
narrowly oblanceolate, narrowly elliptic, elliptic, lanceolate or linear, 3–16 cm long,
0.2–0.7 cm wide, base attenuate, subentire or sinuate-serrate to sinuate-denticulate,
apex obtuse to acute; all leaves scattered-pubescent to glabrous. Capitula 3–14 in lax
to congested or subumbelliform terminal corymbs, discoid; stalk slender, glabrous or
shortly sparsely pubescent; involucre cylindrical, 4–6.5 mm long, 1–2 mm in
diameter; phyllaries 6–9 or 13, usually 8, green tinged purplish in distal part,
glabrous or shortly pubescent at least towards the apex. Disc florets bright to pale
purple, mauve, pink or white, corolla 4–6.2 mm long, lobes 1–2 mm long. Achenes
1.2–1.5 mm long, glabrous; pappus 2.5–4 mm long.

UGANDA. W Nile District: Koboko [Kobboko], no date, *Eggeling* 1847!; Mbale District:
Kapchorwa rest house, Oct. 1952, *G.H.S. Wood* 434!; Mengo District: 20 km S of Nakasongola,
Feb. 1956, *Langdale-Brown* 1957!
KENYA. Trans Nzoia District: Kaptabat, Brockley Primary School, Oct. 1982, *Gilbert & Mesfin*
6455!; Nakuru District: Nakuru–Kericho road just E of Londiani turnoff, Oct. 1971, *Gillett*
19328!; Masai District: Londiani, Aug. 1949, *Maas Geesteranus* 5858!
TANZANIA. Ufipa District: Sumbawanga, June 1968, *Sanane* 189! & 2 km on Tatanda–Mbala
road, Apr. 1997, *Bidgood et al.* 3380!; Njombe District: Ludewa, Itimbo, Nov. 1987, *Mwasumbi
et al.* 13431!
DISTR. **U** 1–4; **K** 3, 5; **T** 4, 7, ?8; Congo (Kinshasa), Zambia, Malawi, Madagascar
HAB. Swamp grassland, moist grassland; may be locally common; 950–2550 m
USES. None recorded on specimens from our area
CONSERVATION NOTES. Least concern

35. **Emilia tenera** (*O.Hoffm.*) *C.Jeffrey* in K.B. 41: 917 (1986). Type: Tanzania,
Njombe District, Ukinga, *Goetze* 939 (B†, holo., BM!, K!, iso.)

Annual or short-lived perennial herb 12–60 cm tall; stems delicate, erect, sparsely
pubescent, at least at the base, or almost glabrous. Leaves sessile; basal leaves
obovate, oblanceolate or broadly elliptic, 1–3.6 cm long, 0.3–1.4 cm wide, cuneate or
attenuate into a petioloid base, margins remotely denticulate, apex obtuse, apiculate;
median and upper leaves oblanceolate-spatulate, oblanceolate to elliptic, 1.2–5.7 cm
long, 0.3–1.3 cm wide, margins sinuate-denticulate, sparsely shortly pubescent.
Capitula 1–3 in lax terminal corymbs, discoid; stalks of the individual capitula

puberulous to pubescent, especially near apex; involucre cylindrical, 5.5–6.5 mm long, 1–2 mm in diameter; phyllaries 8, shortly pubescent especially near the apex, 5–6 mm long. Disc florets mauve or purple, corolla 4–5.3 mm long, lobes 1.2–1.7 mm long. Achenes 1.7–2.1 mm long, glabrous or with a few short hairs; pappus 3–4.5 mm long.

UGANDA. Mt Elgon, Kapchorwa, Sep. 1954, *Lind* 269!
TANZANIA. Morogoro District: Uluguru Mts, Lukwangulo Plateau, Jan. (no year), *Bruce* 692!; Njombe/Mbeya District: Kitulo Plateau above Matamba, Nov. 1986, *Brummitt & Congdon* 18113!; Rungwe District: Rungwe Forest Reserve, S side, Feb. 1986, *Bidgood et al.* 101!
DISTR. U 3; T 6, 7; not known elsewhere
HAB. Stream-banks, swampy grassland or grassland to almost bare ground on lava; 1800–2700 m
USES. None recorded on specimens from our area
CONSERVATION NOTES. One specimen from Uganda, fourteen from Tanzania, but widespread, so probably least concern (LC)

SYN. *Senecio tener* O.Hoffm. in E.J. 30: 434 (1901)

36. **Emilia limosa** (*O.Hoffm.*) *C.Jeffrey* in K.B. 41: 917 (1986); Lisowski, Aster. Fl. Afr. Centr. 2: 369 (1991). Type: Angola, Bie, *Baum* 907 (B†, holo., K!, iso.)

Annual or short-lived perennial herb 15–75 cm tall; stems slender, weak, usually erect, hardly branched, glabrous. Leaves sessile, oblanceolate, narrowly elliptic, lanceolate or almost linear, 1.4–9 cm long, 0.1–1.9 cm wide, remotely obscurely denticulate, sparsely shortly pubescent at least beneath, often glabrescent, sometimes almost glabrous. Capitula 1–4 in lax terminal corymbs, discoid; stalks of the individual capitula glabrous or sparsely pubescent; involucre cylindrical, 5–7 mm long, 1.5–2 mm in diameter; phyllaries 8(–9), green tinged purple, tipped reddish, 4.5–6 mm long, glabrous or sparsely pubescent at least near apex. Disc florets purple or mauve, corolla 5.5–8 mm long, lobes 1.5–2.7 mm long, those of the outer florets the larger. Achenes 1.7–3 mm long, glabrous; pappus 3.5–4 mm long.

TANZANIA. Iringa District: Mufindi, Lake Ngwazi, May 1968, *Renvoize & Abdallah* s.n.!; Njombe District: Kitulo Plateau, Ndumbi Valley, Mar. 1991, *Bidgood et al.* 2136!; Songea District: Mawesu Valley, July 1956, *Semsei* 2422!
DISTR. T 7, 8; Congo (Kinshasa), Angola, Zambia, Malawi, Zimbabwe, South Africa
HAB. Moist grassland, swamp grassland, peat bog; (990–)1650–2800 m
USES. None recorded on specimens from our area
CONSERVATION NOTES. Widespread; least concern (LC)

SYN. *Senecio limosus* O.Hoffm. in Warb., Kunene–Samb. Exped.: 422 (1903)

37. **Emilia pammicrocephala** (*S.Moore*) *C.Jeffrey* in K.B. 41: 917 (1986); Lisowski, Aster. Fl. Afr. Centr. 2: 374 (1991). Type: Congo (Kinshasa), Mt Senga, *Kassner* 2974 (BM!, holo., HBG, iso.)

Short-lived perennial herb 15–70 cm tall; stems decumbent, glabrous except sometimes towards the base. Leaves sessile, green tinged purple; lower broadly ovate, weakly cordate, subrotundate or broadly cuneate and narrowly decurrent onto petioloid base, margins sinuate-serrate, apex rounded; median and upper spatulate, oblanceolate or elliptic-lanceolate, 1.7–4.2 cm long, 0.2–1.3 cm wide, decurrent onto a broadly winged, basally semi-amplexicaul petioloid base; all shortly pubescent beneath, less so to almost glabrous above, purple-tinged. Capitula 7–10 or more in moderately lax corymbs, discoid; stalks of individual capitula glabrous; involucre cylindrical, 2.5–4.5 mm long, 1–1.5 mm in diameter; phyllaries (5–)8(–10), 3–4 mm long, glabrous. Disc florets purple or violet, corolla 2.5–3.5 mm long, lobes 1–1.5 mm long, those of the outer florets the longer. Achenes 1.4–1.5 mm long, shortly hairy; pappus 2–3 mm long.

UGANDA. Kigezi District: Mgahinga–Muhavura saddle, Sep. 1946, *Purseglove* 2188! & Apr. 1970, *Lye & Katende* 5317! & Mt Sabinio, Dec. 1959, *Miller* 602!
DISTR. U 2; Congo (Kinshasa), Rwanda
HAB. Grassland, montane forest or *Hypericum* woodland, bamboo glades; 2200–3000 m
USES. None recorded on specimens from our area
CONSERVATION NOTES. Nine specimens from Uganda; data deficient (DD)

SYN. *Senecio pammicrocephalus* S.Moore in J.L.S. 47: 279 (1925); Maquet in Fl. Rwanda 3: 668, fig. 208/2 (1985)
      S. chiovendeanus sensu auct., *non* Muschl. (1911)

NOTE. The real *S. chiovendeanus* Muschl. in Z.A.E.: 383 (1911) is now *Emilia chiovendeana* (Muschl.) Lisowski and is a taxon close to *E. pammicrocephala*. It occurs just across the border in Congo (Kinshasa) and Rwanda.

38. **Emilia myriocephala** *C.Jeffrey* in K.B. 41: 917 (1986). Type: Tanzania, Kigoma District, Kungwe Mt, Selimweguru, *Newbould & Harley* 4621 (K!, holo.)

Probably annual herb ± 40 cm tall; stems sparsely pubescent towards the base. Leaves pale purplish, sessile; lower ovate, cordate to broadly cuneate and narrowly decurrent onto an exauriculate petioloid base; upper broadly decurrent and auriculate at the base, 1–4 cm long, 0.3–2 cm wide, all coarsely sinuate-dentate, shortly scattered-pubescent. Capitula numerous in congested corymbs, discoid; stalks of individual capitula slender; involucre cylindrical, 2.5 mm long, 1 mm in diameter; phyllaries 5, 2 mm long, glabrous. Disc florets pale magenta, corolla 3 mm long, lobes 1.5 mm long. Achenes 1 mm long, shortly hairy; pappus 1.5 mm long.

TANZANIA. Kigoma District: Kungwe Mt, Selimweguru, July 1959, *Newbould & Harley* 4621!
DISTR. T 4; known from only the type
HAB. 'Erosion gully on steep sandy slope of exposed ridge'; ± 2040 m
USES. None recorded on specimens from our area
CONSERVATION NOTES. Known from only the type; at least vulnerable (VU-D2)

39. **Emilia decipiens** *C.Jeffrey* in K.B. 41: 917 (1986). Type: Tanzania, Songea District, Matagoro Hills just S of Songea, *Milne-Redhead & Taylor* 9868 (K!, holo.)

Annual herb 10–60 cm tall; stems ascending, reddish-purple and shortly pubescent in lower part. Leaves sessile, bright green, glaucous and ± purple-tinged beneath, those on the lower part of the stem ovate or broadly ovate, 2.1–5 cm long, 1.2–2.5 cm wide, cordate and decurrent onto a narrow exauriculate petioloid base, margins sinuate-dentate, apex obtuse, pubescent beneath; those in the upper part of the stem elliptic to oblong, the lower of these often somewhat pinnate-lyrately lobed, 0.8–9 cm long, 0.2–3 cm wide, base expanded and auriculate, margins subentire to sinuate-dentate, apex obtuse. Capitula few in lax corymbs, discoid; stalks of the individual capitula slender, glabrous; involucre cylindrical, 3.5–4 mm long, 1 mm in diameter; phyllaries 8–10, bright green tinged purple, 3–3.5 mm long, glabrous. Disc florets mauve or pink; corolla 5–5.5 mm long, lobes 1.5–1.7 mm long. Achenes 1.3 mm long, hairy; pappus 3 mm long.

TANZANIA. Rungwe District: Vyeja near Rungwe Mission, May 1983, *R. Abdallah* 1356!; Songea District: Matagoro Hills just S of Songea, May 1956, *Milne-Redhead & Taylor* 9868! & NW side of Matagoro Hills, May 1956, *Milne-Redhead & Taylor* 9868a!
DISTR. T 7, 8; Mozambique, Malawi
HAB. Among rock outcrops on poor stony soil or in forest edge; 1200–2000 m
USES. None recorded on specimens from our area
CONSERVATION NOTES. Three specimens from Tanzania; data deficient (DD)

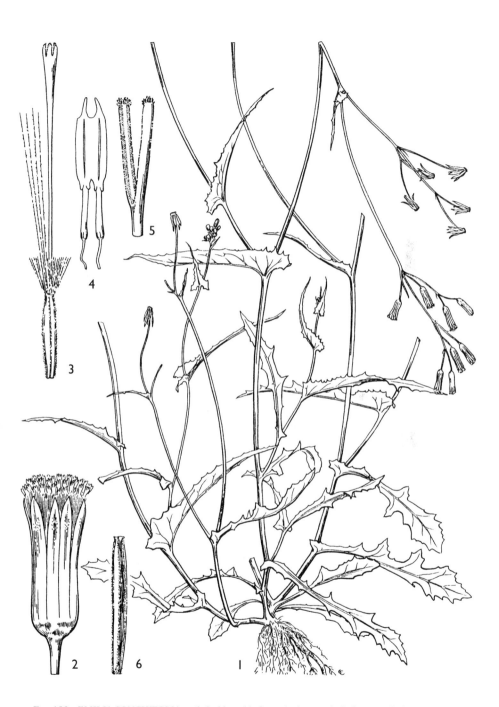

FIG. 123. *EMILIA SONCHIFOLIA* — **1**, habit, × ²/₃; **2**, capitulum, × 4; **3**, floret, × 8; **4**, stamens, × 30; **5**, style branches, × 30; **6**, achene, without pappus, × 5. 1–6 from *Rochecouste* 180. Drawn by Eleanor Catherine, from Flore des Mascareignes.

40. **Emilia sonchifolia** (*L.*) *Wight,* Contrib. Bot. Ind.: 24 (1834); Oliv. & Hiern, F.T.A. 3: (1877); C. Jeffrey in K.B. 41: 917 (1986); Lisowski, Aster. Fl. Afr. Centr. 2: 387 (1991): C. Jeffrey in Fl. Masc. 109: 160, t. 53 (1993). Type: Sri Lanka, *Hermann* s.n. (BM!, lecto.)

Annual herb with basal leaf rosette, 10–45 cm tall; stems erect or semi-erect, pubescent in lower part. Leaves sessile, spreading, subsucculent, often glaucous and/or purplish beneath and on margins, 1.8–9.5 × 0.5–6.5 cm, the lower spatulate, broadly ovate-reniform, broadly ovate or ovate, base cordate, subtruncate or cuneate and decurrent onto an entire or toothed lyrate-pinnately lobed exauriculate petioloid base, margins sinuate-serrate, apex rounded to obtuse, varyingly pubescent especially towards the base; median and upper ovate, lanceolate, oblanceolate or obovate or lyrato-pinnately lobed towards auriculate, semi-amplexicaul base, sinuate-dentate, obtuse, apiculate. Capitula discoid, 2–10 in lax terminal corymbs; peduncles slender, glabrous; involucre cylindrical, sometimes slightly swollen at the base, 7.5–10 mm long, 1.5–2 mm in diameter; phyllaries 8–9, glabrous except sometimes towards the apex or scattered-pubescent throughout, 7–9.5 mm long. Ray florets absent; disc florets reddish-purple, mauve, lavender, pink or pale pink; corollas 6–7.6 mm long, lobes 0.7–1 mm long. Achenes 2.7–3.3 mm long, shortly hairy; pappus 5.5–8 mm long. Fig. 123 (page 595).

KENYA. Lamu District: Mkune Swamp, Mar. 1957, *Rawlins* 418! & Witu, Oct. 1957, *Greenway & Rawlins* 9375!; Kilifi District: Kibarani, Aug. 1950, *Jeffery* 780!
TANZANIA. Uzaramo District: Dar es Salaam, University grounds, May 1968, *Harris & Mpanda* 1702!; Kilosa District: Kifungura km 17, July 1973, *Greenway & Kanuri* 15458!; Lindi District: Nachingwea, June 1953, *Anders* 907!
DISTR. **K** 7; **T** 6, 8; **P**; pantropical
HAB. Swampy grassland, weed of cultivation and gardens; 1–600 m
USES. None recorded on specimens from our area
CONSERVATION NOTES. Least concern (LC)

SYN. *Cacalia sonchifolia* L., Sp. Pl.: 835 (1753)

41. **Emilia mbagoi** *Beentje & Mesfin* in K.B. 59, 2: 325 (2004)
Typus: Tanzania, Nkansi District, 45 km on Namanyere–Karonga road, *Bidgood, Mbago & Vollesen* 2656 (K!, holo., DSM, iso.)

Annual herb 8–20 cm tall, with basal leaf rosette and 1–5 flowering stems; stems erect or semi-erect, little branched, glabrous. Leaves subsessile, the lower broadly elliptic, 1–2.7 cm long, 1–1.8 cm wide, base rounded but near the very base cuneate into a 2 mm petiole, margins entire or minutely undulate, apex rounded with a tiny obtuse mucro; upper leaves very few, more narrowly elliptic and smaller, semi-amplexicaul at base; all leaves subglabrous with some minute hairs on midrib and margins. Capitula 1–few in lax long-stalked terminal cymes, discoid but with the external flowers slightly zygomorphic; stalks of the individual capitula slender, glabrous; involucre cylindrical, sometimes slightly swollen at the base, 4–5 mm long, 2–2.5 mm in diameter; phyllaries 5–6, 4–5 mm long, connate for much of their length, puberulous in upper half or glabrous except for the minutely penicillate apex. Disc florets white; corolla tube 1.6–2.5 mm long, lobes 1–1.4 mm long, glabrous. Achenes to 2.1 mm long, setulose on the ribs; pappus sparse, 1–1.5 mm long, subplumose. Fig. 124 (page 597).

TANZANIA. Ufipa District, 45 km on Namanyere–Karonga road, March 1994, *Bidgood, Mbago & Vollesen* 2656! & 21 km on Kipili–Namanyere road, May 1997, *Bidgood et al.* 3815!
DISTR. **T** 4; not known elsewhere
HAB. *Brachystegia* woodland on sandy soil; 850–950 m
USES. None recorded on specimens from our area
CONSERVATION NOTES. Known from two specimens; at least vulnerable (VU-D2)

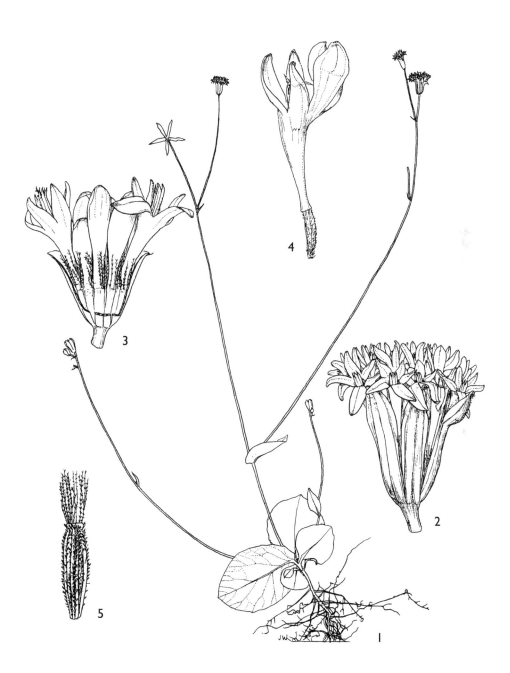

FIG. 124. *EMILIA MBAGOI* — **1**, habit, × 1; **2**, capitulum, × 8; **3**, capitulum, some phyllaries and florets removed, × 10; **4**, outer floret, × 12; **5**, achene and pappus, × 12. 1, 2, 4 from the type; 3, 5 from *Bidgood et al.* 3815. Drawn by Juliet Williamson.

Emilia pumila DC., Prodr. 6: 302 (1838); Oliv. & Hiern, F.T.A. 3: 403 (1877) and *E. humifusa* DC., Prodr. 6: 302 (1838), said to have been described from specimens collected by Bojer on Zanzibar Island, are in fact Madagascan species (Humbert 1963). The same probably applies to the doubtful species *E. gracilis* DC., Prodr. 6: 302 (1838); Oliv. & Hiern, F.T.A. 3: 406 (1877)

Note: *Emilia tessmannii* (Mattf.) C.Jeffrey is reported by Lisowski, in Aster. Fl. Afr. Centr. 2: 366 (1991), to occur in Uganda. We have seen no specimens from Uganda. The distribution area is just over the border in Congo (Kinshasa) and Burundi, so it is not impossible; authenticated specimens are awaited.

93. STENOPS

B.Nord. in Opera Bot. 44: 73 (1978)

Annual herbs. Leaves alternate. Capitula solitary or in few-headed lax cymes; heads small, radiate, with few florets. Involucre ecalyculate; phyllaries connate. Ray florets female; filaments with dilated collar. Achenes 5-ribbed or –winged, glabrous; pappus absent.

Two species in tropical Africa.

Stenops helodes *B.Nord.* in Opera Bot. 44: 73, t. 36 (1978). Type: Tanzania, Ufipa District, road to Katandalo Mission, *Sanane* 241 (K!, holo.)

Annual herb, slightly fleshy, glabrous; stem erect or decumbent, 20–40 cm long, often rooting at lower nodes. Leaves linear to narrowly elliptic, 1–8 cm long, 0.1–0.5 cm wide, base slightly tapering and slightly decurrent, margins entire, apex obtuse. Capitula in lax cymes; involucre campanulate, 1.5–2 mm long; phyllaries 11–13, connate to ± ²/₃, apex acute. Ray florets white or pale mauve, 6–10, tube 1–2 mm long, rays oblong, 4–5.5 mm long, 1.5–2.5 mm wide; disc florets yellow, many, ± 2 mm long. Achenes oblong, ± 1 mm long, 5-angled; pappus absent. Fig. 125 (page 599).

TANZANIA. Ufipa District: between Lake Tanganyika and Lake Rukwa, 1896, *Nutt* s.n.! & Mwimbe dambo on old Sumbawanga road, Apr. 1962, *Richards* 16354! & road to Katandalo Mission, July 1968, *Sanane* 241! (type)
DISTR. **T** 4; Zambia
HAB. In moist grassland or in wet mud; 1500–1950 m
USES. None recorded on specimens from our area
CONSERVATION NOTES. Three specimens from Tanzania, three from Zambia; at least vulnerable (VU-D2)

94. CRASSOCEPHALUM

Moench, Meth.: 516 (1794), nom. rejic. legit.

Senecio subgen. *Gynuropsis* Muschl. in E.J. 43: 38 (1909)

Perennial or annual herbs, erect or scrambling. Leaves alternate, entire to pinnatifid. Capitula solitary or in corymbs, few to many, radiate or discoid, many-flowered. Involucre cylindrical, calyculate; phyllaries narrow; receptacle epaleate. Corolla tube narrow, tapering towards the apex, 5-lobed. Anther collars narrow and elongated, anthers obtuse or slightly sagittate at base. Style arms with short to long

FIG. 125. *STENOPS HELODES* — **1**, habit, × ²/₃; **2**, part of older inflorescence, × ²/₃; **3**, underside of leaf, × 1; **4**, capitulum, × 6; **5**, ray floret, × 12; **6**, ray floret detail, × 20; **7**, disc floret, × 16; **8**, achene, × 24. 1 & 3 from *Richards* 16354, 2 & 4–7 from *Sanane* 241, 8 from *Nutt* s.n. Drawn by Juliet Williamson.

appendages of fused papillae. Achenes cylindrical or oblong, with short undulate or papillose strongly striate epidermal cells and strongly spirally thickened duplex hairs; pappus uniform, of many fine hairs.

A closely knit monophyletic group of 24 species in tropical Africa, Madagascar, the Mascarenes and Yemen.

1. **Crassocephalum splendens** *C.Jeffrey* in K.B. 41: 905 (1986). Type: Tanzania, Iringa District, Mt Imagi, *Richards* 15647 (K!, holo.)

Perennial herb 1.2–1.5 m high; stems thinly pubescent. Leaves ovate to elliptic, 7.5–13 cm long, 3–5.7 cm wide, cuneate or rounded into an exauriculate petioloid base, margins sinuate-serrate or sinuate-dentate, apex obtuse to acute, minutely apiculate, dark green and thinly setulose above, sparsely setulose on veins beneath. Capitula 3–6 in lax terminal corymbs, radiate; stalks of the individual capitula finely pubescent; involucre cylindrical, 10–11.5 mm long, 3.5–4 mm in diameter; bracts of calyculus ± 6, lanceolate, 1.5–5 mm long, ciliate on the margins; phyllaries 12–13, 9–10.5 mm long, glabrous. Ray florets yellow, 8–13, tube 7.5 mm long, glabrous, rays 12–15 mm long, 3–4.5 mm wide, 5–7-veined; disc florets orange or yellow, corolla 9–9.5 mm long, tube glabrous, gradually expanded from above the middle, lobes 0.7 mm long. Achenes 1.5–3.2 mm long, with long hairs in the grooves; pappus 9–10 mm long.

TANZANIA. Iringa District: Dabaga, near Kidabaga village, Aug. 1984, *Bridson* 575!; Mbeya District: Poroto Mts, road to Elton Plateau, Jan. 1961, *Richards* 14135! & Ilembo, Nov. 1975, *Aleljung* 623!
DISTR. **T** 7; not known elsewhere
HAB. Evergreen forest margins; 2000–2100 m
USES. None recorded on specimens from our area
CONSERVATION NOTES. Seven specimens from Tanzania; at least vulnerable (VU-D2)

2. **Crassocephalum ducis-aprutii** (*Chiov.*) *S.Moore* in J.B. 50: 212 (1912); A.V.P.: 222 (1957); Maquet in Fl. Rwanda 3: 650, fig. 202/1 (1985); C. Jeffrey in K.B. 41: 905 (1986); Lisowski, Aster. Fl. Afr. Centr. 2: 330 (1991). Type: Uganda, Ruwenzori, Bujongolo, *Roccati* s.n. (TO, holo.)

Perennial woody herb or shrub, erect to scrambling, 0.6–3(–4) m tall, the whole plant smelling of vanilla; stems pubescent to glabrous. Leaves lanceolate, elliptic to ovate, 3–24 cm long, 1–9 cm wide, base subtruncate, rounded or cuneate and slightly decurrent onto the nearly always auriculate petioloid base, margins sinuate-serrate or serrate-bidentate, apex acute to obtuse, sometimes somewhat attenuate and minutely acuminate-apiculate, glabrous or somewhat pubescent at least on veins beneath; auricles often, usually coarsely, lobed. Capitula discoid, numerous in moderately dense terminal cymes; stalks of the individual capitula pubescent; involucre broadly cylindrical, 4.5–9 mm long, 3–8 mm in diameter; bracts of calyculus 7–10, linear or lanceolate, 1.5–9.5 mm long, glabrous or with a few mostly marginal hairs; phyllaries 8–21, usually 13 or 21, 4–8 mm long, glabrous or thinly pubescent. Ray florets absent; disc florets yellow, corolla 6–11 mm long, tube glabrous, gradually expanded above the middle, lobes 0.7–1.3 mm long. Achenes 1–2 mm long, ribbed, long-hairy between the ribs; pappus 3.5–10 mm long.

UGANDA. Toro District: Ruwenzori, near Nyamileju Hut, Dec. 1968, *Lye* 1266!; Kigezi District: Mt Muhavura, Oct. 1948, *Hedberg* 2115! & Mgahinga [Gahinga] Volcano, Dec. 1930, *Burtt* 2849!
DISTR. **U** 2; Congo (Kinshasa), Rwanda; Ruwenzori and Virunga endemic
HAB. Giant heath or *Hypericum* zone; (2450–)2800–3750 m
USES. None recorded on specimens from our area
CONSERVATION NOTES. More than twenty specimens from Uganda; because of the altitude range, probably least concern (LC)

SYN. *Senecio ducis-aprutii* Chiov. in Ann. Bot. Roma 6: 150 (1907)
    S. gynuroides S.Moore in J.L.S. 38: 203 (1908). Type: Uganda, Ruwenzori, *Wollaston* s.n. (BM!, holo.)
    S. behmianus Muschl. in Z.A.E. 2: 401, t. 43 (1911). Type: Congo (Kinshasa), Ruwenzori, Butagu Valley, *Mildbraed* 2575 (B†, holo., BR!, iso. fragm.)

S. gynuropsis Muschl. in Z.A.E. 2: 404 (1911). Type: Congo (Kinshasa), Ruwenzori, Karisimbi, *Mildbraed* 1614 (B†, holo., BR!, iso., fragm.)
Crassocephalum behmianum (Muschl.) S.Moore in J.B. 50: 212 (1912)
Senecio behmianus Muschl. var. *variostipulatus* De Wild., Pl. Bequaert. 5: 97 (1929). Type: Congo (Kinshasa), Ruanoli Valley, *Bequaert* 4507 & Butagu Valley, *Bequaert* 3860 (both BR!, syn.)

3. **Crassocephalum montuosum** (*S.Moore*) *Milne-Redh.* in K.B. 5: 376 (1951); F.P.S. 3: 21 (1956); Maquet in Fl. Rwanda 3: 652, fig. 202/2 (1985); C. Jeffrey in K.B. 41: 906 (1986); Blundell, Wild Flow. E. Afr.: fig. 361 (1987); Lisowski, Aster. Fl. Afr. Centr. 2: 331, t. 72 (1991); U.K.W.F. ed. 2: 219, t. 91 (1994). Type: Kenya, Machakos to Kikuyu, *Scott Elliot* 6587 (BM!, holo.)

Annual or short-lived perennial herb or soft-wooded shrub, erect or sometimes semi-scandent, rather weak-stemmed, 25–250 cm tall; stems green, sometimes tinged red towards the base, pubescent, glabrescent. Leaves sessile, ovate, lanceolate, elliptic or obovate, unlobed to deeply and narrowly pinnately or pinnato-lyrately 2–8-lobed, 5.5–41 cm long, 1.5–22 cm wide, base rounded, cuneate or attenuate and slightly decurrent onto a petaloid base, margins closely sinuate-serrate to coarsely serrate-bidentate, apex acute to obtuse, often ± attenuate, almost glabrous to scattered-pubescent above and beneath especially on veins, ± glabrescent. Capitula numerous in congested terminal corymbs, discoid; stalks of the individual capitula densely pubescent to glabrous; involucre cylindrical, 5.5–10.5 mm long, 2–4 mm in diameter in bud, to 6 mm at anthesis; bracts of calyculus 6–13, linear to lanceolate, 1.5–7 mm long, dark-tipped, glabrous or ciliate-margined; phyllaries 8–22, usually 13–15, green or yellow-green with dark brown or reddish tips, 5–9.5 mm long, glabrous or sparsely shortly pubescent. Disc florets yellow or sometimes orange, corolla 5.5–9.5 mm long, tube glabrous, gradually expanded above the middle, lobes 0.5–1.5 mm long. Achenes 1.7–2.5 mm long, ribbed, shortly sparsely hairy in the grooves; pappus 5–10 mm long. Fig. 126 (page 603).

UGANDA. Ankole District: Kasyoha-Kitomi Forest Reserve, June 1994, *Poulsen et al.* 534!; Toro District: Kibale National Park, S of Ngogo Camp, June 1997, *Poulsen & Nkuutu* 1292!; Mbale District: N Elgon, Sit R., Jan. 1953, *Dawkins* 785!
KENYA. Northern Frontier District: Warges, Dec. 1958, *Newbould* 3063!; S Nyeri District: Mt Kenya, Karute R. 4 km above Castle Forest Station, Jan. 1985, *Townsend* 2187!; Masai District: Nguruman Escarpment, 9 km NE of Entasekera, Sep. 1977, *Fayad* 191!
TANZANIA. Kilimanjaro, Marangu, May 1961, *Machangu* 74!; Kigoma District: Selimwegeru, July 1959, *Newbould & Harley* 4637!; Iringa District: Luhega Forest Reserve, Feb. 1996, *Frimodt-Møller et al.* 95!
DISTR. U 1–4; K 1–6; T 1–4, 6, 7; from Nigeria to Ethiopia and S to Angola and Zimbabwe, Madagascar
HAB. Moist and evergreen forest, especially in forest edges, clearings etc., generally in disturbed places in the forest and bamboo zone; may be locally common; (360?–)900–3150 m
USES. Minor medicinal for infected eyes (*Tanner*); browsed by donkeys (*Glover*)
CONSERVATION NOTES. Least concern (LC)

SYN. *Senecio montuosus* S.Moore in J.L.S. 35: 354 (1902)
　　S. butaguensis Muschl. in Z.A.E.: 403 (1911). Type: Congo (Kinshasa), Ruwenzori, Butagu Valley, *Mildbraed* 2534 (B†, holo., BR!, iso.)
　　Crassocephalum butaguense (Muschl.) S.Moore in J.B. 50: 211 (1912)
　　Gynura lutea Humbert in Mem. Soc. Linn. Norm. 25: 122, 302 (1923). Type: Madagascar, Andasibe, *Perrier* 2928 (P!, K!, syn.) & Manongarivo, *Perrier* 3213 (P!, syn.)
　　Crassocephalum bumbense S.Moore in J.L.S. 47: 279 (1925). Type: Angola, Sobato de Bumba, *Welwitsch* 3687 (BM!, holo.)
　　C. afromontanum R.E.Fr. in Acta Hort. Berg. 9: 144, t. 6 (1928). Type: Kenya, Nyandarua/Aberdares, Kinangop, *Fries & Fries* 2727 (UPS!, holo.)
　　Senecio rufopilosulus De Wild., Pl. Bequaert. 5: 116 (1929). Type: Congo (Kinshasa), Ruwenzori, Butagu, *Bequaert* 3627 (BR!, lecto., chosen by Jeffrey)
　　Gynura montuosa (S.Moore) Bullock in K.B. 1932: 499 (1932)

Fig. 126. *CRASSOCEPHALUM MONTUOSUM* — **1**, habit, × ²/₃; **2**, leaf, × ¹/₃; **3**, leaf, × ¹/₃; **4**, capitulum, × 4; **5**, floret, × 6; **6**, floret detail, × 12; **7**, achene with pappus, × 6; **8**, achene, without pappus, × 12. 1 & 4–6 from *Frame* 4, 2 from *Geilinger* 4444, 3 from *Geilinger* 4088, 7–8 from *Pocs et al.* 88162. Drawn by Juliet Williamson.

Senecio afromontanum (R.E.Fr.) Humbert & Staner in B.J.B.B. 14: 104 (1936)
Crassocephalum luteum (Humbert) Humbert, Fl. Madag. 189: 836 (1963)

NOTE. Greenway (nr. 8445) observed that on one plant some capitula matured from the centre outwards while others matured from the margins inwards.

4. **Crassocephalum paludum** *C.Jeffrey* in K.B. 41: 906 (1986); Lisowski, Aster. Fl. Afr. Centr. 2: 335 (1991). Type: Uganda, Kigezi District, above Lake Bunyongi, Muko, *Morrison* 251b (K!, holo.)

Herb 30–200 cm tall, erect or sometimes straggling. Stems sometimes purplish, pubescent to glabrous. Leaves sessile, occasionally reddish, linear-lanceolate, lanceolate, elliptic or oblanceolate, 4–19 cm long, 0.4–0.7(–2.5) cm wide, cuneate-attenuate into a winged petioloid or non-petioloid exauriculate or usually auriculate base, margins remotely shallowly sinuate-denticulate to coarsely sinuate-dentate or sinuate-serrate, often with ± recurving teeth, apex obtuse to acute, sometimes attenuate, glabrous or densely to thinly pubescent, often ± glabrescent. Capitula 1–10 in congested to lax terminal corymbs, discoid; stalks of the individual capitula pubescent; involucre cylindrical, 6.5–12 mm long, 2.5–5 mm in diameter; bracts of calyculus 7–16, lanceolate, 1.5–5 mm long, glabrous or ciliate, dark-tipped; phyllaries 13–21, usually 13, green, 6–11 mm long, glabrous or sparsely pubescent. Ray florets 0; disc florets yellow, rarely orange, corolla 7–9.5 mm long, tube glabrous, gradually expanded above the middle, lobes 0.7–1.5 mm long. Achenes 1.5–1.7 mm long, ribbed, shortly hairy between the ribs; pappus 6–10.5 mm long.

UGANDA. Kigezi District: Lake Bunyonyi, Apr. 1970, *Katende & Lye* 149!; Teso District: 2–4 km NW of Serere Agric. Research Station, Nov. 1968, *Lye* 381!; Mengo District: Lutembe Bay, Oct. 1971, *Katende* 1335!
KENYA. Trans Nzoia District: NE Elgon, Kisano swamp, May 1969, *Tweedie* 3646!; Uasin Gishu District: Kipkarren, no date, *Brodhurst Hill* 303!; Kericho District: SW Mau Forest Reserve, Timbilil [Dimbilil] R., Aug. 1949, *Maas Geesteranus* 5672!
TANZANIA. Bukoba District: Karagwe, Nyashozi, Dec. 1931, *Haarer* 2390!; Mwanza District: Mwasonge, Bukumbi, Sep. 1952, *Tanner* 1021!; Buha District: Kaberi swamp, Aug. 1950, *Bullock* 3125!
DISTR. U 2–4; K 3, 5; T 1, 4; Congo (Kinshasa), Rwanda, Burundi, Sudan, Zambia
HAB. Swamps, river-banks, boggy grassland; may grow in water up to 90 cm deep or on floating islands; 1050–2400 m
USES. None recorded on specimens from our area
CONSERVATION NOTES. Over 40 East African specimens in fairly common habitat; least concern (LC)

SYN. *C. sp. A* of Maquet in Fl. Rwanda 3: 652, fig. 201/3 (1985)

5. **Crassocephalum picridifolium** *(DC.)* *S.Moore* in J.B. 50: 212 (1912); F.P.S. 3: 21 (1956); Hilliard, Comp. Natal: 369, fig. 13 (1977); Maquet in Fl. Rwanda 3: 652 (1985); C. Jeffrey in K.B. 41: 906 (1986); Blundell, Wild Flow. E. Afr.: fig. 362 (1987); Lisowski, Aster. Fl. Afr. Centr. 2: 336, t. 73 (1991); U.K.W.F. ed. 2: 220, t. 92 (1994). Type: South Africa, Natal, *Drège* 5161 (G-DC, holo.)

Perennial herb 30–120 cm long or high or scrambling to 180 cm, erect or with the lower stem prostrate and rooting at the nodes, sending up erect branches, sometimes clambering through other plants; stems green or streaked or tinged with purplish, red or brownish-red, glabrous to densely setulose with red hairs. Leaves obovate, oblanceolate, elliptic, lanceolate or ovate, sometimes distinctly pinnately or pinnato-lyrately 2–8-lobed in lower half, 3–15.5 cm long, 0.5–9 cm wide, sessile or cuneate to attenuate into a usually auriculate petioloid base, margins sinuate-dentate or sinuate-serrate to deeply sinuate-lobulate, especially towards the base, apex obtuse to acute, sometimes somewhat attenuate, pubescent, sometimes also setulose on veins

Fig. 127. *CRASSOCEPHALUM PICRIDIFOLIUM* — **1**, habit, × ²/₃; **2**, capitula, × ²/₃; **3**, leaf, × ²/₃; **4**, capitulum, × 3; **5**, phyllary, × 8; **6**, corolla, × 8; **7**, achene, × 6; **8**, achene, without pappus, × 16. 1 & 4 from *Smith et al.* 105, 2 from *Polhill* 143, 3 from *Mwangangi* 2278, 4–8 from *van Someren* 1622. Drawn by Juliet Williamson.

beneath, sometimes ± glabrescent. Capitula 1–5, terminal, lax, discoid, long-stalked; stalks of the individual capitula pubescent; involucre cylindrical, 9–13 mm long, 4–8 mm in diameter, the base wide and truncate; bracts of calyculus linear to lanceolate, 8–15, dark-tipped, 2.5–8 mm long, pubescent or at least ciliate-margined; phyllaries 13–27, usually 21, green with purplish or brownish tips, 8–13 mm long, pubescent or densely glandular, sometimes with the hairs tinged purplish, or rarely glabrous. Disc florets very many, yellow, orange-yellow or orange; corolla 7.2–11.5 mm long, tube glabrous, gradually expanded above the middle, lobes 0.7–1.7 mm long. Achenes 1.5–2.5 mm long, ribbed, shortly hairy between the ribs; pappus 7–12 mm long. Fig. 127 (page 605).

UGANDA. Kigezi District: Nyanja on Kabale–Mbarare road, Jan. 1953, *Norman* 198!; Masaka District: Maramagambo [Malabigambo] Forest 8 km SSW of Katera, Oct. 1953, *Drummond & Hemsley* 4605!; Mengo District: Entebbe, Lake shore, Sep. 1948, *Eggeling* 4931!
KENYA. Naivasha District: NW Lake Naivasha, May 1982, *Mwangangi* 2278!; Machakos District: Kithembe Hill, Kyiamwiitu communal spring, June 1982, *Mwangangi* 2324!; Masai District: Nasampolai Valley, July 1972, *Greenway & Kanuri* 15029!
TANZANIA. Lushoto District: Lushoto Sylviculture Nursery, Oct. 1981, *Shabani* 1292!; Mpwapwa District: Mpwapwa, Apr. 1949, *van Rensburg* 633!; Mbeya District: Poroto Mts, Livingstone Forest Reserve, Sep. 1970, *Thulin & Mhoro* 1236!
DISTR. U 2, 4; K 1, 3–7; T 1–8; W Africa from Gambia to Cameroon, Congo (Kinshasa), Sudan, Ethiopia, and S to South Africa
HAB. Swamps or swampy grassland; may be locally common or even mat-forming; (450–)1050–2700 m
USES. Minor medicinal for wounds (*Koritschoner*)
CONSERVATION NOTES. Over 50 E African specimens in fairly common habitat; least concern (LC)

SYN. *Senecio picridifolius* DC., Prodr. 6: 386 (1838); Oliv. & Hiern, F.T.A. 3: 413 (1877)
    S. acutidentatus A.Rich., Tent. Fl. Abyss. 1: 436 (1848). Type: Ethiopia, Shire, *Dillon* s.n. (P!, holo.)
    S. papaverifolius A.Rich., Tent. Fl. Abyss. 1: 437 (1848). Type: Ethiopia, Adowa, *Dillon* s.n. (P!, holo.)
    Gynura picridifolia (DC.) Burtt Davy in K.B. 1925: 569 (1925)

NOTE. Most likely a hybrid between *C. paludum* and *C. vitellinum.*
    The leaves are aromatic when crushed.

6. **Crassocephalum vitellinum** (*Benth.*) *S.Moore* in J.B. 50: 212 (1912); F.P.S. 3: 21 (1956); F.P.U.: 182, fig. 120 (1962); Maquet in Fl. Rwanda 3: 652, fig. 201/2 (1985); C. Jeffrey in K.B. 41: 907 (1986); Blundell, Wild Flow. E. Afr.: fig. 423 (1987); Lisowski, Aster. Fl. Afr. Centr. 2: 339, t. 74 (1991); U.K.W.F. ed. 2: 220 (1994). Type: Bioko [Fernando Po], *Vogel* s.n. (K!, holo.)

Annual or perennial herb 30–150(–350) cm tall, erect or basally procumbent and rooting at the nodes with erect flowering branches, or semi-scandent; stems green often streaked or tinged with purplish, red or violet, almost glabrous to densely setulose. Leaves sessile, ovate to broadly ovate, elliptic or obovate, 3–12.5 cm long, 1.5–7 cm wide, base cordate, truncate, cuneate or broadly cuneate and slightly decurrent onto sometimes pinnato-lyrately 2–4-lobed, auriculate or exauriculate, petioloid base, margins sinuate-serrate to sinuate-bidentate or broadly sinuate-lobulate, apex obtuse to acute, finely usually densely pubescent above and beneath, sometimes with purplish midrib. Capitula terminal, solitary or up to 3, discoid, long-stalked, sometimes nodding; stalks of the individual capitula densely glandular-pubescent, sometimes purplish; involucre broadly cylindrical, 7–12 mm long, 5–10 mm in diameter; bracts of calyculus 10–18, lanceolate, 2.5–7 mm long, glandular-setulose and ciliate; phyllaries usually 21, rarely 13, green or yellow-green with dark tips, 6.5–11 mm long, densely glandular-setulose with the hairs sometimes purplish. Disc florets orange, less often orange-yellow or yellow; corolla 6.5–10 mm long, tube glabrous, gradually expanded above the middle, lobes 0.7–1.5 mm long. Achenes 2 mm long, ribbed, shortly hairy between the ribs; pappus 6–9 mm long.

UGANDA. Karamoja District: Mt Lonyili [Longili], Apr. 1960, *J. Wilson* 887!; Kigezi District: Kachwekano, July 1945, *A.S. Thomas* 4201!; Mengo District: Makerere Hill, July 1952, *Lind* 104!
KENYA. Uasin Gishu District: Kaptagat, Brockley Primary School, Oct. 1982, *Gilbert & Mesfin* 6456!; Kiambu District: Muguga Forest, Sep. 1965, *Kokwaro & Kabuye* 348!; Masai District: Olekaitorror Escarpment 34 km on Narok–Nairobi road, July 1962, *Glover & Samuel* 3137!
TANZANIA. Bukoba District: Minziro Forest Reserve, Kabobwa, Sep. (no year), *Watkins* 497!; Kigoma District: Ujamba, July 1958, *Mahinde* 100a!; Iringa District: Rufuna Forest, 1982, *R. Abdallah* 1151!
DISTR. U 1–4; K 3–6; T 1, 4, 7; Nigeria, Cameroon, Bioko, Congo (Kinshasa), Rwanda, Burundi, Zambia
HAB. Forest margins and clearings, grassland and bushland in the forest zone, occasionally in swamps (especially in Uganda); may be locally common; 750–2800 m
USES. Minor medicinal for infected eyes (*Tanner*)
CONSERVATION NOTES. Least concern (LC)

SYN. *Gynura aurantiaca* Benth. in Hook., Niger Fl.: 437 (1849), *nom. illegit.*, *non* (Bl.) DC. (1838).
Type as above
G. vitellina Benth. in Hook., Niger Fl.: 438 (1849); Oliv. & Hiern, F.T.A. 3: 402 (1877)

7. **Crassocephalum uvens** (*Hiern*) *S.Moore* in J.B. 50: 212 (1912); C. Jeffrey in K.B. 41: 907 (1986); Lisowski, Aster. Fl. Afr. Centr. 2: 342, t. 75 (1991). Type: Angola, Cacolovar R. between Ivantala and Quilengues, *Welwitsch* 3670 (BM!, holo.)

Perennial herb with short rhizome, lower part of stem creeping but becoming erect and 20–65 cm tall, subscapiform, pale green or slightly reddish-purple tinged, glabrous or pubescent. Leaves sessile, oblanceolate, 1–7 cm long, 0.2–1.2 cm wide, attenuate to an exauriculate base, margins remotely sinuate-denticulate or sinuate-serrate, apex obtuse, apiculate, sparsely shortly setulose or hispid, bright green above, paler beneath, sometimes with purplish margin. Capitula solitary, terminal, erect, long-stalked, discoid; stalks of the individual capitula pilose at least in upper part; involucre cylindrical, 8–11 mm long, 3.5–5 mm in diameter; bracts of calyculus ± 6, lanceolate, green or purple, 2.5–3 mm long; phyllaries 13–18(–21), purple or green tipped purple or brown, 7–10 mm long, glabrous or shortly sparsely pubescent. Ray florets 0; disc florets orange-yellow or orange; corolla 7.5–11 mm long, tube glabrous, gradually expanded above the middle, lobes 0.8–1 mm long. Achenes 3–4 mm long, glabrous; pappus 6.5–10 mm long.

TANZANIA. Mbeya District: Kitulo Plateau, between Nwela and Ishinga Mt, Feb. 1979, *Cribb et al.* 11371!; Iringa District: Lake Ngwazi, May 1968, *Renvoize & Abdallah* 2002!; Songea District: Mbinga, Peramiho, May 1977, *Mhoro* 2884!
DISTR. T 7, 8; Congo (Kinshasa), Angola, Zambia, Malawi, Zimbabwe
HAB. Marshy or riverine grassland; 950–2100 m
USES. None recorded on specimens from our area
CONSERVATION NOTES. Eight specimens from Tanzania, but widespread; least concern (LC)

SYN. *Senecio uvens* Hiern, Cat. Welw. Afr. Pl. 1(3): 602 (1898)
S. kyimbilensis Mattf. in E.J. 59, Beibl. 133: 34 (1924). Type: Tanzania, Rungwe District, Kyimbila, *Stolz* 2325 (B†, holo., C!, EA!, K!, iso.)

8. **Crassocephalum effusum** (*Mattf.*) *C.Jeffrey* in K.B. 41: 907 (1986). Type: Tanzania, Rungwe District, Kyimbila, *Stolz* 1456 (B†, holo., K!, M!, iso.)

Annual or short-lived perennial herb 90–150 cm tall, erect; stems pale green with reddish tinge, sparsely shortly pubescent. Leaves sessile, elliptic, 3–21 cm long, 0.5–4 cm wide, attenuate to an exauriculate petioloid base or (especially the upper) sessile, margins shallowly and remotely to (especially the upper) coarsely sinuate-serrate, unlobed or (especially the upper) lyrato-pinnately or pinnately 2–8-lobed, apex ± attenuate and obtuse to acute, shortly pubescent. Capitula discoid, several to numerous in lax terminal cymes, erect; stalks of the individual capitula glabrous or pubescent; involucre cylindrical, 6.5–7 mm long, 2.5–3 mm in diameter; bracts of calyculus 9–12, lanceolate, dark-tipped, 1.5–3.5 mm long, ciliate; phyllaries 13, pale green with dark tips, 6–6.5 mm long, sparsely shortly pubescent. Disc florets mauve or pinkish-mauve (label of type says white), corolla 8–9 mm long, tube glabrous, gradually expanded above the middle, lobes 1–1.2 mm long. Achenes 1.8–2 mm long, ribbed, hairy between the ribs; pappus 6–8.5 mm long.

TANZANIA. Njombe District: Lukumburu, July 1956, *Milne-Redhead & Taylor* 10972!; Iringa District: Usagara/Mazombe area, July-Aug. 1936, *Ward* U48!; Songea District: R. Mukuluzi 3 km NE of Kigonsera, Apr. 1956, *Milne-Redhead & Taylor* 9594!
DISTR. **T** 7, 8; Malawi
HAB. Riverine grassland in *Brachystegia-Uapaca* woodland zone; 950–1150 m
USES. None recorded on specimens from our area
CONSERVATION NOTES. Five specimens from Tanzania; data deficient (DD)

SYN. *Senecio effusus* Mattf. in E.J. 59, Beibl. 133: 34 (1924)

9. **Crassocephalum radiatum** *S.Moore* in J.B. 56: 227 (1918); C. Jeffrey in K.B. 41: 907 (1986); Lisowski, Aster. Fl. Afr. Centr. 2: 345, t. 76 (1991). Type: Congo (Kinshasa), Lake Mweru, *Kassner* 2825 (BM!, holo., HBG, iso.)

Annual herb 10–240 cm tall, erect; stems pale green, often speckled or tinged reddish-purple, sparsely setulose. Leaves sessile, elliptic, narrowly elliptic or lanceolate, 1.2–14 cm long, 0.3–6.5 cm wide, cuneate or attenuate into an exauriculate petioloid base, margins remotely sinuate-serrate to serrate-laciniate, unlobed or shortly to deeply 2-lobed towards the base, apex obtuse, apiculate, shortly pilose-setulose. Capitula numerous to rarely solitary in lax terminal corymbs, erect, radiate or discoid; stalks of the individual capitula slender, glabrous or almost so; involucre cylindrical, slightly swollen at the base, 5.5–7.5 mm long, 2–3 mm in diameter; bracts of calyculus 8–12, dark-tipped, 1–3.5 mm long, glabrous or ciliate; phyllaries 13 or sometimes 8, green with purple tips, 5–7 mm long, shortly hispid or glabrous. Ray florets 13, 8, 5 or 0, tube 4 mm long, glabrous, rays when present deep pink or pinkish-mauve, 1–3.5 mm long (6–8 mm fide Lisowski), 1 mm wide, 3–4-veined; disc florets dull brick-red, magenta or white, corolla 5.5–8 mm long, tube glabrous, gradually expanded above the middle, lobes 0.5–0.7 mm long. Achenes 1.5–1.6 mm long, ribbed, shortly hairy in the grooves; pappus 4.5–6 mm long.

TANZANIA. Ufipa District: Nkansi, Chala Mt, May 1997, *Bidgood et al.* 3696!; Rungwe District: Kiwira Forest Reserve, June 1962, *Mgaza* 484!; Songea District: Matengo Hills, Kihuru Hill, May 1956, *Milne-Redhead & Taylor* 10288a!
DISTR. **T** 4, 7, 8; Congo (Kinshasa), Burundi, Malawi
HAB. Grassland or bushed grassland, mist forest; (1000–)1650–2200 m
USES. None recorded on specimens from our area
CONSERVATION NOTES. Ten Tanzanian specimens, but due to distribution probably least concern (LC)

SYN. *Senecio heteromorphus* Hutch. & Burtt in Rev. Zool. Bot. Afr. 23: 42 (1932). Type: Congo (Kinshasa), Kitendwe–Kasiki, *de Witte* 411 (BR!, holo.)
    Crassocephalum heteromorphum (Hutch. & Burtt) C. Jeffrey in K.B. 41: 906 (1986)

NOTE. Lisowski reduced *C. heteromorphum* to a synonym of *C. radiatum*.

10. **Crassocephalum rubens** (*Jacq.*) *S.Moore* in J.B. 50: 212 (1912); F.P.S. 3: 21 (1956); Maquet in Fl. Rwanda 3: 650 (1985); C. Jeffrey in K.B. 41: 907 (1986); Lisowski, Aster. Fl. Afr. Centr. 2: 350, t. 78 (1991); C. Jeffrey in Fl. Masc. 109: 157, t. 52 (1993). Type: from a plant cultivated in Austria, Vienna (not traced)

Annual herb 20–150 cm tall, erect; stems green striate with purple, densely to sparsely pubescent or setulose. Leaves sessile, obovate, oblanceolate, elliptic or lanceolate, rarely ovate, unlobed or (especially the upper) lyrato-pinnately or pinnately 2–8-lobed, 1.2–20 cm long, 0.5–7.5 cm wide, cuneate or attenuate into an exauriculate petioloid base or (especially the upper) sessile, margins remotely sinuate-denticulate to coarsely sinuate-serrate, apex rounded to obtuse or acute, scattered-pubescent at least on veins beneath. Capitula 1–8, long-stalked, usually at first nodding then ± erect, but sometimes erect throughout, discoid; stalks of the individual capitula ± pubescent and purplish-tinged; involucre cylindrical, 8–13 mm long, 2.5–8 mm in diameter; bract of calyculus 5–23, purplish or dark green with purple tips, lanceolate or narrowly lanceolate, 3–6(–8) mm long, glabrous or ciliate; phyllaries 13–25, commonly 21 or 13, pale green to green, often tinged purple especially towards the apex and purple-tipped, 7.5–12 mm long, glabrous or sparsely shortly pubescent. Disc florets blue, purple or mauve, less often pink or red; corolla 6.2–10.5 mm long, tube glabrous, gradually expanded in upper third, lobes 0.4–1.5 mm long. Achenes 2–2.5 mm long, ribbed, hairy or shortly so in the grooves; pappus 7–12 mm long.

DISTR. U 1–4; K 2–7; T 1–8; Ethiopia, Burundi, Sudan, Zambia, Malawi, Mozambique, Zimbabwe, South Africa; Madagascar, Comoro Is., Yemen

HAB. Grassland or swampy to muddy sites in the woodland zone or woodland, but usually in disturbed situations such as waste ground or in cultivations; may be locally common; (1–)800–2500 m

var. **rubens**

Leaves often narrowly and deeply 2–8-lobed. Capitula 1–4; involucre broadly cylindrical, 9.5–13 mm long, 5–8 mm in diameter; bract of calyculus 12–23; phyllaries 13–25, commonly 21; disc florets blue, purple, mauve, magenta, pink or red.

UGANDA. W Nile District: Kisomoro, Terego, Apr. 1940, *Eggeling* 3877!; Toro District: Nyakatonzi, Aug. 1961, *Wilson* 1189!; Masaka District: Lwera on Masaka–Kampala road, Feb. 1971, *Kabuye* 330!

KENYA. Elgon, Oct.–Nov. 1930, *Lugard* 439!; Fort Hall District: Thika to Muranga, Mar. 1948, *Bogdan* 1539!; Nairobi, Sep. 1915, *Dowson* 304!

TANZANIA. Mpanda District: Tumba, Jan. 1952, *Siame* 113!; Rungwe District: Rungwe Forest Reserve, May 1983, *Macha* 317!; Njombe District: Luilo, Sep. 1970, *Thulin & Mhoro* 1175!

DISTR. U 1–4; K 3–5; T 1, 4, 6–8; W Africa from Liberia to Cameroon, Sudan, Ethiopia and S to South Africa; Madagascar, Comoro Is., Mauritius, Reunion

HAB. As for species

USES. None recorded on specimens from our area

CONSERVATION NOTES. Least concern (LC)

SYN. *Senecio rubens* Jacq., Hort. Vindob. 3: 50, t. 98 (1777)
    S. cernuus L.f., Suppl.: 370 (1782), *nom. illegit. superfl.* Type as above
    Gynura cernua Benth. in Hook., Niger Fl.: 437 (1849); Oliv. & Hiern, F.T.A. 3: 402 (1877), *nom. illegit. superfl.*
    G. rubens (Jacq.) Muschl. in F.R. 11: 119 (1912)

var. **sarcobasis** (*DC.*) *C.Jeffrey & Beentje* **comb. nov.** Type: Madagascar, without locality, *Bojer* s.n. (G-DC, holo.)

Leaves usually rather broadly 2–8-lobed. Capitula 1–12; involucre cylindrical, 8–12 mm long, 2.5–7 mm in diameter; bract of calyculus 5–17; phyllaries 13–25, commonly 13; disc florets purple, mauve, magenta or pink, very rarely blue.

UGANDA. Mbale District: Kapcherwa, Sep. 1954, *Lind* 282! & Sebei, Kyesoweri, Oct. 1955, *Norman* 298! & Bulago, Bugishu, Aug. 1932, *A.S. Thomas* 335!

KENYA. W Suk District: km 67 on Kitale–Moroto road, Suk scarp, Oct. 1952, *Verdcourt* 746!; Ravine District: 6 km N of Eldama Ravine on Torongo road, Oct. 1981, *Gilbert & Mesfin* 6432!; Nairobi, Thompson's estate, June 1930, *Napier* 192a!

TANZANIA. Mbulu District: Pienaar's Heights, May 1962, *Polhill & Paulo* 2345!; Dodoma District: Dodoma–Iringa road, Matumbulu reservoir, July 1974, *Mhoro & Backeus* s.n.!; Iringa District: edge of Lake Ngwazi, May 1968, *Renvoize & Abdallah* 2006!

Distr. **U** 3; **K** 2–7; **T** 1–8; Congo (Kinshasa), Rwanda, Burundi, Sudan, Ethiopia, Zambia, Malawi, Mozambique, Zimbabwe, South Africa; Madagascar, Comoro Is., Yemen
Hab. As for species
Uses. None recorded on specimens from our area
Conservation notes. Least concern (LC)

Syn. *Gynura sarcobasis* DC., Prodr. 6: 300 (1838)
    Crassocephalum sarcobasis (DC.) S.Moore in J.B. 50: 211 (1912); Maquet in Fl. Rwanda 3: 650, fig. 201/1 (1985); C. Jeffrey in K.B. 41: 907 (1986); Blundell, Wild Flow. E. Afr.: fig. 678 (1987); Lisowski, Aster. Fl. Afr. Centr. 2: 349 (1991); U.K.W.F. ed. 2: 219, t. 92 (1994)

Note. *C. sarcobasis* is brought into synonymy here. The two taxa are very close but there seem to be consistent, if minor, differences.

11. **Crassocephalum bauchiense** (*Hutch.*) *Milne-Redh.* in K.B. 5: 376 (1951); C. Jeffrey in K.B. 41: 908 (1986); Lisowski, Aster. Fl. Afr. Centr. 2: 354 (1991). Type: Nigeria, Vom, *Young* 158 (K!, holo.)

Annual herb up to 120 cm tall; stems erect, shortly pubescent. Leaves sessile, obovate-lanceolate in outline, deeply pinnate-lyrately lobed with 5–20 lanceolate sinuate-dentate or sinuate-lobulate lateral lobes, 2–15 cm long, 2–5.5 cm wide, apex obtuse, minutely acuminate-apiculate, scattered-pubescent especially on veins. Capitula numerous in copious compound terminal corymbs, discoid; stalks of the individual capitula pubescent; involucre cylindrical, 7.5–8.5 mm long, 3 mm in diameter; bracts of calyculus 11–18, narrowly lanceolate, 2–3.5 mm long, ciliate; phyllaries 12–18, 7–8 mm long, glabrous. Disc florets pale blue; corolla 6–7 mm long, tube glabrous, gradually expanded in upper third, lobes 0.7 mm long. Achenes 2.5–3 mm long, ribbed, very shortly sparsely hairy between the ribs; pappus 6–7 mm long.

Uganda. Mubende District: Kakumiro, Oct. 1945, *A.S. Thomas* 4320!
Distr. **U** 4; Nigeria, Cameroon, Congo (Kinshasa), Zambia
Hab. 'Rocky outcrop'; 1290 m
Uses. None recorded on specimens from our area
Conservation notes. One specimen from Uganda, but due to distribution probably least concern (LC)

Syn. *Gynura caerulea* Hutch. & Dalz., F.W.T.A. 2: 147, 148 (1931), *nom. illegit.*, *non G. coerulea* O. Hoffm. (1893). Type as above
    G. bauchiensis Hutch. in F.W.T.A. 2: 608 (1936)

12. **Crassocephalum crepidioides** (*Benth.*) *S.Moore* in J.B. 50: 211 (1912); F.P.S. 3: 21 (1956); F.P.U.: 182 (1962); Maquet in Fl. Rwanda 3: 652 (1985); C. Jeffrey in K.B. 41: 908 (1986); Lisowski, Aster. Fl. Afr. Centr. 2: 354 (1991); C. Jeffrey in Fl. Masc. 109: 159 (1993); U.K.W.F. ed. 2: 219 (1994). Type: Sierra Leone, without locality, *Don* s.n. (BM!, holo.)

Annual herb 25–120(–150) cm tall, rarely a short-lived perennial; stems erect, green, often flecked with dull purplish-red, pubescent. Leaves sessile, obovate, broadly elliptic, rhombic or ovate, unlobed or pinnato-lyrately 2–8-lobed, 5–26 cm long, 2–10 cm wide, cuneate or attenuate and slightly decurrent onto an exauriculate petioloid base, margins usually coarsely and sometimes irregularly sinuate-serrate or sinuate-serrato-bidentate, apex obtuse to acute, pilose-setulose, sometimes with purplish margins or purplish-tinged beneath especially on main veins. Capitula few to numerous in dense to fairly lax terminal corymbs, discoid, drooping in bud and at anthesis, becoming erect in flower and fruit; stalks of the individual capitula pubescent; involucre cylindrical or ± so above a somewhat bulging base, 9–13.5 mm long, 3–5 mm in diameter; bracts of calyculus 6–21, purplish or blackish with darker tips, 2–6 mm long, scattered-pubescent or at least ciliate; phyllaries 13–21, usually 21 or less frequently 13, yellow-green or green with purplish or blackish tips, 8–12.5 mm long, scattered-setulose, or rarely glabrous. Ray florets 0; disc florets many, with orange-red or brick-red, rarely yellow corolla-lobes and red style-arms, tube often yellow, corolla 7–12 mm long, glabrous, slender, gradually expanded in upper third, lobes 0.5–1 mm long. Achenes 1.8–2.7 mm long, ribbed, shortly hairy between the ribs; pappus 7–13 mm long.

UGANDA. Kigezi District: Ruhinda, Jan. 1951, *Purseglove* 3552!; Busoga District: 5 km W of Busesa, Oct. 1962, *Lewis* 6029!; Mengo District: Kisubi, Entebbe, Sep. 1948, *Eggeling* 5830!
KENYA. Kiambu District: Muguga, Oct. 1976, *Mungai* 109!; Kericho District: SW Mau Forest, Timbilil, Aug. 1961, *Kerfoot* 2879!; Kwale District: Kwale, 1934, *McCraig* 9171!
TANZANIA. Lushoto District: Mafi Hill near headwaters of Kwalukonge stream, Jan. 1985, *Borhidi et al.* 85/279!; Kigoma District: Gombe National Park [Stream Reserve], Jan. 1964, *Pirozynski* 305!; Iringa District: Mufindi, Kibao, Mar. 1962, *Polhill & Paulo* 1755!; Zanzibar: without further locality, 1901, *Lyne* 33!
DISTR. **U** 1–4; **K** 1, 3–7; **T** 1–8; **Z**; **P**, W Africa from Senegal to Cameroon, Sudan, Ethiopia, and S to South Africa; naturalized in large parts of tropical Asia and the Pacific
HAB. In moist sites along rivers or lakes, in forest margins and secondary vegetation, a common weed of disturbed places and cultivation; 0–2500 m
USES. Young leaves used as vegetable in **K** 3 and **T** 6; minor medicinal against sleeping sickness and swollen lips (*Tanner*)
CONSERVATION NOTES. Least concern (LC)

SYN. *Senecio diversifolius* A.Rich., Tent. Fl. Abyss. 1: 437 (1848), *nom. illegit., non* Dum. (1827). Type: Ethiopia, Adowa, *Dillon* s.n. (P!, holo., iso.)
Gynura crepidioides Benth. in Hook., Niger Fl.: 438 (1849); Oliv. & Hiern, F.T.A. 3: 403 (1877)
Crassocephalum diversifolium Hiern, Cat. Welw. Afr. Pl. 3(1): 594 (1898), *nom. illegit. superfl.*, type as for *G. crepidioides*

NOTE. Greenway reports that on a single plant the florets may mature from the margins of the capitulum inwards, or from the centre of the capitula outwards.

95. AUSTROSYNOTIS*

C. Jeffrey in K.B. 41: 878 (1986); Bremer, Asterac., Cladistics & Class.: 498 (1994)

Perennial scandent herb. Leaves alternate, with thickened prehensile petiole; blade ovate-cordate, auriculate, palmately veined. Capitula in corymbose panicles, heterogamous, radiate; involucre calyculate; anthers caudate; style arms without appendages, fringed with short hairs, but in ray florets virtually without hairs. Achenes oblong, ribbed; pappus of fine bristles.

Monotypic.

Austrosynotis rectirama (*Baker*) *C.Jeffrey* in K.B. 41: 878 (1986). Type: Malawi, Nyika, *Whyte* 110 (K!, holo.)

Climber to 7 m; stems tomentose to puberulent, glabrescent, often with purple spots. Leaves cordate, 3–8 cm long, 2–6 cm wide, base cordate, margins dentate to denticulate, apex broadly acuminate or cuspidate, sparsely arachnoid and glabrescent above, tomentose or sometimes pubescent beneath, veins palmate, conspicuous from beneath; petioles 2.6–3.5 cm long, sparsely tomentose, auriculate. Capitula in axillary or terminal corymbs, radiate; peduncle tomentose; involucre 3.2–3.9 mm long, 1–1.4 mm in diameter; bracts of calyculus 5, lanceolate, 1.5–3.2 mm

* by Italo Malombe, East African Herbarium, P.O. Box 45166, Nairobi, Kenya. malombe@avu.org/plants@africaonline.co.ke

FIG. 128. *AUSTROSYNOTIS RECTIRAMA* — **1**, habit, × ²/₃; **2**, capitulum, × 4; **3**, ray floret, × 10;
4, disc floret, × 10; **5**, achene, most pappus removed, × 12. 1–4 from *Richards* 15048, 5 from
Salubeni 1690. Drawn by Juliet Williamson.

long; phyllaries 12–13, tomentose, bearded at the apex. Ray florets 5, yellow, 4.5–7 mm long, 1.2–1.5 mm broad, 4–5-veined; disc florets ± 15, yellow, 4.5–8 mm long, lobes 1.2–1.5 mm long. Achenes 0.5–2.3 mm long, 4-ribbed, strigose; pappus 5–8 mm long, creamy white. Fig. 128 (page 612).

TANZANIA. Ufipa District: Mt Kito, Apr. 1961, *Richards* 15048!; Rungwe District: Bulambia [Ulambya], Aug. 1974, *Leedal* 2039!
DISTR. **T** 4, 7; Malawi
HAB. Dense riverine forest; 1400–1800 m
USES. None recorded on specimens from our area
CONSERVATION NOTES. Two specimens from Tanzania; data deficient (DD)

SYN. *Senecio rectiramus* Baker in K.B. 1898: 155 (1898)

NOTE. Specimens belonging to this genus have long been identified as *Mikaniopsis* Milne-Redh. because of the habit and most of the morphological characters. But *Austrosynotis* is distinct in the presence of auriculate leaves and the radiate capitulum. I have also noticed that most parts such as the stem, peduncle, bracts and involucre have purple spots.

96. **MIKANIOPSIS***

Milne-Redh. in Exell, Suppl. Cat. Vasc. Pl. S. Tomé: 27 (1956); C. Jeffrey in K.B. 41: 878 (1986); Bremer, Asterac., Cladistics & Class.: 505 (1994)

Scandent herbs or sub-shrubs with fluted stems. Leaves alternate, simple, cordate, 5–7-palmately veined; petioles prehensile with basal thickening, which remains after leaf abscission. Capitula racemose, cymose or paniculate, disciform, heterogamous; involucre calyculate. Outer florets female, inner hermaphrodite; all florets with narrow cylindrical tube and campanulate 5-lobed upper limb; anthers sagittate at base, caudate; style arms long, slightly flattened, apex truncate or rounded. Achenes cylindrical, 8–10-ribbed, glabrous, rarely tomentellous; pappus 1–3-seriate with very many bristles.

About 15 species in tropical Africa.

1. Leaves tomentose beneath 2
   Leaves thinly floccose to glabrous 3
2. Involucre 7.5–8 mm long; florets ± 23 . . . .   1. *M. tedliei*
   Involucre 4–6 mm long; florets ± 12 . . . .   2. *M. bambuseti*
3. Corolla orange-yellow, white inside; florets up to 18 . . . .   3. *M. usambarensis*
   Corolla white; florets up to 12 4
4. Inflorescence axis without glandular hairs; **T** 4 . . . .   4. *M. tanganyikensis*
   Inflorescence axes with glandular hairs; Uganda . . . .   5. *M. vitalba*

1. **Mikaniopsis tedliei** (*Oliv. & Hiern*) C.D.Adams in J.W. Afr. Sci. Assoc. 6: 153 (1961); C.D.Adams in F.W.T.A. ed. 2, 2: 243 (1963); Lisowski, Aster. Fl. Afr. Centr. 2: 441 (1991). Type: Ghana, Ashanti, *Tedlie* s.n. (K!, holo.)

Woody climber reaching several m in length; branches striate, tomentose-pubescent and glandular on young parts, becoming puberulous with age. Leaves cordate, 6–10.8 cm long, 4.5–10.1 cm wide, base subcordate, margins sparsely dentate, apex acuminate, dark green and puberulous above, grey-green and tomentose beneath; palmately 5–7-veined from the base; petiole 3–6.5 cm long. Capitula numerous in terminal and axillary corymbs to 22 cm long, disciform; stalk

* by Italo Malombe, East African Herbarium, P.O. Box 45166, Nairobi, Kenya. malombe@avu.org/plants@africaonline.co.ke

of individual heads 5–7.3 cm long; involucre hemispherical, 7.5–8 mm long; bracts of calyculus 3–4, lanceolate, 2.5–4 mm long; phyllaries 8–14, lanceolate, 7.5–8 mm long, acute, puberulous, margins scarious. Outer florets female, to 8 mm long; inner hermaphrodite, corolla 6–8 mm long, lobes 1.3–2 mm long; style-arms with short hairy subulate appendages. Achenes 0.6–2.1 mm long, 5-ribbed, glabrous; pappus 6.5–8 mm long, cream or dirty white.

UGANDA. Masaka District: Bugoma, Sese, June 1925, *Maitland* 779!
DISTR. **U** 4; Ghana, Cameroon, Congo (Kinshasa)
HAB. Along paths in dense equatorial forest; ± 1150 m
USES. None recorded on specimens from our area
CONSERVATION NOTES. Widespread; least concern (LC)

SYN. *Senecio tedliei* Oliv. & Hiern, F.T.A.. 3: 420 (1877)
    Gynura tedliei (Oliv. & Hiern) Hutch. & Dalz., F.W.T.A. 2: 608 (1936)
    Mikaniopsis sp. A of C.Jeffrey in K.B. 41: 878 (1986)

NOTE. *M. tedliei* was initially thought to be confined to West Africa, but was reported to occur in Central Africa by Lisowski (1991). Therefore its presence in FTEA is not a surprise because the habitat (**U** 4) is congruent to the known distribution.

2. **Mikaniopsis bambuseti** (*R.E.Fr.*) *C.Jeffrey* in K.B. 41: 878 (1986); K.T.S.L.: 558, fig. (1994); U.K.W.F. ed. 2: 219 (1994). Types: Kenya, Mt Kenya, *Fries & Fries* 1203 (UPS, syn., K! photo.) & Nyandarua/Aberdare Mts, *Fries & Fries* 2749 (UPS, syn.)

Scandent subshrub to 8 m long; stems white-tomentose, glabrescent, glandular. Leaves membranous, dark green above, ovate, 5–13 cm long, 3–11 cm wide, base cordate (the incision acute), margins sinuate-denticulate with glandular-apiculate teeth, apex acute to abruptly acuminate, glabrous to lanate above, densely white-tomentose beneath; palmately 5–7-veined from the base; petiole 1–8 cm long. Capitula in dense terminal and upper axillary paniculoid corymbs, disciform; stalk of individual heads 3.5–9 mm long; involucre obconic, to 6 mm long and 2–3 mm wide; bracts of calyculus 3–7, ± 3 mm long; phyllaries 7–10, oblong-lanceolate, 4–5.5 mm long, acute to acuminate, tomentellous to glabrous. Florets ± 12, outer female, inner hermaphrodite, corollas 5–8.2 mm long, pale yellow to almost white; lobes 1.2–1.5 mm long. Achenes cylindrical, 1–2.4 mm long, ribbed, glabrous; pappus 5–11 mm long, creamy white. Fig. 129 (page 615).

UGANDA. Mbale District: Mt Elgon, Sasa trail, 28 Dec. 1996, *Wesche* 474! & same locality, 6 Jan. 1997, *Wesche* 691!
KENYA. Elgeyo District: Cherangani Hills between Kaptalamwa and Tangul, Feb. 1985, *Townsend* 2393!; Nakuru District: Mt Londiani 12 km ENE of Londiani, Dec. 1967, *Perdue & Kibuwa* 9225!; Masai District: Nasampolai Valley W side, Sep. 1972, *Greenway & Kanuri* 15056!
DISTR. **U** 2; **K** 2–4, 6; not known elsewhere
HAB. Bamboo forest or thicket, less often in dry forest; 2100–3150 m
USES. None recorded on specimens from our area
CONSERVATION NOTES. Restricted in distribution but with rather wide altitude range; least concern (LC)

SYN. *Senecio bambuseti* R.E.Fr. in Act. Hort. Berg. 9: 155 (1928)
    Mikaniopsis clematoides sensu Agnew, U.K.W.F.: 471 (1974), *non* (A. Rich.) Milne-Redh.

NOTE. Reported to flower only rarely, but observed to flower during the first and the last quarter of the year.
    Luke reports he has seen this in the Bale Mts in Ethiopia, in 2003.

3. **Mikaniopsis usambarensis** (*Muschl.*) *Milne-Redh.* in Exell, Cat. Vasc. Pl. S. Tomé Suppl.: 30 (1956); U.K.W.F.: 471 (1974); C. Jeffrey in K.B. 41: 878 (1986); Lisowski, Aster. Fl. Afr. Centr. 2: 443 (1991); K.T.S.L.: 559, fig. (1994); U.K.W.F. ed. 2: 219 (1994). Type: Tanzania, Lushoto District, Amani, *Braun* 768 (B†, holo., EA!, iso.)

FIG. 129. *MIKANIOPSIS BAMBUSETI* — **1**, habit, × ¹/₂; **2**, capitulum, × 4 ¹/₂; **3**, female floret, × 6; **4**, central floret, × 6. 1–4 from *Jonsson* 644. Drawn by Juliet Williamson.

Robust perennial climber several m long, branched; stem semi-succulent, sparsely tomentellous in young parts, glabrescent, glandular. Leaves fleshy, broadly ovate, 5–9.5 cm long, 3–8.1 cm wide, base often cordate, rarely obtuse, margins usually irregularly sinuate-dentate, rarely entire, apex acute to broadly acuminate, glabrous or with remnants of floccose indumentum at least at the base and along midrib; palmately 5–6-veined from the base; petiole 2.8–5.9 cm long. Capitula in terminal and axillary simple or paniculate corymbs, subsessile, disciform; stalk of individual heads 6–20 mm long; involucre obconic, 4.5–6(–9) mm long; bracts of calyculus 3–5, lanceolate, 1.6–4 mm long, glabrous; phyllaries 7–10, linear-oblong, 4.3–6 mm long, glabrous but shortly hairy at the apex. Florets ± 8–12, outer 4–6 female, with corollas often shorter than the inner, inner hermaphrodite; corollas orange-yellow, white inside, 5.5–7(–10) mm long; lobes 1.3–2 mm long; style-arms with very short obtuse appendages. Achenes subcylindrical, 1–2.4 mm long, ribbed, glabrous; pappus 5.4–11 mm long, silvery white.

KENYA. Kericho District: Mau forest, Sambret, Oct. 1961, *Kerfoot* 2978!; Kiambu District; Kereita forest, northward road, June 2001, *Malombe et al.* 986!
TANZANIA. Lushoto District: Shagai Forest Reserve, Aug.1955, *Semsei* 2188! & Amani, July 1914, *Grote* 7997! & Amani to Bomole, June 1914, *Peter* K712!
DISTR. **K** 4, 5; **T** 3; Rwanda, Burundi, Congo (Kinshasa)
HAB. Moist forest margins; 850–950(–2350) m
USES. Minor medicinal for stomach ulcers (*Braun*)
CONSERVATION NOTES. Wide distribution, least concern (LC)

SYN. *Senecio usambarensis* Muschl. in E.J. 43: 63 (1909); T.T.C.L.: 160 (1949)
    Mikaniopsis paniculata sensu Troupin (1982: 145) pro parte *quoad specim. Troupin* 15817, *non* Milne-Redh.
    M. tedliei sensu Troupin (1982: 145) pro parte *quoad specim. Bouxin* 476 & *Troupin* 15249, *non* (Oliv. & Hiern) C.D.Adams
    M. clematoides sensu Troupin (1982: 145); Maquet in Fl. Rwanda 3: 656 (1985) pro parte *quoad specim. Runyinya* 453, *non* (A. Rich.) Milne-Redh.
    M. sp. A of Maquet in Fl. Rwanda 3: 656, fig. 204/2 (1985)

NOTE. Recent collections by *Malombe* et al. from Kereita forest, West Nyandarua/Aberdares (**K** 4) are sterile. Efforts are being made to collect fertile material.

4. **Mikaniopsis tanganyikensis** (*R.E.Fr.*) *Milne-Redh.* in Exell, Cat. Vasc. Pl. S. Tomé Suppl.: 30 (1956); C. Jeffrey in K.B. 41: 879 (1986); Lisowski, Aster. Fl. Afr. Centr. 2: 444 (1991). Type: Zambia, Luvingu, *Fries* 1103 (UPS, holo., K!, fragm. & photo.)

Scandent subshrub or herbaceous liane, 2.5–6.3 m high; stem glabrous. Leaves broadly ovate, 3–12 cm long, 2.5–9 cm wide, base cordate, margins entire or with two teeth on each side near the base, apex acute-cuspidate, usually glabrous, rarely minutely pilose or sparsely tomentellous; palmately 5–7-veined from the base; petiole 3–6 cm long. Capitula in terminal and upper axillary simple or paniculate corymbs with glabrous axes, disciform; stalk of individual heads 5.3–6.3 mm long; involucre obconic, 4.2–5.5 mm long; bracts of calyculus 3, lanceolate, 1.5–2 mm long, sparsely hairy; phyllaries 8–11, oblong-lanceolate, 4.2–5.5 mm long, glabrous but shortly hairy at the apex. Florets female on the periphery, hermaphrodite towards the centre; corolla white, 5.5–7.5 mm long, lobes 0.5–1.4 mm long; stigma bifid, flattened, apex truncate or rounded. Achenes subcylindrical, 1.5–4 mm long, ribbed, glabrous. Pappus 6–7 mm long, creamy white.

TANZANIA. Ufipa (?Mpanda) District: between Lake Tanganyika and Lake Rukwa, anno 1896, *Nutt* s.n.!
DISTR. **T** 4; Congo (Kinshasa), Zambia, Malawi
HAB. no data (in southern Africa in forest edges, gallery forest or swamp forest); ± 1800 m
USES. None recorded on specimens from our area
CONSERVATION NOTES. Widespread; least concern (LC)

SYN. *Senecio tanganyikensis* R.E.Fr., Wiss. Ergebn. Schwed. Rhod.–Kongo-Exped. 1911–12, 2: 345 (1916); T.T.C.L.: 160 (1949)

NOTE. Remains of floccose and tomentellous indumentum were observed on the collection by *Nutt.* This feature was not recorded before.

5. **Mikaniopsis vitalba** (*S.Moore*) *Milne-Redh.* in Exell, Cat. Vasc. Pl. S. Tomé Suppl.: 30 (1956); C. Jeffrey in K.B. 41: 879 (1986); Lisowski, Aster. Fl. Afr. Centr. 2: 445 (1991). Type: Uganda, Mengo District: Entebbe, *Bagshawe* 729 (BM!, holo.)

Woody climber 3–8 m long; stem glabrous, rarely with remains of floccose indumentum. Leaves ovate, 4–9 cm long, 2.8–6 cm wide, base subcordate, margins entire or sparsely denticulate near the base, apex acuminate, glabrous or sometimes floccose above when young; palmately 5–7-veined from the base; petioles 2.5–4 cm long. Capitula in terminal and upper axillary racemes or panicles with glandular axes, disciform; stalks of individual capitula 4–14 mm long; involucre obconic, 5–7.5 mm long; bracts of calyculus 2–3, lanceolate, 2 mm long, ciliate; phyllaries 7–8, linear-oblong, 4.4–7.5 mm long, acute and short-hairy at the apex, otherwise glabrous to puberulous. Florets 8–12, the outer female, the inner hermaphrodite; corolla white, 4.5–7.5 mm long, lobes 0.4–1.2 mm long; style arms slightly narrowed towards apex, tip truncate or rounded. Achenes cylindrical, 2–3 mm long, ribbed, glabrous; pappus 4–8 mm long, creamy white.

UGANDA. Kigezi District: Kabale, Rubanda, Omubwindi swamp, July 1989, *Katende* 3777!; Mengo District: Kipayo, Aug. 1915, *Dummer* 1012! & Entebbe, Aug.1905, *Bagshawe* 729! (type)
DISTR. **U** 2, 4; Cameroon, Gabon, Congo (Kinshasa), Angola
HAB. Forest; 1200–2250 m
USES. None recorded on specimens from our area
CONSERVATION NOTES. Widespread; least concern (LC)

SYN. *Senecio vitalba* S.Moore in J.B. 44: 85 (1906)

NOTE. Lisowski (1991) states that *M. vitalba* has 14–16 florets per capitulum, but the range of the florets was found to be 8–12.

97. SENECIO

L., Sp. Pl.: 866 (1753) & Gen. Pl., ed. 5: 373 (1754)

Annual or perennial herbs or shrubs, sometimes scandent. Leaves usually alternate, very variable. Inflorescence from solitary capitula to very compound many-headed cymes or panicles. Capitula discoid or radiate, usually stalked, solitary or in groups which often form corymbose cymes; involucre usually with a single row of phyllaries subtended by a small calyculus of much smaller bracts; phyllaries free or rarely connate, with membranous or scarious margin. Flowers mostly yellow, sometimes purple or reddish; anthers ecaudate, obtuse or sagittate at base, appendiculate at apex; style branches obtuse or truncate with a ring of hairs and stigmatic lines. Achenes cylindrical, ellipsoid or obovoid, with a pappus of fine bristles.

About 1000 species, worldwide.

1. Capitula discoid; ray florets absent 2
   Capitula radiate 30
2. Lianas 3
   Herbs or shrubs, not scandent 4
3. Stem glabrous; corolla 6.5–9 mm long . . . . 1. *S. syringifolius* (p. 621)
   Stem pubescent; corolla 3–5 mm long . . . . 5. *S. deltoideus* (p. 625)

4. Leaves peltate . . . . 38. *S. oxyriifolius* (p. 646)
    Leaves not peltate 5
5. Annual herb 6
    Perennials, or herb with annual shoots from
      perennial rootstock 7
6. Involucre 3.5–5.5 mm long; ray florets usually
      present; pappus 3–4 mm 85
    Involucre 7–8 mm long; ray florets absent; pappus
      5–5.5 mm . . . . 72. *S. vulgaris* (p. 669)
7. Plants with flowers before leaves appear 8
    Plants with flowers and leaves simultaneously 9
8. Capitula solitary; involucre 15–22 mm long . . . . . 6. *S. pachyrhizus* (p. 626)
    Capitula 7–20 together; involucre 6–7 mm long . 21. *S. spartareus* (p. 635)
9. Leaves succulent or sub-succulent (may not be
      clear indried specimens) 10
    Leaves not succulent 11
10. Leaves viscid and glandular . . . . 55. *S. schweinfurthii* (p. 659)
    Leaves tomentose beneath . . . . 54. *S. telekii* (p. 657)
11. Florets purple, mauve or red 12
    Florets yellow, cream or white 15
12. Plant subscapigerous, with more basal leaves than
      stem leaves and few unbranched stems 13
    Plant with branched stems with many stem leaves 14
13. Achenes and ovary glabrous; Ruwenzori . . . . 62. *S. mattirollii* (p. 663)
    Achenes and ovary hairy in grooves; Tanzania . . . 68. *S. erubescens* (p. 667)
14. Achenes shortly hairy; **T** 2, 6 . . . . 69. *S. cyaneus* (p. 667)
    Achenes glabrous . . . . 70. *S. purpureus* (p. 668)
15. Leaves glandular and viscid 16
    Leaves not glandular 19
16. Achenes and ovary hairy 17
    Achenes and ovary ± glabrous . . . . 57. *S. purtschelleri* (p. 660)
17. Herb to 30 cm; leaves 5.5–9 × 0.4–0.7 cm; **U** 4 . . . 56. *S. navugabensis* (p. 659)
    Herb to 120 cm; basal leaves 7–38 × 0.9–10 cm 18
18. Involucre 4–8 mm long; widespread . . . . 66. *S. hochstetteri* (p. 666)
    Involucre 8.5–9 mm long; **T** 4, 7 . . . . 68. *S. greenwayi* (p. 666)
19. Capitulum solitary 20
    Capitula several together 21
20. Dwarf rosette herb; involucre 15–22 mm long . . . 6. *S. pachyrhizus* (p. 626)
    Leaves cauline; involucre 8–11.5 mm long . . . . . . 7. *S. kayomborum* (p. 626)
21. Capitula 1–5 together 22
    Capitula 6 or more together 23
22. Basal leaves 27–32 cm long; involucre 10 mm long 28. *S. discokaraguensis* (p. 638)
    Basal leaves 1–8 cm long; involucre 3.5–5.5 mm long 43. *S. mesogrammoides* (p. 649)
23. Herb with annual shoots from perennial rootstock 33. *S. urundensis* (p. 642)
    Perennials with perennial above-ground parts 24
24. Achenes and ovary hairy; **U** 2 . . . . 65. *S. transmarinus*
                                                              × *hochstetteri* (p. 665)
    Achenes and ovary ± glabrous 25
25. Involucre 3.5–4 mm in diameter 26
    Involucre more than 5.5 mm in diameter 27
26. Stem slightly glandular-pubescent; basal leaves
      5.5–13 cm wide . . . . 31. *S. exarachnoideus* (p. 641)
    Stem glabrous; basal leaves 0.6–5 cm wide . . . . . . 34. *S. semiamplexifolius* (p. 642)
27. Median leaves 4.5–32 × 0.5–10.5 cm; scapigerous
      plant 28
    Median leaves 3.5–6.5 × 0.4–0.6 cm (unknown for
      S. doryphoroides); subscapigerous plant 29

28. Pappus 4.5–7 mm long; leaves arachnoid-pubescent
    when young .... 29. *S. gramineticola* (p. 640)
    Pappus 9 mm long; leaves glabrous but for the
    midrib .... 30. *S. immixtus* (p. 640)
29. Phyllaries ± 18 .... 20. *S. doryphoroides* (p. 635)
    Phyllaries 10–14 .... 21. *S. spartareus* (p. 635)
30. (*plants with ray florets*) Florets purple, mauve or red   61. *S. roseiflorus* (p.662)
    Florets yellow 31
31. Leaves noticeably glandular, and often viscid 32
    Leaves not glandular or minutely and hardly
    visibly glandular 39
32. Ray less than 10 mm long 33
    Ray more than 10 mm long 36
33. Achenes/ovary glabrous 34
    Achenes/ovary hairy 35
34. Rays 0–7, less than 2.2 mm wide; pappus less than
    6.5 mm long .... 57. *S. purtschelleri* (p. 660)
    Rays 8–14, more than 2.5 mm wide; pappus more
    than 7 mm long .... 63. *S.* × *pirottae* (p. 663)
35. Leaves 5–11.5 × 0.4–3 mm; rays 11–13, 6–10 mm long   44. *S. moorei* (p. 651)
    Leaves 2.5–4 × 0.3–0.5 mm; "rays" 9, 4 mm long   46. *S. mooreoides*
                                                          × *hochstetteri* (p. 652)
36. Achenes/ovary glabrous; pappus less than 8 mm long 37
    Achenes/ovary hairy; pappus 8.5–10.5 mm long   58. *S. snowdenii* (p. 660)
37. Rays 6–12; median leaves to 24 × 6.5 cm; basal
    leaves present .... 64. *S. transmarinus* (p. 664)
    Rays 10–23; median leaves to 11.5 × 4.5 cm; basal
    leaves absent 38
38. Rays 17–23, 2.5–3 mm wide; Mt Elgon .... 59. *S. sotikensis* (p. 661)
    Rays 10–13, 3.7–4 mm wide; Virunga Mts .... 60. *S. polyadenus* (p. 662)
39. Leaves succulent (may not be clear in dried specimens) 40
    Leaves not succulent 46
40. Creeping herb 41
    Erect herb or shrub, or scandent 42
41. Leaves sparsely pilose or glabrous; rays 7–8.5 mm long   49. *S. meyeri-johannis* (p. 654)
    Leaves white-tomentose to archnoid, at least
    below; rays 5.5–6 mm long .... 53. *S. keniophytum* (p. 656)
42. Plants from above 3000 m; stem arachnoid-pubescent 43
    Plants from below 3000 m; stem glabrous 44
43. Achenes/ovary shortly hairy; phyllaries glabrous .   50. *S. aequinoctalis* (p. 654)
    Achenes/ovary glabrous; phyllaries hairy .... 54. *S. telekii* (p. 657)
44. Petiole absent 45
    Petiole 0.4–3.5 cm long; leaves elliptic or obovate,
    more than 1 cm wide .... 39. *S. hadiensis* (p. 646)
    (Leaves peltate .... 37. *S. milanjianus*) (p. 644)
45. Leaves subcylindric, 2–6 mm across .... 40. *S. margaritae* (p. 648)
    Leaves flat, 1.2–5.5 cm wide .... 36. *S. ruwenzoriensis* (p. 643)
    (or rarely 25. *S. sororius* with leaves 0.7–1 cm wide)
46. Annual herb 47
    Perennial herb or shrub 48
47. Leaves glabrous; calyculus of 16 bracts; **U** 1, Moroto   42. *S. morotonensis* (p. 649)
    Leaves glabrous or pubescent; calyculus of 1–7
    bracts; Kenya or Tanzania .... 43. *S. mesogrammoides* (p. 649)
48. Plant (sub-)scapigerous 49
    Plant with most leaves cauline 55
49. Achenes/ovary hairy 50
    Achenes/ovary glabrous 52

Note: *S. macroglossus* DC. which is reported as occurring in Kenya (Lisowski, Aster. Fl. Afr. Centr. 2: 261 (1991)) has not been seen by us from that country, apart from the following specimen: Kenya, Nairobi, in Jan & Bert Gillett's garden; Dec. 1977, *Gillett* 21645! A herb with sub-succulent deltoid 5-lobed leaves with cordate base and large ray florets.

Other cultivated Senecio: *S. tamoides* DC. with orange florets, in the gardens of the National Museums of Kenya, Aug. 1965, *Kirika* 463!

S. viravira Hieron., City Park, Nairobi, Feb. 1961, *Perkins* H42/61! with much-divided grey leaves.

1. **Senecio syringifolius** *O.Hoffm.* in E.J. 20: 236 (1894); T.T.C.L.: 160 (1949); U.K.W.F.: 476, fig. (1974); Maquet in Fl. Rwanda 3: 666 (1985); C. Jeffrey in K.B. 41: 885 (1986); Blundell, Wild Flow. E. Afr.: fig. 232 (1987); Lisowski, Aster. Fl. Afr. Centr. 2: 261 (1991); U.K.W.F. ed. 2: 221, t. 93 (1994). Type: Tanzania, Lushoto District, Usambara Mts, Lutindi, *Holst* 3264 (B†, holo., M!, iso.)

FIG. 130. *SENECIO SYRINGIFOLIUS* — **1**, habit, × ²/₃; **2**, capitulum, × 3; **3**, floret, × 6; **4**, achene, × 3. 1–3 from *Drummond & Hemsley* 2703, 4 from *Haarer* 1515. Drawn by Juliet Williamson.

Liana twining to 1.5–10 m high or long, rather succulent; stems up to 1.5 cm in diameter, glabrous, developing a grey bark when old. Leaves glossy, triangular, rhombic-triangular, oblong-triangular or ovate in outline, 3.8–10 cm long, 1.5–9 cm wide, rounded to truncate or weakly cordate and shortly decurrent onto the petiole, margins entire or remotely paucidentate or paucilobulate in lower half to coarsely sinuate-dentate with apiculate teeth, apex obtuse to acuminate, glabrous to moderately densely pubescent, 5–7-veined from base; petiole glabrous or with a few hairs on the margins near the base, exauriculate, 1.4–5.6 cm long. Capitula discoid, many in usually dense, spreading, terminal compound corymbs, sweet-scented; involucre cylindrical, 5.5–9 mm long, 2.5–3.5 mm in diameter; bracts of calyculus 3–5, lax, oblong-linear or oblong-lanceolate, 1.5–5 mm long, obtuse, shortly ciliate at the apex or glabrous; phyllaries 8, green, 5–8.5 mm long, glabrous or hairy at apex. Ray florets 0; disc florets 15–20, yellow or creamy yellow; corolla 6.5–9 mm long, expanded from below the middle, glabrous; lobes 1–1.9 mm long. Achenes 3–4 mm long, ribbed, glabrous; pappus 5–9 mm long. Fig. 130 (page 622).

UGANDA. Kigezi District: Luhiza, Kinaba, Mar. 1947, *Purseglove* 2358! & Luhiza Road, June 1951, *Purseglove* 3685!
KENYA. S Nyeri District: Kiandogoro, Aug. 1963, *Mathenge* 203!; Masai District: S summit Ngong Hills, Oct. 1977, *Gillett & Stearn* 21589!; Teita District: Vuria Hill, May 1985, *Taita Hills Expedition* 134!
TANZANIA. Lushoto District: Mazumbai Forest, July 1966, *Semsei* 4056!; Morogoro District: Uluguru Mts, Lukwangulu Plateau, Sep. 1970, *Thulin & Mhoro* 1021!; Mbeya District: Mporoto Ridge at Igale Pass, June 1992, *Mwasumbi* 16363!
DISTR. **U** 2; **K** 1, 3–7; **T** 2–3, 6–7; Rwanda, Malawi
HAB. Evergreen forest, forest margins, bamboo zone, secondary bushland in forest zone; (900–)1500–3300 m
USES. Root used for emetic (*Koritschoner*)
CONSERVATION NOTES. Widespread and fairly common in East Africa; least concern (LC)

SYN. *S. exsertiflorus* Baker in K.B. 1898: 154 (1898). Type: Malawi, Zomba, *Whyte* s.n. (K!, holo.)
    S. nyikensis Baker in K.B. 1898: 154 (1898), *nom. illegit.*, *non* Baker (1897). Type: Malawi, Nyika, *Whyte* 238 (K!, holo.)
    S. subpetitianus Baker in K.B. 1898: 303 (1898). Type: Malawi, Zomba, *Whyte* 238 (K!, holo.)
    S. elliotii S.Moore in J.L.S. 35: 360 (1902). Type: Uganda, Ruwenzori, *Scott Elliot* 7826 (BM!, holo.)

2. **Senecio maranguensis** *O.Hoffm.* in P.O.A. C: 418 (1895); T.T.C.L.: 158 (1949); Maquet in Fl. Rwanda 3: 664, fig. 206/2 (1985); C. Jeffrey in K.B. 41: 885 (1986); Lisowski, Aster. Fl. Afr. Centr. 2: 262 (1991); K.T.S.L.: 562 (1994); U.K.W.F. ed. 2: 221 (1994). Type: Tanzania, Moshi District, Marangu, *Volkens* 1127 (B†, syn.) & 1491 (B†, syn., BM!, isosyn.)

Woody herb or shrub to 3 m or semi-scandent to 6 m; stems leafy, long and whippy, sometimes tinged purplish or red, pubescent or thinly arachnoid, glabrescent. Leaves petiolate, leathery, lanceolate, ovate or oblong, 3–16.5 cm long, 1–6 cm wide, base cuneate, rounded, truncate or weakly cordate, margins finely to coarsely sinuate-serrate, dentate or bidentate, apex obtuse to acute, glabrous to pubescent especially on veins beneath, glabrous above except for midrib and sometimes the main veins; somewhat marcescent when old; petiole sparsely pubescent or arachnoid above to glabrous, 0.2–3 cm long, very narrowly winged, exauriculate or with coarsely dentate stipuliform auricles at the base. Capitula radiate, numerous in copious congested to lax spreading terminal compound corymbs; stalk of individual heads pubescent or arachnoid; involucre 3–5 mm long, 1.7–2.5 mm in diameter; bracts of calyculus 3–6, 1–3 mm long, arachnoid or pubescent at least on margins at apex; phyllaries 7–8, green with brown tips, 2.5–5 mm long, glabrous or pilose or arachnoid. Ray florets 5–8, corolla tube 2.5–3 mm long, hairy at apex or glabrous, rays pale to bright yellow, 3–5 mm long, 1–2.2 mm wide, 4-veined, spreading; disc florets yellow turning red-brown, corolla 4–6 mm long, tube expanded above the middle, glabrous, lobes 0.5–0.7 mm long. Achenes ribbed, 1.5–2 mm long, glabrous; pappus 3.5–6 mm long.

UGANDA. Ruwenzori Mts, Jan. 1967, *Magogo* 26!; Kigezi District: Mafuga, Aug. 1949, *Purseglove* 3079! & Bufumbira, Apr. 1970, *Katende* 193!
KENYA. Nyandarua [Aberdare] Mts, Kinangop, Apr. 1922, *Fries & Fries* 2699!
TANZANIA. Kilimanjaro, Barankata, Aug. 1971, *Shabani* 746!; Lushoto District: Chambogo Forest, July 1987, *Kisena* 524!; Rungwe District: Iringu Forest, Nov. 1982, *Leliyo* 312!
DISTR. U 2; K 3; T 2–3, 6–7; Congo (Kinshasa), Rwanda, Burundi, Malawi
HAB. Montane forest or forest margins, secondary bushland replacing forest, heath zone, bamboo zone; may be locally common; 1800–3250 m
USES. None recorded on specimens from our area
CONSERVATION NOTES. Fairly widespread, many recent specimens; least concern (LC)

SYN. *S. scrophulariifolius* O.Hoffm. in E.J. 24: 474 (1898). Type: Tanzania, Morogoro District, Uluguru Mts, *Stuhlmann* 9172 (B†, holo.)
　　S. psiadioides O.Hoffm. in E.J. 30: 436 (1901); T.T.C.L.: 158 (1949). Type: Tanzania, Njombe District, Ukinga Mts, *Goetze* 1201 (B†, EA, syn.) & 1259 (B†, syn., BM!, isosyn.)
　　S. roccatii Chiov. in Ann. Bot. Roma 6: 151 (1907). Type: Uganda, Toro District, Mubuku Valley, *Roccati* s.n. (TO, syn.)
　　S. jugicola S.Moore in J.L.S. 28: 264 (1908). Type: Uganda, Toro District, Ruwenzori, *Wollaston* s.n. (BM!, holo.)
　　S. hageniae R.E.Fr. in Acta Hort. Berg. 9: 151 (1928). Type: Kenya, Nyandarua [Aberdare] Mts, Kinangop, *Fries & Fries* 2699 (UPS!, holo.)
　　S. subcarnosulus De Wild., Pl. Bequaert. 5: 118 (1929). Type: Congo (Kinshasa), Ruwenzori, Butagu, *Bequaert* 3728 (BR!, holo.)

3. **Senecio mariettae** *Muschl.* in Z.A.E. 2: 387, t. 42 (1911); Maquet in Fl. Rwanda 3: 666, fig. 206/1 (1985); C. Jeffrey in K.B. 41: 886 (1986); Lisowski, Aster. Fl. Afr. Centr. 2: 265 (1991). Type: Congo (Kinshasa), Sabinio [Sabinjo], *Mildbraed* 1700 (B†, holo., BR!, iso.)

Semi-scandent shrub 2–4 m high. Stems leafy, finely pubescent. Leaves shortly petiolate, narrowly elliptic, 4.8–9.5 cm long, 1–4.1 cm wide, base cuneate, subtruncate or weakly cordate, margins coarsely sinuate-dentate or sinuate-serrate, apex rounded to obtuse, apiculate, densely grey-tomentose beneath except for midrib, green and glabrous above; petiole finely shortly pubescent, exauriculate, 0.5–1.1 cm long. Capitula radiate, numerous in copious dense terminal compound corymbs; stalk of individual heads floccose-tomentose; involucre 4.5–6 mm long, 2.5 mm in diameter; bracts of calyculus 3–5, lanceolate, 1.5–2 mm long, arachnoid-pubescent on margins and also in lower part; phyllaries 8–12, 4–6 mm long, thinly arachnoid near base. Ray florets 7–13, corolla tube 2.7–3 mm long, glabrous or hairy, rays yellow, 5.5–12 mm long, 1.7–2.5 mm wide, 4-veined; disc florets yellow, corolla 4.2–5.2 mm long, tube glabrous, lobes 1 mm long. Mature achenes unknown but ovary 1.3 mm long, glabrous; pappus 4.2–5.2 mm long.

UGANDA. Kigezi District: Mgahinga, 1932, *Eggeling* 1065! & Aug. 1938, *A.S. Thomas* 2477! & Muhavura, Oct. 1948, *Hedberg* 2059!
DISTR. U 2; Congo (Kinshasa), Rwanda, Burundi
HAB. *Hypericum*, heath or bamboo zone; 2550–3400 m
USES. None recorded on specimens from our area
CONSERVATION NOTES. Four Ugandan specimens; data deficient (DD)

4. **Senecio lyratus** *Forssk.*, Fl. Aegypt-Arab.: 148 (1775); C. Jeffrey in K.B. 41: 886 (1986); U.K.W.F. ed. 2: 221 (1994). Type: Yemen, *Forsskål* s.n. (C!, syn.)

Perennial woody herb or soft shrub, erect or scrambling, often scandent, stems 1–5 m long, angular, green, sparsely to densely pubescent. Leaves spreading, triangular, rhombic, ovate-triangular or triangular-lanceolate, 4.4–9 cm long, 1.3–5.2 cm wide, base cuneate, subtruncate or cordate, margins coarsely apiculate-dentate or bidentate, apex acute to subacute and minutely apiculate, often with 2 or rarely 4 much smaller ± retrorsely-directed oblong lateral lobes on the petiole/winged base, pubescent to finely tomentose especially on veins beneath, scattered-pubescent above, narrowly petiolate with the petiole auriculate at the base. Capitula radiate, numerous in divaricately-branched terminal corymbs; stalks of individual capitula pilose or arachnoid, often glabrescent; involucre cylindrical, 4.5–5.5 mm long, 2.5–3 mm in diameter; bracts of calyculus 3–4, lanceolate, 1–2.5 mm long, ciliate on margins or at apex or ± glabrous; phyllaries 8–13, green, sometimes tinged reddish, 4–5 mm long, pubescent or glabrous. Ray florets 5–8, corolla tube 1.7–2.5 mm long, hairy at apex, ray bright yellow, 4–9 mm long, 1.7–2.2 mm wide, 4-veined; disc florets orange-yellow, corolla 4.5–5.8 mm long, tube expanded from just below the middle, glabrous or with a very few hairs, lobes 0.5–0.7 mm long. Achenes 2–2.5 mm long, hairy on ribs or glabrous; pappus 4.5–5 mm long.

Kenya. Northern Frontier District: Mt Kulal, June 1960, *Oteke* 125! & Marsabit, Aug. 1968, *Faden* 68/497!; Elgeyo District: Tambach, Jan. 1962, *Tweedie* 2283!
Tanzania. Ufipa District: Nsanga Forest, Aug. 1960, *Richards* 13038!; Iringa District: Dabaga, Sep. 1932, *Geilinger* 1899!; Njombe District: Nyumbanyito, 13 km WNW of Njombe, July 1956, *Milne-Redhead & Taylor* 11112!
Distr. **K** 1, 3–5; **T** 4, 7; Sudan, Ethiopia, Somalia, Yemen
Hab. Evergreen dry forest margins or bushed grassland; 1500–2250 m
Uses. None recorded on specimens from our area
Conservation notes. Widespread; least concern (LC)

Syn. *S. auriculatus* Vahl, Symb. Bot. 1: 72, t. 18 (1790), *nom. illegit. superfl.* Type as for *S. lyratus* Forssk.
  S. appendiculatus Poir., Encycl. 7: 102 (1806), *nom. illegit. superfl.* Type as for *S. lyratus* Forssk.
  S. lyratipartitus A.Rich., Tent. Fl. Abyss. 1: 439 (1848). Type: Ethiopia, Selleuda Mts near Adowa, *Schimper* 184 (K!, P!, isosyn.) & 1184 (K!, isosyn.), Orbata, *Dillon* s.n. (P!, syn.), Ouodgerate Province, *Petit* s.n. (P!, syn.)
  Cineraria schimperi Oliv. & Hiern, F.T.A. 3: 404 (1877). Type: Ethiopia, without further locality, *Schimper* s.n. & *Hildebrandt* s.n. (not traced) – possibly the same as *S. lyratipartitus*
  Senecio basipinnatus Baker in K.B. 1895: 217 (1895). Type: Somalia, Golis Range, *Cole* s.n. & *Lort Phillips* s.n. (both K!, syn.)
  S. masonii De Wild., Pl. Bequaert. 5: 128 (1929). Type: Kenya, Nairobi, *Mason* s.n. (BR!, holo.)
  S. lyratipartita (A.Rich.) Cuf., E.P.A.: 1152 (1967); *nom superfl.*

Note. Resembles *Cineraria deltoidea* but distinct in its achenes.

5. **Senecio deltoideus** *Less.*, Syn. Comp.: 392 (1832); Oliv. & Hiern, F.T.A. 3: 420 (1877); Hilliard, Comp. Natal: 492 (1977); C. Jeffrey in K.B. 41: 886 (1986); U.K.W.F. ed. 2: 221 (1994). Type: South Africa, Cape of Good Hope, *Sonnerat* s.n. (P, holo.) (At P HB saw a sheet marked *Senecio deltoideus* Less. DC! Cap., pre-1837, which is possibly this specimen - there are no other Sonnerat sheets of this taxon)

Scandent perennial herb or liana; stems 2–7 m long, green, ribbed, finely pubescent, often glabrescent. Leaves ovate-triangular or triangular, 4.3–12 cm long, 2–6.5 cm wide, base truncate to cordate, margins coarsely apiculate-dentate or bidentate, apex obtuse to subacute or acuminate, sometimes with 2 (or very rarely 4) much smaller lateral lobes, finely pubescent, mostly on veins, especially beneath, to glabrous or almost so, narrowly petiolate with the petiole auriculate at the base. Capitula discoid, numerous in divaricately-branched terminal and axillary corymbs forming large, lax, pendent thyrsoid panicles; involucre cylindrical, 4–5.5 mm long, 2.5–3.5 mm in diameter; bracts of calyculus 3–5, lanceolate to broadly lanceolate, 1.5–2 mm long, floccose or ciliate or glabrous; phyllaries 8–13, green, 3.5–5 mm long, sparsely pubescent or glabrous. Ray florets absent; disc florets pale creamy yellow to yellow, corolla 3–5 mm long, tube expanded above the middle, glabrous, lobes 1–1.2 mm long. Achenes 2.5–3 mm long, ribbed, hairy; pappus white, 3–5.5 mm long.

KENYA. Kiambu District: Muguga Forest, Sep. 1965, *Kokwaro & Kabuye* 341!; Machakos District: Kilungu 4 km N of Nunguni, Aug. 1971, *Mwangangi* 1658!; Masai District: Olekaitorror Escarpment, Sep. 1962, *Glover & Wateridge* 3279!

TANZANIA. Mbulu District: Ngorongoro, Lositete, Aug. 1981, *Chuwa & Baraza* 2280!; Lushoto District: Mtowanguue, Sep. 1955, *Sangiwa* 71! & W Usambara Mts, Shagayu Forest Reserve SE of Shagein, Oct. 1986, *Borhidi et al.* 86/094!

DISTR. **K** 3–7; **T** 2, 3; Malawi, Mozambique, Zimbabwe, Swaziland, South Africa

HAB. Moist or dry forest or forest margins, secondary bushland in the forest zone; 700–2200 m

USES. Minor medicinal use as an emetic (*Koritschoner, Semsei*)

CONSERVATION NOTES. Wide distribution; least concern (LC)

SYN. *Eupatorium auriculatum* Lam., Encycl. 2: 410 (1788), *non Senecio auriculatus* N.L.Burm. (1768). Type as for *S. deltoideus*

    Senecio sarmentosus O.Hoffm. in E.J. 20: 236 (1894), *nom. illegit., non* (Bl.) Schweinf. (1867); T.T.C.L.: 159 (1949). Type: Tanzania, Lushoto District: Usambara Mts, *Holst* 439, 621 (B†, syn.), 3256a (B†, M!, syn.) & 8914 (B†, K!, syn.)

    S. rooseveltianus De Wild., Pl. Bequaert. 5: 129 (1929). Type: Kenya, Nairobi, *Mearns* 984 (BR!, holo., BM!, iso.)

    S. tanzaniensis Cuf., E.P.A.: 1159 (1967). Type as for *S. sarmentosus* O. Hoffm.

NOTE. No ray florets have been seen for this taxon in East Africa, but in Southern Africa some specimens have 1(–4) ray florets.

6. **Senecio pachyrhizus** *O.Hoffm.* in E.J. 30: 435, t. 19/a–g (1901); C. Jeffrey in K.B. 41: 887 (1986); Lisowski, Aster. Fl. Afr. Centr. 2: 270 (1991). Type: Tanzania, Mbeya/Chunya District: Usafwa [Usafua], *Goetze* 1121 (B†, holo.)

Dwarf perennial herb with woody rootstock, precocious; rootstock 1–2.5 cm in diameter, crown often woolly and with conspicuous fibrous remnants of old leaf-bases. Stems usually several from each crown, forming small cushions 5–16 cm high, pale green, minutely glandular-pubescent and sometimes also slightly arachnoid, glabrescent. Leaves all basal, oblong to elliptic, 13–25 cm long, 1–4.5 cm wide, long-attenuate into petioloid base, margins crenate-serrate, apex acute to attenuate, arachnoid-pubescent above and beneath; stem-leaves scale-like, sessile, concave, pale green or brown, oblanceolate or lanceolate, 1–2.5 cm long, 0.2–0.5 cm wide, erose-denticulate, acuminate or acute, glabrous or sparsely arachnoid beneath, the uppermost constituting the calyculus. Capitula erect, discoid, terminal, usually solitary; involucre broadly cylindrical, brownish, 15–22 mm long, 11–18 mm in diameter, glabrous or sparsely arachnoid at base; bracts of calyculus 5–8, broadly ovate to lanceolate, 10–20 mm long, acuminate or acute, shortly erose-denticulate; phyllaries 16–25, long-attenuate, 15–22 mm long, shortly ciliate at margins near apex. Ray florets absent; disc florets white or cream, anthers fully exserted, corolla 15.5–20 mm long, tube abruptly expanded and campanulate in upper one-sixth, shortly hairy on the narrow cylindrical lower part, lobes 2–2.5 mm long. Achenes when mature 11 mm long, densely long-hairy; pappus 8–22 mm long, white.

UGANDA. Acholi District: Imatong Mts, Apr. 1938, *Eggeling* 3610!

TANZANIA. Kigoma District: Mahali Mts, Kabesi, Sep. 1958, *Newbould & Jefford* 2309!; Mbeya District: Mbeya Range, Oct. 1956, *Richards* 6451!; Njombe District: Chimala Plateau, Oct. 1963, *E.S. Brown* 483!

DISTR. **U** 1; **T** 4, 7; Cameroon, Congo (Kinshasa), Burundi, Angola, Zambia, Malawi

HAB. Pyrophytic, in upland grassland, flowering about a month after fire; in large clumps or cushion-forming; 1500–2600 m

USES. None recorded on specimens from our area

CONSERVATION NOTES. Widespread, in fairly common habitat; least concern (LC)

7. **Senecio kayomborum** *Beentje* in K.B. 58: 233 (2003). Type: Tanzania, Iringa District, Mufindi, Igowole, *C.J. & M.J. Kayombo* 220 (K!, holo., MO, NHT, iso.)

Erect or creeping herb with annual shoots up to 40 cm high or long from a thickened undergound part; stems in the upper part reddish purple and sparsely pilose with multicellular hairs, otherwise glabrous. Leaves narrowly obovate to almost linear, 1.5–4.5 cm long, 0.2–1.6 cm wide, base attenuate, margins entire or denticulate but slightly thickened and minutely revolute, apex obtuse and somewhat thickened, sparsely pilose-scabridulous on both surfaces, glabrescent (no basal leaves or root crown visible or indicated). Capitula solitary, erect, discoid; involucre cylindrical, dark purple-red, 8–12 mm long, 6–9 mm in diameter, densely pilose-scabridulous at base; bracts of calyculus 5–10, narrowly lanceolate and up to 5 mm long, acute, pilose; phyllaries 15–20, attenuate, sparsely pilose-scabridulous, penicillate-bearded at apex. Ray florets absent; disc florets orange-yellow or yellow, corolla 8.5–10.2 mm long, tube slightly and gradually widening in the upper half, lobes 1.5–2.5 mm long; anthers 3 mm long; style arms obtuse, papillate at apex. Achenes ± 3 mm long, 8–9-ribbed, pilose between the ribs; pappus 7.5–8.5 mm long, white.

TANZANIA. Iringa District: Lake Ngwazi S, Nov. 1986, *Goldblatt et al.* 8058! & Mufindi, Igowole, Mar. 1989, *C.J. & M.J. Kayombo* 220! (type) & Udzungwa Mountain National Park, Camp 295 to Exit Gully, June 2002, *Luke et al.* 8630!
DISTR. **T** 7; Zambia
HAB. Wet meadow grassland; 1800–2100 m
USES. None recorded on specimens from our area
CONSERVATION NOTES. Known from only three Tanzanian specimens and one Zambian one; at least vulnerable (VU-D2)

8. **Senecio subsessilis** *Oliv. & Hiern*, F.T.A. 3: 415 (1877); A.V.P.: 242 (1957); Maquet in Fl. Rwanda 3: 664, fig. 205/2 (1985); C. Jeffrey in K.B. 41: 892 (1986); Lisowski, Aster. Fl. Afr. Centr. 2: 271, t. 58 (1991); U.K.W.F. ed. 2: 222 (1994). Type: Ethiopia, Gaffat, *Schimper* 1532 (K!, holo., BM!, iso.)

Perennial herb, woody herb or weakly woody shrub 45–300 cm high; stems densely to thinly arachnoid or sparsely pubescent to glabrescent; roots thick. Basal leaves petiolate, upper leaves sessile, leaves lanceolate, ovate, ovate-triangular to broadly elliptic, 11–50 cm long, 3–13 cm wide or up to 27 cm wide in basal leaves, base cordate, subtruncate, rounded or cuneate, shortly decurrent onto the petiole, margins sinuate-dentate to coarsely bidentate, apex obtuse or rounded and minutely apiculate, subacute or acuminate; petiole unwinged, exauriculate at the base in basal leaves, broadly auriculate and semi-amplexicaul to exauriculate in stem leaves, 11–67 cm long; all leaves coriaceous to chartaceous, thinly arachnoid or floccose-tomentose and glabrescent above, tomentose or arachnoid or pubescent beneath. Capitula radiate, erect, usually numerous in a copious congested to lax sometimes divaricately-branched terminal corymbiform cyme; stalks of the individual capitula arachnoid-tomentose or shortly pubescent; involucre cylindrical, 6–11 mm long, 3–7 mm in diameter, arachnoid or woolly at least at base or glabrescent; bracts of calyculus (2–)6–12, lanceolate or linear-lanceolate, 1.5–10 mm long, thinly arachnoid to glabrous and sometimes bearded at the apex; phyllaries 8–20, green, 5.5–10 mm long, thinly arachnoid or pubescent to glabrous, bearded and shortly tapered or sometimes slightly expanded at the apex. Ray florets pale to bright yellow, 5–8(–12), corolla tube glabrous or sometimes shortly glandular-hairy or with a few long straight multicellular eglandular hairs in lower part, 4.2–5 mm long, rays 6–18 mm long, 2–5 mm wide, 4–10-veined; disc florets yellow, corolla 6.5–9 mm long, tube gradually expanded above the middle or from just below the middle, glabrous or sometimes with long multicellular hairs near the base, lobes 1–1.5 mm long. Achenes 2–5 mm long, glabrous; pappus 5–9 mm long, white. Fig. 131 (page 628).

UGANDA. Karamoja District: Morongole, Nov. 1939, *A.S. Thomas* 3313!; Kigezi District: Mt Mgahinga [Gahinga], Apr. 1970, *Katende* 176!; Mbale District: Elgon, Bulambuli, Nov. 1933, *Tothill* 2394!

FIG. 131. *SENECIO SUBSESSILIS* — **1**, habit, × ²/₃; **2**, leaf from near base of plant, × ²/₃; **3**, ray floret, × 6; **4**, disc floret, × 8; **5**, achene, × 6. 1 & 3–5 from *Richards* 21706, 2 from *Richards* 23857. Drawn by Juliet Williamson.

KENYA. Nakuru District: Mau Forest, Endabarra, Jan. 1946, *Bally* 4900!; Kericho District: SW Mau Forest, Sambret, Oct. 1961, *Kerfoot* 2915!; Masai District: Nasampolai Valley, Jan. 1972, *Greenway & Kanuri* 14973!

TANZANIA. Kilimanjaro: Marangu, May 1961, *Machangu* 72!; Lushoto District: W Usambara, Mtumbi Forest Reserve, Feb. 1985, *Borhidi et al.* 85/650!; Mbeya District: Poroto Mts, May 1957, *Richards* 9743!

DISTR. U 1–3; **K** 3–6; **T** 2, 3, 6, 7; Congo (Kinshasa), Rwanda, Burundi, Sudan, Ethiopia

HAB. Montane bushland, giant heath, *Hagenia* woodland/forest, bamboo, often in clearings and possibly a pioneer there; 1800–3600 m

USES. None recorded on specimens from our area

CONSERVATION NOTES. Widespread in a reasonably common habitat; least concern (LC)

SYN. *S. denticulatus* Engl. in Abh. Preuss. Akad. Wiss. 1891: 442 (1892); T.T.C.L.: 158 (1949), *nom. illegit., non* DC. (1838). Type: Tanzania, Kilimanjaro, *Meyer* 72 (B†, holo.)
  S. bagshawei S.Moore in J.L.S. 38: 264 (1906). Type: Uganda, Ruwenzori, *Wollaston* s.n. (BM!, holo.)
  S. wollastonii S.Moore in J.L.S. 38: 264 (1906). Type: Uganda, Ruwenzori, *Wollaston* s.n. (BM!, holo.)
  S. thomsianus Muschl. in E.J. 43: 59 (1909). Type: Tanzania, E Usambara Mts, *Keil* 180 (B†, holo.)
  S. trichopterygius Muschl. in Z.A.E. 2: 394 (1911); F.P.S. 3: 49 (1956). Type: Congo (Kinshasa), Rugege, Rukarara, *Mildbraed* 905 (B†, holo.)
  S. denburgianus Muschl. in Z.A.E. 2: 394 (1911). Type: Congo (Kinshasa), Karisimbi, *Mildbraed* 1613 (B†, holo., BR!, iso.)
  S. fistulosus De Wild., Pl. Bequaert. 5: 105 (1929). Type: Congo (Kinshasa), Ruwenzori, *Bequaert* 3830 (BR!, holo.)
  S. fistulosus De Wild. var. *mukuleensis* De Wild., Pl. Bequaert. 5: 106 (1929). Type: Congo (Kinshasa), Ruwenzori, *Bequaert* 5925 (BR!, holo.)
  S. latealatopetiolatus De Wild., Pl. Bequaert. 5: 109 (1929). Type: Congo (Kinshasa), Lanuri, *Bequaert* 4538 (BR!, holo.)

9. **Senecio mabberleyi** *C.Jeffrey* in K.B. 41: 893 (1986). Type: Tanzania, Kilosa District, Ukaguru Mts, Mamiwa Forest Reserve, *Mabberley, Pocs & Salehe* 1300 (K!, holo.)

Perennial herb 0.8–2 m high; stems erect, weakly woody, sparsely arachnoid or pubescent to glabrous. Basal leaves petiolate, ovate-triangular to triangular, 13–16 cm long, 10–12.5 cm wide, base cordate, margins coarsely dentate, apex obtuse, apiculate; petiole unwinged, glabrous, 18–38 cm long; median and upper stem leaves ovate, broadly elliptic or oblong-lanceolate, 7.5–19 cm long, 2.8–7 cm wide, base of median leaves rounded to cuneate into a broadly winged dentate semi-amplexicaul base, or cordate and petiolate, base of upper leaves sometimes sessile and sometimes slightly narrowed towards an auriculate and semi-amplexicaul base, margins coarsely dentate or denticulate, apex obtuse; all leaves rather coriaceous, green, sparsely arachnoid, glabrescent or glabrous on both surfaces or glabrous above and sparsely pilose on veins beneath. Capitula radiate, erect, 6 to numerous in lax terminal corymbose cymes; stalks of the individual capitula slightly arachnoid or glabrous; involucre cylindrical, 7.5–9 mm long, 6 mm in diameter, slightly woolly at the base; bracts of calyculus 6–12, lanceolate or linear-lanceolate, reflexing, 4.5–9 mm long, slightly arachnoid to glabrous, bearded at the apex; phyllaries 12–16, sometimes long-tapered, 6.5–9 mm long, sparsely pubescent or arachnoid in upper part, bearded at apex. Ray florets 5–8, yellow, corolla tube 4–5 mm long, glabrous, rays 10–13 mm long, 3.5–4 mm wide, 4–7-veined; disc florets yellow, corolla 7–9 mm long, slender, gradually slightly expanded above the middle, lobes 1–1.5 mm long. Mature achenes unknown; ovary of disc florets 2–4 mm long, glabrous, ribbed; pappus 7–8 mm long, white.

TANZANIA. Kilosa District: Ukaguru Mts, Mamiwa Forest Reserve, July 1972, *Mabberley, Pocs & Salehe* 1300! & W slopes of Mt Mnyera, May 1978, *Thulin & Mhoro* 2751!; Morogoro District: Uluguru Mts, W of Lukwangule Plateau, May 1988, *Pocs & Minja* 88/065u!

DISTR. **T** 6; not known elsewhere

HAB. Upland forest or forest margins; may be locally common; 1800–2300 m

Uses. None recorded on specimens from our area
Conservation notes. Restricted to two mountain ranges in a vulnerable habitat; vulnerable
    (VU-B1abiii)

10. **Senecio crispatopilosus** *C.Jeffrey* in K.B. 41: 893 (1986); U.K.W.F. ed. 2: 222, as
crispatipilosus (1994). Type: Uganda, Mt Elgon, *Dummer* 3598 (K!, holo.)

Coarse perennial herb, 60–150 cm high; stems erect, pubescent. Basal and stem
leaves broadly ovate, ovate or ovate-lanceolate, 7.5–14.5 cm long, 2.5–12 cm wide,
base cordate, margins dentate to bidentate, apex rounded; petiole unwinged,
exauriculate, 7–18 cm long, the upper stem leaves gradually becoming sessile,
auriculate, semi-amplexicaul; upper stem leaves 4–12.5 cm long, 1.2–6 cm wide; all
leaves membraneous, sparsely pubescent on veins beneath. Capitula radiate, erect,
numerous in divaricately-branched terminal corymbs; stalks of the individual
capitula arcuate-divaricate, pubescent; involucre cylindrical, 9 mm long, 4.5 mm in
diameter, slightly pubescent at the base; bracts of calyculus 8–12, linear to lanceolate,
3.5–7 mm long, pubescent on margins or sometimes glabrous, bearded at the apex;
phyllaries 18–22, 7–8.5 mm long, pubescent or glabrous, bearded at the apex. Ray
florets 10–13, yellow, corolla tube 4.5–5 mm long, glabrous, rays 15–18 mm long,
3.5–4.5 mm wide, 4–7-veined; disc florets yellow, corolla 6.5–8 mm long, the tube
slightly expanded from just below the middle, glabrous, lobes 1 mm long. Achenes
3–4.5 mm long, slightly hairy or glabrous; pappus 6.5–7 mm long.

Uganda. Mbale District: Elgon, Jan. 1918, *Dummer* 3598! & Nov. 1933, *Tothill* 2290! & Bulambuli,
    Nov. 1933, *Tothill* 2372!
Kenya. S Nyeri District: Queen's Cave waterfall gorge, June 1962, *Coe* 754! & idem, July 1960,
    Polhill 12024; NW Mt Kenya, Aug. 1931, *Slade* 19!
Distr. U 3; **K** 4; not known elsewhere
Hab. Clearings in bamboo zone, *Juniperus* zone or just below; 2700–3050 m
Uses. None recorded on specimens from our area
Conservation notes. Three specimens from Uganda, all from Elgon; three from Kenya;
    vulnerable (VU-D2)

Syn. *S. sp. F* of U.K.W.F. ed. 1: 480 (1974)

11. **Senecio volcanicola** *C.Jeffrey* in K.B. 41: 893 (1986). Type: Tanzania, Arusha
District, Mt Meru, *Bally* 11581 (K!, holo., BR!, iso.)

Coarse perennial herb or weakly woody shrub, 1–2 m high; stems erect, sparsely
arachnoid or glabrous. Basal leaves "large", details unknown. Stem leaves ovate or
ovate-lanceolate, 7–21 cm long, 3–8 cm wide, cordate at the base into a broadly
winged, auriculate petioloid base to 7 cm long in lower leaves, upper leaves sessile,
auriculate, slightly narrowed above the auricles or semi-amplexicaul in uppermost
leaves, margins strongly dentate, apex shortly obtusely acuminate, apiculate; all
leaves firmly membraneous, glabrous above, moderately to thinly arachnoid-
pubescent beneath. Capitula radiate, erect, very numerous in copious, congested
terminal corymbs; stalks of the individual capitula sparsely shortly pubescent;
involucre cylindrical, 5–6.5 mm long, 3–4 mm in diameter, slightly woolly at the base;
bracts of calyculus 3–7, lanceolate to linear, narrow, 1.5–4 mm long, ± arachnoid or
glabrous except at the margins, bearded and sometimes slightly expanded at the
apex; phyllaries 7–12, 5–6 mm long, glabrous, shortly tapered to bearded apex. Ray
florets 4–5, pale to golden yellow, corolla tube 4.5 mm long, glabrous, rays 6–10 mm
long, 2.5–3 mm wide, 4–6-veined; disc florets yellow, corolla 7–9 mm long, gradually
expanded above the middle, glabrous, lobes 1–1.3 mm long. Achenes unknown,
ovary of disc florets 1.5 mm long, glabrous; pappus 6.5–7 mm long, white.

Tanzania. Masai District: Ngorongoro Crater rim, Dec. 1989, *Chuwa* 3069!; Arusha District: Mt
    Meru, Sep. 1932, *B.D. Burtt* 4107!; Kilimanjaro, above Engare Nairobi, Nov. 1948, *Salt* 37!

DISTR. **T** 2; known from Ngorongoro, Mt Meru and Kilimanjaro
HAB. Montane forest clearing, *Hagenia* zone; 2100–2750 m
USES. None recorded on specimens from our area
CONSERVATION NOTES. Small area of occupancy, but habitat quite extensive; least concern (LC)

12. **Senecio dentatoalatus** *C.Jeffrey* in K.B. 41: 893 (1986). Type: Tanzania, Morogoro District, S Uluguru Forest Reserve, *Semsei* 2062 (K!, holo., EA, iso.)

Perennial herb, ± 1 m high; stems erect, slightly arachnoid or pubescent, glabrescent. Basal leaves unknown; lower stem leaves petiolate, upper stem leaves sessile and semi-amplexicaul, blade ovate or ovate-triangular, 11–21 cm long, 6.8–12.5 cm wide, base truncate or subcordate, margins coarsely dentate, apex obtuse; petiole winged, 24–33 cm long, the wings coarsely dentate; all leaves coriaceous, glabrous except for a few glandular hairs on margins and mid-rib beneath. Capitula radiate, numerous in copious lax to dense, conspicuously bracteate terminal corymbs; inflorescence bracts resembling reduced upper stem leaves; stalks of the individual capitula sparsely pilose or arachnoid in upper part; involucre broadly cylindrical, ± 9 mm long, 8 mm in diameter, pilose at base; bracts of calyculus 6–10, oblong-oblanceolate, reflexing, 6–12 mm long, sparsely ciliate at margins, obtuse and bearded at the apex; phyllaries 14–18, 7–8 mm long, slightly pilose to glabrous. Ray florets 8, yellow, corolla tube 4–4.5 mm long, glabrous, rays 12–15 mm long, 4 mm wide, 6–9-veined; disc florets yellow, corolla 7–7.5 mm long, gradually expanded above the middle, glabrous, lobes 1.2–1.5 mm long. Achenes unknown, but ovary of disc florets 2 mm long, glabrous; pappus 7 mm long, white.

TANZANIA. Morogoro District: N Uluguru Mts, Oct. 1938, *Schlieben* 2781! & Lupanga Peak, Dec. 1931, *B.D. Burtt* 3466! & S Uluguru Forest Reserve, Lukwangule Plateau, Mar. 1955, *Semsei* 2062!
DISTR. **T** 6; known from only the Uluguru Mts
HAB. Clearing in upland forest or on rock wall; 1800–2150 m
USES. None recorded on specimens from our area
CONSERVATION NOTES. Not collected for almost fifty years – the cited material is all we have seen; at least vulnerable (VU-D2)

13. **Senecio subfractiflexus** *C.Jeffrey* in K.B. 41: 894 (1986). Type: Tanzania, Morogoro District, Morogoro–Lupanga Peak track, *Greenway & Eggeling* 8619 (K!, holo.)

Perennial herb, 40–100 cm high; stems flexuous, erect or straggling, weak-stemmed, wiry, tinged reddish, obscurely arachnoid, glabrescent. Leaves dark green above, often bright purple when old beneath, petiolate, triangular-lanceolate, 3.5–11.5 cm long, 1.1–5.4 cm wide, base cordate or subtruncate to cuneate, margins coarsely remotely sinuate-dentate, apex attenuate, shortly acuminate-apiculate, glabrous except for sparse arachnoid indumentum on mid-rib beneath; petiole unwinged, almost glabrous, 0.5–11 cm long, absent in uppermost leaves. Capitula radiate, erect, 3–12 in divaricate terminal and axillary corymbs; stalks of the individual capitula slightly arachnoid; involucre broadly cylindrical, 6.5 mm long, 4.5–5.5 mm in diameter; bracts of calyculus reflexing, 6–12, lanceolate, 2.5–6 mm long, slightly arachnoid, acute; phyllaries 8–12, 6–6.5 mm long, sparsely arachnoid, glabrescent, shortly tapered, apically bearded. Ray florets 5–8, bright golden yellow, corolla tube 3 mm long, glabrous, rays 6–8 mm long, 2.5–3 mm wide, 6–7-veined; disc florets yellow, corolla 7 mm long, abruptly expanded above the middle, glabrous, lobes 1 mm long. Achenes 3.5 mm long, glabrous, ribbed; pappus 6–6.5 mm long, white.

TANZANIA. Morogoro District: Uluguru Mts, June 1933, *Schlieben* 4010! & Salaza Forest 5 km S of Bunduki, Mar. 1953, *Drummond & Hemsley* 1614! & ridge above Maunga Valley, W of Bondwa, July 1972, *Mabberley* 1196!
DISTR. **T** 6; known from only the Uluguru Mts
HAB. Evergreen forest; 1600–2100 m

Uses. None recorded on specimens from our area

Conservation notes. Known from four collections in a vulnerable habitat; at least vulnerable (VU-D2)

14. **Senecio pseudosubsessilis** *C.Jeffrey* in K.B. 41: 894 (1986); U.K.W.F. ed. 2: 222 (1994). Type: Kenya, Elgeyo District: 1.7 km W of Lobot cross-roads, *Tweedie* 4193 (K!, holo.)

Perennial herb, 60–300 cm high; stems pubescent to glabrous; rhizome creeping? Basal leaves sessile with petioloid base, lanceolate or oblanceolate to oblong, 7.5–78 cm long (including an up to 20 cm petioloid base), 1.6–18 cm wide, base of basal leaves gradually attenuate into narrowly winged petioloid base, that of stem leaves cordate and semi-amplexicaul, not or somewhat expanded and semi-amplexicaul at the base, margins dentate to coarsely bidentate, apex acute to rounded and sometimes obscurely acuminate, coriaceous to chartaceous, thinly pubescent above and especially on mid-rib beneath to glabrous. Capitula radiate, erect, numerous in a congested terminal cymes; stalks of the individual capitula sparsely pilose or arachnoid-pubescent to glabrous; involucre cylindrical, 7–9 mm long, 4–5 mm in diameter, slightly arachnoid or glandular-pubescent at base; bracts of calyculus 4–8, yellow-green with brown tips, lanceolate, 3–7 mm long, glabrous or sparsely pubescent at margins and/or base; phyllaries 10–14, yellow-green with brown tips, 5.5–9 mm long, glabrous or sparsely pubescent. Ray florets 6–8, bright yellow, corolla tube 3–5.5 mm long, glabrous or glandular-pubescent, rays 7–9.5 mm long, 3–4 mm wide, 4–8-veined; disc florets somewhat darker yellow, corolla 6–7.5 mm long, tube gradually expanded above the middle, glabrous, lobes 0.8–1.3 mm long. Achenes 4.7 mm long, glabrous, ribbed; pappus 6–7.5 mm long, white.

Kenya. Elgeyo District: Kaibwibich, Nov. 1966, *Tweedie* 3379!; Fort Hall District: Kimakia Forest Station, Dec. 1960, *Lucas, Polhill & Verdcourt* 4!; Kericho District: Itare River, Feb. 1940, *Copley* 780!
Distr. **K** 3–5; not known elsewhere
Hab. Upland forest margins, grassland, swamps, riverine; 1900–3050 m
Uses. None recorded on specimens from our area
Conservation notes. Common habitat; least concern (LC)

Syn. *S. subsessilis* sensu U.K.W.F. ed. 1, *non* Oliv. & Hiern

15. **Senecio sabinjoensis** *Muschl.* in Z.A.E. 2: 395 (1911); Maquet in Fl. Rwanda 3: 664 (1985); C. Jeffrey in K.B. 41: 894 (1986); Lisowski, Aster. Fl. Afr. Centr. 2: 274 (1991). Type: Congo (Kinshasa), Sabinio [Sabinjo], *Mildbraed* 1718 (B†, holo., BR!, iso.)

Perennial herb or shrubby herb, rhizomatous, 30–120(–200) cm high; stems stout, glabrous. Basal leaves sessile, oblanceolate, 10–31 cm long (including 3.5–10.5 cm petioloid base), 2–7 cm wide, gradually attenuate into winged petioloid base, margins sinuate-dentate, apex obtuse to acute; stem leaves sessile, oblanceolate or lanceolate to ovate, 3.5–28 cm long, 1–5 cm wide, base ± expanded-auriculate and semi-amplexicaul at base, margins crenate-dentate to sinuate-dentate, obtuse to subacute; all leaves coriaceous, glabrous except sometimes for a few hairs on margins and mid-rib beneath. Capitula radiate, erect, numerous in copious terminal corymbs; stalks of the individual capitula glabrous or almost so; involucre cylindrical, 9–10 mm long, 5–7 mm in diameter, glabrous or with a few hairs at the base; bracts of calyculus 6–15, 4–9 mm long; phyllaries 10–21, lanceolate or linear-lanceolate, 5–10 mm long, glabrous except for slightly bearded apex. Ray florets 8–13, yellow, corolla tube 4–5.5 mm long, glandular-hairy, rays 11–13 mm long, 2.5–3.5 mm wide, 4–5-veined; disc florets yellow, corolla 7–8.5 mm long, tube gradually expanded above the middle, glabrous, lobes 1–1.5 mm long. Achenes 4.5–5 mm long, blackish, sparsely shortly hairy, many-ribbed; pappus 6.5–8 mm long, white.

UGANDA. Kigezi District: Mt Sabinio, no date, *Eggeling* 1110! & Mt Mgahinga [Gahinga], Apr. 1970, *Katende* 180! & Mt Muhavura, Oct. 1948, *Hedberg* 2058!
DISTR. U 2; Congo (Kinshasa), Rwanda
HAB. Heath zone, grassland, mossy crevice on bare lava slope; 2900–3950 m
USES. None recorded on specimens from our area
CONSERVATION NOTES. Restricted to the Virunga Mts; twelve Ugandan specimens, but quite a lot from Congo; least concern or near threatened (LC/NT)

SYN. *S. subsessilis* sensu Hedb. in A.V.P.: 242, 364 (1957), *non* Oliv. & Hiern pro parte

16. **Senecio laticorymbosus** *Gilli* in Ann. Naturh. Mus. Wien 78: 159 (1974), as *latecorymbosus*; C. Jeffrey in K.B. 41: 894 (1986). Type: Tanzania, Njombe District, Uwemba, *Zerny* 634 (W!, holo.)

Coarse perennial herb 1.5–2.4 m high; stems erect, leafy, sparsely pubescent or glabrous. Stem leaves sessile, elliptic, narrowly elliptic or lanceolate, 20–27 cm long, 3–5 cm wide, base rounded and semi-amplexicaul to tapered and broadly expanded, very shortly narrowly decurrent, margins regularly dentate, apex attenuate, acute, glabrous above, sparsely minutely pubescent beneath. Capitula radiate, erect, numerous in a copiously branched terminal corymb to 90 cm long; involucre cylindrical, 9–12 mm long, 5–6 mm in diameter, glabrous except for a few glandular hairs at the base; bracts of calyculus 8–12, lanceolate, 3.5–8 mm long, shortly pubescent especially on the margins; inner phyllaries 12–16, 9–11 mm long, glabrous, bearded at the apex. Ray florets 7–8, yellow, corolla tube 5–7 mm long, glandular-hairy or glabrous, rays 12–13 mm long, 3.5–5 mm wide, 4–6-veined; disc florets yellow, corollas 9–10 mm long, expanded above the middle, glabrous, lobes 1.5–2 mm long. Achenes 5–5.5 mm long, very sparsely hairy, many-ribbed; pappus 8–9 mm long, white.

TANZANIA. Mbeya District: Poroto Mts, Livingstone Forest Reserve, Sep. 1970, *Thulin & Mhoro* 1234!; Njombe District: Iboma Forest on Njombe–Uqemba road, July 1982, *Magogo* 2239!; Rungwe District: Rungwe, Sep. 1932, *Geilinger* 2307!
DISTR. T 7; Malawi
HAB. Upland forest, bamboo forest or plantation forest; 2150–2550 m
USES. None recorded on specimens from our area
CONSERVATION NOTES. Six Tanzanian specimens; data deficient (DD)

17. **Senecio inornatus** *DC.*, Prodr. 6: 385 (1838); Hilliard, Comp. Natal: 471 (1977); C. Jeffrey in K.B. 41: 894 (1986); Lisowski, Aster. Fl. Afr. Centr. 2: 277, t. 59 (1991). Type: South Africa, Cape Province, *Drège* 5151 (G-DC, holo.)

Erect perennial herb 1–2 m high; stems winged in the lower part, glabrous; rhizome creeping. Basal leaves sessile, elliptic-lanceolate, 20–73 cm long (including 4–25 cm long petioloid base), 1.5–10 cm wide, gradually attenuate basally into a narrowly-winged petioloid base, margins dentate, apex acute; stem leaves sessile, ovate-lanceolate to oblanceolate, 6–33 cm long, 1–7.5 cm wide, basally attenuate into an entire-winged, not or slightly expanded and decurrent petioloid base, margins denticulate, apex acute or shortly acuminate; all leaves bright green, with thick pallid midrib, glabrous. Capitula radiate, erect, numerous in copious lax to congested terminal corymbs; stalks of the individual capitula glabrous or shortly glandular-pubescent below capitula; involucre shortly hairy at the base, cylindrical, 6–7.5 mm long, 2.5–4 mm in diameter; bracts of calyculus 2–6, lanceolate, 1.5–5 mm long; phyllaries 10–16, yellow-green, sometimes with black tips, 4–7.5 mm long, glabrous or almost so, bearded at the apex. Ray florets 5–8, golden yellow to yellow-orange, corolla tube 2.5–4 mm long, glandular-hairy, rays 5–9.5 mm long, 2–3.5 mm wide, 4–9-veined; disc florets bright yellow, with yellow styles and brown or yellow anthers, corolla 5–8 mm long, gradually expanded above the middle, glabrous or almost so, lobes 1–1.5 mm long. Achenes ellipsoid, 4 mm long, angled, glabrous or slightly hairy; pappus 5–7.5 mm long.

TANZANIA. Njombe District: Ndumbi Forest, Feb. 1954, *Paulo* 258! & Kitulo Plateau, Ndumbi
    Valley, Mar. 1991, *Bidgood et al.* 2128!; Rungwe District: Ngozi Crater, July 1991, *Kayombo* 1102!
DISTR. **T** 7; Congo (Kinshasa), Burundi, Angola, Zambia, Malawi, Zimbabwe, South Africa
HAB. Upland grassland, forest margins, may be abundant near streams; once recorded as a
    weed in pyrethrum; 1600–2700 m
USES. None recorded on specimens from our area
CONSERVATION NOTES. Least concern (LC)

SYN. *S. lygodes* Hiern, Cat. Welw. Afr. Pl. 1(3): 599 (1898), *non* sensu Hilliard, Comp. Natal: 453
    (1977). Type: Angola, Huila, *Welwitsch* 3676 (BM!, holo., BR!, iso.)
    S. stolzii Mattf. in E.J. 59, Beibl. 133: 37 (1924). Type: Tanzania, Rungwe District, Kyimbila,
    Stolz 1116 (B†, holo., K!, M!, iso.)

18. **Senecio microalatus** *C.Jeffrey* in K.B. 41: 894 (1986). Type: Tanzania, Mbeya
District, Mbeya Mt, *Milne-Redhead & Taylor* 10214 (K!, holo.)

Perennial herb ± 1 m high; stems solitary, green, glabrous, winged. Basal leaves
sessile, narrowly elliptic, 48.5 cm long (including 18 cm long petioloid base), 2.7 cm
wide, long-tapered into a petioloid base, margins finely sinuate-denticulate, apex
attenuate, subacute; stem leaves sessile, lanceolate or oblanceolate, 10–14.5 cm long,
1.1–1.9 cm wide, pseudopetiolate, with slightly decurrent margins forming ridges on
the stem; all leaves green, glabrous. Capitula radiate, erect, 7–19 in divaricate lax
corymbs; involucre cylindrical, 6–8 mm long, 4–5 mm wide; bracts of calyculus
16–18, imbricate, 3.5–5 mm long, somewhat arachnoid especially at margins;
phyllaries ± 14, bright green, tipped brown, 6–6.5 mm long, bearded at apex, scantily
finely arachnoid. Ray florets ± 3, yellow, corolla tube 4 mm long, hairy, rays 8 × 3 mm,
4-veined; disc florets dull yellow, with yellow styles and anthers, corolla 7 mm long,
tube gradually expanded above the middle, glabrous, lobes ovate-triangular, 1.2 mm
long. Achenes 4.5 mm long, glabrous, many-ribbed; pappus 6–6.5 mm long, white.

TANZANIA. Mbeya District: Mbeya Mt, May 1956, *Milne-Redhead & Taylor* 10214!
DISTR. **T** 7; known from only the type
HAB. Boggy grassland by streamside; 2310 m
USES. None recorded on specimens from our area
CONSERVATION NOTES. Known from one collection; at least vulnerable (VU-D2)

19. **Senecio doryphorus** *Mattf.* in E.J. 59, Beibl. 133: 35 (1924); C. Jeffrey in K.B.
41: 894 (1986); Lisowski, Aster. Fl. Afr. Centr. 2: 280 (1991). Type: Tanzania, Rungwe
District, Kyimbila, *Stolz* 2419 (B†, holo., BR!, EA!, K!, iso.)

Perennial herb, 30–90 cm high, subscapigerous; stem solitary, erect, slender,
glabrous. Basal leaves sessile, oblanceolate to linear, 9.8–23 cm long (including
3.2–7.2 cm long petioloid base), 0.3–1.8 cm wide, long-attenuate into petioloid base,
margins obscurely denticulate and revolute, apex shortly obtusely acuminate; stem
leaves smaller and becoming progressively smaller up the stem, sessile, narrowly
lanceolate, with slightly expanded base; all leaves glabrous, coriaceous. Capitula
radiate, erect, 1–4 in lax terminal corymb; stalk of individual heads slightly glandular-
pubescent just below capitula, otherwise glabrous; involucre cylindrical, 7–9 mm long,
5–6 mm in diameter, glabrous except for a few glandular hairs at base; bracts of
calyculus 4–10, lanceolate, 3–7 mm long; phyllaries 10–22, pale green with yellow-green
tips, 6.2–10 mm long, apically bearded. Ray florets 8–13, clear yellow, corolla tube
4–5 mm long, with a few hairs at the apex, rays 10–15 mm long, 2–4 mm wide, 4–7-
veined; disc florets darker yellow, with yellow styles and brown anthers, corolla 6–7 mm
long, tube slightly expanded in upper half, glabrous or almost so, lobes 1 mm long,
erect. Achenes 3.5–4 mm long, glabrous, many-ribbed; pappus 5–7 mm long, white.

TANZANIA. Iringa District: Lake Ngowasi, Mar. 1962, *Polhill & Paulo* 1836! & Mufindi, Lake
    Nsonso, May. 1983, *Magogo* 2463!; Njombe District: Njombe, Dec. 1931, *Lynes* 142!

Distr. **T** 7; Congo (Kinshasa)
Hab. Moist or marshy grassland; 1700–2000 m
Uses. None recorded on specimens from our area
Conservation notes. Twelve specimens from a restricted area, but status of habitat unclear: data deficient (DD)

20. **Senecio doryphoroides** *C.Jeffrey* in K.B. 41: 895 (1986). Type: Tanzania, Mpanda District, Silkcub Highlands, *Richards* 7170 (K!, holo.)

Perennial herb, 50 cm high, scapigerous; stem solitary, erect, glabrous. Basal leaves sessile, oblanceolate-linear, 16–25.5 cm long (including 8–13 cm long petioloid base), 0.9–1.2 cm wide, long-attenuate into a petioloid base, margins minutely denticulate, apex acute, glabrous. Stem leaves (if any) unknown. Capitula discoid, erect, 6 in a very lax terminal corymb; stalks of the individual capitula slightly arachnoid near apex; involucre broadly cylindrical, 7.5 mm long, 6.5 mm in diameter; bracts of calyculus 2–3, lanceolate, 3 mm long, slightly arachnoid; phyllaries ± 18, 7 mm long, almost glabrous, slightly tapered and white-bearded at apex. Ray florets 0; disc florets yellow, corolla 5 mm long, tube slightly expanded above the middle, glabrous, lobes 0.8 mm long. Mature achenes unknown, but ovary of disc florets 2.5 mm long, glabrous; pappus 4 mm long.

Tanzania. Mpanda District: Silkcub Highlands, Dec. 1956, *Richards* 7170!;
Distr. **T** 4; known from only the type
Hab. Grassland on red soil; 1650 m
Uses. None recorded on specimens from our area
Conservation notes. Known from one specimen; at least vulnerable (VU-D2)

21. **Senecio spartareus** *S.Moore* in J.L.S. 35: 358 (1902); C. Jeffrey in K.B. 41: 895 (1986), as *spartaceus*; Lisowski, Aster. Fl. Afr. Centr. 2: 283 (1991); U.K.W.F. ed. 2: 221 (1994), as *spartaceus*. Type: Kenya, 'Kavirondo', *Scott Elliot* 7029 (BM!, holo., K!, iso.)

Perennial herb, 45–60 cm high, precocious, subscapigerous; stems erect, solitary or paired, appearing before the basal leaves, dull dark green or maroon, slightly arachnoid to almost glabrous. Basal leaves unknown; stem leaves narrow, lanceolate, 2.5–6.5 cm long, 0.4–0.6 cm wide, attenuate, scale-like, slightly arachnoid to almost glabrous. Capitula discoid, erect, 7–20 in terminal corymbs; branches and stalks of the individual capitula slightly arachnoid to almost glabrous; involucre broadly cylindrical, 6–7 mm long, 6–7 mm in diameter, somewhat pubescent or arachnoid at base; bracts of calyculus 3–8, lanceolate, 2–3 mm long, sparsely arachnoid; phyllaries 10–14, dull dark green, 6–7 mm long, glabrous, abruptly tapered and ciliate-bearded at the apex. Ray florets 0; disc florets lemon yellow to yellow, corolla 4.7–6 mm long, glabrous, the tube gradually expanded above the middle, lobes 0.7–1 mm long. Mature achenes unknown, but ovary of disc florets 1 mm long, glabrous; pappus 4–6 mm long, white.

Uganda. W Nile District: Lindu, Mar. 1945, *Greenway & Eggeling* 7224!
Kenya. Trans Nzoia District: Kitale, Apr. 1962, *Tweedie* 2336! & Milimani, Jan. 1970, *Tweedie* 3750!; Uasin Gishu District: Kipkarren, 1932, *Brodhurst Hill* 741!
Distr. **U** 1; **K** 3, 5; Congo (Kinshasa), Burundi
Hab. Grassland; 1450–1900 m
Uses. None recorded on specimens from our area
Conservation notes. Distribution area is fairly large and habitat common; least concern (LC)

Syn. *S.* sp. D of U.K.W.F. ed. 1

Note. Flowering with early rains.
    The protologue has '*spartareus*', a name that filled HB with suspicion, especially as Jeffrey (1986) has '*spartaceus*', which looked more likely. However, in the Oxford Latin Dictionary, 'spartarius' stands for 'of Spanish broom'; and Spencer Moore does say 'distinguished by the spartioid habit'. We believe '*spartareus*' must stand.

22. **Senecio strictifolius** *Hiern*, Cat. Welw. Afr. Pl. 1, 3: 600 (1895); C. Jeffrey in K.B. 41: 895 (1986); Lisowski, Aster. Fl. Afr. Centr. 2: 284 (1991). Type: Angola, near Lopollo and Lake Ivantala, *Welwitsch* 3675 (BM!, holo.)

Perennial herb (Lisowski says annual), 40–120 cm high, (?precocious); stems erect, slightly arachnoid. Stem leaves sessile, lanceolate to oblanceolate, 4–10 cm long, 0.4–0.8 cm wide, attenuate to an exauriculate base, margins denticulate, apex obtuse and apiculate, glabrous except for slightly arachnoid midrib. Capitula erect, radiate, numerous in a copious terminal corymb; involucre cylindrical, 5 mm long, 3.5–4 mm in diameter, almost glabrous; bracts of calyculus 6–7, 1.5–2.5 mm long, glabrous; phyllaries 12–13, 4.5–5 mm long, glabrous. Ray florets 9–12, yellow, corolla tube 3.2 mm long, glabrous, rays 3–6.5 mm long, 3 mm wide, 4-veined; disc florets yellow, corolla 5–6 mm long, tube gradually expanded above the middle, glabrous, lobes 1 mm long. Achenes unknown, but ovary of disc florets 1.2 mm long, glabrous; pappus 4–5 mm long.

Tanzania. Iringa District: 25 km from Mafinga on Mbeya road, Apr. 1983, *R. Abdallah* 1286! & 30 km from Mafinga [Sao Hill] on Mbeya road, Mar. 1988, *Bidgood et al.* 797! & Uhafiwa, Aug. 1989, *Kayombo* 709!
Distr. **T** 7; Congo (Kishasa), Angola, Zambia, Malawi, Zimbabwe, Botswana, Namibia
Hab. Marshy grassland; 1700–2150 m
Uses. None recorded on specimens from our area
Conservation notes. Least concern (LC)

23. **Senecio depauperatus** *Mattf.* in E.J. 59, Beibl. 133: 36 (1924); C. Jeffrey in K.B. 41: 895 (1986). Type: Tanzania, Rungwe District, Kyimbila, *Stolz* 2336 (B†, holo., BR!, EA!, K!, iso.)

Perennial herb, pyrophytic, 16–75 cm high; stems one to several, erect, virgate, leafy, sparsely arachnoid and glabrescent or glabrous, from a creeping woody rootstock. Stem leaves numerous, sessile, linear-lanceolate, 1.3–3 cm long, 0.1–0.2 cm wide, tapered at the base, margins entire and revolute, apex obtusely to acutely acuminate-apiculate, glabrous except for a few hairs at the very base. Capitula erect, radiate, few to numerous in lax to subumbelliform terminal corymbs; stalks of the individual capitula glabrous or almost so; involucre cylindrical, 3.5–5 mm long, 2.5–3.5 mm in diameter, glabrous or slightly pubescent at the base; bracts of calyculus 2–6, lanceolate or linear-lanceolate, 1–3 mm long, sparsely arachnoid or ciliate on margins or glabrous except at apex; phyllaries 8–10, 3–4.5 mm long, tapered to rounded and bearded at the apex. Ray florets 4–5, yellow, corolla tube 2.5 mm long, glabrous, rays 4.5–6.5 mm long, 1.5–2.5 mm wide, 4–7-veined; disc florets yellow to light brown, corolla 4–5 mm long, glabrous, gradually expanded above the middle. Achenes (immature) 2 mm long, glabrous; pappus 3.5–4.5 mm long, white.

Tanzania. Mbeya District: Elton Plateau, Nov. 1963, *Richards* 18435!; Njombe District: Njombe, Dec. 1931, *Lynes* 59! & Ndumbi Forest Reserve, Nov. 1986, *Brummitt & Mwasumbi* 18126!
Distr. **T** 7; not known elsewhere
Hab. Grassland, especially where subject to burning; 1550–2900 m
Uses. None recorded on specimens from our area
Conservation notes. Twenty specimens at EA and Kew; not uncommon in a common habitat; probably least concern (LC)

24. **Senecio lelyi** *Hutch.* in K.B. 1921: 382 (1921); C. Jeffrey in K.B. 41: 895 (1986). Type: Nigeria, between Hepham & Ropp, *Lely* 356 (K!, holo.)

Perennial herb, 50–100 cm high; stems 1–2, erect, shortly white-woolly to almost glabrous, pale green; roots rather fleshy. Basal leaves unknown, represented by only fibrous remnants; stem leaves sessile, ± erect, subequal, lanceolate, 5–20 cm long,

0.6–2.5 cm wide, tapered and shortly decurrent to slightly expanded and semi-amplexicaul at the base, margins denticulate, apex acute, glabrous except for a few arachnoid hairs near the base. Capitula erect, radiate, 4–20 in lax terminal corymb; stalks of the individual capitula slightly arachnoid; involucre cylindrical, 8–10 mm long, 5–7 mm in diameter, slightly woolly at the base; bracts of calyculus 4–8, lanceolate, 2.5–5 mm long, ± arachnoid especially at margins; phyllaries 10–14, bright green tipped black, 7.5–9 mm long, apically bearded, oblong, glabrous or sparsely lanate. Ray florets 3–13, bright yellow, corolla tube 4–6 mm long, glabrous, rays 8–15 mm long, 2.5–4 mm wide, 8–9-veined; disc florets dull yellow with yellow styles and yellow or brown anthers, corolla 6–7.5 mm long, tube gradually expanded above the middle, glabrous, lobes 1–1.5 mm long. Achenes not known mature but ovary of disc florets 1.5–2 mm long, hairy; pappus 6–8 mm long.

TANZANIA. Iringa District: Ngowasi [Ngwazi], Jan. 1987, *Lovett* 1252!; Njombe District: Lihogosa Swamp, Jan. 1957, *Richards* 7893!; Songea District: Matengo Hills, Lupembe Hill, Jan. 1956, *Milne-Redhead & Taylor* 8216!
DISTR. **T** 7, 8; Guinea, Nigeria, Cameroon, Sudan
HAB. Grassland, secondary woodland, swamp edge; 950–2100 m
USES. None recorded on specimens from our area
CONSERVATION NOTES. Least concern (LC)

SYN. *S. graciliserra* Mattf. in E.J. 59, Beibl. 133: 26 (1924). Type: Cameroon, Baja Highlands, Kunde, *Mildbraed* 9217 (B†, K!, syn.) & Bamenda, Bambulue, *Thorbecke* 288 (B†, syn.)

25. **Senecio sororius** *C.Jeffrey* in K.B. 41: 895 (1986). Type: Tanzania, Songea District, R. Luhimba, *Milne-Redhead & Taylor* 8571b (K!, holo., BR!, iso.)

Perennial herb, 60–65 cm high; stems 1–2, erect, yellow-green tinged dull purple, almost glabrous. Basal and stem leaves sessile, rather fleshy, green, with fine reddish-purple margins, oblanceolate-linear, 5–26 cm long (including 3.5–10 cm long petioloid base), 0.7–1 cm wide, long-attenuate into a petioloid base, margins dentate, apex obtuse, ± woolly and vinous at very base, otherwise glabrous, held erect with margins folded upwards; stem leaves becoming progressively smaller up the stem and becoming semi-amplexicaul. Capitula erect, radiate, 2–4 in lax terminal corymb; stalks of the individual capitula slightly arachnoid near apex; involucre broadly cylindrical, 10–12 mm long, 10 mm in diameter, slightly arachnoid-lanate at base; bracts of calyculus 10–12, lanceolate, 5–9.5 mm long, sparsely arachnoid; phyllaries 20–22, tinged dull purple, 8–10.5 mm long, slightly arachnoid or glabrous, bearded at apex. Ray florets 8–11, bright yellow, corolla tube 5 mm long, sparsely hairy, rays irregular and somewhat reflexed, 17 mm long, 3 mm wide, 8-veined; disc florets paler yellow, with brown anthers, corolla 8 mm long, tube gradually expanded above the middle, glabrous, lobes ovate, 1 mm long. Achenes 5 mm long, shortly pubescent, many-ribbed; pappus 7.5–8.5 mm long.

TANZANIA. Songea District: R. Luhimba, Jan. 1956, *Milne-Redhead & Taylor* 8571b!
DISTR. **T** 8; known only from the type
HAB. Boggy grassland, where it is locally dominant; 1000 m
USES. None recorded on specimens from our area
CONSERVATION NOTES. Known from a single specimen; at least vulnerable (VU-D2)

26. **Senecio tabulicola** *Baker* in K.B. 1895: 155 (1895); C. Jeffrey in K.B. 41: 895 (1986). Type: Malawi, Nyika, *Whyte* 162 (K!, holo.)

Coarse perennial herb, subscapigerous, 60–210 cm high; stems solitary, erect, at first slightly arachnoid, glabrescent, pale green with reddish longitudinal ridges. Basal leaves sessile, narrowly lanceolate to elliptic, 19–49 cm long (including 10–17 cm long petioloid base), 2–7 cm wide, gradually attenuate into a petioloid base, margins sinuate-dentate, apex shortly acuminate, green and glabrescent above, whitish to pale green and arachnoid but often glabrescent beneath; stem leaves smaller and becoming progressively smaller up the stem, sessile, not or somewhat expanded-auriculate and semi-amplexicaul at the base, sinuate-dentate to subentire. Capitula erect, radiate, 3–35 in a lax to crowded terminal corymb; involucre cylindrical, 8–13 mm long, 6–10 mm in diameter, ± woolly at the base; bracts of calyculus 6–16, lanceolate, 2–7 mm long, ± arachnoid-tomentose; phyllaries ± 10–22, green with reddish or brownish tips, 7.5–11 mm long, shortly lanate or glabrous except near the base, apically bearded. Ray florets 8–13, bright yellow, corolla tube glabrous, rays 8–11 mm long, 3–4 mm wide, 6–9-veined; disc florets deep yellow to orange, with yellow styles and brown anthers, corolla 6–10 mm long, tube gradually expanded above the middle, lobes 0.8–1 mm long. Achenes not known mature but ovary of disc florets 1.7 mm long, glabrous, many-ribbed; pappus 6.5–10 mm long.

TANZANIA. Mpwapwa District: Kiboriani Mts, Oct. 1938, *Greenway* 5792!; Iringa District: Lake Ngwazi, May 1983, *R. Abdallah* 1373!; Songea District: NW side of Matogoro Hills, May 1956, *Milne-Redhead & Taylor* 9878!
DISTR. **T** 5, 7, 8; Malawi, Mozambique
HAB. *Brachystegia* or *Uapaca* woodland, grassland; 1200–2350 m
USES. None recorded on specimens from our area
CONSERVATION NOTES. Ten Tanzanian specimens, but from a widespread habitat; probably least concern (LC)

27. **Senecio karaguensis** *O.Hoffm.* in P.O.A. C: 147 (1895); Maquet in Fl. Rwanda 3: 662, fig. 205/1 (1985); C. Jeffrey in K.B. 41: 895 (1986); Lisowski, Aster. Fl. Afr. Centr. 2: 284 (1991). Type: Tanzania, Lake Province, *Stuhlmann* 1679 (B†, syn.), 1978 (B†, syn.), 2122 (B†, syn., K!, isosyn.)

Coarse perennial herb, 40–120(–200) cm high, subscapigerous; stems solitary, erect, whitish, arachnoid. Basal leaves sessile, lanceolate, narrowly lanceolate, elliptic or oblanceolate, 26–45 cm long (including 8.5–13.5 cm long petioloid base), 1.2–4 cm wide, attenuate into a petioloid base, margins sinuate-dentate, apex acute, green and glabrescent to glabrous above, white-tomentose except sometimes for glabrescent midrib beneath; stem leaves smaller and becoming progressively smaller up the stem, the lower ± expanded-auriculate and semi-amplexicaul at the base. Capitula erect, radiate, 5–15 in a lax terminal corymb; stalks of the individual capitula tomentose; involucre broadly cylindrical, 9–15 mm long, arachnoid-tomentose; bracts of calyculus 10–22, imbricate, lanceolate, 6–12 mm long, arachnoid-tomentose; phyllaries ± 16–30, 10–12 mm long, arachnoid-tomentose, apically bearded. Ray florets 8–20, lemon to deep yellow, corolla tube 5–6.5 mm long, glabrous, rays 11–14 mm long, 3–4 mm wide, 7–12-veined; disc florets yellow, corolla 7–10 mm long, tube slightly expanded above the middle, glabrous, lobes 0.8–1 mm long, ovate-triangular, erect. Achenes 4.5 mm long, glabrous; pappus white, 7–9.5 mm long. Fig. 132 (page 639).

UGANDA. Kigezi District: Rukungire, Jan. 1945, *Purseglove* 1626! & Nyakageme, May 1947, *Purseglove* 2413!; Ankole District: Buhweju, Isingiro, Sep. 1992, *Rwaburindore* 3456!
TANZANIA. Bukoba District: Karagwe Hills, Mar. 1894, *Scott Elliot* 7478!; Ngara District: Bugufi, Kirushya, Feb. 1960, *Tanner* 4727!; Masai District: Loliondo, Ngosaro Sambu, July 1956, *Williams* 697!
DISTR. **U** 2; **T** 1, 2; Rwanda, Burundi
HAB. Grassland, may be locally common; 1200–2100 m
USES. None recorded on specimens from our area
CONSERVATION NOTES. Distribution fairly wide, in a common habitat; least concern (LC).

28. **Senecio discokaraguensis** *C.Jeffrey* in K.B. 41: 895 (1986). Type: Tanzania, Mbeya District, Mbeya Airfield, *Milne-Redhead & Taylor* 10044 (K!, holo.)

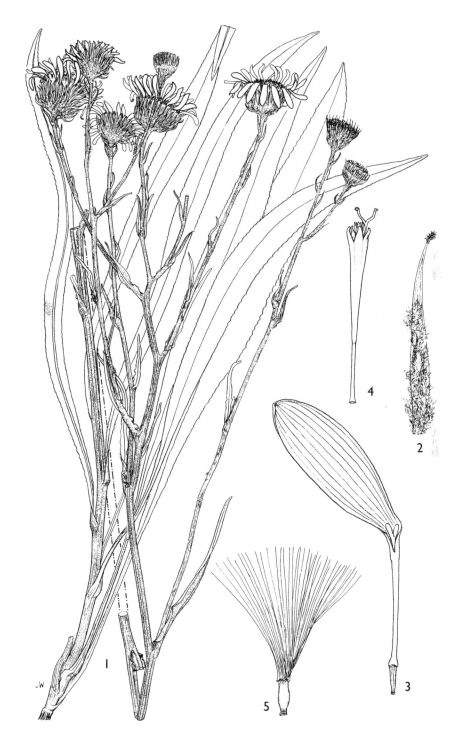

Fig. 132. *SENECIO KARAGUENSIS* — **1**, habit, × ²/₃; **2**, phyllary, × 5; **3**, ray floret, × 5; **4**, disc floret, × 5; **5**, young achene, × 5. 1–4 from *Purseglove* 1626, 5 from *Purseglove* 2413. Drawn by Juliet Williamson.

Perennial herb to 2 m high, subscapigerous; stems paired, erect, furrowed, greyish-green, arachnoid. Basal leaves sessile, elliptic, 27–32 cm long (including 7.5–11.5 cm long petioloid base), 2.5–6 cm wide, attenuate into a petioloid base, sinuate-dentate, acute, coriaceous, green and glabrous above, greyish-green and arachnoid beneath, the petioloid base thinly arachnoid; stem leaves smaller and becoming progressively smaller up the stem, sessile, linear-lanceolate, auriculate, ± entire, 5.5–11 cm long, 0.7–1 cm wide. Capitula discoid, erect, 4–5 in a lax terminal corymb; stalks of the individual capitula arachnoid; involucre globose, arachnoid, ± 10 mm long, 10 mm in diameter; bracts of calyculus ± 12–14, broadly lanceolate, arachnoid, apically bearded, 8 mm long; phyllaries ± 26–28, arachnoid, greyish-green tipped reddish, apically long-attenuate and glabrous except at bearded tip. Ray florets 0; disc-florets, styles and anthers orange-yellow, corolla 7.5 mm long, tube gradually expanded upwards, glabrous, lobes 0.5 mm long. Mature achenes unknown, but ovary of disc florets ± 1.5 mm long, glabrous; pappus 6 mm long.

TANZANIA. Mbeya District: Mbeya Airfield, May 1956, *Milne-Redhead & Taylor* 10044!
DISTR. **T** 7; known only from the type
HAB. Grassland; 1620 m
USES. Noné recorded
CONSERVATION NOTES. Known from one specimen; at least vulnerable (VU-D2)

29. **Senecio gramineticola** *C.Jeffrey* in K.B. 41: 895 (1986). Type: Tanzania, Ufipa District: Nsanga Mt, *Richards* 12130 (K!, holo.)

Perennial herb, 40–150 cm high; stems erect, with white arachnoid tomentum, glabrescent, reddish-brown. Basal leaves sessile, elliptic, 31–60 cm long (including 10–22 cm long petioloid base), 7–17 cm wide, attenuate into a membraneous petioloid base, margins closely to coarsely dentate or bidentate, apex obtuse; stem leaves sessile, elliptic to lanceolate, 4.5–32 cm long, 0.5–10.5 cm wide, attenuate into an exauriculate or scarcely auriculate petioloid base in lower leaves, expanded and semi-amplexicaul in upper leaves, margins strongly dentate or bidentate, obtuse, shortly acuminate; all leaves arachnoid-pubescent and ± glabrescent. Capitula discoid, erect, numerous in congested terminal corymbs; involucre cylindrical, 7–9 mm long, 5.5–6.5 mm in diameter, arachnoid at least at the base; bracts of calyculus 8–10, lanceolate, 3–8 mm long, ± arachnoid; phyllaries 16–18, 7–8 mm long, arachnoid or thinly so, long-tapered and bearded at the apex. Ray florets 0; disc florets yellow to orange or brownish, corolla 4.5–7 mm long, tube gradually expanded above the middle, glabrous, lobes 0.7–1 mm long. Achenes (immature) 4 mm long, glabrous; pappus 4.5–7 mm long, white.

TANZANIA. Ufipa District: Mbizi Mts, *Bidgood et al.* 3579!; Iringa District: Ngowasi [Ngwazi], Mar.
1989, *Kayombo & Kayombo* 99!; Njombe District: near Kifanya, Feb. 1963, *Richards* 17667!
DISTR. **T** 4, 7; Malawi
HAB. Grassland or bushed grassland, ?only in unburnt areas; 1500–2250 m
USES. None recorded on specimens from our area
CONSERVATION NOTES. Fairly wide distribution in common habitat; least concern (LC)

30. **Senecio immixtus** *C.Jeffrey* in K.B. 41: 896 (1986); Lisowski, Aster. Fl. Afr. Centr. 2: 287 (1993). Type: Tanzania, Njombe District, Kipengere Mts, *Richards* 7618 (K!, holo.)

Perennial herb to 1.2 m high; stems erect, arachnoid-floccose, glabrescent. Basal leaves narrowly elliptic, 15–32 cm long, 2–4.4 cm wide, base petioloid and attenuate, margins slightly dentate, apex acute; median stem leaves sessile, oblong, 23–29 cm long, 3.6–5 cm wide, attenuate into a petioloid base, margins strongly sinuate-dentate, apex shortly acuminate-apiculate, coriaceous, sparsely arachnoid on midrib beneath, glabrescent; upper stem leaves sessile, ovate-lanceolate, 7 cm long, 0.6–2 cm wide, with expanded auriculate semi-amplexicaul base. Capitula erect, discoid, numerous in dense rounded terminal corymbs; stalks of the individual capitula densely arachnoid, with stalked glands; involucre broadly cylindrical, 7–8 mm long, 6–8 mm in diameter; bracts of calyculus 8–11, lanceolate, 2.5–7 mm long, acute, arachnoid; phyllaries 18–21, 8 mm long, arachnoid, also glandular-pubescent, long-tapered at apex. Ray florets 0; disc florets yellow, corolla 5–9 mm long, tube glabrous, abruptly expanded above the middle, lobes 1–1.3 mm long. Achenes 4–4.5 mm long, glabrous; pappus 9 mm long, white.

TANZANIA. Njombe District: Kipengere Mts, Jan. 1957, *Richards* 7618!; Njombe/Mbeya District: Makete, Kitulo Plateau just E of Ndumbi R. along Matamba–Kitulo road, Feb. 1989, *Gereau et al.* 3098!
DISTR. **T** 7; Congo (Kinshasa)
HAB. Upland grassland; 2400–2600 m
USES. None recorded on specimens from our area
CONSERVATION NOTES. Known from only three specimens; at least vulnerable (VU-D2)

NOTE. Very close to *S. gramineticola.*

31. **Senecio exarachnoideus** *C.Jeffrey* in K.B. 41: 896 (1986). Type: Tanzania, Songea District, 5.5 km E of Songea, *Milne-Redhead & Taylor* 8685 (K!, holo., BR!, iso.)

Perennial herb to 1.5 m high; stems solitary, erect, pale green with darker green striations, sometimes reddish-purple below, slightly glandular-pubescent in upper part. Basal leaves sessile, ovate to broadly elliptic, 12–51 cm long (including 11.5–19 cm long petioloid base), 5.5–13.5 cm wide, attenuate into a very sparsely pubescent petioloid base, margins strongly dentate, apex obtuse or obtuse and shortly acuminate; stem leaves sessile, broadly ovate, lanceolate or elliptic, 7–32 cm long, 1.4–8.5 cm wide, cuneate into a petioloid base, the upper leaves with broadly winged, ± auriculate base, margins sinuate-dentate, apex obtuse; all leaves deep green and glabrous above, pale with prominent veins and glabrous except for a few arachnoid hairs on mid-rib beneath. Capitula discoid, erect, numerous in a copious but lax terminal corymb; stalks of the individual capitula slender, glandular-pubescent or setose; involucre cylindrical, 6–7 mm long, 4 mm in diameter; bracts of calyculus 4–6, lanceolate, 2.5–6 mm long, slightly ciliate; phyllaries 12–14, pale yellowish green, 5–7 mm long, almost glabrous, long-attenuate, bearded at the apex. Ray florets 0; disc florets and styles pale greenish yellow, anthers pale greenish yellow to sulphur, turning brown, corolla 6.3–7 mm long, tube glabrous, gradually expanded above the middle, lobes 1.5 mm long. Achenes (immature) 3 mm long, glabrous; pappus 6–7 mm long, white.

TANZANIA. Songea District: Matengo Hills, Lupembe Hill, Jan. 1956, *Milne-Redhead & Taylor* 8220! & 5.5 km E of Songea, Feb. 1956, *Milne-Redhead & Taylor* 8685! & Mar. 1956, *Milne-Redhead & Taylor* 8685a!
DISTR. **T** 8; not known elsewhere
HAB. Deciduous woodland and grassland; 1100–1900 m
USES. None recorded on specimens from our area
CONSERVATION NOTES. Known from three Tanzanian collections from two sites; at least vulnerable (VU-D2)

32. **Senecio coronatus** (*Thunb.*) *Harv.*, Fl. Cap. 3: 369 (1865); Hilliard, Comp. Natal: 462 (1977); C. Jeffrey in K.B. 41: 896 (1986); Lisowski, Aster. Fl. Afr. Centr. 2: 268 (1991). Type: South Africa, Cape, *Thunberg* s.n. (UPS, holo.)

Perennial herb, subscapigerous; crown of rootstock densely woolly and covered with remnants of old leaf-bases; stems 1–2, erect, 20–54 cm high, arachnoid-lanate. Basal leaves sessile, oblanceolate, 5–16 cm long, 0.4–3 cm wide, gradually tapered into a basally slightly expanded and sheathing petioloid base, margins subentire or indistinctly crenate-dentate in distal part, apex shortly acuminate, green and almost glabrous above, green and sparsely villous beneath; stem leaves smaller and becoming progressively smaller up the stem, sessile, oblong-lanceolate, base semi-amplexicaul, apex shortly acuminate. Capitula radiate, ± 4 in a lax terminal corymb; stalks of the individual capitula woolly; involucre broadly hemispherical, 9–11 mm long, 10–15 mm in diameter, woolly at base; bracts of calyculus ± 10, narrowly lanceolate, 6–7 mm long; phyllaries ± 18–20, lanceolate, 8–12 mm long, laxly villous, shortly bearded at apex. Ray florets 10–12, yellow, corolla tube 5–6 mm long, strongly pubescent, rays 11–18 mm long, 3.5 mm wide, 10-veined; disc florets deep yellow, corolla 9 mm long, tube slightly expanded above the middle, pubescent at and below the middle, lobes 1–1.5 mm long, triangular. Achenes 4–5 mm long, shortly hairy; pappus 8 mm long, white.

TANZANIA. Ufipa District: Old Sumbawanga road, Jan. 1962, *Richards* 15882!
DISTR. **T** 4, ?8 (see Note); Angola, Zambia, Malawi, Zimbabwe, Botswana, Swaziland, South Africa
HAB. Marshy grassland; 1500 m
USES. None recorded on specimens from our area
CONSERVATION NOTES. Least concern (LC)

SYN. *Cineraria coronata* Thunb., Fl. Cap.: 670 (1825)

NOTE. A specimen from **T** 8, Songea District: Luhimba R., 1000 m, Jan. 1956, *Milne-Redhead & Taylor* 8571a! is very similar but lacks the characteristic pubescence on the tube of the ray florets. It is included here provisionally.

33. **Senecio urundensis** *S.Moore* in J.L.S. 35: 355 (1902); C. Jeffrey in K.B. 41: 896 (1986); Lisowski, Aster. Fl. Afr. Centr. 2: 288 (1991). Type: "Urundi" (see note), *Scott Elliot* 8181 (BM!, holo., K!, iso.)

Herb with annual shoots 20–100 cm high from a perennial rootstock; crown of rootstock woolly; stems erect, leafy, arachnoid to glabrescent. Basal leaves sessile, elliptic, 20–30 cm long (including 4–14 cm long petioloid base), 4–7.5 cm wide, long-attenuate into a petioloid base, margins sinuate-dentate, apex rounded; lower stem leaves with expanded membraneous exauriculate base; median and upper stem leaves sessile, broadly ovate to lanceolate, elliptic or oblanceolate-oblong, 3–12 cm long, 1–5 cm wide, tapered and exauriculate to an expanded, auriculate and semi-amplexicaul base, margins denticulate, apex shortly acuminate; all leaves thinly arachnoid, glabrescent except for a few hairs on midrib beneath. Capitula discoid, numerous in copious congested to lax terminal corymbiform clusters; stalks of the individual capitula slender, thinly arachnoid, often glabrescent; involucre cylindrical, 4.5–7.5 mm long, 3–4 mm in diameter; bracts of calyculus 2–6, lanceolate to linear, 2–7 mm long, usually thinly arachnoid and ciliate; phyllaries 8–14, 4.5–7 mm long, thinly arachnoid, glabrescent, bearded at apex. Ray florets 0; disc florets yellow, corolla 4.7–6.5 mm long, tube rather abruptly expanded in upper third, glabrous, lobes 0.8–1.2 mm long. Achenes 2–3 mm long, ribbed, hairy; pappus 5–6 mm long, white.

TANZANIA. Buha District: Kalinzi, Nov. 1962, *Verdcourt* 3413!; Mpwapwa, Feb. 1930, *Hornby* 189!; Mbeya District: N Usafwa Forest, Oct. 1959, *Procter* 1497!
DISTR. **T** 1? (see note), 4, 5, 7; Congo (Kinshasa), Rwanda, Burundi, Zambia, Malawi
HAB. Grassland and sparsely wooded grassland, especially where burned; 1500–2700 m
USES. None recorded on specimens from our area
CONSERVATION NOTES. Ten Tanzanian specimens; data deficient (DD) but probably least concern

NOTE. It is unclear whether the type is from Tanzania – Scott Elliot's label only has 'Urundi, 4–5000 feet, Sep.'. It seems most likely this is in Burundi.

34. **Senecio semiamplexifolius** *De Wild.* in B.J.B.B. 5: 86 (1915); C. Jeffrey in K.B. 41: 896 (1986); Lisowski, Aster. Fl. Afr. Centr. 2: 296 (1991). Type: Congo (Kinshasa), Little Luemba Valley, *Hock* s.n. (BR!, holo.)

Perennial herb, 50–90 cm high; crown of rootstock shortly woolly; stems 1–2, erect, leafy, glabrous except at the nodes. Leaves slightly glaucous, lower stem leaves smaller than the median; median and upper stem leaves sessile, oblong or lanceolate-elliptic, 3–17 cm long, 0.6–5 cm wide, cordate and semi-amplexicaul at the base, margins entire, subentire or remotely minutely denticulate, apex shortly acutely acuminate-apiculate, glabrous. Capitula discoid, numerous in a copious terminal corymb; stalks of the individual capitula slender, glabrous; involucre cylindrical, 5–6 mm long, 3.5–4 mm in diameter, glabrous; bracts of calyculus 1–4, lanceolate, 0.5–2.5 mm long, glabrous; phyllaries 6–9, 4–6 mm long, obtusely tapered and slightly bearded at the apex. Ray florets 0; disc florets pale yellow to yellow, corolla 5–6.5 mm long, distinctly but gradually expanded in upper half, glabrous, lobes somewhat unequal, with two rather shorter than the other three, 1–2 mm long. Achenes (immature) 4 mm long, glabrous; pappus 4–6 mm long, white.

TANZANIA. Mbulu District: Mt Hanang, Feb. 1946, *Greenway* 7574!; Morogoro District: Nguru Mts near Maskati Mission, Mgundwilo Mt, June 1978, *Thulin & Mhoro* 3130!; Njombe District: Poroto Mts, Kitulo Plateau, Ndumbi Valley, Mar. 1991, *Bidgood et al.* 2138!
DISTR. T 2, 6, 7; Congo (Kinshasa)
HAB. Grassland, usually on hillslopes, or shrubby vegetation; 1600–2600 m
USES. None recorded on specimens from our area
CONSERVATION NOTES. Probably least concern (LC)

35. **Senecio pergamentaceus** *Baker* in K.B. 1898: 154 (1898); C. Jeffrey in K.B. 41: 896 (1986). Type: Malawi, Zomba, *Whyte* s.n. (K!, holo.)

Perennial herb, 60–80 cm high; crown of rootstock woolly; stems erect, leafy, pale green, rather slender, glabrous. Stem leaves sessile, ovate to lanceolate, the median largest 10–15 cm long, 3.3–4.5 cm wide, the upper and lower smaller, base semi-amplexicaul, margins entire or remotely denticulate, apex obtuse, shortly acuminate-apiculate, glabrous, minutely glandular. Capitula radiate, numerous in a divaricate lax terminal corymb; involucre cylindrical, 5.5–6 mm long, 3.5 mm in diameter, glabrous; bracts of calyculus 2, 1–2 mm long; phyllaries 6–8, pale green, 5–6 mm long, shortly tapered, slightly bearded at margins at and towards apex. Ray florets 4–5, yellow, corolla tube 3.5 mm long, glabrous, rays 8–13 × 2–2.5 mm, 4-veined; disc florets yellow with yellow styles and orange anthers, corolla 5.5–6 mm long, tube abruptly expanded above the middle, glabrous, lobes 1–1.2 mm long. Achenes 5 mm long, glabrous, ribbed, pappus 5 mm long, white.

TANZANIA. Ufipa District: Ufipa Plateau 1 km S of Mutimmbwa, Jan. 1987, *Moyer & Sanane* 145!; Songea District: Kigonsera, Dec. 1973, *Mhoro* 1700! & ± 7 km W of Songea, Jan. 1956, *Milne-Redhead & Taylor* 8260!
DISTR. T 4, 7, 8; Zambia, Malawi
HAB. Secondary woodland or grassland with *Protea*; 950–2250 m
USES. None recorded on specimens from our area
CONSERVATION NOTES. Four Tanzanian specimens, but because of extra-EA distribution and the habitat probably least concern (LC)

36. **Senecio ruwenzoriensis** *S.Moore* in J.L.S. 35: 355 (1902); Verdcourt & Trump, Common Poison. Pl. E.Afr.: 156, fig. 14 (1969); Hilliard, Comp. Natal: 475 (1977); Maquet in Fl. Rwanda 3: 662, fig. 208/4 (1985); C. Jeffrey in K.B. 41: 897 (1986); Lisowski, Aster. Fl. Afr. Centr. 2: 296, t. 63 (1991); U.K.W.F. ed. 2: 224, t. 93 (1994). Type: Uganda, Toro District, Ruwenzori, *Scott Elliot* 8043 (BM!, holo., K!, iso.)

Perennial herb, 15–100 cm high, with creeping fleshy rhizome; stems leafy, erect, pale green, sometimes purplish at the base, glabrous. Leaves cauline, light green, usually with glaucous bloom beneath, succulent, sessile, obovate to almost rhombic, oblanceolate, elliptic or lanceolate, 2.5–11.5 cm long, 1–5.3 cm wide, base rounded to cuneate, margins entire or few-denticulate, apex rounded to obtuse and shortly apiculate, glabrous; pinnately veined but the veins strongly ascending and often with two especially prominent ascending lateral veins from near the base. Capitula radiate, 1–20 in a lax, long-stalked leafless terminal corymb; involucre cylindrical, slightly swollen at the base, 7–10 mm long, 4–8 mm in diameter, glabrous; bracts of calyculus 1–2, ovate-lanceolate or oblong-lanceolate, 1–2 mm long, glabrous; phyllaries 10–14, pale green, 6–9 mm long, glabrous, shortly tapered and bearded at the apex. Ray florets (5–)8, pale to bright yellow, corolla tube 3.5–4.5 mm long, glabrous or with a few hairs in upper part, rays 5–11 mm long, 1.5–3.5 mm wide, 4-veined; disc florets yellow with yellow styles and brown anthers, corolla 4.5–7.5 mm long, gradually expanded above the middle, glabrous, lobes 0.5–1 mm long. Achenes 2.7–4 mm long, ribbed, glabrous or sometimes sparsely hairy; pappus 4–7 mm long, white. Fig. 133 (page 645).

UGANDA. Karamoja District: Mt Kadam, Apr. 1959, *Wilson* 778!; Ankole District: 3 km E of Kamatarisi, Sep. 1969, *Lye et al.* 4321!; Mubende District: Kakumiro, May 1945, *A.S. Thomas* 4121!
KENYA. Trans Nzoia District: slopes of Elgon above Endebess, Apr. 1961, *Polhill* 401!; N Nyeri District: Nanyuki, June 1943, *Moreau & Moreau* 52!; Kisumu-Londiani District: Londiani, May 1932, *Graham* 2801!
TANZANIA. Bukoba District: Mabira, Sep. 1958, *Procter* 1005!; Masai District: Lerai, Jan. 1989, *Chuwa* 89/035F!; Ulanga District: Ngongo Mt, Jan. 1979, *Cribb et al.* 11128!
DISTR. **U** 1, 2, 4; **K** 1–6; **T** 1, 2, 4, 6–8; Nigeria to Sudan and S to South Africa
HAB. Shallow soil among rocks, rocky grassland or bushland, occasionally in wooded grassland, sometimes by streams or a weed of cultivation; 1150–3000 m
USES. None recorded on specimens from our area; possibly poisonous to livestock
CONSERVATION NOTES. Least concern (LC)

SYN. *S. paucifolius* DC., Prodr. 6: 403 (1838); Oliv. & Hiern, F.T.A. 3: 414 (1877), *nom. illegit., non* S.G. Gmel. (1774). Type: South Africa, Cape Province, *Ecklon* 1253 & 140-8, *Drege* 5847 (G-DC, syn.)
S. othonniformis Fourcade in Trans. Roy. Soc. S. Afr. 21: 89 (1943). Type as for *S. paucifolius* DC.

NOTE. Comes up after the first rains, also flowers after fires.
"Leaves thick and juicy and if growing near pools or lakes generally has a small green frog sitting in it" (*Irwin* 67 from Mt Elgon).

37. **Senecio milanjianus** *S.Moore* in J.L.S 35: 359 (1902); C. Jeffrey in K.B. 41: 897 (1986). Type: Malawi, Mt Mlanje, *Whyte* s.n. (BM!, holo.)

Perennial herb; stem succulent, pale green, purple-tinged, from which arise erect leafy flowering stems 18–38 cm high, glabrous. Basal and median stem leaves succulent, dull green with mauve margins, paler and slightly glaucous beneath, petiolate, peltate, circular, 3.8–7.5 cm in diameter, shallowly and regularly sinuate; petiole 2–7.5 cm long, mauve, exauriculate; upper stem leaves sessile, few, oblong to ovate, 3.5–4.1 cm long, 1.2–2.2 cm wide, base semi-amplexicaul, margins entire, apex obtuse. Capitula radiate, erect in congested terminal corymbs; stalks of the individual capitula short; involucre cylindrical, 7 mm long, 3–5 mm in diameter; bracts of calyculus 6–7, linear, 6–7 mm long; phyllaries 10–13, yellow-green with purple tips, 6.5 mm long. Ray florets 5–6, tube 4 mm long, glabrous, rays bright yellow, 6 mm long, 3 mm wide, 4–5-veined; disc florets yellow, 6.5 mm long, expanded from below the middle, glabrous, lobes 1 mm long. Mature achenes unknown; ovary 1.5 mm long, glabrous; pappus 4–4.5 mm long.

TANZANIA. Songea District: Lupembe Hills, May 1956, *Milne-Redhead & Taylor* 10464!
DISTR. **T** 8; Malawi, Zimbabwe
HAB. In rock crevices with water seep on steep rocky outcrop, stems creeping between rocks; 1950 m
USES. None recorded on specimens from our area
CONSERVATION NOTES. A single collection from Tanzania; data deficient (DD)

Fig. 133. *SENECIO RUWENZORIENSIS* — **1**, habit, × ²/₃; **2**, capitulum, × 4; **3**, ray floret, × 8; **4**, disc floret, × 8; **5**, detail of style arms, × 16. 1–5 from *Friis & Hansen* 2557. Drawn by Juliet Williamson.

SYN. *S. tropaeolifolius* O.Hoffm. in E.J. 30: 437, t. 21 (1901), *nom. illegit.*, *non* F. Muell. (1867).
    Type: Tanzania, Njombe District, Ukinga Mts, *Goetze* 964 (B†, holo., BM!, iso.)
    S. conradii O.Hoffm. in E.J. 43: 43 (1909). Type as for *S. tropaeolifoius* O. Hoffm.

38. **Senecio oxyriifolius** *DC.*, Prodr. 6: 405 (1838); Hilliard, Comp. Natal: 473 (1977); C. Jeffrey in K.B. 41: 897 (1986); Lisowski, Aster. Fl. Afr. Centr. 2: 303, t. 66 (1991). Type: South Africa, Cape Province, Fish River, *Drège* 5653 (G-DC, holo.)

Perennial herb 40–110 cm high with horizontal, fleshy tuberous rhizome (or woody, fide *Kerfoot* 1779) and erect flowering stems; stems glabrous, succulent or rather succulent. Lower and median stem leaves succulent or somewhat succulent, pale green or glaucous, the lower small, soon disappearing, the median functional, 4–8, petiolate, peltate, circular, subreniform-circular to almost rounded-triangular or rounded-subhastate in outline, 2.5–6 cm long, 3.8–9.5 cm wide, margins closely sinuate-dentate to broadly sinuate-apiculate-dentate, apex rounded to obtuse and apiculate; petiole 2.5–8 cm long; upper leaves few, remote, small, lanceolate, scale-like. Capitula discoid, pendulous to erect, numerous in long-stalked dense to spreading or lax terminal corymbs; stalks of the individual capitula slender, often glaucous; involucre cylindrical, pale green or glaucous, 6.5–8.5 mm long, 2–3 mm in diameter; bracts of calyculus 0–2, lanceolate, 1.5–5.5 mm long; phyllaries 5–8, pale green or glaucous, 6–8 mm long. Ray florets 0; disc florets pale to golden yellow, corolla 4.8–6 mm long, expanded above the middle into infundibuliform to campanulate upper part, glabrous, lobes 1 mm long. Achenes 2.5–3.2 mm long, ribbed, shortly hairy; pappus 4.7–5 mm long.

TANZANIA. Ufipa District: Mbisi, no date, *Bullock* 2800! & Mbisi Mts, 07°54'S 31°43'E, Feb. 1994, *Bidgood et al.* 2519!; Rungwe District: Rungwe Mts N of Rungwe Mission, May 1983, *Magogo* 2447!
DISTR. T 4, 7, 8; Congo (Kinshasa), Angola, Zambia, Malawi, Mozambique, Zimbabwe, South Africa
HAB. Grassland or bushed grassland, often in rocky sites, sometimes in cultivated areas or wooded grassland; 950–2800 m
USES. None recorded on specimens from our area
CONSERVATION NOTES. Least concern (LC)

39. **Senecio hadiensis** *Forssk.*, Fl. Aegypt.-Arab.: 149, t. 19 (1773); Maquet in Fl. Rwanda 3: 666, fig. 207/3 (1985); C. Jeffrey in K.B. 41: 898 (1986); Blundell, Wild Flow. E. Afr.: 176, fig. 382 (1987); Lisowski, Aster. Fl. Afr. Centr. 2: 305 (1991); U.K.W.F. ed. 2: 221, t. 93 (1994). Type: Yemen, *Forsskål* s.n. (C!, holo.)

Succulent herb or shrub, weakly erect to ± 60 cm but more usually scandent and to 15 m long; stems up to 2.5 cm in diameter, rather succulent, glabrous, developing a grey warty bark when older, with pungent smell when cut. Leaves rather succulent, petiolate, glossy pale green, elliptic, broadly elliptic or obovate, 3–13.5 cm long, 1.5–6 cm wide, base cuneate or attenuate into the petiole, entire or minutely to coarsely serrate or dentate-apiculate, apex obtuse to rounded, minutely acuminate-apiculate; petiole narrowly winged, slightly expanded at the base, exauriculate, 0.4–3.5 cm long. Capitula radiate or occasionally some heads without rays, numerous in conspicuous copious open to dense terminal compound corymbs; involucre 3.5–5.5 mm long, 2–2.5 mm in diameter; bracts of calyculus 2–4, lax, broadly linear, ciliate near apex, 1–3 mm long; phyllaries 7–11, green or yellow-green, 3–5 mm long. Ray florets 5, yellow, tube 3–4 mm long, glabrous or with a few hairs at apex, rays 3–5.5 mm long, 1.2–2 mm wide, 4-veined; disc florets yellow, corolla 4–7 mm long, expanded upwards from below the middle, glabrous, lobes 0.7–1 mm long. Achenes 2–3 mm long, ribbed, shortly hairy in the grooves; pappus 5–6 mm long. Fig. 134 (page 647).

Fig. 134. *SENECIO HADIENSIS* — **1**, habit, × ²/₃; **2**, capitulum, × 4; **3**, ray floret, × 6; **4**, disc floret, × 6; **5**, achene, × 6. 1–4 from *Robertson* 1550, 5 from *Rammell* 3485. Drawn by Juliet Williamson.

UGANDA. Karamoja District: Mt Moroto, Jan. 1959, *J. Wilson* 659!; Kigezi District: Nkanda, foot of Mfumbira Mts, Oct. 1970, *Katende* 596!; Masaka District: 3 km NW of Sembabule, May 1971, *Lye & Katende* 6109!

KENYA. Northern Frontier District: Kulal, June 1960, *Oteke* 108!; Nakuru District: 1 km N of Lake Nakuru, Aug. 1967, *Mwangangi* 87!; Kiambu District: Muguga relict forest, Feb. 1967, *Kirrika* 488!

TANZANIA. Musoma District: Moru Kopjes, Feb. 1968, *Greenway & Kanuri* 13176!; Arusha District: Mt Meru, Feb. 1973, *Richards & Arasululu* 28920!; Same District: Mkomazi Game Reserve, Kisima Hill, June 1996, *Abdallah et al.* 96/137!

DISTR. U 1, 2, 4; **K** 1, 3, 4, 6, 7; **T** 1–3; Congo (Kinshasa), Rwanda, Burundi, Ethiopia, Somalia, Madagascar, Yemen, Saudi Arabia

HAB. Dry forest or forest margins, evergreen or semi-deciduous bushland, riverine thicket, termite mound thicket, grassland; may be locally common; 500–2600 m

USES. In W Uganda used as a hedge plant; roots used medicinally, but plant possibly poisonous

CONSERVATION NOTES. Least concern (LC)

SYN. *S. petitianus* A.Rich., Tent. Fl. Abyss. 1: 442 (1848); Oliv. & Hiern, F.T.A. 3: 419 (1877); T.T.C.L.: 158 (1949); Verdcourt & Trump, Common Poison. Pl. E.Afr.: 155 (1969). Type: Ethiopia, Shoa, *Petit* s.n. (P, holo., not found)

　　S. subcrassifolius De Wild., Pl. Bequaert. 5: 120 (1929). Type: Congo (Kinshasa), Bukombo, *Bequaert* 5266 (BR!, holo.)

NOTE. Can make striking masses of gold colour in the landscape (fide *E. Polhill* 118, Naivasha). Flowered heavily on Kilimanjaro in 1992, but hardly flowered in 1993 (fide *Grimshaw*).

　　A form with white-margined leaves has been recorded in cultivation in Nairobi (*Graham Bell* in EA 12192).

　　S. brachypodus DC. from Mozambique to South Africa may not be distinct; see Hilliard, Comp. Natal: 490–491 (1977).

　　Greenway & Kanuri 13176 is reported to lack the ray florets.

40. **Senecio margaritae** *C.Jeffrey* in K.B. 41: 898 (1986); U.K.W.F. ed. 2: 222 (1994). Type: Kenya, Naivasha District, Kedong Valley, Mt Margaret, *Bally* 1013 (K!, holo.)

Succulent shrub 90–150 cm high, with straggly branches; stems glabrous, with pungent smell when cut. Leaves sessile, succulent, subcylindrical, grooved above, 5–11.5 cm long, 0.2–0.6 cm wide, entire, apiculate, glabrous. Capitula radiate, very numerous in copious long-stalked compound terminal corymbs; involucre cylindrical, 10–12 mm long, 3–5 mm in diameter; bracts of calyculus 2–5, lax, lanceolate, 2–4 mm long; phyllaries 5–8, 6.5–12 mm long. Ray florets 5, bright yellow, tube 6.6–9.5 mm long, glabrous; rays 1.7–3.5 mm long, rather irregularly lobed; disc florets yellow, corolla 8.3–11.5 mm long, expanded in upper two-thirds, glabrous, lobes 1 mm long. Achenes 4 mm long, glabrous or sparsely shortly hairy; pappus 7.5–9 mm long.

KENYA. Naivasha District: Kedong Valley, Mt Margaret, June 1940, *Bally* 1013!

DISTR. **K** 3; not known elsewhere

HAB. Not mentioned; possibly rocky bushland; 1800–1950 m

USES. None recorded on specimens from our area

CONSERVATION NOTES. Known from Mt Margaret (a small hill on the former Mt Margaret Estate) and, according to Bally, on the old Kajiado road NW of the Ngong Hills; no specimens from the second locality. Cultivated in Nairobi (1960, *Bally* s.n.; 1963, *Verdcourt* 3591; 1965, *Kirrika* 464). Data deficient (DD), but possibly extinct (Ex)

SYN. *Kleinia barbertonicus* sensu U.K.W.F. ed. 1

41. **Senecio plantagineoides** *C.Jeffrey* in K.B. 41: 898 (1986); U.K.W.F. ed. 2: 222 (1994). Type: Kenya, Nakuru District, Ndoroto, *Bally* 7427 (K!, holo., EA!, iso.)

Perennial subscapigerous herb 12–60 cm high; stems 1–3, erect, or ± decumbent near base, flecked dingy purple, shortly glandular-hairy or glabrous. Basal leaves sessile, oblanceolate to narrowly oblanceolate-linear, 6.5–16 cm long (including

3–6.5 cm long petioloid base), 0.4–1.7 cm wide, attenuate into fibrous petioloid base, margins remotely coarsely to obscurely dentate, apex subobtuse, shortly acuminate-apiculate; stem leaves sessile, lanceolate, oblong-lanceolate or linear, 3.2–5.5 cm long, 0.3–0.6 cm wide, little expanded at the base, margins coarsely remotely serrate to entire, apex acute or attenuate; all leaves glabrous or sparsely pubescent on margins and mid-rib near base, glaucous beneath. Capitula radiate, erect, 1–15 in terminal and upper axillary, usually lax corymbs; inflorescence bracts and stalk of individual heads glandular-hairy or pilose; involucre cylindrical, 7–9 mm long, 3–5 mm in diameter; bracts of calyculus 2–6, lanceolate, 1.5–6 mm long, remote, shortly pilose, especially on margins; phyllaries 10–16, pilose or sometimes glabrous, 6–8 mm long, brownish and bearded at the apex. Ray florets golden yellow, 6–9, tube 4.5 mm long, glabrous, rays 7–9 mm long, 2–2.5 mm wide, 4–7-veined; disc florets yellow, corolla 5.5–7 mm long, expanded above the middle, glabrous, lobes 0.8–1 mm long. Achenes 2.7–3.5 mm long, shortly hairy, many-ribbed; pappus 5.5–6.5 mm long, white.

KENYA. Elgeyo District: Kaisungur, July 1961, *Tweedie* 2169!; Naivasha District: Gilgil, Nov. 1933, *Mainwaring* 6076!; Kisumu-Londiani District: 6 km SSE of Timboroa Station, July 1949, *Maas Geesteranus* 5516!
DISTR. **K** 3, 5; not known elsewhere
HAB. Grassland; 2000–3150 m
USES. None recorded on specimens from our area
CONSERVATION NOTES. The habitat is widespread; twelve Kenyan specimens seen; least concern (LC)

SYN. *S.* sp. E of Agnew, U.K.W.F.: 480 (1974)

42. **Senecio morotonensis** *C.Jeffrey* in K.B. 41: 898 (1986). Type: Uganda, Karamoja District, Moroto Mts, *Eggeling* 2910 (K!, holo.)

Annual herb 30–90 cm high; stems erect, glabrous. Leaves oblanceolate-linear to linear, 1.5–9 cm long, 0.1–0.3 cm wide, the lower petiolate, exauriculate and slightly decurrent, the median and upper sessile, slightly expanded and auriculate at the base, margins entire or remotely denticulate, apex acute, glabrous. Capitula radiate, in small lax terminal corymbs; stalks of the individual capitula slender, glabrous; involucre 4.5–5 mm long, 2–2.5 mm in diameter; bracts of calyculus ± 16, imbricate, slightly pubescent on the margins, 1.5–2 mm long; phyllaries ± 20, glabrous, 4–4.5 mm long. Ray florets ± 11, golden yellow, tube 2 mm long, hairy, rays ± 5 × 2 mm, 4-veined; disc florets yellow, corolla 3.7 mm long, expanded above the middle, glabrous, lobes 0.5 mm long. Achenes 1.8 mm long, slightly ribbed, shortly hairy in the grooves; pappus 5 mm long.

UGANDA. Karamoja District: Mt Moroto, July 1930, *Liebenberg* 363! & Feb. 1936, *Eggeling* 2910! & Sep. 1958, *J. Wilson* 513!
DISTR. **U** 1; only known from Mt Moroto
HAB. Once recorded as forest margins; ± 2900 m
USES. None recorded on specimens from our area
CONSERVATION NOTES. Three specimens known from one site; at least vulnerable (VU-D2)

43. **Senecio mesogrammoides** *O.Hoffm.* in P.O.A. C: 417 (1895); C. Jeffrey in K.B. 41: 899 (1986); U.K.W.F. ed. 2: 222, t. 93 (1994). Type: Tanzania, *Fischer* 357 (B†, syn.) & Moshi District, Marangu, *Volkens* 553 (B†, syn., BM!, isosyn.)

Annual or perennial herb 5–60 cm high; stems branched, decumbent or ascending, then erect, arachnoid-pubescent to glabrous except at the nodes, sometimes purple-tinged. Stem leaves sessile, linear to elliptic, lanceolate or narrowly ovate in outline, entire to strongly pinnately toothed or lobed, pinnatifid into narrow

FIG. 135. *SENECIO MESOGRAMMOIDES* — **1**, habit, × ²/₃; **2**, habit of tiny plant × ²/₃; **3**, capitulum, × 5; **4**, ray floret, × 8; **5**, disc floret, × 12; **6**, achene, × 8. 1 from *Bogdan* 1989, 2 & 5 from Gillett 13941, 3–4 from *Grant* 4/5, 6 from *Richards* 23713. Drawn by Juliet Williamson.

triangular-ascending acute lobes, or deeply pinnately divided into remote narrowly oblong segments, 1–8 (including petioloid base when present) cm long, 0.1–1.3(–3 across the lobes) cm wide, base exauriculate or slightly auriculate or the lower narrowed into a basally expanded but exauriculate petioloid base; all leaves sparsely arachnoid-pubescent above and often more densely so beneath, or glabrous. Capitula radiate or less often discoid, erect, 1–5 in small lax corymbs terminal on the branches, when numerous forming as a whole a lax terminal compound corymb; stalks of the individual capitula slender, sparsely arachnoid-pubescent, especially at the apex, or glabrous; involucre cylindrical or broadly so, 3.5–5.5 mm long, 2–6 mm in diameter; bracts of calyculus 1–9, lanceolate or broadly so, sometimes remote, 1.2–2 mm long, thinly arachnoid or glabrous; phyllaries 11–16, 3–5 mm long, thinly arachnoid or glabrous, shortly tapered to bearded apex. Ray florets 7–14 or sometimes 0, bright yellow, tube 2–2.5 mm long, glandular-hairy, rays 3.5–6 × 1.5–2.2 mm, 4–5-veined; disc florets yellow or orange-yellow, corolla 3–4.5 mm long, tube expanded above the middle, glabrous or with a few glandular hairs in lower part, lobes 0.3–0.5 mm long. Achenes 1.7–3 mm long, ribbed, shortly hairy; pappus 3–4 mm long, white. Fig. 135 (page 650).

Kenya. Northern Frontier District: Furrole, Sep. 1952, *Gillett* 13491!; Nairobi, Langata, Nov. 1981, *Gilbert* 6818!; Masai District: Ngerendei, June 1961, *Glover et al.* 1577!
Tanzania. Musoma District: Serengeti, Engare Nanyuki, Mar. 1962, *Greenway* 10485!; Masai District: Serengeti, Oldiang'arangar, Nov. 1962, *Oteke* 224!; Mbulu/Singida District: Yaida Swamp, Jan. 1970, *Richards* 25148!
Distr. **K** 1, 3, 4, 6; **T** 1, 2, 5; Ethiopia
Hab. Open grassland, often on black cotton soil, occasionally a weed in cultivation; 1200–2500 m
Uses. Grazed by all domestic stock, and frequented by bees (*Glover et al.*)
Conservation notes. The habitat is widespread; least concern (LC)

Syn. *S.* sp. C of Agnew, U.K.W.F.: 479 (1974)

44. **Senecio moorei** *R.E.Fr.* in Act. Hort. Berg. 9: 152 (1928); Verdcourt & Trump, Common Poison. Pl. E. Afr.: 154 (1969); C. Jeffrey in K.B. 41: 899 (1986); Blundell, Wild Flow. E. Afr.: fig. 381 (1987); U.K.W.F. ed. 2: 223 (1994). Type: Kenya, Mt Kenya, *Fries* 1323 (UPS!, holo.)

Perennial woody herb or shrub 45–180 cm high; stems much branched, erect, sparsely arachnoid but glabrescent, sometimes also glandular, often purplish especially towards the base. Leaves oblanceolate or elliptic to narrowly oblanceolate or narrowly elliptic, 5–11.5 cm long, 0.4–3 cm wide, narrowed into a basally auriculate petioloid base, margins closely sinuate-denticulate or sinuate-serrate to coarsely acutely dentate, apex acute to obtuse, shortly apiculate, thinly arachnoid or floccose-tomentose, especially beneath, to almost glabrous, scattered-glandular beneath; auricles coarsely toothed or lobed; occasionally somewhat fleshy. Capitula radiate, erect, numerous in congested or less often lax terminal corymbs; stalks of the individual capitula almost glabrous to sparsely glandular, and slightly arachnoid just beneath the capitulum; involucre cylindrical, 6.5–9 mm long, 3–4.5 mm in diameter; bracts of calyculus 2–6, lanceolate, 1–4.5 mm long, acute to obtuse, sparsely arachnoid or glabrous except at the margins; phyllaries 12–15, green, dark brown or black at the tips, 6–8.5 mm long, sparsely arachnoid to glabrous. Ray florets bright lemon yellow to golden yellow, 11–13, tube 4–5.5 mm long, sparsely hairy, rays 6–10 mm long, 2–3.3 mm wide, 4-veined; disc florets yellow, corolla 5.5–8.3 mm long, expanded above the middle, glabrous or with a few hairs at the base, lobes 0.7–1 mm long. Achenes 2.5–4 mm long, ribbed, shortly sparsely hairy in the grooves; pappus 5–8 mm long.

Uganda. Mbale District: Bulambuli, Bugishu, Sep. 1932, *Thomas* 519! & Elgon near Sasa stream, Mar. 1951, *G. Wood* 153!

KENYA. Naivasha District: South Kinangop, the Elephant, June 1968, *Mwangangi* 1016!; Kiambu
    District: Kinale Forest Station, Oct. 1976, *Mungai* 115!; Masai District: Enesambulai Valley,
    Mar. 1969, *Greenway & Kanuri* 13594!
DISTR. U 3; **K** 3–6; not known elsewhere
HAB. Open spaces in upland forest, glades in bamboo, heath zone, near swamps or moist sites,
    montane grassland; 1750–3500 m
USES. None recorded on specimens from our area
CONSERVATION NOTES. Fairly widespread in a common habitat; least concern (LC)

NOTE. Abundant all over the Kinangop (fide *Verdcourt* 1953).

45. **Senecio mooreioides** *C.Jeffrey* in K.B. 41: 899 (1986). Type: Tanzania, Mbulu
District, Mt Hanang, *B.D. Burtt* 4043 (K!, holo.)

Shrubby perennial herb 10–120 cm high; stem much branched, white-tomentose
to slightly arachnoid and glabrescent. Leaves greyish green to silvery white, sessile,
oblanceolate-linear, linear, narrowly elliptic or narrowly lanceolate, 2.8–8 cm long,
0.25–6 cm wide (excluding lobes when present), auriculate and semi-amplexicaul at
the base, margins remotely serrulate or denticulate to narrowly pinnately
paucilobulate, apex obtuse to acute and apiculate, thinly arachnoid to glabrous
above, arachnoid-tomentose (except sometimes for midrib) beneath. Capitula
radiate, erect in short- to long-stalked usually lax terminal corymbs; stalks of the
individual capitula lanate to slightly arachnoid-glabrescent; involucre cylindrical or
suburceolate-campanulate, 5–7 mm long, 3–5 mm in diameter, lanate or arachnoid
in the lower half or at the base; bracts of calyculus 8–18, lanceolate, 1–2.5 mm long,
acute, imbricate, dark-tipped; phyllaries 10–14 or 18–22, 4–6 mm long, glabrous,
dark-tipped. Ray florets 12–15, bright yellow, tube 2.5–3 mm long, hairy or sparsely
so, rays 5–7 mm long, 2–3 mm wide, 4-veined; disc florets yellow or orange, corolla
4.5–6 mm long, slightly expanded above the middle, glabrous, lobes 0.5–0.7 mm
long. Achenes 2–2.8 mm long, hairy; pappus 4–5.5 mm long.

TANZANIA. Masai District: Lemagrut cone, Oct. 1988, *Chuwa* 2711! & Oldoinyo Lengai, July
    1931, *St. Clair-Thompson* 197!; Mbulu District: Hanang, Feb. 1946, *Greenway* 7645!
DISTR. **T** 2; not known elsewhere
HAB. In rocky sites, often with *Helichrysum*, in the giant heath zone, also in montane grassland;
    once found near steaming crack on Lengai; (1750–)2400–3450 m
USES. None recorded on specimens from our area
CONSERVATION NOTES. Restricted in distribution, but with wide altitude range, and habitat not
    under threat? Fourteen herbarium specimens seen; least concern (LC)

46. **Senecio mooreoides** × **Senecio hochstetteri**; C. Jeffrey in K.B. 41: 899 (1986)

Much branched perennial herb to 50 cm high; stems shortly glandular-setose.
Leaves sessile, oblanceolate, 2.5–4 cm long, 0.35–0.5 cm wide, tapered into a slightly
expanded exauriculate base, coarsely serrate, obtuse, dark brownish green, shortly
glandular-setose above, sparsely arachnoid and setose beneath. Capitula subradiate
with irregularly lobulate enlarged outer disc florets, erect, in small terminal stalked
corymbs; stalks of the individual capitula glandular, also thinly arachnoid; involucre
7 mm long, 4 mm in diameter; bracts of calyculus 3–5, lanceolate, 2–3 mm long;
phyllaries 12–14, 6–7 mm long, finely glandular-pubescent. Florets pale lemon
yellow; outer disc florets ± 9, ray-like with enlarged outer lobe ± 4 mm long, ±
bilabiate; inner disc florets regular, corolla 5.5 mm long, glabrous, lobes 0.5 mm
long. Achenes unknown, but ovary of inner disc florets 1.5 mm long, hairy; pappus
5–5.5 mm long.

TANZANIA. Mbulu District: Mt Hanang, Werther's Peak, Feb. 1946, *Greenway* 7712!
DISTR. **T** 2; only known from one collection
HAB. In more open parts of giant heath/bushland zone on rocky ridge; ± 3050 m

NOTE. This plant has the habit of *S. mooreoides* and the indument of *S. hochstetteri*; the florets are of a form intermediate between these two.

47. **Senecio rhammatophyllus** *Mattf.* in E.J. 59, Beibl. 133: 32 (1924); A.V.P.: 239 (1957); C. Jeffrey in K.B. 41: 899 (1986); U.K.W.F. ed. 2: 222 (1994). Type: Uganda, Elgon, *Dummer* 3333 (B†, holo., K!, iso.)

Shrub 60–100 cm high; stems erect, branched, leafy, sparsely arachnoid-pubescent, glabrescent. Leaves sessile, crowded, linear, 2.6–8 cm long, 0.07–0.15 cm wide, slightly expanded but exauriculate at base, remotely serrulate with revolute margins, apex slightly attenuate and obtuse to acute, glabrous or nearly so. Capitula radiate, erect, terminal, solitary or more often in small cymes; stalks of the individual capitula shortly pubescent or arachnoid-pubescent; involucre broadly cylindrical, 6–8 mm long, 4–5 mm in diameter; bracts of calyculus 14–16, lanceolate, 1.5–3.5 mm long, obtuse, dark-coloured, shortly ciliate on margins or at least near the apex; phyllaries 18–22, 5–7.5 mm long, glabrous, dark-tipped. Ray florets 12–14, yellow, tube 2.8 mm long, hairy, rays ± 13 × 2.5 mm; disc florets yellow, corolla 4–4.8 mm long, expanded above the middle, glabrous, lobes 0.5 mm long. Achenes 2–2.5 mm long, ribbed, shortly hairy in the grooves; pappus 3.5–4.5 mm long.

UGANDA. Mbale District: Elgon, 1930, *Saunders & Hancock* 57! & 6 June 1949, *Osmaston* 4000! & Caldera track, 6 Dec. 1967, *Hedberg* 4513!
KENYA. Elgon, Dec. 1930, *Lugard* 306! & E slope of Koitoboss, 10 May 1948, *Hedberg* 866!
DISTR. **U** 3; **K** 3; endemic to Mt Elgon
HAB. Moorlands with grasses and giant heath; 3000–4150 m
USES. None recorded on specimens from our area
CONSERVATION NOTES. Restricted in distribution, but with wide altitude range and habitat not threatened except by fire; five Ugandan specimens, two Kenyan ones seen; near threatened (NT)

48. **Senecio hedbergii** *C.Jeffrey* in K.B. 41: 899 (1986); U.K.W.F. ed. 2: 222 (1994). Type: Tanzania, Kilimanjaro, above Marangu, *Hedberg* 1422 (K!, holo., EA!, iso.)

Subshrub 17–50 cm high; stems white with dense to thin arachnoid tomentum or densely pubescent. Leaves crowded or rather crowded, sessile, narrowly elliptic, elliptic or oblanceolate in outline, 1.5–2.6 cm long, 0.1–0.25 cm wide (or 0.2–0.7 cm across the lobes), ± expanded but exauriculate at base, margins remotely pinnate-serrate or narrowly pinnately lobulate, revolute, apex obtuse, thinly arachnoid and glabrescent, or pilose but glabrescent above, white with arachnoid tomentum beneath. Capitula radiate, erect, solitary and terminal on the branches or in small terminal corymbs; stalks of the individual capitula arachnoid-tomentose or shortly pubescent; involucre broadly cylindrical, 5–9 mm long, 3.5–5 mm in diameter; bracts of calyculus 12–20, imbricate, lanceolate, acute, dark-coloured or with dark-coloured tips, 1.5–6 mm long, thinly arachnoid or pubescent; phyllaries ± 20–21, 5–8 mm long, thinly arachnoid at least in lower part, glabrescent, dark-tinged and/or with dark tips, sometimes long-attenuate. Ray florets 12–14, yellow, tube 2–3.5 mm long, hairy, rays 6.5–11 × 2–2.5 mm, 4-5-veined; disc florets yellow, corolla 3.5–6.2 mm long, expanded above the middle, glabrous, lobes 0.5–1 mm long. Achenes 2–2.5 mm long, ribbed, hairy in the grooves; pappus 3–6.2 mm long.

KENYA. West Suk District: Mt Kachakulon, Feb. 1933, *Champion* in *Mortimer* s.n.!; Elgeyo District: Chepkotet, Aug. 1968, *Thulin & Tidigs* 217!
TANZANIA. Kilimanjaro, above Marangu, June 1948, *Hedberg* 1422! (type)
DISTR. **K** 2, 3; **T** 2; not known elsewhere
HAB. Giant heath zone; 3300–3650 m
USES. None recorded on specimens from our area
CONSERVATION NOTES. Known from only three specimens; possibly vulnerable (VU-D2)

49. **Senecio meyeri-johannis** *Engl.* in Abh. Preuss. Akad. Wiss. 1891: 444 (1892); A.V.P.: 240, 361 (1957); C. Jeffrey in K.B. 41: 899 (1986). Type: Tanzania, Kilimanjaro, *Meyer* 257 (B†, holo.), neotype: Kilimanjaro, *Hedberg* 1208 (UPS, neo., EA, K!, S , isoneo., chosen by Hedberg, 1957)

Perennial herb, subscapigerous with creeping, mat-forming stems and flowering stems 4–30 cm high, basally decumbent, then erect, laxly arachnoid or sparsely pilose, glabrescent. Basal and lower stem leaves sessile, somewhat succulent, grey-green, oblanceolate to elliptic, 1–8 cm long (including 0.25–1.6 cm long petioloid base), 0.3–1.6 cm wide, tapered into a basally slightly expanded but exauriculate petioloid base, margins slightly to strongly serrate or crenulate or quite deeply pinnately-lyrately lobulate with few-dentate obtuse lobes, apex obtuse, apiculate, sparsely pilose to almost glabrous; upper stem leaves few, smaller, sessile, lanceolate, with ± auriculate base. Capitula radiate, erect, terminal, solitary or rarely paired; stalks of the individual capitula arachnoid or sparsely so, slightly thickened below the capitulum; involucre bright green, 6.5–9 mm long, 4–5.5 mm in diameter; bracts of calyculus 6–12, lanceolate, 2–5 mm long, often thinly arachnoid, sometimes ciliate; phyllaries 18–22, 6–8.5 mm long, thinly arachnoid to glabrous. Ray florets 16–22, brilliant yellow but purple in bud, tube 4 mm long, hairy or sparsely so, rays 7–8.5 × 2–2.7 mm, 4-veined; disc florets deeper yellow, corolla 5.5–6.2 mm long, expanded above the middle, glabrous, lobes 0.5–0.7 mm long. Achenes 2.5 mm long, ribbed, hairy in the grooves; pappus 5–6 mm long, white.

Tanzania. Kilimanjaro, saddle between Kibo and Mawenzi, June 1948, *Hedberg* 1208! & Shira Plateau, Feb. 1969, *Richards* 23980!; Masai District: Ololmoti [Olomoti], Sep. 1932, *B.D. Burtt* 4399!
Distr. **T** 2; known from Kilimanjaro and Ololmoti (Ngorongoro)
Hab. On moist peaty soil in upper moorlands, afro-alpine grassland, seepage areas; 2650–4500 m
Uses. None recorded on specimens from our area
Conservation notes. Restricted distribution but wide altitude range; not threatened except by fire; least concern (LC)

Syn. *S. deaniensis* Muschl. in E.J. 43: 58 (1909). Type: Tanzania, Kilimanjaro, *Jaeger* 405 (B†, holo.)
    S. uhligii Muschl. in E.J. 43: 65 (1909). Type: Tanzania, Kilimanjaro, *Uhlig* 351 (B†, holo.)
    S. meyeri-johannis Engl. subsp. *olomotiensis* C.Jeffrey in K.B. 41: 900 (1986). Type: Tanzania, Masai District, Ololmoti [Olmoti], *Greenway* 9121 (K!, holo.)

Note. Subsp. *olomotiensis* was distinguished by Jeffrey based on the leaves being more robust, oblanceolate-elliptic, 3–8 cm long, 0.8–1.6 cm wide, margins slightly serrulate or crenate-denticulate. On perusal of the material come to light since 1986 it was noticed that in Kilimanjaro populations the leaf margins indentations, which formed the main distinguishing character, varied from the pinnately lobulate to almost entire; leaf size also overlapped the distinctions between the types. The subspecies is here put into synonymy as no distinguishing characters remain.
    It would be interesting to know if this taxon occurs on Mt Meru.

50. **Senecio aequinoctalis** *R.E.Fr.* in Acta Hort. Berg. 9: 152 (1928); A.V.P.: 242, 362 (1957); C. Jeffrey in K.B. 41: 900 (1986); U.K.W.F. ed. 2: 223 (1994). Type: Kenya, Nyandarua/Aberdare Mts, Sattima, *Fries & Fries* 2360 (UPS!, holo., K!, iso.)

Woody herb, tufted, with often numerous ascending flowering stems 5–37 cm high, stems purple-tinged, thinly arachnoid and glabrescent. Lower leaves subsucculent, light green with purple-tinged veins, sessile, linear-lanceolate to oblanceolate or elliptic in outline, 2–7.5 cm long, 0.15–0.7 cm wide (–1.5 cm across the lobes), attenuate into a not or hardly expanded membraneous base, margins entire or denticulate to conspicuously pinnately serrate or lobulate or with a few remote pinnate teeth or narrow ascending lobes, apex obtuse to subacute, apiculate, sparsely arachnoid to glabrous except at very base; upper leaves sessile, smaller, ± serrate, subauriculate at the base. Capitula radiate, erect, 1–5 in a terminal corymb on each flowering stem; stalks of the individual capitula arachnoid, at least at apex, ± slender; involucre cylindrical-campanulate or shortly so, green, 6–7.5 mm long, 3.5–5 mm in diameter; bracts of calyculus 8–16, lanceolate to broadly so, imbricate or spreading, green or purplish with dark tips or dark-coloured in upper half, 1–3.5 mm long, slightly arachnoid or ciliate at least at base; phyllaries 15–20, green or purplish, dark in upper part or at the apex, 5–7 mm long, glabrous or almost so. Ray florets 10–14, yellow, tube 3.2–3.5 mm long, hairy, rays 6.5–9 mm long, 2–3 mm wide, 4-veined; disc florets yellow, corolla 4–5 mm long, expanded above the middle, glabrous or with 1 or 2 hairs at the base only, lobes 0.5 mm long. Achenes 2–2.5 mm long, ribbed, shortly hairy in the grooves; pappus 3.5–4.5 mm long.

KENYA. Mt Kenya, Teleki Valley, Aug. 1948, *Hedberg* 1880! & Hinde Valley, Jan. 1985, *Townsend* 2262! & NW slopes, Mar. 1968, *Mwangangi & Fosberg* 581!
DISTR. **K** 3, 4; endemic to Nyandarua/Aberdares and Mt Kenya
HAB. Moist sites in afro-alpine grassland, shallow soil over rocks; 3000–4250 m
USES. None recorded on specimens from our area
CONSERVATION NOTES. Restricted in distribution but wide altitude range, and habitat not threatened; least concern (LC)

NOTE. Townsend found both the entire-leaved form and the pinnatilobed form in the same population.

51. **Senecio amplificatus** *C.Jeffrey* in K.B. 41: 900 (1986); U.K.W.F. ed. 2: 223 (1994). Type: Kenya, Naivasha District, Kinangop, *Hedberg* 4326 (K!, holo., EA!, iso.)

Perennial herb, creeping or decumbent with erect flowering stems 9–30 cm high; stems sparsely pilose to arachnoid, dark-coloured. Leaves sessile, oblanceolate, 2.5–5.6 cm long, 0.2–0.4(–0.6 across the teeth) cm wide, reducing in size somewhat up the stems, entire or with a few remote teeth, apex subacute or shortly acuminate, apiculate; glabrous or pilose especially above and on midrib and veins beneath. Capitula radiate, erect, terminal, solitary or few; stalks of the individual capitula pilose or almost glabrous; involucre broadly cylindrical, 10 mm long, 7 mm in diameter; bracts of calyculus 12–14, lanceolate, 3.5–4.5 mm long, ciliate or ciliate-arachnoid at least in upper part; phyllaries 20–22, green or purple, 8.5–9.5 mm long. Ray florets 21, yellow, tube 3.5 mm long, hairy, rays 13 × 3 mm, 4-veined; disc florets yellow, corolla 5.5 mm long, expanded above the middle, glabrous, lobes 0.7 mm long. Mature achenes unknown; ovary 2 mm long, appressed-hairy; pappus 5 mm long.

KENYA. Naivasha District: Kinangop, Sep. 1967, *Hedberg* 4326! & N Aberdares, Apr. 1959, *Alexander* 11635
DISTR. **K** 3; known from only the Nyandarua/Aberdare Mts
HAB. On a small summit in the upper giant heath zone; 3500 m
USES. None recorded on specimens from our area
CONSERVATION NOTES. Two specimens known; at least vulnerable (VU-D2)

52. **Senecio jacksonii** *S.Moore* in J.L.S. 35: 358 (1902); A.V.P.: 240, 361 (1957); C. Jeffrey in K.B. 41: 900 (1986); U.K.W.F. ed. 2: 223 (1994). Type: Kenya, Elgon, 'Sotik', *Jackson* s.n. (BM!, holo.)

Creeping perennial herb, subscapigerous, rhizomatous, tufted with erect flowering stems 4–15 cm high; stems green or purplish, arachnoid, glabrescent. Leaves ± crowded, sessile, dark green, narrowly oblanceolate, oblanceolate-spatulate or oblanceolate, 2–7 cm long, 0.1–0.6 cm wide, tapered into a slightly expanded membraneous base, margins entire or remotely denticulate and revolute, apex obtuse, subapiculate, arachnoid especially on midrib or sparsely arachnoid or pilose especially towards or only at the base and on margins, glabrescent, slightly purplish. Capitula radiate, erect, terminal, solitary or 2–3; stalks of the individual capitula arachnoid to glabrous; involucre obconic-cylindrical, 7.5–9 mm long, 4.5–6 mm in diameter; bracts of calyculus 6–20, imbricate, lanceolate or broadly so, dark or purplish at the apex, 1–4.5 mm long, thinly arachnoid or ciliate or glabrous; phyllaries 16–22, green or purple-tinged with dark purplish tips, 7–8.5 mm long, thinly arachnoid to glabrous. Ray florets 15–20, yellow, tube 3–3.8 mm long, hairy, rays 9–12 × 2.5–3 mm, 4–5-veined; disc florets yellow, corolla 4.7–5.8 mm long, slightly expanded above the middle, glabrous, lobes 0.7–0.8 mm long. Achenes 2.5–2.8 mm long, ribbed, hairy or sparsely so in the grooves; pappus 4–5 mm long, white.

UGANDA. Elgon, Nov. 1933, *Tothill* 2416! & below Jackson's summit, 1930, *Liebenberg* 1583! & track through Caldera, Dec. 1967, *Hedberg* 4485!
KENYA. Elgon, S slopes, Dec. 1967, *Mwangangi* 315!; Nyandarua/Aberdares, Kinangop, Jan. 1932, *Dale* 2693! & track from Wanderi's camp to Sattima, Feb. 1985, *Townsend* 2407!
DISTR. U 3; K 3/4; endemic to Elgon and Nyandarua/Aberdares
HAB. On thin soil over rocks and rocky ground, swampy grassland; 3250–4150(–4500) m
USES. None recorded on specimens from our area
CONSERVATION NOTES. Restricted distribution area and range, but habitat not threatened; 9 Ugandan specimens, 13 Kenyan specimens; near threatened (NT)

SYN. *S. caryophyllus* Mattf. in E.J. 59, Beibl. 133: 33 (1924); C. Jeffrey in K.B. 41: 900 (1986); U.K.W.F. ed. 2: 223 (1994). Type: Uganda, Elgon, *Dummer* 3308 (B†, holo., K!, iso.)
    S. sympodialis R.E.Fr. in Acta Hort. Berg. 9: 153 (1928). Type: Kenya, Nyandarua/Aberdare Mts, Sattima, *Fries & Fries* 2581 (UPS!, holo.)
    S. jacksonii S.Moore subsp. *sympodialis* (R.E.Fr.) Hedberg, A.V.P.: 241 (1957)
    S. jacksonii S.Moore subsp. *caryophylla* (Mattf.) Hedberg, A.V.P.: 241 (1957)

NOTE. *S. caryophyllus* is here put in synonymy. Hedberg downgraded the taxon to subspecific level, but Jeffrey in the 1986 treatment of East African *Senecio* reinstated full specific status, with the key character 'Leaves linear/oblanceolate or narrowly so'. After study of the material available the junior author believes this is the only difference, and one that has several intermediates. Habitat and altitude range are the same and the species are united here.

53. **Senecio keniophytum** *R.E.Fr.* in Acta Hort. Berg. 9: 154, t. 9 (1928); A.V.P.: 241, 362 (1957); C. Jeffrey in K.B. 41: 900 (1986); U.K.W.F. ed. 2: 223 (1994). Type: Kenya, Mt Kenya, *Fries & Fries* 1414 (UPS!, holo., BR!, K!, S, iso.)

Perennial creeping herb with erect flowering branches 4–13(–20) cm high, branching, rhizomatous, tough; stems densely white-lanate, though eventually glabrescent. Leaves sessile, somewhat succulent, oblanceolate to oblong or oblong-linear, 1.4–7 cm long, 0.5–1.8 cm wide, ± attenuate into a membraneous often purplish slightly expanded but exauriculate petioloid base, margins obscurely serrulate to remotely coarsely serrate, apex obtuse, apiculate, white-tomentose to green and glabrous above, densely white-tomentose to thinly arachnoid and with glabrous midrib beneath. Capitula radiate, erect, solitary or rarely 2–3 together terminal on the branches; stalks of the individual capitula thinly to densely lanate; involucre 9–12 mm long, 6–7 mm in diameter; bracts of calyculus 8–12, lanceolate, dark-tipped, 5–9 mm long, thinly to densely arachnoid-lanate; phyllaries 12–20, black-tipped, 7–11 mm long, densely white-lanate to thinly arachnoid and glabrescent, sometimes also minutely puberulous. Ray florets 12–20, bright yellow, tube 3 mm long, glabrous, rays 5.5–6 × 2.5–3 mm, 4-veined; disc florets dull yellow, becoming brown, corolla 5–6 mm long, glabrous, expanded above the middle, lobes 0.7–1 mm long. Achenes 3 mm long, ribbed, glabrous; pappus 5–5.5 mm long.

Kenya. Mt Kenya: upper Hausberg Valley, Jan. 1965, *Allt* 14! & head of Teleki Valley, Sep. 1948,
  J.G. Williams in *Bally* 6441! & small lake just below Tyndall Glacier, July 1948, *Hedberg* 1743!
Distr. **K** 4; endemic to Mt Kenya
Hab. Boggy ground near streams or lakes, may grow close to snow line or glacier lakes in
  shelter of rocks; (3700–)4050–4500(–5000) m
Uses. None recorded on specimens from our area
Conservation notes. Limited distribution but over 20 specimens; near threatened (NT)

Syn. *S. keniophytum* R.E.Fr. var. *decumbens* R.E.Fr. in Acta Hort. Berg. 9: 155 (1928), *nom. non rite
    publ.* Type as for *S. keniophytum*
  S. keniophytum R.E.Fr. var. *candido-lanatus* R.E.Fr. in Acta Hort. Berg. 9: 155 (1928). Type:
    Kenya, Mt Kenya, *Fries & Fries* 1408 (K!, iso.)
  S. keniophytum R.E.Fr. var. *glabrior* R.E.Fr. in Acta Hort. Berg. 9: 155 (1928). Type: Kenya,
    Mt Kenya, *Fries & Fries* 1312 (K!, iso.)

Note. *Luke et al* 4765 from Mt Keya, Lake Alice, is close to this and *S. telekii.*

54. **Senecio telekii** (*Schweinf.*) *O.Hoffm.* in P.O.A. C: 417 (1895); A.V.P.: 239, 360, t.
1b (1957); C. Jeffrey in K.B. 41: 900 (1986). Type: Tanzania, Kilimanjaro, *von Höhnel*
s.n. (B†, holo.), neotype: Kilimanjaro, SW slope of Mawenzi, *Hedberg* 1275 (UPS,
neo., EA, K!, S, isoneo., chosen by Hedberg)

Woody herb 7–30(–60) cm high, much branched, rhizomatous; stems leafy,
arachnoid-pubescent; roots up to 40 cm long. Leaves sessile, rather succulent, oblong
to oblanceolate, 1.2–3.5 cm long, 0.2–1 cm wide, slightly expanded and sometimes
auriculate at the base, margins obscurely crenulate-serrulate, apex obtuse, minutely
apiculate, densely white arachnoid-tomentose to sparsely arachnoid or glabrous
above, densely white arachnoid-tomentose beneath except for base of midrib.
Capitula radiate or discoid, erect, 1–4 together terminal on the branches, sessile or
shortly stalked; stalks of the individual capitula swollen, arachnoid-pubescent;
involucre 7–9 mm long, 4.5–6 cm in diameter; bracts of calyculus 6–10, lanceolate to
oblong, dark-tipped, 3–6 mm long, obtuse to acute, arachnoid especially on margins
to densely lanate-tomentose; phyllaries 18–20, sometimes purplish, 7–8.5 mm long,
thinly to densely arachnoid or lanate, especially in lower half. Ray florets absent or
10–14, when present bright yellow, tube 2.5–3 mm long, hairy throughout or at base
only, rays 8–12 mm long, 3–4 mm wide, 4–5-veined; disc florets dull yellow, corolla
4.5–6 mm long, expanded above the middle, glabrous, lobes 0.5–1 mm long.
Achenes 2.5 mm long, ribbed, glabrous; pappus 5–6 mm long, white.

Tanzania. Kilimanjaro, Shira Ridge, Aug. 1968, *Carmichael* 1463! & Kibo Hut, Oct. 1952, *Bally*
  8363!; Arusha District: Mt Meru, near summit, Sep. 1932, *Burtt* 4056!
Distr. **T** 2; endemic to Kilimanjaro and Mt Meru
Hab. Dry stony slopes and scree, stony gravel, in rock crevices, in the lower part of its range in
  open giant heath, in the upper part of its range in the shelter of rocks or even next to
  glaciers; the highest phanerogam in Africa; (3350–)3600–5400 m
Uses. None recorded on specimens from our area
Conservation notes. Restricted distribution but habitat not specifically threatened; > 20
  specimens; least concern (LC)

Syn. *Erigeron telekii* Schweinf. in Hoehnel, Zum Rudolph- und Stephanie-See: 861 (1892)
    Senecio volkensii O.Hoffm. in P.O.A. C: 418 (1895); T.T.C.L.: 159 (1949). Type: Tanzania,
    Kilimanjaro, *Volkens* 1367 (B†, holo., BM!, K!, iso.)

Note. 'Spectacular' (*Burtt*).
  Grimshaw 93/635 from Kilimanjaro, 4200 m near Moir's Hut, is an erect shrublet to 10 cm
  high with lobed subglabrous leaves. It seems intermediary between *S. telekii* and *S. meyeri-
  johannis*, and Grimshaw believes it is indeed a hybrid; both 'parent' species of this Grimshaw's
  Groundsel occur in the locality as well.
  Grimshaw 93/637 from the Lent Valley at 4200 m, above Moir's Hut, looks more like *S.
  telekii* itself but lacks the dense indument typical of this taxon. It is a low shrub to 25 cm high
  and formed large colonies on rock ledges. The lobes of the disc florets are 0.5 mm long,
  much shorter than in typical *S. telekii.*

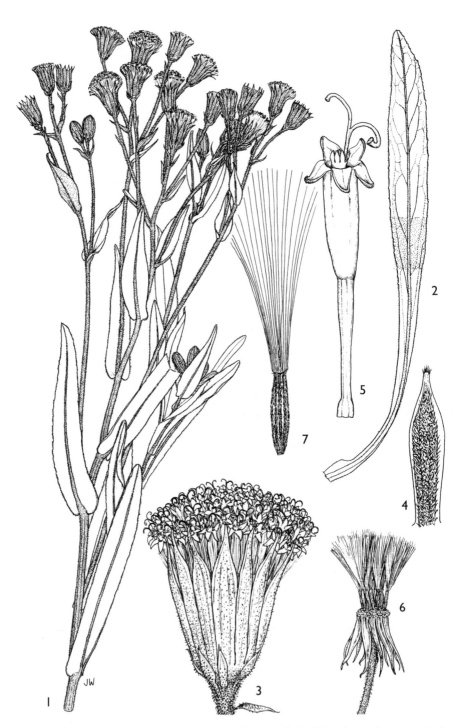

FIG. 136. *SENECIO SCHWEINFURTHII* — **1**, habit, × ²/₃; **2**, leaf from base of plant, × ²/₃; **3**, capitulum, × 4; **4**, phyllary, × 4; **5**, floret, × 6; **6**, old capitulum with mature achenes, × 1 ¹/₂; 7, achene, × 6. 1 from *Greenway & Kanuri* 13588, 2–3 & 5 from *Hedberg* 1881, 4 from *Glover et al.* 1036, 6–7 from *Townsend* 2218. Drawn by Juliet Williamson.

55. **Senecio schweinfurthii** *O.Hoffm.* in P.O.A. C: 417 (1895); A.V.P.: 244, 363, t. 7a (1957); C. Jeffrey in K.B. 41: 901 (1986); Blundell, Wild Flow. E. Afr.: fig. 383 (1987); Lisowski, Aster. Fl. Afr. Centr. 2: 307 (1991); U.K.W.F. ed. 2: 222 (1994). Type: Tanzania, Kilimanjaro, *von Höhnel* s.n. (B†, holo.), neotype: Kilimanjaro, Horombo [Peters] Hut, *Hedberg* 1179 (UPS, neo., EA, K!, S, isoneo., chosen by Hedberg)

Perennial herb with creeping woody rhizome and sometimes with the main stem prostrate, and ascending branching leafy flowering stems 20–150 cm high; stems pale to bright green, sparsely glandular-pubescent. Leaves crowded in terminal rosettes, with smaller ones on the inflorescence, somewhat succulent, sticky, oblanceolate-linear, oblanceolate, oblong or narrowly lanceolate, 6.5–26 cm long, 0.6–3 cm wide, tapered into a slightly expanded subauriculate semi-amplexicaul base, margins minutely to strongly sinuate-dentate or sinuate-serrate, revolute, apex subacute to rounded, minutely apiculate, bright green with paler midrib, densely glandular-pubescent above, glandular and minutely ciliate on the margins, glandular-pubescent and sometimes also slightly arachnoid beneath; stem leaves few, sessile, semi-amplexicaul, auriculate. Capitula discoid, scented with a sweet or musky odour, erect, 3–many in lax terminal cymose corymbs or panicles; stalks of the individual capitula glandular; involucre suburceolate-campanulate, cylindrical in lower $^2/_3$, very slightly contracted just below, then expanded at, the apex, 9–14 mm long, 4–7 mm in diameter; bracts of calyculus 3–6, lax, lanceolate to broadly so, 3–6 mm long, pale green with dark tips, glandular; phyllaries 10–20, green, brown and bearded at the apex, 8–12 mm long, glandular, rarely also sparsely arachnoid, with hyaline margins and paler main veins translucent towards the apex. Ray florets 0; disc florets pale to bright yellow, corolla 7.5–11 mm long, tube expanded above the middle, glabrous or with a few hairs at the base, lobes 0.7–1.2 mm long. Achenes 2.7–4 mm long, ribbed, shortly hairy; pappus 10 mm long. Fig. 136 (page 658).

KENYA. Trans Nzoia District: Cherangani, Kaisungor, Feb. 1970, *Tweedie* 3763!; Nakuru District: above Mau Narok, Dec. 1974, *Williams Sangai* 51!; Mt Kenya, Sirimon track, Mar. 1968, *Fosberg & Mwangangi* 49878!

TANZANIA. Masai District: Ngorongoro, Mt Lemagarut, Oct. 1988, *Chuwa* 2695!; Kilimanjaro, Horombo [Peters] Hut, Feb. 1934, *Greenway* 3744!; Arusha District: Mt Meru, foot of ash cone, Mar. 1968, *Greenway & Kanuri* 13202!

DISTR. **K** 3–6; **T** 2; Congo (Kinshasa)

HAB. Open moorland, lava flows and dry stone slopes, giant heath/*Protea* zone, clearings in upper montane forest; 2300–4500 m

USES. None recorded on specimens from our area

CONSERVATION NOTES. Common within much of its habitat; least concern (LC)

SYN. *S. serra* Schweinf. in Hoehnel, Zum Rudolph- und Stephanie-See: 864 (1892), *nom. illegit.*, *non* Hook. (1834) *nec* Sond. (1850); type as for *S. schweinfurthii*

S. massaiensis Muschl. in E.J. 43: 67 (1909). Type: Kenya, Nakuru District: Mau Plateau, *Baker* s.n. (B†, holo.)

S. melanophyllus Muschl. in E.J. 43: 68 (1909). Type: Tanzania, Kilimanjaro, *Uhlig* 1083 (B†, holo., EA, iso.)

S. theodori K.Afzel. in Svensk Bot. Tidskr. 19: 422 (1925). Type: Kenya, Mt Kenya, *Fries & Fries* 665 (UPS, holo., BR!, K!, S, iso.)

S. roberti-friesii K.Afzel. in Svensk Bot. Tidskr. 19: 420 (1925). Type: Kenya, Mt Kenya, *Fries & Fries* 1355 (UPS, holo., BM!, BR!, S, iso.)

S. roberti-friesii K.Afzel. var. *subcanescens* K.Afzel. in Svensk Bot. Tidskr. 19: 420 (1925). Type: Kenya, Mt Kenya, *Fries & Fries* 2355 (UPS, holo.)

S. sattimae K.Afzel. in Svensk Bot. Tidskr. 19: 420 (1925). Type: Kenya, Nyandarua/Aberdare Mts, Sattima, *Fries & Fries* 2355a (UPS, holo.)

S. roberti-friesii K.Afzel. var. *kilimanjaricus* K.Afzel. in Acta Hort. Berg. 15: 71 (1949). Type: from plants in cultivation from seeds collected on Kilimanjaro by *Ambjörn* (SBT, syn.)

56. **Senecio navugabensis** *C.Jeffrey* in K.B. 41: 901 (1986). Type: Uganda, Masaka District, Lake Navugabo, *Synge* 1947 (BM!, holo.)

Perennial herb 25–30 cm high, ? rhizomatous; flowering stems decumbent at base, then erect, glandular-hairy. Leaves sessile, narrowly oblong or oblanceolate-oblong, 5.5–9 cm long, 0.4–0.7 cm wide, exauriculate, margins denticulate, apex subobtuse, apiculate, shortly glandular-hairy above and beneath. Capitula discoid, terminal, erect, 4–6 together in dense terminal cymes; stalk of individual heads glandular; involucre 10 mm long, 7–8 mm in diameter; bracts of calyculus ± 6, lanceolate, 4 mm long, glandular; phyllaries 13, narrowly oblong-lanceolate, 9–10 mm long, attenuate, glandular, dark-coloured and slightly bearded at the apex, with narrow scarious margins. Ray florets 0; disc florets yellow, corolla 8.5–9 mm long, glabrous, lobes 0.7 mm long. Mature achenes unknown, but ovary pubescent; pappus 11 mm long, white.

UGANDA. Masaka District: Lake Navugabo, Apr. 1935, *Synge* 1947!
DISTR. **U** 4; known from only the type
HAB. Edge of swamp; ± 1200 m
USES. None recorded on specimens from our area
CONSERVATION NOTES. Known from only one specimen in a fairly well-collected area; at least vulnerable (VU-D2)

57. **Senecio purtschelleri** *Engl.* in Abh. Preuss. Akad. Wiss. 1891, 2: 443 (1892); A.V.P.: 237, 360 (1957); C. Jeffrey in K.B. 41: 901 (1986); U.K.W.F. ed. 2: 222 (1994). Type: Tanzania, Kilimanjaro, *Meyer* 213 & 262 (B†, syn.), neotype: Kilimanjaro, *Hedberg* 1314 (UPS, neo., EA!, K!, isoneo., chosen by Hedberg)

Perennial herb 7–60(?–90?) cm high, rhizomatous; stems erect, branched, ± glandular-pubescent, sometimes also arachnoid-glabrescent, leafy. Leaves sessile, oblong to oblanceolate, 2–13.5 cm long, 0.4–2.7 cm wide, tapered to an auriculate or subauriculate somewhat expanded base, margins sinuate-crenate or obscurely to coarsely sinuate-dentate or bidentate with rounded apiculate teeth, apex obtuse to rounded, minutely apiculate, dark green, sticky, glandular-pubescent and sometimes also thinly arachnoid-glabrescent above, glandular-pubescent beneath especially on midrib and veins and sometimes also thinly arachnoid, marcescent when old. Capitula radiate, subradiate with irregularly enlarged outer disc florets, or discoid, erect, 2–12 in usually congested terminal cymes; stalks of the individual capitula glandular, rarely also slightly arachnoid; involucre cylindrical, 8–11.5 mm long, 4–5 mm in diameter; bracts of calyculus 4–8, lax, lanceolate, 3–7 mm long, dark-tipped, ± glandular, at least on margins; phyllaries 10–14, 7–11 mm long, thinly finely glandular-pubescent and sometimes also very thinly arachnoid. Ray florets 5–7 or 0, variable within populations, when present yellow, tube 3–4 mm long, sparsely hairy, rays 4–6 × 1.5–2.2 mm; disc florets yellow, corolla 5–6.7 mm long, tube expanded above the middle, glabrous, lobes 0.5–0.8 mm long. Achenes 2.5–3.5 mm long, ribbed, glabrous or almost so; pappus 5.5–6.5 mm long, white.

KENYA. Mt Kenya, above head of Nithi Valley, Aug. 1944, *Le Pelley* in *Bally* 3445! & head of Sirimon Valley, Nov. 1970, *Mabberley* 417! & N slope near Kami Hut, June 1986, *Linder* 3672!
TANZANIA. Kilimanjaro, saddle, June 1948, *Hedberg* 1193! & Mawenzi hut, Mar. 1956, *Trump* s.n.!: Mt Meru, summit, Mar. 1946, *Fuggles Couchman* in AH 9817!
DISTR. **K** 4; **T** 2; restricted to Mt Kenya, Kilimanjaro and Mt Meru
HAB. Damp sites in afro-alpine zone, usually in rock crevices or in shelter of rocks, may be only flowering plant on rock screes, also in upper giant heath belt; (3350–)3700–4800(–5000) m
USES. None recorded on specimens from our area
CONSERVATION NOTES. Limited distribution area but habitat not threatened; least concern (LC)

SYN. *S. platzii* Muschl. in E.J. 43: 67 (1909). Type: Tanzania, Kilimanjaro, *Meyer* 51 (B†, syn.) & Meru, *Uhlig* 604 (B†, EA!, syn.)

58. **Senecio snowdenii** *Hutch.* in K.B. 1920: 24 (1920); A.V.P.: 236, t. 7b (1957); C. Jeffrey in K.B. 41: 901 (1986); U.K.W.F. ed. 2: 222 (1994). Type: Uganda, Elgon, *Snowden* 463 (K!, holo., BM!, iso.)

Perennial herb or subshrub 30–100 cm high, viscid; stems erect, glandular-pubescent and viscid (smelling of cinnamon), purple-tinged, leafy. Leaves sessile, oblong-linear to oblong-lanceolate or oblanceolate, 4.5–17 cm long, 0.5–2 cm wide, tapered to an auriculate or subauriculate ± semi-amplexicaul base, margins remotely to coarsely sinuate-denticulate, apex subacute to rounded, apiculate, light green, sticky, ± glandular-pubescent above and beneath, marcescent when old. Capitula radiate, erect, 1–7 in lax terminal cymes; stalks of the individual capitula glandular; involucre cylindrical, 9–16 mm long, 5–8 mm in diameter; bracts of calyculus 4–6, lax, lanceolate, 3.5–9 mm long, densely glandular; phyllaries 12–14, 8.5–14 mm long, densely glandular. Ray florets 10–13, golden yellow, tube 5.5–6.5 mm long, sparsely hairy, rays 10–17 mm long, 3–4 mm wide, 4–5-veined; disc florets yellow, corolla 8–10.2 mm long, tube expanded above the middle, glabrous, lobes 0.5–1 mm long. Achenes 3.5–4.2 mm long, ribbed, hairy; pappus 8.5–10.5 mm long, white.

UGANDA. Elgon, near Bulambuli, Nov. 1933, *Tothill* 2314! & Gabaralome, Dec. 1938, *A.S. Thomas* 2681! & W slope above Butadiri, near Mountain Hut, Dec. 1967, *Hedberg* 4455!
KENYA. Elgon, Feb. 1930, *Gardner* 2268! & 13 km SW of Suam Saw Mills, Dec. 1967, *Mwangangi* 320! & W of Kipsare Hill, Dec. 1967, *Gillett* 18460!
DISTR. **U** 3; **K** 3; endemic to Mt Elgon
HAB. From clearings in the bamboo zone almost to the summit, in grassland or afro-alpine bushland; forms patches and can be locally common; 2700–4250 m
USES. None recorded on specimens from our area
CONSERVATION NOTES. Restricted in area but in a wide range of habitats and altitude; much destruction has disturbed Mt Elgon habitats; ?near threatened (NT)

SYN. *S. elgonensis* Mattf. in E.J. 59, Beibl. 133: 32 (1924), *nom. illegit.*, *non* T.C.E.Fr. (1923). Type: Uganda, Elgon, *Dummer* 3323 (B†, holo., K!, iso.)
    S. mattfeldii R.E.Fr. in Acta Hort. Berg. 9: 151 (1928). Type as for *S. elgonensis*

59. **Senecio sotikensis** *S.Moore* in J.L.S. 35: 357 (1902); A.V.P.: 236 (1957); C. Jeffrey in K.B. 41: 902 (1986); U.K.W.F. ed. 2: 222 (1994). Type: Kenya, Elgon, Sotik [?Sudek], *Jackson* s.n. (BM!, holo.)

Perennial woody herb or weak-stemmed shrub 45–200 cm high, viscid; stems erect, branched, glandular-setose. Leaves sessile, broadly oblong, oblong or elliptic-oblong, 3.2–9 cm long, 0.8–3 cm wide, base auriculate with toothed auricles and semi-amplexicaul, pinnately lobulate with acute to rounded entire or few-denticulate or apiculate-lobulate lobes, apex obtuse, minutely apiculate, viscid, glandular-pubescent, sometimes also sparsely arachnoid beneath, crowded at the ends of the branches, weakly marcescent when old. Capitula radiate, erect, 3–20 in lax to dense terminal corymbs; stalks of the individual capitula glandular-hairy, slightly expanded below the capitula; involucre 8–10 mm long, 4.5–6 mm in diameter; bracts of calyculus 5–9, lax, narrowly lanceolate, 2.5–5.5 mm long, glandular; phyllaries 16–21, 7.5–10 mm long, glandular, dark-tipped. Ray florets 17–23, yellow, tube 3.5–4 mm long, hairy, rays 10–17 mm long, 2.5–3 mm wide, 4-veined; disc florets yellow, corolla 6–6.5 mm long, tube expanded above the middle, glabrous, lobes 0.7 mm long. Achenes 3–4 mm long, ribbed, glabrous; pappus 6.5–7 mm long.

UGANDA. Elgon crater, Jan. 1918, *Dummer* 3324! & W slope above Butadiri, near Mountain Hut, Dec. 1967, *Hedberg* 4453! & Sasa Hut, June 1970, *Lye et al.* 5726!
KENYA. Elgon, Dec. 1930, *Lugard & Lugard* 416! & Dec. 1970, *Kokwaro* 2464! & Mbere R. Valley S of Koitcut, Dec. 1967, *Gillett* 18435!
DISTR. **U** 3; **K** 3; endemic to Mt Elgon
HAB. Marshy ground, grassland; may be locally common; 3000–4500 m
USES. None recorded on specimens from our area
CONSERVATION NOTES. Restricted in area but in a common habitat; much destruction has disturbed Mt Elgon habitats; ?near threatened (NT)

SYN. *S. transmarinus* of U.K.W.F.: 479 (1974), *non* S.Moore

60. **Senecio polyadenus** *Hedb.*, A.V.P.: 237, 365, fig. 17 (1957); C. Jeffrey in K.B. 41: 902 (1986); Lisowski, Aster. Fl. Afr. Centr. 2: 308 (1991). Type: Uganda, Kigezi District, Muhavura, *Hedberg* 2091 (UPS, holo., EA!, K!, iso.)

Perennial herb or spreading shrub 30–100 cm high, rhizomatous; stems erect, glandular-pubescent, leafy. Lower and median stem leaves ± equal in size, sessile and ovate-oblong or oblong, 4.5–11.5 cm long, 1.5–4.5 cm wide, base auriculate, margins sinuate-denticulate to strongly dentate-lobulate, apiculate-bidentate or lobulate, apex obtuse, apiculate; upper stem leaves sessile, similar to the lower, margins dentate-lobulate, apex acute, 3–5 cm long, 1.2–1.5 cm wide; all leaves finely glandular above and beneath. Capitula radiate, erect, numerous in copious often dense terminal corymbs; stalk of individual heads densely glandular-hairy; involucre cylindrical, 7–11 mm long, 4.5–6.5 mm in diameter; bracts of calyculus 4–6, lanceolate to linear, glandular and 5–8 mm long; phyllaries 12–13, 8.5–10 mm long, glandular. Ray florets 10–13, citrus yellow, corolla tube 4–5.5 mm long, glabrous, rays 10–15 mm long, 3.7–4 mm wide, 5–10-veined; disc florets yellow, corolla 6.5–9 mm long, tube expanded or abruptly expanded above the middle, glabrous, lobes 1–1.2 mm long. Achenes not known mature, immature achenes 2–4 mm long, glabrous; pappus 5–7 mm long, white.

Uganda. Kigezi District: Mt Muhavura, Dec. 1933, *A.S. Thomas* 1119! & June 1939, *Purseglove* 768! & Oct. 1948, *Hedberg* 2229!
Distr. U 2; Congo (Kinshasa); endemic to Mts Muhavura and Karisimbi
Hab. Heath zone, grassland or *Senecio* forest; 3500–4000 m
Uses. None recorded on specimens from our area
Conservation notes. Four sheets seen from Uganda and one from Congo; at least vulnerable (VU-D2)

61. **Senecio roseiflorus** *R.E.Fr.* in Acta Hort. Berg. 9: 150 (1928); A.V.P.: 235 (1957); C. Jeffrey in K.B. 41: 902 (1986); Blundell, Wild Flow. E. Afr.: fig. 682 (1987); U.K.W.F. ed. 2: 222, t. 94 (1994). Type: Kenya, Nyandarua/Aberdare Mts, Sattima, *Fries & Fries* 2442 (UPS, holo., K!, iso.)

Perennial woody herb or weak-stemmed shrub 75–150(–240) cm high; stems erect or occasionally scrambling, branched, glandular, leafy, viscid, disagreeably aromatic. Leaves sessile, oblong-lanceolate to lanceolate, 5–9 cm long, 0.4–2.4 cm wide, base somewhat auriculate and semi-amplexicaul, margins crenate to coarsely sinuate-serrate, bidentate or lobulate, apex obtuse to acute, apiculate, crowded, green, viscid, glandular-hairy, marcescent when old. Capitula radiate, erect, ± 6–20 in lax to dense terminal corymbs; stalks of the individual capitula glandular-hairy, viscid; involucre 7.5–11 mm long, 4–7 mm in diameter; bracts of calyculus 1–4, lax, narrow, lanceolate, 2.5–6 mm long, acute, glandular-hairy; phyllaries 8–14, deep dull purple, 7–10 mm long, densely glandular. Ray florets 12–14, violet, bright to pale purple or mauve, corolla tube 3.5–4.5 mm long, glabrous, rays 8–14 mm long, 3–4.5 mm wide, 4–5-veined; disc florets purple to dark red, corolla 7–8 mm long, tube expanded from below the middle, glabrous, lobes 1 mm long. Achenes 3.2–3.5 mm long, ribbed, shortly sparsely hairy; pappus 7.5–8 mm long.

Kenya. Naivasha District: Nyandarua/Aberdares National Park, 00°15'S, 36°35'E, Mar. 1987, *Beentje* 3239!; Mt Kenya, Urumandi, Aug. 1944, *Le Pelley* in *Bally* 3414! & Sirimon Track, Mar. 1968, *Fosberg & Mwangangi* 49878!
Distr. K 3, 4; endemic to Nyandarua/Aberdares and Mt Kenya
Hab. Giant heath and *Stoebe-Protea* zone, rocky moorland, streamsides; (2900–)3200–3600 (–4200) m
Uses. None recorded on specimens from our area
Conservation notes. Three specimens from the Nyandarua/Aberdares, > 20 from Mt Kenya; low risk except through fire; near threatened (NT)

Note. Chromosome number n=11 according to *Lewis* 5918.

62. **Senecio mattirolii** *Chiov.* in Ann. Bot. Roma 6: 150 (1907); A.V.P.: 235, 363 (1957); C. Jeffrey in K.B. 41: 902 (1986); Lisowski, Aster. Fl. Afr. Centr. 2: 309, t. 67 (1991). Type: Congo (Kinshasa), Ruwenzori, *Roccati* s.n. (TO, holo.)

Perennial herb 10–30(–40) cm high, subscapigerous; stems reddish, crispate-glandular. Basal leaves often reddish, sessile, oblanceolate or oblanceolate-spatulate, 5.5–13 cm long, 1–2 cm wide, narrowed into a basally expanded membraneous exauriculate petaloid base, margins sinuate-denticulate or bidenticulate, apex obtuse, minutely apiculate, glandular; stem leaves sessile, smaller, oblong, 2–4.8 cm long, 0.3–1 cm wide, subauriculate. Capitula discoid, erect, 2–6 in lax terminal corymbs; stalks of the individual capitula glandular; involucre 9–11 mm long, 5.5–7.5 mm in diameter; bracts of calyculus 4–6, lax, narrowly lanceolate, 2.5–7 mm long, glandular; phyllaries 14–20, purplish, 8–12 mm long, glandular. Ray florets 0; disc florets purple or mauve, corolla 6–7.5 mm long, tube expanded above the middle, glabrous, lobes 0.7–1 mm long. Achenes 3.5–4 mm long, ribbed, glabrous; pappus 6.5–7 mm long.

UGANDA. Toro District: Ruwenzori, Lake Kitandara, Aug. 1960, *Richardson & Kendall* 26! & below Bujuku Hut, June 1968, *Manum* 104! & between Bujuku Hut and Lake Irene Hut, Jan. 1969, *Lye* 1308!
DISTR. U 2; Congo (Kinshasa); endemic to Ruwenzori
HAB. Open stony and grassy areas, often near stream sides or glaciers; 3600–4500 m
USES. None recorded on specimens from our area
CONSERVATION NOTES. Restricted in area but habitat not threatened except through fire; near threatened (NT)

63. **Senecio × pirottae** *Chiov.* in Ann. Bot. Roma 6: 149 (1907), *pro spec.*; C. Jeffrey in K.B. 41: 902 (1986); Lisowski, Aster. Fl. Afr. Centr. 2: 311 (1991). Type: Congo (Kinshasa), Ruwenzori, *Roccati* s.n. (TO, holo.)

Perennial herb 10–30 cm high. Basal leaves sessile, oblanceolate in outline, 5.7–11 cm long, 1.2–2.5 cm wide, tapered into a petioloid base, pinnately lobed with rounded denticulate lobes, apex obtuse; stem leaves sessile, oblong, 1.7–7.5 cm long, 0.4–1.7 cm wide, base auriculate and semi-amplexicaul; all leaves glandular above, glandular and thinly arachnoid beneath. Capitula radiate or subradiate with enlarged irregularly lobed outer disc florets, 6–30 in lax to congested terminal corymbs; stalks of the individual capitula finely glandular-hairy; involucre 9–10 mm long, 5–6 mm in diameter; bracts of calyculus 2–6, lax, lanceolate, 1–7 mm long, acute; phyllaries 12–23, 7–10 mm long, glandular. Ray florets or enlarged outer disc florets 8–14, very variable in colour, from white and pink to pale red and bright yellow, tube 4–7 mm long, glabrous, rays 4.5–10 mm long, 2.5–3.5 mm wide, 5–8-veined; disc florets yellow to dark red, corolla 6.8–7.5 mm long, lobes 1 mm long. Achenes 5 mm long, glabrous; pappus 7–7.5 mm long.

UGANDA. Toro District: Ruwenzori, Lake Kitandara, Apr. 1948, *Hedberg* 733! & Apr. 1948, *J. Adamson* 37–41! & Feb. 1954, *Davis* 97!
DISTR. U 2; Congo (Kinshasa); endemic to Ruwenzori
HAB. Open stony areas; (3000–)3800–4200 m
USES. None recorded on specimens from our area
CONSERVATION NOTES. Narrow endemic, but hybrids do not deserve conservation ratings

SYN. *S. pirottae* Chiov. var. *infundibuliformis* Chiov. in Ann. Bot. Roma 6: 149 (1907). Type: Congo (Kinshasa), Ruwenzori, *Roccati* s.n. (TO, holo.)
    S. humphreysii R.D.Good in J. B. 66: 40 (1928). Type: Congo (Kinshasa), Ruwenzori, *Humphreys* 547 & 548 (BM!, syn.)
    S. mattirolii × *S. transmarinus* of A.V.P.: 235, 363 (1957)

NOTE. A hybrid between *S. mattirolii* and *S. transmarinus*. *Davis* 97 is a series of specimens showing the range of ray floret number and colour.

64. **Senecio transmarinus** *S.Moore* in J.L.S. 35: 356 (1902); A.V.P.: 243, 362 (1957); Maquet in Fl. Rwanda 3: 664 (1985); C. Jeffrey in K.B. 41: 902 (1986); Lisowski, Aster. Fl. Afr. Centr. 2: 312 (1991). Type: Uganda, Ruwenzori, *Scott Elliot* 7730 (BM!, holo., K!, iso.)

Perennial herb, sometimes ± straggling with erect stems 30–240 cm high, glabrous, sparsely glandular or sparsely setose. Basal leaves sessile, oblanceolate, 3–7.6 (including 1.5–2.5 cm petioloid base) cm long, 0.7–2 cm wide, narrowed into a slender, basally expanded petioloid base, deeply pinnately lobed with ovate strongly denticulate-lobulate apiculate lobes; stem leaves sessile, oblong-lanceolate to elliptic or oblanceolate, 2–24(–40) cm long, 0.7–6.5(–9) cm wide, ± auriculate at the base, margins sinuate-denticulate, crenulate to remote-dentate or pinnately lobed with ovate lobes, apex obtuse to acute; all leaves glabrous or almost so to ± glandular-pubescent above and especially on veins beneath, sometimes reddish-purple beneath. Capitula radiate, erect, 1–many in lax to sometimes dense terminal corymbs; stalks of the individual capitula glabrous to thinly glandular or finely pubescent; involucre 7–13 mm long, 3.5–6.5 mm in diameter; bracts of calyculus 2–8, narrow, lanceolate, 2–10 mm long, setaceous, glabrous, sparsely glandular, or sparsely hairy on margins; phyllaries 11–14(–21), 6–12 mm long, glabrous or sparsely glandular or thinly pubescent, green, sometimes purple-striped. Ray florets 6–12, yellow, tube 3–6 mm long, glabrous, rays 10–16 × 2–5 mm, 5-veined; disc florets yellow, corolla 6.5–9 mm long, tube expanded above the middle, glabrous, lobes 0.7–1 mm long. Achenes 4–5 mm long, glabrous; pappus 5–8 mm long (achenes and pappus unknown in vars. *major* and *virungae*).

1. Stem leaves pinnately lobed or deeply dentate 2
   Stem leaves crenulate to denticulate 3
2. Stem leaves dentate to double-dentate . . . . a. var. **transmarinus**
   Stem leaves pinnatipartite with dentate lobes . . . . b. var. **sycephyllus**
3. Involucre 13 mm long; ray florets ± 12 . . . . c. var. **major**
   Involucre 7–10 mm long; ray florets 8–9 . . . . d. var. **virungae**

a. var. **transmarinus**; A.V.P.: 362 (1957); C. Jeffrey in K.B. 41: 902 (1986); Lisowski, Aster. Fl. Afr. Centr. 2: 312 (1991)

Stem leaves ± auriculate, dentate to double-dentate, to 24 cm long and 5 cm wide; involucre 7–10 mm long, 3.5–5.5 mm in diameter; calyculus bracts 2–7.5 mm long; phyllaries 6–9 mm long. Ray florets 8–9.

UGANDA. Toro District: Ruwenzori, Kivata, May 1893–94, *Scott Elliot* 7730! & ?Yerua, *Scott Elliot* 7877! & Ruwenzori, Aug. 1938, *Purseglove* 223!
DISTR. U 2; Congo (Kinshasa), Rwanda; endemic to Ruwenzori and Virunga
HAB. Forest; 2400–2700 m
USES. None recorded on specimens from our area
CONSERVATION NOTES. Three specimens known from Uganda; data deficient (DD)

SYN. *S. confertoides* De Wild., Pl. Bequaert. 5: 101 (1929). Type: Congo (Kinshasa), Ruwenzori, Lanuri, *Bequaert* 4672 (BR!, syn.)
   S. lanuriensis De Wild., Pl. Bequaert. 5: 107 (1929). Type: Congo (Kinshasa), Ruwenzori, Lanuri, *Bequaert* 4545 (BR!, holo.)

b. var. **sycephyllus** (*S.Moore*) Hedb., A.V.P.: 244, 362 (1957); C. Jeffrey in K.B. 41: 902 (1986), as *sycephalus*; Lisowski, Aster. Fl. Afr. Centr. 2: 314, t. 68 (1991). Type: Congo (Kinshasa), Ruwenzori, *Scott Elliot* 7965 (BM!, holo., K!, iso.)

Basal leaves sessile; stem leaves ± auriculate, pinnately lobed with ovate, rounded, ± apiculate-lobulate lobes, to 8 cm long and 6.5 cm wide; involucre 6.5–10 mm long, 3.5–5.5 mm in diameter; calyculus bracts 2–7.5 mm long; phyllaries 6–10 mm long. Ray florets 8–9.

UGANDA. Toro District: Ruwenzori, Bujuku Valley near Bigo Camp, Mar. 1948, *Hedberg* 357! & Freshfield Pass, July 1951, *Osmaston* 3888! & near Nyamileju Hut, Dec. 1968, *Lye* 1248!
DISTR. **U** 2; Congo (Kinshasa); endemic to Ruwenzori
HAB. Moorlands; 2850–4200 m
USES. None recorded on specimens from our area
CONSERVATION NOTES. Seventeen specimens from Uganda; data deficient (DD)

SYN. *S. sycephyllus* S.Moore in J.L.S. 37: 324 (1906)
    S. coreopsoides Chiov. in Ann. Bot. Roma 6: 149 (1907). Type: Congo (Kinshasa), Ruwenzori, *Roccati* s.n. (TO, holo.)
    S. cortesianus Muschl. in Z.A.E.: 399 (1911). Type: Congo (Kinshasa), Ruwenzori, Butagu Valley, *Mildbraed* 2552 (B†, holo., BR!, iso.) – note – BR label has the number 2582

NOTE. Leaf margins vary considerably in the material seen and *Osmaston* 1252 from 2400 m has the stem leaves conforming to the description of var. *transmarinus* but lower leaves are pinnatilobed as in var. *sycephyllus*. *Magogo* 31 from 2600 m has most leaves dentate but some seem to be almost pinnatilobed.

c. var. **major** *C.Jeffrey* in K.B. 41: 902 (1986). Type: Uganda, Ruwenzori, Buturungu Valley, *Hedberg* 682 (K!, holo.)

Leaves to 8 cm long and 5 cm wide, leaf margins sinuate-denticulate; involucre 13 mm long, 6.5 mm in diameter; calyculus bracts to 10 mm long; phyllaries 6–12 mm long. Ray florets ± 12.

UGANDA. Toro District: Ruwenzori, Buturungu Valley, *Hedberg* 682!
DISTR. **U** 2; known from only the type
HAB. *Dendrosenecio* 'forest'; 4100 m
USES. None recorded on specimens from our area
CONSERVATION NOTES. Known from only a single specimen; at least vulnerable (VU-D2)

SYN. *S. dewildemanianus* Muschl. in B.S.B.B. 49: 229 (1913). Type: Congo (Kinshasa), Ruwenzori, *Kassner* 3123 (B†, holo., BM!, K!, P, iso.)

d. var. **virungae** *C.Jeffrey* in K.B. 41: 903 (1986). Type: Uganda, Kigezi District, Muchoza valley, *Kutland* 181 (K!, holo.)

Leaves to 8 cm long and 5 cm wide, leaf margins crenulate to remote-dentate; involucre 7–10 mm long, 3.5–5.5 mm in diameter; calyculus bracts 2–7.5 mm long; phyllaries 6–9 mm long. Ray florets 8–9.

UGANDA. Kigezi District: summit of mountain road, June 1938, *Tothill* 2730! & Muchoza valley S of Kabale, May 1953, *Kutland* 181! & Muko, above Lake Bunyonyi, Jan. 1962, *Morison* 222!
DISTR. **U** 2; Congo (Kinshasa); endemic to Virunga
HAB. Bamboo zone and/or floating swamp; 2250–2400 m
USES. None recorded on specimens from our area
CONSERVATION NOTES. Data deficient (DD)

SYN. *S. gwinnerianus* Muschl. in Z.A.E. 2: 397 (1911). Type: Congo (Kinshasa), Karisimbi, *Mildbraed* 1606 (BR!, fragm.) & 1616 (B†, syn.) put into synonymy of *S. sabinjoensis* by Jeffrey, but Lisowski believes they are *S. transmarinus* var. *virungae* (fide Lisowski 1991), and we agree.

65. **Senecio transmarinus × hochstetteri**; C. Jeffrey in K.B. 41: 903 (1986)

Perennial herb 60 cm high; stems finely pubescent. Leaves sessile, oblong-lanceolate in outline, bidentate-lobulate, 8–10.5 cm long, 1.2–3.2 cm wide, apex acuminate-apiculate. Capitula subradiate with irregularly enlarged marginal disc florets; stalks of the individual capitula glandular; involucre 8.5 mm long, 5 mm in diameter. Ray florets 0; disc florets yellow, corolla 8 mm long, expanded above the middle, lobes 1.2 mm long. Mature achenes unknown, but ovary 1.7 mm long, hairy; pappus 8 mm long.

UGANDA. Kigezi District: Luhiza, Kanaba [Kinaba], Mar. 1947, *Purseglove* 2351! & Virunga Mts,
   Kanaba Gap, Nov. 1934, *G. Taylor* 1845!
DISTR. **U** 2; known from only the cited specimens
HAB. Hill-side 'scrub'; 2250–2400 m
USES. None recorded on specimens from our area
CONSERVATION NOTES. Known from two specimens; at least vulnerable (VU-D2)

66. **Senecio hochstetteri** *A.Rich.*, Tent. Fl. Abyss. 1: 435 (1848); Oliv. & Hiern,
F.T.A. 3: 414 (1877); Maquet in Fl. Rwanda 3: 664 (1985); C. Jeffrey in K.B. 41: 903
(1986); Lisowski, Aster. Fl. Afr. Centr. 2: 316, t. 69 (1991); U.K.W.F. ed. 2: 222, t. 93
(1994). Type: Ethiopia, Mt Kubbi [Koubi], *Schimper* 268 (BR!, K!, M!, iso.)

Perennial herb 26–120 cm high, rhizomatous, rosette-forming, subscapigerous,
viscid; stems 1–several from rosette, glandular, purplish near base. Basal leaves
sessile, narrowly to broadly oblanceolate to narrowly elliptic in outline, 8–38 cm long,
1–10 cm wide, narrowed into a petioloid base, margins pinnately-lyrately sinuate-
dentate or bidentate to lobulate with denticulate lobes or rarely merely sinuate-
denticulate, apex obtuse to acute, apiculate, glandular-pubescent or densely so; stem
leaves sessile, oblong, ovate-oblong, elliptic or oblanceolate, 5–13 cm long, 0.4–3.7 cm
wide, subauriculate, pinnately lobed or dentate to (especially the uppermost) entire.
Capitula discoid, 3–many in lax to congested usually long-stalked terminal corymbs;
stalk of individual heads glandular; involucre 4.5–8 mm long, 2.5–5 mm in diameter;
bracts of calyculus 2–5, lax, narrow, lanceolate, 1.5–5 mm long; phyllaries 8–14,
4–8 mm long, glandular, green with purple to black tips. Ray florets 0; disc florets
white, cream or pale yellow, corolla 5–10 mm long, campanulate in upper half,
glabrous, lobes 0.7–1.6 mm long. Achenes 2–3.5 mm long, ribbed shortly hairy or
sparsely so in the grooves; pappus 5–8 mm long, white.

UGANDA. Karamoja District: Mt Morongole, July 1965, *J. Wilson* 1651!; Kigezi District: Behungi,
   Dec. 1933, *A.S. Thomas* 1079! & N slope of Mgahinga [Gahinga]–Muhavura saddle, Apr. 1970,
   Lye & Katende 5321!
KENYA. Northern Frontier District: Ndoto Mts, Sirwan, Jan. 1959, *Newbould* 3355!; Nakuru
   District: NE of Menengai Crater, Aug. 1967, *Mwangangi* 197!; S Nyeri District: Nanyuki Forest
   Station, Aug. 1932, *Napier* 2179!
TANZANIA. Lushoto District: Chambogo Forest reserve, July 1987, *Kisena* 521!; Morogoro
   District: Uluguru Mts, Morningside to Bondwa, Dec. 1970, *Harris & Wingfield* 2216!; Mbeya
   District: Mbeya Range, N face, Mar. 1960, *Kerfoot* 1785!
DISTR. **U** 1, 2; **K** 1, 3–6; **T** 1–3, 6, 7; W Africa from Sierra Leone to Cameroon, Congo
   (Kinshasa), Rwanda, Burundi, Sudan, Ethiopia, Malawi, Zimbabwe, South Africa
HAB. Grassland, occasionally in forest margins or wooded grassland; 900–2800(–3350) m
USES. None recorded on specimens from our area
CONSERVATION NOTES. Least concern (LC)

SYN. *S. chlorocephalus* Muschl. in Z.A.E. 2: 393 (1911). Type: Rwanda, SE of Karisimbi, *Mildbraed*
      1637 (B†, holo., BR!, iso.)
   S. polygonoides Muschl. in Z.A.E. 2: 398 (1911). Type: Congo (Kinshasa), E of Karisimbi,
      Mildbraed 1571 (BR!, iso.)
   S. rusisiensis R.E.Fr. in Wiss. Ergebn. Schwed. Rhod.–Kongo Exped. 1: 344 (1911). Type:
      Congo (Kinshasa), S of Lake Kivu along Rusisi R., *Fries* 1550 (UPS, holo.)
   S. tshitirungensis De Wild., Pl. Bequaert. 5: 125 (1929). Type: Congo (Kinshasa),
      Tshitirunge, *Bequaert* 5999 (BR!, holo.)
   S. longoensis De Wild., Pl. Bequaert. 5: 122 (1929). Type: Congo (Kinshasa), Tongo-Mukule,
      Bequaert 5864 (BR!, holo.)
   S. lugardae Bullock in K.B. 1932: 499 (1932). Type: Kenya, Elgon, *Lugard & Lugard* 541 (K!,
      holo., EA!, iso.)

67. **Senecio greenwayi** *C.Jeffrey* in K.B. 41: 903 (1986); Lisowski, Aster. Fl. Afr.
Centr. 2: 319 (1991). Type: Tanzania, Mbeya, *Greenway* 6182 (K!, holo.)

Perennial herb 28–120 cm high, viscid and somewhat foetid; stems glandular. Basal leaves sessile, oblanceolate to oblanceolate-elliptic, 7–30 cm long, 0.9–5 cm wide, attenuate into a petioloid base, margins pinnately sinuate-dentate or lobate with broadly ovate, apiculate-denticulate lobes; stem leaves few, sessile, ovate-lanceolate or oblong to oblanceolate, 3–7 cm long, 0.4–0.9(–2) cm wide, semi-amplexicaul; all leaves crispate-glandular. Capitula discoid, 4–many in lax to congested terminal corymbs; stalks of the individual capitula glandular; involucre 8.5–9 mm long, 4.5–5 mm in diameter; bracts of calyculus 1–4, lax, narrow, 0.5–3 mm long, glandular; phyllaries 11–15, 8–9 mm long, glandular. Ray florets 0; disc florets white to yellow, corolla 6–10 mm long, glabrous, lobes 1–1.2 mm long. Achenes 3 mm long, ribbed, pubescent; pappus 6–8 mm long.

TANZANIA. Ufipa District: foot of Mt Mbaa opposite Tatanda Mission, Nov. 1986, *Brummitt et al.* 18026! & Mbizi Forest, Nov. 1958, *Napper* 1028!; Mbeya District: Ipinda, N Usafwa Forest Reserve, Oct. 1959, *Procter* 1432!
DISTR. T 4, 7; Congo (Kinshasa), Zambia, Malawi
HAB. Grassland, especially in after-burn flora; 1650–2600 m
USES. None recorded on specimens from our area
CONSERVATION NOTES. Rather uncommon in its distribution area; eleven Tanzanian collections; data deficient (DD)

68. **Senecio erubescens** *Ait.*, Hort. Kew 3: 190 (1789); Hilliard, Comp. Natal: 421 (1977); C. Jeffrey in K.B. 41: 903 (1986); Lisowski, Aster. Fl. Afr. Centr. 2: 318 (1991). Type: South Africa, Cape of Good Hope, cult. Kew, *Masson* s.n. (BM!, holo.)

Erect or decumbent perennial herb 30–60 cm high, rosette-forming, subscapigerous; stems reddish-brown, glandular or sparsely so. Basal leaves sessile, oblanceolate, broadly oblanceolate or oblanceolate-elliptic in outline, 7–21 cm long, 0.9–7 cm wide, attenuate into a petioloid base, margins often reddish, sinuate-serrate to sinuate-dentate, bidentate-lobate or pinnately lobed with rounded, denticulate lobes, apex rounded to obtuse, apiculate, glandular above and on margins and especially or at least on mid-rib beneath; stem leaves few, sessile, oblong or oblanceolate-oblong, 4–5 × 0.6–1 cm, semi-amplexicaul, sinuate-denticulate. Capitula discoid, 3–many in a lax terminal corymb; inflorescence axes glandular; involucre cylindrical, 8–12 mm long, 4–6 mm in diameter; bracts of calyculus 2–4, narrow, lanceolate, 1.5–4 mm long, glandular; phyllaries 12–21, dark green with purplish tips, 7.5–10.5 mm long, glandular. Ray florets 0; disc florets purple, mauve or rarely yellow, corolla 5.5–8.8 mm long, tube campanulate in upper third, glabrous, lobes 1–1.2 mm long. Achenes 3–3.7 mm long, ribbed, shortly hairy in the grooves; pappus 6–8.5 mm long.

TANZANIA. Ufipa District: Malonje Plateau, Mar. 1957, *Richards* 8430!; Mbeya District: Elton Plateau, Jan. 1961, *Richards* 14170!; Songea District: 12 km E of Songea by Nonganonga stream, Dec. 1955, *Milne-Redhead & Taylor* 7765!
DISTR. T 4, 7, 8; Congo (Kinshasa), Angola, Zambia, Malawi, Zimbabwe, Botswana, Swaziland, South Africa
HAB. Grassland, often in boggy sites; (990–)1600–2400 m
USES. None recorded on specimens from our area
CONSERVATION NOTES. Least concern (LC)

SYN. *S. ianthinus* Mattf. in E.J. 59, Beibl. 133: 39 (1924). Type: Tanzania, Rungwe District: Kyimbila, *Stolz* 2418 (K!, iso.)

69. **Senecio cyaneus** *O.Hoffm.* in E.J. 20: 235 (1895); C. Jeffrey in K.B. 41: 904 (1986); U.K.W.F. ed. 2: 222 (1994). Type: Tanzania, Kilimanjaro, *Volkens* 1162 (K!, lecto., chosen by Jeffrey)

Perennial ± erect weak-stemmed herb or straggling woody herb 30–180 cm high; stems glandular-pubescent or sparingly so, or sometimes sparsely arachnoid, brownish, leafy. Leaves green and often glossy, sessile, oblanceolate to elliptic or broadly elliptic, broadly lanceolate or ovate in outline, 5.5–26.5 cm long, 0.9–7.5 cm wide, base ± auriculate and semi-amplexicaul, margins denticulate or more usually bidentate or lobate, shortly to deeply pinnately or pinnato-lyrately lobed, or ± laciniate, especially towards the base, the lobes ovate or oblong, denticulate, with rounded and apiculate apices, and almost glabrous to sparsely glandular-hairy or thinly arachnoid above, sometimes purple-tinged and sparsely glandular-hairy especially on margins, main veins or at least the midrib, or pilose to arachnoid beneath. Capitula discoid, numerous in copious upper axillary and terminal lax to dense corymbs; stalks of the individual capitula thinly to densely glandular-hairy; involucre 7–10 mm long, 3–4.5 mm in diameter, greenish-brown or purple; bracts of calyculus 3–6, lax, narrow, lanceolate or narrowly so, 2–7 mm long, somewhat glandular-hairy at least on the margins; phyllaries 10–13, densely crispate-glandular to glabrous, 6.5–9.5 mm long. Ray florets 0; disc florets purple, mauve or rarely blue, corolla 5.5–9 mm long, tube campanulate or narrowly so in upper half or third, glabrous, lobes 1–1.3 mm long. Achenes 3.2–5 mm long, ribbed, sparsely shortly hairy in the grooves; pappus 6–9.5 mm long.

TANZANIA. Arusha District: E Mt Meru, Jan. 1971, *Mabberley* 602!; Kilimanjaro, S slope between Umbwe and Weru-Weru Rs., Aug. 1932, *Greenway* 3146!; Morogoro District: Uluguru Mts, Lukwangule Plateau, May 1988, *Pocs & Minja* 88/070/J!
DISTR. **T** 2, 6; endemic to Kilimanjaro, Mt Meru, Hanang and the Lukwangule Plateau
HAB. Upland grassland and heath zone, less often in forest margin; 2300–3350 m
USES. None recorded on specimens from our area
CONSERVATION NOTES. Locally common in its restricted distribution area; least concern (LC)
NOTE. *Luke et al.* 8575 from **T** 7, Udzungwa Mts, is close.

70. **Senecio purpureus** *L.*, Syst. Ed. 10: 1214 (1759); Hilliard, Comp. Natal: 419 (1977); C. Jeffrey in K.B. 41: 904 (1986); Lisowski, Aster. Fl. Afr. Centr. 2: 320 (1991). Type: *Senecio viscosus aethiopicus, flore purpureo* Breyne, Exotic Pl. Cent. Prima t. 67 (1678), lecto.!, chosen by Hilliard

Erect perennial herb or woody herb with numerous stems from woody base, 60–240 cm high; stems pale green, glandular-hairy. Basal leaves sessile, oblong-elliptic or lyrate, 35–73 cm long, 9.7–12.5 cm wide, tapered into a petioloid base, margins pinnately lobed with large, oblanceolate, bidentate-lobulate terminal lobe and smaller ovate or broadly ovate dentate lateral lobes with rounded and apiculate apex; median and upper leaves sessile, lanceolate to broadly elliptic in outline, 5.5–17 cm long, 2.2–7 cm wide, auriculate at the base, pinnately lobed especially in lower part with ovate, denticulate lobes, obtuse to acute, acuminate-apiculate; all leaves sparsely shortly glandular-hairy to thinly arachnoid above, ± pubescent to arachnoid-tomentose beneath. Capitula discoid, numerous in copious congested terminal and upper axillary corymbs; stalks of the individual capitula densely glandular; involucre 7–9 mm long, 3–3.5 mm in diameter; bracts of calyculus 4–9, lax, narrow, lanceolate, 2–4.5 mm long; phyllaries 10–13, green with purplish tips to entirely purple, 7–8.5 mm long, shortly glandular-hairy. Ray florets absent; disc florets purple or mauve, rarely blue or white, corolla 5–7.2 mm long, glabrous, lobes 1–1.2 mm long. Achenes (immature) 3.5–5 mm long, glabrous; pappus 5–7 mm long.

TANZANIA. Ufipa District: near Mmemya Mt, Feb. 1951, *Bullock* 3668!; Rungwe District: Kiwira, Rungwe Forest Reserve, Apr. 1983, *Magogo* 2413!; Songea District: Matengo Hills, Jan. 1956, *Milne-Redhead & Taylor* 8233!
DISTR. **T** 4, 7, 8; Congo (Kinshasa), Angola, Zambia, Malawi, Zimbabwe, South Africa
HAB. Marshy ground by streams or in swamp edges, in ruderal sites such as road-sides and cultivation; (1200–)1500–2700 m
USES. None recorded on specimens from our area
CONSERVATION NOTES. Least concern (LC)

Syn. *S. bussei* Muschl. in E.J. 43: 67 (1909). Type: Tanzania, Songea District, Matengo, *Busse* 1315 (B†, holo., EA!, iso.)

     S. lubumbashiensis De Wild., Pl. Bequaert. 5: 111 (1929). Type: Congo (Kinshasa), Lubumbashi R. source, *De Giorgi* 299 (BR!, holo.)

71. **Senecio madagascariensis** *Poir.*, Encycl. Suppl. 5: 130 (1804); U.K.W.F. ed. 2: 222 (1994). Type: Madagascar, *Commerson* s.n. in herb. Desfontaines (P, holo.)

Resembles *S. mesogrammoides*, but the leaf midrib is scabrid, while the leaves are glabrous otherwise; and the number of phyllaries is higher (± 20–29).

Kenya. Naivasha District: farm near Gilgil, Oct. 1986, *Taylor* s.n.! and idem, Kwetu Farm, Mar. 1987, *Armstrong* s.n.!
Distr. **K** 3; southern Africa, Madagascar
Hab. Weed in farmland; ± 2400 m
Uses. None recorded on specimens from our area; toxic to livestock
Conservation notes. Least concern (LC)

Note. Resembles *S. mesogrammoides*; but while that species is grazed, *S. madagascariensis* is toxic to livestock.

72. **Senecio vulgaris** *L.*, Sp. Pl.: 867 (1753); Oliv. & Hiern, F.T.A. 3: 411 (1877); C. Jeffrey in K.B. 41: 904 (1986); U.K.W.F. ed. 2: 222 (1994). Type: *Senecio foliis pinnatifidis denticulatis, laciniis aequalibus patentissimis rachi lineari* L., Hort. Cliff.: 406 (1737), lecto. chosen by Jeffrey & Chen 1984, BM-HC!

Annual herb 14–37 cm high. Stems at first white-pubescent, glabrescent. Basal leaves sessile, oblanceolate or oblanceolate-elliptic in outline, tapered into an expanded exauriculate petioloid base, pinnately toothed or lobed, rounded, soon disappearing; stem leaves sessile, oblong to broadly oblong in outline, auriculate, semi-amplexicaul, shortly to deeply pinnately lobulate with oblong to oblong-linear, apiculate-dentate lobes, 2.3–6 cm long, 0.15–0.6 (–1–1.7 cm across the lobes) cm wide; all leaves with scattered long white hairs above on midrib and beneath especially on midrib. Capitula discoid, in congested terminal corymbs; stalk of individual heads with long white pubescence, glabrescent; involucre cylindrical, 7–8 mm long, 2.5 mm in diameter; bracts of calyculus ± 8, imbricate, lanceolate, glabrous, up to 2 mm long; phyllaries ± 19, glabrous, 6.5–7.5 mm long. Ray florets 0; disc florets yellow, corolla 5 mm long, tube expanded in the upper $^{1}/_{3}$, glabrous, narrowly tubular below, lobes 0.5 mm long. Achenes 2.5 mm long, ribbed, hairy in the grooves; pappus 5–5.5 mm long.

Kenya. Nakuru District: Molo, Nov. 1954, *Lacey* in EA 10585!; Kericho District: SW Mau, Farm 1645, 1956, *Whitall* 163! & Timbilil, Aug. 1961, *Kerfoot* 2875!
Tanzania. Pare District: Kilomeni, June 1915, *Peter* 53951!
Distr. **K** 3, 5; **T** 3; originally from temperate Eurasia and N Africa, now widely naturalized
Hab. Very little information, but probably in ruderal sites and a weed of cultivation; 1800–3000 m
Uses. None recorded on specimens from our area
Conservation notes. Least concern (LC)

Excluded species

Verdcourt & Trump, Common Poison. Pl. E.Afr.: 155 (1969) mention *Haarer* 1596 from **T** 7 as *S. latifolius* DC., a species not found in our area; it occurs further to the South. We have not been able to trace the specimen.

98. **SOLANECIO**

(Sch. Bip.) Walp., Rep. 6: 273 (1846)

Senecio L. subgen. *Solanecio* Sch. Bip. in Flora 25: 441 (1842)
Senecio L. sect. *Crassuli* Muschl. in E.J. 43: 38 (1909)
Senecio L. sect. *Tuberosi* Muschl. in E.J. 43: 38 (1909)

Perennial herbs, often scandent, rarely shrubs or small trees. Leaves alternate, entire to pinnatilobed, often somewhat succulent. Capitula discoid, in paniculate-corymbose inflorescences. Involucre calyculate; receptacle epaleate. Florets yellow; anthers obtuse to sagittate at base, appendaged at apex; style branches truncate, sometimes with a tuft of fused papillae. Achenes ribbed; pappus of many fine bristles.

About 16 species in tropical Africa, Yemen and Madagascar.

1. Scrambling or prostrate herbs (the ones that are sometimes
   erect have leaves that are hairy on the lower surface) 2
   Shrubs, trees, erect herbs or epiphytes 5
2. Young stems and leaves hairy 3
   Young stems and leaves glabrous or nearly so 4
3. Leaves coarsely toothed and lobed, thinly tomentose beneath   8. *S. nandensis*
   Leaves only slightly denticulate or serrate, densely tomentose
   beneath . . . .   9. *S. cydoniifolius*
4. Leaves triangular to sagittate; achenes glabrous . . . .   10. *S. biafrae*
   Leaves lyrately lobed or compound; achenes short-hairy in
   the grooves . . . .   11. *S. angulatus*
5. Perennial herbs from root tubers 6
   Shrubs, trees or epiphytic herbs without tubers 8
6. Ovary and achenes glabrous . . . .   1. *S. gymnocarpus*
   Ovary and achenes hairy 7
7. Phyllaries 8–13 . . . .   2. *S. goetzei*
   Phyllaries 5 . . . .   3. *S. tuberosus*
8. Epiphytes 9
   Terrestrial shrubs or trees 11
9. Capitula few, in small corymbs; **U** 2 . . . .   6. *S. kanzibiensis*
   Capitula many, in groups of corymbs 10
10. Leaves spread along the stem; involucre 6–7 mm long;
    T 2, 3 . . . .   4. *S. mirabilis*
    Leaves crowded towards the apex of the stem; involucre
    7–10 mm long; **T** 6, 7 . . . .   5. *S. epidendricus*
11. Shrub or tree, 1–8 m high; involucre 4–7.5 mm long;
    achenes hairy in the grooves; widespread . . . .   12. *S. mannii*
    Shrubs to 3.3 m; involucre 8–11.5 mm long; achenes
    glabrous 12
12. Leaf margins entire; capitula in compound corymbose
    inflorescences; pappus 5–7 mm long . . . .   7. *S. buchwaldii*
    Leaf margins bidentate; capitula in corymbs; pappus 10 mm
    long . . . .   13. *S. gynuroides*

1. **Solanecio gymnocarpus** *C.Jeffrey* in K.B. 41: 920 (1986). Type: Tanzania, Mwanza District, Igalukiro, *Tanner* 1599 (K!, holo.)

Perennial herb 0.6–1.2 m high, glabrous, with erect stems arising from a fleshy tuberous rootstock to 2.5 cm in diameter; upper part of stem yellow. Leaves sessile, subsucculent; the basal narrowly oblanceolate to obovate, 16–30 cm long, 1–3 cm wide, attenuate into a petioloid base, remotely dentate; median and upper leaves lanceolate or narrowly oblong-lanceolate, 2–15 cm long, 0.5–1 cm wide, expanded at the base, entire or usually narrowly laciniate-serrate. Capitula discoid, 5–12 clustered in small terminal subumbelliform corymbs; involucre cylindrical, 8.5–9.5 mm long, 3 mm in diameter; bracts of calyculus 3–4, lax, lanceolate, 2–3 mm long; phyllaries 8, 8–9 mm long. Ray florets absent; disc florets yellow, corolla 5.5–6.5 mm long, tube glabrous, expanded from about or below the middle, lobes 0.7–1 mm long. Mature achenes ± 5 mm long, glabrous; pappus 4.5–7 mm long.

TANZANIA. Mwanza District: Mwana, ca. 1926, *R.L. Davis* 231!; Tabora District: Malongwe–Nyahua, Jan. 1926, *Peter* 34548!; Ufipa District: km 7 on Namanyere–Karonga road, Mar. 1994, *Bidgood et al.* 2626!
DISTR. **T** 1, 4; not known elsewhere
HAB. Swampy grassland in seasonally inundated sites or in hardpan depressions; 1100–1500 m
USES. None recorded on specimens from our area
CONSERVATION NOTES. Four Tanzanian specimens seen; at least vulnerable (VU-D2).

NOTE. A specimen from Tanzania, Ufipa District: km 8 on Mkasama road from Namanyere–Chala road, May 1997, *Bidgood et al.* 3768! is close to *S. gymnocarpus*, but has the basal leaves obovate, to 5.5 cm wide, the margins dentate and almost lobed, and above all thinly pubescent with multicellular hairs. The flowers are said to be pinkish purple. More material of this might prove this taxon to be distinct.

2. **Solanecio goetzei** (*O.Hoffm.*) *C.Jeffrey* in K.B. 41: 921 (1986); U.K.W.F. ed. 2: 220 (1994). Type: Tanzania, Iringa District, Iringa, *Goetze* 654 (B†, holo., K!, iso.)

Perennial herb with several erect stems 0.3–2 m high arising from a tuberous rootstock to 10 cm long and 5 cm in diameter; stem glabrous, sometimes reddish near base. Leaves sessile, somewhat succulent, oblanceolate, elliptic or lanceolate, 5.5–22 cm long, 2–7.5 cm wide, the lower tapered into an exauriculate petioloid base, the median and upper tapered and exauriculate to ± cordate, semiamplexicaul auriculate base, margins remotely obscurely to coarsely sinuate-dentate, sinuate-bidentate to sinuate-crenate, apex obtuse to rounded, shortly acuminate-apiculate, glabrous. Capitula discoid, several to many in usually dense, long-stalked terminal compound corymbs; involucre cylindrical, 5.5–9.5 mm long, 2.5–3 mm in diameter; bracts of calyculus 2–4, narrowly lanceolate, 2–4 mm long; phyllaries 8–13, yellow-green or yellow distally, 5–9 mm long. Ray florets 0; disc florets yellow or orange, corolla 5–7 mm long, tube glabrous, expanded above the middle, lobes 0.7–1.2 mm long. Achenes 3.5–4 mm long, ribbed, hairy; pappus 4.5–6 mm long.

KENYA. Machakos District: Athi River, Nov. 1960, *Ossent* 528! & Loitokitok–Emali road, *Kokwaro et al.* 3545; Masai District: Selengai Game Post, Dec. 1969, *Kanure Kibui* 120!
TANZANIA. Masai District: Kakessio, Apr. 1961, *Newbould* 5836! & Ngorongoro Crater wall, Jan. 1989, *Pocs & Chuwa* 89/010/N!; Chunya District: Mbangala, Feb. 1994, *Bidgood et al.* 2298!
DISTR. **K** 4, 6; **T** 1–5, 7, 8; not known elsewhere
HAB. Marshy or seasonally inundated grassland, woodland or wooded grassland; 750–2050 m
USES. None recorded on specimens from our area
CONSERVATION NOTES. In a fairly common habitat; probably least concern (LC)

SYN. *Senecio goetzei* O. Hoffm. in E.J. 28: 507 (1901)

3. **Solanecio tuberosus** (*A.Rich.*) *C.Jeffrey* in K.B. 41: 921 (1986); U.K.W.F. ed. 2: 220 (1994). Type; Ethiopia, Djeladjeranne, *Schimper* 1610 (K!, P!, syn.), *Dillon* s.n. (P, syn.) & Schoata, *Schimper* 1361 (K!, P!, syn.)

Perennial herb with glabrous erect stems 20–60 cm high arising from a tuberous rootstock. Leaves mostly basal, sessile, rather succulent, variable, oblanceolate in outline, 9.5–21 cm long, 0.8–3(–7) cm wide, tapered into a petioloid base, usually with remote obscure to prominent marginal teeth or narrow forward-pointing lobes, apex obtuse, minutely and obtusely acuminate-apiculate, glabrous, shiny green with paler mid-rib and main veins and with mid-rib of petioloid base purple. Capitula discoid, numerous in erect compound terminal corymbs; stalks of individual capitula sparsely glandular-pubescent; involucre cylindrical, 7–10 mm long, 2.5–4 mm in diameter; bracts of calyculus 1–3, lanceolate, acute, almost glabrous, pale green with yellowish tips, 1.5–3 mm long; phyllaries 5, glabrous or almost so, 6.5–7 mm long. Ray florets 0; disc florets yellow, corolla 5–7.5 mm long, glabrous, tube expanded above the middle, lobes 1–1.5 mm long. Achenes 4.5 mm long, ribbed, hairy; pappus 5–5.5 mm long.

Uganda. Karamoja District: Kokumongole, May 1939, *A.S. Thomas* 2886!
Kenya. Elgon, SW slopes, 1934, *Tweedie* 162!
Distr. U 1; **K** 3; Sudan, Ethiopia
Hab. Swampy grassland or thin seasonally marshy soil pockets on rock outcrop; 1200–2000 m
Uses. None recorded on specimens from our area
Conservation notes. Certainly very uncommon in our area; data deficient (DD)

Syn. *Senecio tuberosus* A.Rich., Tent. Fl. Abyss. 1: 434, t. 58 (1848); Oliv. & Hiern, F.T.A. 3: 413 (1877); F.P.S. 3: 49 (1956)
    S. solanoides Aschers. in Schweinf., Beitr. Fl. Aethiop.: 159 (1867). Type: Ethiopia, *Schimper* 1361 pro parte (B†, holo., K!, iso., P!, iso.)

Note. Mesfin has described a *S. tuberosus* (A. Rich.) C.Jeffrey var. *pubescens* Mesfin from Ethiopia, in K.B. 49: 140 (1995).

4. **Solanecio mirabilis** (*Muschl.*) *C.Jeffrey* in K.B. 41: 921 (1986). Type: Tanzania, Lushoto District, Amani, *Engler* 569, 575 (B†, syn.)

Epiphytic succulent herb; stems slender, leafy, glabrous. Leaves succulent, sessile, oblanceolate-spatulate, obovate or elliptic, 1.4–6.2 cm long (including 0.4–1.4 cm petioloid base), 0.6–2.6 cm wide, cuneate or attenuate into an exauriculate petioloid base, margins entire, apex obtuse to rounded, glabrous. Capitula discoid, numerous in terminal thyrses of stalked congested subumbelliform corymbs; stalks of individual capitula slender; involucre ± cylindrical, 6–7 mm long, 2–3 mm in diameter; bracts of calyculus 1–3, lanceolate or oblanceolate, 1–1.5 mm long, ciliate at margins, obtuse; phyllaries 5–7 mm long. Ray florets 0; disc florets yellow, corolla 7 mm long, tube glabrous, expanded in upper third, lobes 2 mm long. Achenes 2.5 mm long, ribbed, glabrous; pappus 4.5–5 mm long.

Kenya. Teita District: Mbololo hill, Mraru ridge, Oct. 1970, *Faden & Githui* 70/712
Tanzania. Arusha District: Mt Meru, E slope, Jan. 1970, *Greenway* in *EA* 14277!; Lushoto District: Amani–Maramba, Oct. 1935, *Greenway* 4150! & W Usambara, Ambangulu Tea Estate, Sep. 1997, *Luke* 4763!
Distr. **K** 7; **T** 2, 3; not known elsewhere
Hab. Moist upland forest; 950–2100 m
Uses. None recorded on specimens from our area
Conservation notes. Known from six specimens only, in a diminishing habitat; at least vulnerable (VU-D2)

Syn. *Senecio mirabilis* Muschl. in E.J. 59, Beibl. 43: 63 (1909)

5. **Solanecio epidendricus** (*Mattf.*) *C.Jeffrey* in K.B. 41: 921 (1986). Type: Tanzania, Rungwe District, Kyimbila, *Stolz* 857 (B†, holo., K!, M!, W!, iso.)

Epiphytic succulent herb; stems short, creeping, glabrous, flowering stems erect. Leaves succulent, sessile, crowded towards the apex of the stem, oblanceolate to narrowly oblanceolate-elliptic, 3.5–10.5 cm long, 1.2–2.5 cm wide, gradually attenuate into a petioloid base, margins entire, apex obtuse, minutely acuminate-apiculate, glabrous. Capitula discoid, numerous in long-stalked terminal compound corymbs; involucre cylindrical but the phyllaries soon spreading, 7–10 mm long;

bracts of calyculus 4–5, lanceolate, pubescent on margins or at apex or glabrous, 1.5–2.5 mm long; phyllaries 8, green, 7–8 mm long. Ray florets 0; disc florets dull yellow, corolla 5–8 mm long, tube glabrous, expanded above the middle, lobes 1.2 mm long. Achenes 2.5 mm long, ribbed, glabrous; pappus 4.5–5.5 mm long.

TANZANIA. Iringa District: Mwanihana Forest Reserve above Sanje, Oct. 1984, *D.W. Thomas* 3840!; Iringa District: Mt Image, Mar. 1962, *Polhill & Paulo* 1643!; Rungwe District: Kyimbila, Aug. 1911, *Stolz* 857!
DISTR. **T** 7; not known elsewhere
HAB. Moist forest, on mossy trees; 1400–2400 m
USES. None recorded on specimens from our area
CONSERVATION NOTES. Only four specimens seen, from a diminishing habitat; at least vulnerable (VU-D2)

SYN. *Senecio epidendricus* Mattf. in E.J. 59, Beibl. 133: 41 (1924)

6. **Solanecio kanzibiensis** (*Humbert & Staner*) *C.Jeffrey* in K.B. 41: 921 (1986); Lisowski, Aster. Fl. Afr. Centr. 2: 412 (1991). Type: Congo (Kinshasa), Kahuzi–Kivu, *Humbert* 7725 (BR!, holo.)

Epiphytic rather succulent herb; stems 20–60 cm long, slender, rooting at lower nodes, leafy, pendulous, glabrous. Leaves remote, petiolate; blade elliptic, 1–4 cm long, 0.5–1.7 cm wide, base cuneate to attenuate, margins entire, apex obtuse, glabrous; petiole slender, exauriculate, 0.5–1.6 cm long. Capitula discoid, in small terminal and upper axillary corymbs; involucre cylindrical to obconic, 5–7.5 mm long; bracts of calyculus 2–3, lanceolate, 1–2 mm long, glabrous; phyllaries ± 8, 5–7.5 mm long. Ray florets 0; disc florets ± 10, yellow or orange, corolla 5–6 mm long; tube glabrous, expanded above the middle, lobes 0.8–1.2 mm long. Mature achenes unknown, ovary 1.5–2 mm long, glabrous; pappus 5–6 mm long.

UGANDA. Kigezi District: Impenetrable Forest, Nyebeya, Oct. 1940, *Eggeling* 4136! & Kabale, Rubanda, July 1989, *Katende* 3784!
DISTR. **U** 2; Bioko, Congo (Kinshasa), Rwanda
HAB. Rain-forest; ± 2250 m
USES. None recorded on specimens from our area
CONSERVATION NOTES. Widespread; least concern (LC)

SYN. *Senecio kanzibiensis* Humbert & Staner in B.J.B.B. 14: 108, t. 5/5–7 (1936); Maquet in Fl. Rwanda 3: 666, fig. 209/2 (1985)

7. **Solanecio buchwaldii** (*O.Hoffm.*) *C.Jeffrey* in K.B. 41: 921 (1986). Type: Tanzania, Lushoto District, Usambara Mts, *Buchwald* 183 (B†, holo., BM!, iso.)

Succulent shrub 1–3.3 m high, ± erect; stem thick, succulent, often red-tinged, glabrous. Leaves sessile, succulent or subsucculent, crowded towards the apex of the stem, dark green to grey-green, often with ± purple-tinged or reddish midrib, obovate to oblanceolate, 10–20 cm long, 3.2–9.2 cm wide, cuneate or attenuate-decurrent onto wide exauriculate petioloid base, margins entire, apex obtuse to rounded, minutely broadly apiculate, glabrous. Capitula discoid, numerous in copious long-stalked terminal cymes of congested subumbelliform corymbs; inflorescence branches often reddish-purple, slender, bracteolate, 9–15 mm long; involucre cylindrical, 8–10 mm long, 3–5 mm in diameter; bracts of calyculus 1–3, lanceolate, 1.5–2 mm long, glabrous, lax; phyllaries 5–6, olive-green or reddish-purple, 7.5–9 mm long. Ray florets 0; disc florets dull brownish or pinkish yellow or dull yellow, corolla 7–8.2 mm long, tube glabrous, slightly expanded in upper part, lobes 1 mm long. Achenes 4.5 mm long, obtusely ribbed, glabrous; pappus 5–7 mm long.

KENYA. Teita District: Ngaongao, Feb. 1966, *Gillett et al.* 17108! & Mbololo, Feb. 1969, *Archer* in EA 14092! & Kasigau, Nov. 1994, *Luke* 4222!

TANZANIA. Lushoto District: Usambara Mts, Matondwe Hill, head of Kwai Valley, Feb. 1953, *Drummond & Hemsley* 1352! & Kwa-Mashai, Feb. 1964, *Semsei* 3643!
DISTR. **K** 7; **T** 3; not known elsewhere
HAB. Rocky outcrop crevices in moist forest zone; 800–1900 m
USES. None recorded on specimens from our area
CONSERVATION NOTES. Restricted to a small area in a specialized habitat; vulnerable (VU-B1ab)

SYN. *Senecio buchwaldii* O.Hoffm. in E.J. 24: 474 (1898)

8. **Solanecio nandensis** (*S.Moore*) *C.Jeffrey* in K.B. 41: 921, fig. 6d (1986); Lisowski, Aster. Fl. Afr. Centr. 2: 413 (1991); U.K.W.F. ed. 2: 220 (1994). Type: Kenya, Nandi, *Scott Elliot* 6987 (BM!, holo., K!, iso.)

Scrambling perennial herb, occasionally erect, 1–10.5 m long, rather succulent. Stems succulent, with purplish markings, usually floccose or finely pubescent, glabrescent. Leaves semi-succulent, aromatic, petiolate; blade sublyrate, oblong to triangular or ovate-triangular in outline, 3–15 cm long, 3–10 cm wide, base weakly cordate to subtruncate and shortly narrowly decurrent onto the petiole, margins very coarsely sinuate-bidentate or sinuate-laciniate, apex obtuse to acute, shortly acuminate-apiculate, thinly arachnoid or minutely puberulous to glabrous except for main veins above, finely densely to thinly shortly white-tomentose beneath; petiole narrowly winged in upper part, 2–5.7 cm long, thinly arachnoid, exauriculate, in the upper leaves often bearing 1–2 small, oblong lateral lobes. Capitula discoid, many in dense terminal compound corymbs; stalks of individual capitula thinly arachnoid, sparsely shortly pubescent or glabrous; involucre cylindrical, slender, 5.5–7 mm long, 2 mm in diameter; bracts of calyculus 2–5, lanceolate, shortly pubescent or at least ciliate, 1.2–3 mm long; phyllaries 8, light green with pale margins and tips, glabrous or almost so, 5–6.5 mm long. Ray florets absent; disc florets dull white, cream or pale yellow, corolla 5–9 mm long, tube glabrous, expanded above the middle, lobes 0.7–1.2 mm long. Achenes 2.3–3 mm long, ribbed, hairy; pappus 5.5–8 mm long.

KENYA. Naivasha District: Lake Naivasha, Crescent Is., Sep. 1976, *Hayes* 55!; Machakos District: Lukenya, Aug. 1953, *Bally* 9053!; Masai District: Nasampolai Valley, Oct. 1969, *Greenway & Kanuri* 13834!
TANZANIA. Kondoa District: Beriku Ridge near Salanga, Jan. 1928, *B.D.Burtt* s.n.!
DISTR. **K** 3–4, 6; **T** 5; Rwanda, Ethiopia
HAB. Forest margins, woodland, bushland, bushed grassland or on bare rock; 1500–2700 m
USES. None recorded on specimens from our area
CONSERVATION NOTES. Fifteen East African collections, the most recent one from 1976; distribution wide, so least concern (LC)

SYN. *Senecio nandensis* S.Moore in J.L.S. 35: 360 (1902); T.T.C.L.: 159 (1949); Maquet in Fl. Rwanda 3: 668 (1985)
    S. adolfi-fridericii Muschl. in Z.A.E. 2: 385 (1911), as *adolfi-friderici*. Type: Rwanda, between Lake Luhondo and Muhavura, *Mildbraed* 1832 (B†, holo., BR!, iso.)
    S. ceranianus Chiov. in Ann. Bot. Roma 10: 386 (1912). Type: Ethiopia, Galla Arussi, *Negri* 812, 1420 (FT!, syn.)

NOTE. Looks quite like *Gynura scandens* but distinct in the flower colour (orange or bright yellow in *G. scandens*) and the shorter pappus (9–11 mm in *G. scandens*). The *Gynura* also has 13 phyllaries and an unpleasant smell.

9. **Solanecio cydoniifolius** (*O.Hoffm.*) *C.Jeffrey* in K.B. 41: 921, fig. 6e (1986); Lisowski, Aster. Fl. Afr. Centr. 2: 414 (1991); U.K.W.F. ed. 2: 220 (1994). Type: Tanzania, Lushoto District, Usambara Mts, *Holst* 2106 (B†, holo., K!, iso.)

Erect to scrambling semi-succulent shrubby herb with foetid smell, 0.6–3.6(–7) m high or long; stems succulent, white-tomentose, ± glabrescent. Leaves petiolate; blade subsucculent, silvery-white or grey-green to light pale green, ovate, lanceolate, ovate-rhombic or oblong, 4.5–25 cm long, 2.3–16 cm wide, base rounded to cuneate and shortly decurrent onto the petiole, margins remotely sinuate-denticulate or sinuate-serrate, apex obtuse to rounded, apiculate, tomentose to thinly floccose above, densely tomentose beneath; petiole densely tomentose, exauriculate, 1.5–8 cm long. Capitula discoid, numerous in large terminal stalked loose to contracted thyrses of stalked congested subumbelliform corymbs; stalks of individual capitula short, arachnoid-tomentose; involucre cylindrical, 6.5–14 mm long, ± 2.5 mm in diameter; bracts of calyculus 2–6, narrowly lanceolate, 1.5–6 mm long, arachnoid at least at the base and on margins; phyllaries 8–13, green, 6–142 mm long, glabrous or thinly hairy and glabrescent. Ray florets absent; disc florets yellow, corolla 6–12 mm long, tube glabrous, gradually expanded from below the middle, lobes 1–1.7 mm long. Achenes 3–3.7 mm long, glabrous or shortly sparsely hairy; pappus 6–12 mm long.

UGANDA. Ankole District: Igara, Mar. 1939, *Purseglove* 541!; Teso District: Serere, Mar. 1932, *Chandler* 528!; Mengo District: 4 km N of Bale, June 1956, *Langdale-Brown* 2117!
KENYA. Nairobi, Karura Forest, Nov. 1976, *Kahurananga & Mungai* 132!; Masai District: Trans Mara, Olosendo area, June 1961, *Glover et al.* 1839!; 'Mombasa', Nov. 1884, *Wakefield* s.n.!
TANZANIA. Mwanza District: Igombe, Sep. 1952, *Tanner* 950!; Tanga District: Sawa, Feb. 1965, *Faulkner* 3457!; Uzaramo District: Ruvu Thicket, July 1969, *Harris* 2957!
DISTR. U 2–4; K 1, 4, 6, 7; T 1–3, 6; Congo (Kinshasa), Rwanda, Burundi, Zambia
HAB. Forest margin, cultivation edges, thicket, coral rocks near the sea; 0–2100 m
USES. Minor medicinal against sores (*Tanner*); bark used like cotton thread (*Jarrett*); possibly poisonous; often planted near houses (*Tanner*)
CONSERVATION NOTES. Widespread and common; least concern (LC)

SYN. *Senecio cydoniifolius* O.Hoffm. in Engl., Glied. Veg. Usambara: 19 (1894); Maquet in Fl. Rwanda 3: 668, fig. 207/2 (1985)
  S. stuhlmannii Klatt in Leopoldina 31: 10 (1895); T.T.C.L.: 159 (1949); Verdcourt & Trump, Common Poison. Pl. E.Afr.: 153 (1969). Type: Tanzania, Pangani, *Stuhlmann* s.n. (B†, holo.)
  S. dewevrei Dur. & De Wild. in B.S.B.B. 39: 35 (1900). Type: Congo (Kinshasa), near Lake Tanganyika, *Dewèvre* 935 (BR!, holo.)
  S. elskensii De Wild., Pl. Bequaert. 5: 103 (1929). Type: Burundi, Kitete, *Elskens* 111 (BR!, lecto., chosen by C. Jeffrey)
  S. mearnsii De Wild., Pl. Bequaert. 5: 114 (1929). Type: Uganda, ?Ankole District: near Kisingo, *Mearns* 2593 (BR!, holo.)
  S. giorgii De Wild., Contib. Fl. Katanga suppl. 3: 154 (1930). Type: Congo (Kinshasa), Kalemie [Albertville], *Giorgi* 13 (BR!, holo.)
  S. simulans Chiov., Racc. Bot. Miss. Consol. Kenya: 70 (1935). Type: Kenya, Meru, *Balbo* 31 (FT!, holo.)

10. **Solanecio biafrae** (*Oliv. & Hiern*) *C.Jeffrey* in K.B. 41: 922 (1986). Type: Cameroon, Mt Cameroon, *Mann* 1325 (K!, holo.)

Scrambling subsucculent herb, 'very high'; stems glabrous. Leaves petiolate; blade triangular-subhastate or narrowly so or saggitate, 2.2–9 cm long, 1.3–6.2 cm wide, base weakly cordate or subtruncate and shortly decurrent onto the petiole, margins remotely sinuate-paucidentate in lower half, apex ± attenuate, acute and apiculate, glabrous; petiole slender, glabrous, exauriculate, 0.8–5.5 cm long. Capitula discoid, numerous in terminal loose thyrses of stalked congested subumbelliform corymbs; stalks of the individual capitula short, shortly sparsely pubescent; involucre cylindrical, 7–10 mm long, 2–4 mm in diameter; bracts of calyculus 2–4, lax, glabrous or with a few marginal hairs, 1.5–3 mm long; phyllaries 8, green, glabrous, 6.5–10 mm long. Ray florets absent; disc florets pale yellow, corolla 6–8.5 mm long, tube glabrous, slightly expanded from above the middle, lobes 1.2–1.5 mm long. Achenes 3 mm long, ribbed, glabrous; pappus 6–8 mm long.

UGANDA. Bunyoro District: Bujenje, Budongo Forest, Mar 1971, *Synnott* 530!; Kigezi District: Virunga, Muhavura N slopes, Nov. 1954, *Stauffer* 974!; Mengo District: Sai, Apr. 1915, *Dummer* 2440!

DISTR. **U** 2, 4; Sierra Leone, Liberia, Ghana, Nigeria, Cameroon, Congo (Kinshasa)
HAB. Moist forest; 1150–2700 m
USES. None recorded on specimens from our area
CONSERVATION NOTES. Widespread; least concern (LC)

SYN. *Senecio biafrae* Oliv. & Hiern, F.T.A. 3: 420 (1877)
    Crassocephalum biafrae (Oliv. & Hiern) S.Moore in J.B. 50: 211 (1912)

11. **Solanecio angulatus** (*Vahl*) *C.Jeffrey* in K.B. 41: 922 (1986); Lisowski, Aster. Fl. Afr. Centr. 2: 415, t. 88 (1991); K.T.S.L.: 562, map. (1994); U.K.W.F. ed. 2: 220, t. 92 (1994). Type: Yemen, Hadie, *Forsskahl* s.n. (C!, syn.)

Scrambling perennial succulent herb, occasionally prostrate, 0.6–3(–10) m long or high; stems rather succulent, pale green, often variegated with pale yellow green and purple, slightly angular, glabrous. Leaves rather succulent, pinnate-lyrately lobed or compound, 3.7–22 cm long, 1–14 cm wide, with 1–4 pairs of spreading or ± recurving ovate to oblong lobes and a larger ± triangular, sometimes basally deeply 2-lobulate, terminal lobe, the very base with conspicuous basal semi-amplexicaul rounded auricles, margins coarsely sinuate-dentate or sinuate-laciniate to sinuate-lobulate, apex acute, often attenuate, glabrous or almost so (rarely slightly puberulous), often glaucous, especially beneath. Capitula discoid (see note), strongly scented, numerous in terminal stalked loose thyrses of stalked, usually congested subumbelliform corymbs, the whole inflorescence often pendent, the distal part sometimes turned up, strongly scented; stalks of capitula glabrous or shortly pubescent; involucre cylindrical, 6.5–10.5 mm long, 2–5 mm in diameter; bracts of calyculus 2–4, lanceolate, 1.2–2 mm long, glabrous or shortly pubescent on margins and apex; phyllaries 5–8, pale green, often with dull purple tinge at apex and base, 6–10 mm long. Ray florets nearly always absent; disc florets dull golden yellow, corolla 5.5–9.5 mm long, tube glabrous, gradually slightly expanded from below the middle, lobes 1.3–2 mm long. Achenes 3–3.5 mm long, ribbed, shortly hairy in the grooves; pappus 5.5–8 mm long. Fig. 137 (page 677).

UGANDA. Karamoja District: Moroto, Dec. 1956, *J. Wilson* 328!; Ankole District: Ruizi R., Nov. 1950, *Jarrett* 128!; Mengo District: Kisinsi Point opposite Kaazi, Feb. 1970, *Lye et al.* 5087!
KENYA. Naivasha District: Crater Lake, Aug. 1992, *Luke* 3255!; Embu District: Thuchi R. 5 km W of Ishiara, Feb. 1995, *Miringu & Beentje* 28!; Kisumu, Nov. 1939, *Opiko* in *Bally* 665!
TANZANIA. Pare District: Mkomazi Game Reserve, E end of Zange Ridge, June 1996, *Abdallah et al.* 96/56!; Buha District: Gombe National Park, Momgono, Apr. 1964, *Pirozynski* 706!; Rungwe District: Mwaksi R. near Tukuyu–Ipinda road, Sep. 1971, *Perdue & Kibuwa* 11623!
DISTR. **U** 1–4; **K** 1–7; **T** 1–8; Cameroon, Equatorial Guinea, Gabon, Congo (Kinshasa), Rwanda, Burundi, Sudan, Ethiopia, Somalia, Angola, Zambia, Malawi, Mozambique, Zimbabwe, South Africa, Comoro Is., Madagascar; Yemen
HAB. Forest, especially in riverine or lakeshore forest, bushland, bushed grassland, thickets; (250–)600–1800(–2500) m
USES. Minor medicinal against fever (*Jeffery*) or colds (*Jarrett*), childrens' sore eyes or throat infection (*Meyerhoff*)
CONSERVATION NOTES. Widespread and common; least concern (LC)

SYN. *Cacalia sonchifolia* sensu Forssk., Fl. Aegypt.-Arab. 119, 485 (1775), *non* L.
    C. angulata Vahl, Symb. Bot. 3: 92 (1794)
    Senecio bojeri DC., Prodr. 6: 376 (1838); T.T.C.L.: 159 (1949). Type: Madagascar, near Antananarivo, *Bojer* s.n. (G-DC, holo.)
    S. subscandens A.Rich., Tent. Fl. Abyss. 1: 434 (1848); Oliv. & Hiern, F.T.A. 3: 421 (1877). Type: Ethiopia, Shoa, *Dillon & Petit* s.n. & sine loc., *Schimper* 1926 (both P!, syn.)
    S. gabonicus Oliv. & Hiern, F.T.A. 3: 421 (1877). Type: Gabon, Gabon R., *Mann* 995 (K!, holo.)
    Crassocephalum subscandens (A.Rich.) S. Moore in J.B. 50: 211 (1912)
    C. bojeri (DC.) Robyns, Fl. Sperm. Parc Nat. Albert 2: 544 (1947); F.P.U.: 182 (1962)
    Senecio angulatus sensu Maquet in Fl. Rwanda 3: 666, fig. 209/1 (1985), misapplied name

NOTE. *Greenway* 10493 from the Serengeti says that a few pale yellow ray florets were present but this is extremely unusual.

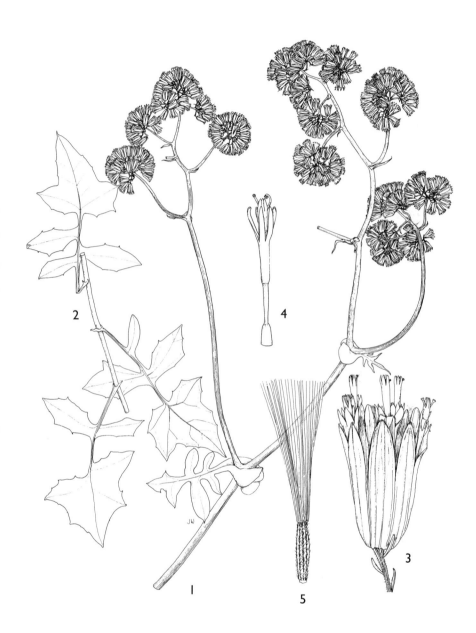

Fig. 137. *SOLANECIO ANGULATUS* — **1**, habit, × ¹/₂; **2**, leafy stem, × ¹/₂; **3**, capitulum, × 4; **4**, floret, × 6; **5**, achene, × 6. 1 from *Peter* 41630, 2 from *Archbold* 3015, 3 from *Whittall* 128, 4 from *Archbold* 2700, 5 from *Bally* 11583. Drawn by Juliet Williamson.

12. **Solanecio mannii** (*Hook.f.*) *C.Jeffrey* in K.B. 41: 922, fig. 6f (1986); Lisowski, Aster. Fl. Afr. Centr. 2: 418 (1991); K.T.S.L.: 562, fig., map (1994); U.K.W.F. ed. 2: 220, t. 92 (1994). Type: Bioko [Fernando Po], *Mann* 282 (K!, holo.)

Erect softly woody shrub or small tree 1–8(–10) m high, with foul smell; stems rather thick and succulent, green, with prominent persistent leaf-scars, branching in a dichotomous manner (like frangipani) to form a rounded crown; trunk up to 15 cm in diameter and with thin, smooth greyish or greenish bark. Leaves crowded towards the apices of the branches, slightly succulent, obovate, oblanceolate, or elliptic, 10–55 cm long, 2–16 cm wide, attenuate into an exauriculate petioloid base, margins shallowly to prominently sinuate-serrate or sinuate-laciniate, apex acute, acuminate-apiculate, glabrous or pubescent to thinly floccose especially on main veins above, paler and glabrous to floccose or tomentose beneath; sometimes the lower midrib purplish. Capitula discoid, very numerous in large thyrses of ± congested subumbelliform corymbs; stalks of individual capitula glabrous or shortly pubescent; scented; involucre cylindrical, 4–7.5 mm long, 1.5–2 mm in diameter; bracts of calyculus 3–4, lanceolate or oblong-lanceolate, pubescent at least on the margins, 1–3 mm long; phyllaries 5–8, green, sometimes with purple tips, 3.5–7 mm long, glabrous or thinly pubescent. Ray florets absent; disc florets yellow, less often orange or white, corolla 4.5–8.5 mm long, tube expanded above the middle, glabrous or shortly hairy in upper part, lobes 0.7–2 mm long. Achenes 2–3.5 mm long, ribbed, shortly hairy in the grooves; pappus 3.5–8.5 mm long. Fig. 138 (page 679).

UGANDA. Karamoja District: Napak Mts, Dec. 1958, *Langdale-Brown* 88!; Kigezi District: Luhiza Road, June 1951, *Purseglove* 3686!; Mengo District: Entebbe, July 1922, *Maitland* 9!
KENYA. Northern Frontier District: Mt Kulal, Apr. 1959, *T. Adamson* K9b!; Naivasha District: Kipipiri, Feb. 1964, *Karanja* 45!; Masai District: Trans Mara, Abossi Hill, Aug. 1961, *Glover et al.* 2343!
TANZANIA. Masai/Mbulu District: Ngorongoro, Rotian glade, Jan. 1989, *Pocs & Chuwa* 89/025b!; Ulanga District: Ujiji near Mahenge, May 1960, *Haerdi* 522/0!; Mbeya District: Mporoto Ridge, Galijembe Forest Reserve, Nov. 1992, *Mwasumbi* 16525!
DISTR. U 1–4; K 1–6, 7; T 1–8; Nigeria, Bioko, Cameroon, Congo (Kinshasa), Rwanda, Burundi, Sudan, Ethiopia, Angola, Zambia, Malawi, Mozambique, Zimbabwe
HAB. Forest or forest glades or margins, bushland or even grassland on rocky slopes, riverine, near cultivated areas; 80–2700 m
USES. Planted as a hedge around homesteads (*Tanner*), or in crops to keep root-eating rodents away (*Davies*); stems used as light rafters in huts; minor medicinal as a purgative (*Peter*), against fever and gonorrhoea (*Tanner*), stomach trouble (*Maitland*), against bad circulation in babies (*Meyerhoff*). Wood of lower stem used to make stands for gourds (*Napier*); crushed leaves are put in water to prevent lightening strike or death (*Katz*); nectaries poisonous (*Ichikawa*)
CONSERVATION NOTES. Widespread and common; least concern (LC)

SYN. *Senecio mannii* Hook.f. in J.L.S. 6: 14 (1861); Oliv. & Hiern, F.T.A. 3: 418 (1877); Maquet in Fl. Rwanda 3: 668, fig. 207/1 (1985)
    S. multicorymbosus Klatt in Ann. Naturhist. Mus. Wien 7: 103 (1892); T.T.C.L.: 159 (1949). Type: Angola, Malanga, *von Mechow* 187 (WU, holo., BR!, iso.)
    S. congolensis De Wild. in Ann. Mus. Congo ser. 5: 86 (1904). Type: Congo (Kinshasa), Kisantu, *Gillet* 1356 (BR!, holo.)
    S. morrumbalensis De Wild., Pl. Nov. Herb. Hort. Ten. 2: 135, t. 101 (1910). Type: Mozambique, Morrumbala Mts, *Luja* 32 (BR!, holo.)
    S. acervatus S.Moore in J.L.S. 40: 121 (1911). Type: Zimbabwe, Chirinda Forest, *Swynnerton* 665 (BM!, holo.)
    Crassocephalum multicorymbosus (Klatt) S.Moore in J.B. 50: 211 (1912)
    Senecio bogoroensis De Wild., Pl. Bequaert. 5: 99 (1929). Type: Congo (Kinshasa), Bogoro, *Bequaert* 4866 (BR!, holo.)
    S. mannii Hook. f. var. *kikuyensis* Chiov., Racc. Bot. Miss. Consol. Kenya: 71 (1935). Type: Kenya, Fort Hall District, Tuso, *Balbo* 97 (FT, holo.)
    Crassocephalum mannii (Hook.f.) Milne-Redh. in K.B. 5: 377 (1951); I.T.U.: 95 (1952); F.P.S. 3: 22 (1956); F.W.T.A. ed. 2, 2: 246, t. 252 (1963); Blundell, Wild Flow. E. Afr.: fig. 360 (1987)

Fig. 138. *SOLANECIO MANNII* — **1**, habit; **2**, involucre; **3**, floret; **4**, anthers; **5**, style arms. Drawn by W.E. Trevithick; from the Flora of West Tropical Africa.

13. **Solanecio gynuroides** *C.Jeffrey* in K.B. 41: 923 (1986). Type: Uganda, Kigezi District, Ishasha Gorge, *Purseglove* 2935 (K!, holo.)

Shrub to 3 m high; stems finely pubescent. Leaves sessile, ovate, 30 cm long, 13 cm wide, cuneate and decurrent onto a petioloid base, margins coarsely sinuate-bidentate, apex acuminate, thinly floccose especially on mid-rib above, glabrescent. Capitula discoid, in lax terminal corymbs; stalks of individual capitula arachnoid; involucre cylindrical, 11.5 mm long, 5–7 mm in diameter; bracts of calyculus ± 13, lanceolate, floccose towards the base, 3–4 mm long; phyllaries 13, glabrous, 10.5 mm long. Ray florets absent; disc florets yellow, corolla 10 mm long, tube glabrous, gradually expanded in upper half, lobes 0.7 mm long. Mature achenes unknown, but ovary 2 mm long, glabrous; pappus 10 mm long.

UGANDA. Kigezi District, Ishasha Gorge, Mar. 1947, *Purseglove* 2935!
DISTR. U 2; known from only the type
HAB. 'Forest'; 1500 m
USES. None recorded on specimens from our area
CONSERVATION NOTES. Known from only the type; at least vulnerable (VU-D2)

99. KLEINIA

Mill., Gard. Dict. abbr. ed. 4 (1754).

Senecio subgen. *Kleinia* (Mill.) O.Hoffm. in E. & P. Pf. IV, 5 (54): 301 (1890); P. Halliday, Noteworthy sp. *Kleinia*, in Hookers Ic. Pl. 29 (4): 1–135 (1988)

Perennial herbs or shrubs, erect or prostrate; stem often succulent. Leaves alternate, simple, succulent, sometimes reduced or rudimentary. Capitula discoid, solitary or in corymbs or panicles, campanulate; involucre usually calyculate, but sometimes bracts obscure or absent; receptacle epaleate. Flowers tubular, the tube widened at base, of various colours. Anthers obtuse at base, appendaged at apex. Style branches with conical to elongated terminal appendages. Achenes cylindrical to oblong, ribbed; pappus of fine white bristles.

About 40 species, mostly in tropical Africa but a few in North Africa, the Canary Is., South Africa, Madagascar, Arabia, India and Sri Lanka.

Note: many species *may flower when leafless*.

1. Herbs or shrubs, the stems fleshy but less than 8(–10) mm across; achenes hairy; florets white, cream, pale yellow, purple or pink 2
   Herbs with succulent thick stems, usually well over 8 mm in diameter; achenes glabrous or hairy; florets red, orange or bright yellow 8
2. Involucre 19–24 mm long; bracts of calyculus 6–9 mm long; fruiting pappus 36–50 mm long . .   4. *K. dolichocoma* (p. 683)
   Involucre < 19 mm long; bracts of calyculus < 4 mm long; fruiting pappus < 32 mm long 3
3. Flowers pink or purple 4
   Flowers white, cream or yellow-green 5
4. Phyllaries 8–13; fruiting pappus 10–17.5 mm . . . .   3. *K. breviflora* (p. 683)
   Phyllaries 5; fruiting pappus 20–32 mm . . . .   6. *K. squarrosa* (p. 684)
5. Capitula in stalked cymes, raised well above the stem apex; phyllaries 3–4 . . . .   1. *K. triantha* (p. 682)
   Capitula in clusters near the stem apex; phyllaries 5–8 6

6. Corolla lobes 1.3–2 mm long; fruiting pappus 26–29 mm long; central tuft of style-arm papillae prominent, 0.5 mm long .... 7. *K. odora* (p. 684)
   Corolla lobes 0.8–1.3 mm long; fruiting pappus 10–20 mm long; central tuft of style-arm papillae short or absent 7
7. Involucre 9–12 mm long, 3–4 mm in diameter; pappus 11–14 mm .... 2. *K. scottioides* (p. 682)
   Involucre 14–19 mm long, 5–12 mm in diameter; pappus14.5–20 mm .... 5. *K. negrii* (p. 683)
8. Leaves present and always wider than 4 mm 9
   Leaves absent, scale-like (< 12 mm long) or cylindrical, but always < 4 mm wide 16
9. Stem and leaves hairy; fruiting pappus to 35 mm .. 8. *K. implexa* (p. 685)
   Stem and leaves glabrous; fruiting pappus shorter (except *in K. grantii*, 50 mm) 10
10. Leaves (at least some of them) dentate, serrate or laciniate; plant erect 11
    Leaves with entire margin; plant erect or creeping 12
11. Phyllaries 13; flowering stems precocious, arising before the leaves, bearing small sessile serrate scale-like leaves .... 9. *K. schweinfurthii* (p. 687)
    Phyllaries 5–8; flowering stems arising from a basal rosette of leaves, bearing linear bracts in the upper part .... 10. *K. oligodonta* (p. 687)
12. Vegetative stem prostrate, with many crowded leaves along its length; pappus 18–25 mm ..... 11. *K. petraea* (p. 688)
    Vegetative stem erect or prostrate, the leaves mostly near the apex or at base of the flowering stem; pappus < 20 mm 13
13. Leaves thin; capitula solitary; achenes 7 mm long, pappus 16–25 mm .... 13. *K. mweroensis* (p. 689)
    Leaves succulent; capitula 1–many; achenes > 3 mm; pappus < 16.5 mm (–50 mm in fruit in *K. grantii*) 14
14. Capitula pendulous, in a many-headed panicle; flowers orange-yellow .... 12. *K. amaniensis* (p. 688)
    Capitula erect in flower, solitary or up to 10; flowers red or orange-red 15
15. Phyllaries prominently striate longitudinally ..... 21. *K. grantii* (p. 692)
    Phyllaries not striate .... 22. *K. abyssinica* (p. 694)
16. Leaves absent or scale-like and 2–12 mm long 17
    Leaves cylindrical or subulate, over 20 mm long 19
17. Stem erect; leaves (when present) 2–6 mm; corolla tube glabrous .... 18. *K. gregorii* (p. 691)
    Stem creeping; leaves 3–12 mm long; corolla tube puberulous 18
18. Stems with few leaves < 9 mm long .... 19. *K. pendula* (p. 692)
    Stems densely beset with leaves 9–12 mm long ... 20. *K. vermicularis* (p. 692)
19. Leaves 1–3 mm in diameter; florets 17–24 mm long; pappus 15–22 mm 20
    Leaves 3–4 mm in diameter; florets 14–15 mm long; pappus 10–14 mm 21

1. **Kleinia triantha** *Chiov.*, Fl. Somal. 2: 208 (1932). Type: Somalia, Garossellei, *Senni* 649 (FT!, holo.)

Succulent herb or shrub, erect or subscandent and 60–220 cm high; stems fleshy, glabrous, with prominent leaf-scars. Leaves succulent, obovate, 4.5–8 cm long, 1.5–3 cm wide, cuneate into a petioloid base, margins entire, apex obtuse, glabrous. Capitula discoid, numerous in long-stalked terminal corymbs of subumbelliform cymes; involucre cylindrical, 8–10 mm long, ± 2 mm in diameter; bracts of calyculus 1–2, subulate-lanceolate, 1 mm long; phyllaries 3–4, yellow, 8–10 mm long, glabrous. Ray florets absent; disc florets white, corolla 9.5–12.5 mm long, tube glabrous, slightly expanded in upper part, lobes 1 mm long. Achenes 3.2–4.2 mm long, hairy; pappus 9–18 mm long, white.

KENYA. Northern Frontier District: El Wak, May 1952, *Gillett* 13391! & 5 km NE of El Wak, Dec. 1971, *Bally & Radcliffe-Smith* 14551!
DISTR. **K** 1; Somalia
HAB. Overgrazed *Acacia-Commiphora* bushland or open *Acacia* bushland on limestone; 350–400 m
USES. None recorded on specimens from our area
CONSERVATION NOTES. Two specimens from Kenya; in 1952 common at El Wak; data deficient (DD)

2. **Kleinia scottioides** *C.Jeffrey* in K.B. 41: 925 (1986); U.K.W.F. ed. 2: 224 (1994). Type: Kenya, Turkana District, Kongelai Escarpment, *Tweedie* 3129 (K!, holo., M!, iso.)

Succulent herb or shrub, erect or scandent, 120–180 cm high; stems fleshy, glaucous, glabrous. Leaves succulent, oblanceolate to broadly obovate-spatulate, 2–8.5 cm long, 1.2–3.5 cm wide, cuneate or attenuate into petioloid base, margins entire, apex rounded to acuminate-apiculate, glabrous, deciduous. Capitula discoid, several to many in congested terminal subumbelliform cymose corymbs; stalks of individual capitula stout, erect; involucre cylindrical, 9–12 mm long, 3–4 mm in diameter; bracts of calyculus 1–3, lanceolate, lax, 2–4 mm long; phyllaries 5–8, 9–11 mm long. Ray florets absent; disc florets white or cream, corolla 11–14 mm long, tube glabrous, slightly expanded in upper part, lobes 0.8–1 mm long. Achenes 5–7 mm long, ribbed, hairy, pappus 11–14 mm long.

UGANDA. Karamoja District: Kakamari, June 1930, *Liebenberg* 357! & Moroto R., Feb. 1936, *Eggeling* 2953! & Mt Debasien, Jan. 1937, *A.S. Thomas* 2212!
KENYA. Northern Frontier District: 38 km N of Maralal on Baragoi road, June 1979, *Gilbert et al.* 5466!; West Suk District: Wei Wei, Sep. 1978, *Meyerhoff* 112!; Masai District: Morijo, Feb. 1979, *Bamps* 6606!
DISTR. **U** 1; **K** 1, 2, 6; not known elsewhere
HAB. On or between rocks in bushland or wooded grassland area; 900–1900 m
USES. Stem and root as minor medicinal
CONSERVATION NOTES. Three specimens from Uganda, all collected before 1940, and three specimens from Kenya; at least vulnerable (VU-D2)

SYN. *K. sp. C* of U.K.W.F. ed. 1: 483 (1974)

3. **Kleinia breviflora** *C.Jeffrey* in K.B. 41: 925 (1986); U.K.W.F. ed. 2: 224 (1994). Type: Kenya, Kiambu District, 27 km on Kikuyu–Narok road, *Verdcourt* 3561 (K!, holo.)

Succulent herb or shrub 40–150 cm high, erect or straggling, stiff and tough, often much branched; stems succulent, at first smooth, green with pallid streaks, sometimes glaucous, becoming brownish and grooved when old. Leaves succulent, oblanceolate to broadly obovate or rotundate-spatulate, 2.5–5.3 cm long, 1–3.3 cm wide, attenuate to broadly cuneate into petioloid base, margins entire, apex obtuse to rounded, minutely acuminate-apiculate, glabrous, deciduous. Capitula discoid, few or several in terminal congested subumbelliform cymes; involucre cylindrical, 9–14 mm long, 4–5 mm in diameter; bracts of calyculus 1–4, lanceolate, 1.5–4 mm long; phyllaries 8 or 13, green, maroon-tipped, 8–14 mm long. Ray florets absent; disc florets pink, purple or magenta, corolla 10–18 mm long, tube glabrous, expanded in upper part, lobes 1–1.3 mm long. Achenes 4.5–5.5 mm long, hairy; pappus 10–17.5 mm long.

KENYA. Northern Frontier District: Ndoto Mts, Ndigri-Alori, Jan. 1959, *Newbould* 3589!; Naivasha District: Hell's Gate, Jan. 1966, *Magius* in *E. Polhill* 159!; Masai District: Mt Suswa, Oct. 1962, *Glover & Samuel* 3295!
TANZANIA. Masai District: Serengeti, Oldiang'arangar, Nov. 1962, *Newbould* 6260! & Mt Longido, Jan. 1969, *Richards* 23663!; Pare District: Mbalu Hill near Mkomazi, Jan. 1948, *Bally* 5752!
DISTR. **K** 1, 3, 4, 6; **T** 2, 3; not known elsewhere
HAB. Rocky sites in dry bushland zone, where it may form thickets; (650–)1500–2100 m
USES. None recorded on specimens from our area
CONSERVATION NOTES. Fairly widespread in common habitat: least concern (LC)

SYN. *K.* sp. *B* of U.K.W.F. ed. 1

NOTE. Many specimens were named *K. kleinioides* or *K. longiflorus*, taxa not known from our area.

4. **Kleinia dolichocoma** *C.Jeffrey* in K.B. 41: 925 (1986). Type: Kenya, Northern Frontier District, Danissa Hills, *Bally & Smith* 14600 (K!, holo.)

Succulent erect shrub; stems succulent, striate, glabrous. Leaves succulent, obovate-spatulate, 1.5–2.7 cm long, 0.8–1.4 cm wide, attenuate into petioloid base, margins entire, apex rounded, glabrous. Capitula discoid, terminal in small 2–3-headed cymes; stalks of individual capitula stout; involucre cylindrical, 19–24 mm long, 5–9 mm in diameter; bracts of calyculus 2–3, oblanceolate, lax, 6–9 mm long; phyllaries 6, glaucous, 19–24 mm long. Ray florets absent; disc florets colour not indicated, probably white or yellow, corolla 20–21 mm long, tube glabrous, slightly expanded in upper part, lobes 1 mm long. Achenes 4 mm long, hairy; pappus in flower 19–20 mm long, in fruit 36–50 mm long.

KENYA. Northern Frontier District, Danissa Hills 70 km SSW of Ramu, Dec. 1971, *Bally & Smith* 14600!
DISTR. **K** 1; Somalia
HAB. Open bushland; ± 630 m
USES. None recorded on specimens from our area
CONSERVATION NOTES. Only one specimen known from Kenya; data deficient (DD)

5. **Kleinia negrii** *Cufod.* in Nuov. Giorn. Bot. Ital. n.s. 50: 113 (1943); C. Jeffrey in K.B. 41: 925 (1986). Type: Ethiopia, Kaskei R., *Corradi* 1934 (FT!, holo.)

Succulent semi-erect or scandent shrub; stems succulent, striate, glabrous. Leaves sub-succulent, oblanceolate, 3.7–5 cm long, 1–2 cm wide, attenuate into petioloid base, margins entire, apex obtuse, apiculate, glabrous, deciduous. Capitula discoid, in few-headed terminal corymbs, with few axes; stalks of individual capitula stout; involucre cylindrical, 14–19 mm long, 5–12 mm in diameter; bracts of calyculus 1–3, lax, 2–4 mm long; phyllaries 8, 14–19 mm long. Ray florets absent; disc florets yellow-green, corolla 16.5–21 mm long, tube glabrous, slightly expanded in upper part, lobes 1–1.3 mm long. Achenes unknown, but ovary 2 mm long, hairy; pappus 14.5–20 mm long.

UGANDA. Kodish (?=Kodit?), Karasuk, Jan. 1957, *J. Wilson* 312! (see note)
KENYA. Northern Frontier District: Moyale, July 1952, *Gillett* 13522!
DISTR. U 1; K 1, 2? (see note); Ethiopia
HAB. Degraded bushland; ± 1100 m
USES. None recorded on specimens from our area
CONSERVATION NOTES. Only two East African specimens; data deficient (DD)

NOTE. The *Wilson* specimen may very well be from Kenya, as the local name is in Turkana. But the hand-written label specifically says Uganda.

6. **Kleinia squarrosa** *Cufod.* in Nuov. Giorn. Bot. Ital. n.s. 50: 114 (1943); C. Jeffrey in K.B. 41: 925 (1986); Blundell, Wild Flow. E. Afr.: fig. 772 (1987); U.K.W.F. ed. 2: 224 (1994). Type: Ethiopia, Asile, *Corradi* 1654 (FT!, lecto., chosen by Jeffrey)

Succulent suberect or more usually straggling to ascending stiff herb or shrub 0.6–5 m high, usually much branched, sometimes forming dense thickets; once reported to have an underground rootstock; stems thick, succulent, glabrous, green, often white-dotted and sometimes with purplish or reddish bands; sap with strong acrid or resinous smell. Leaves few, near stem apex, or absent at flowering time, succulent, ovate or elliptic, 1.5–4.5 cm long, 0.3–2 cm wide, broadly cuneate to attenuate into petioloid base, margins entire, apex rounded, shortly acuminate-apiculate, short-lived. Capitula discoid, 2–16 in congested terminal subumbelliform cymes; involucre cylindrical, 10–16 mm long, 3–4 mm in diameter; bracts of calyculus 1–3, lanceolate, 0.5–1.5 mm long; phyllaries 5, greenish, sometimes glaucous, usually tinged with purple or brown at least near the apex, 10–16 mm long. Ray florets absent; disc florets pink to mauve or purple, rarely white, corolla 14.5–22 mm long, tube glabrous, expanded in upper part, lobes 1.2–1.7 mm long. Achenes 4–6.5 mm long, hairy; pappus 14–21.5 mm long in flower, 20–32 mm long in fruit.

UGANDA. Karamoja District: Katakekile, Jan. 1957, *J. Wilson* 314! & mile 124 on Kitale–Moroto road, Dec. 1959, *Napper* 1506! & Pian, Lodoketeminit, June 1962, *Kerfoot* 3826!
KENYA. Northern Frontier District: 1–3 km NE of Mado Gashi, June 1970, *Gillett & Newbould* 19181!; Masai District: 01°55'S 36°38'E, Jan. 1991, *Newton* 3743!; Teita District: foot of Mbololo Hill, Sep.–Oct. 1938, *Joanna* in *Bally* 8984!
TANZANIA. Arusha District: Ngaserai Plain, Jan. 1971, *Richards & Arasululu* 26471!; Pare/Lushoto District: Mkomazi, Kalamba–Mnazi, Jan. 1930, *Greenway* 2053!; Mpwapwa District: 15 km S of Gulwe on Kibakwe track, Apr. 1988, *Bidgood et al.* 980!
DISTR. U 1; K 1–7; T 2, 3, 5, 6; Ethiopia, Somalia
HAB. Dry bushland or sub-desert, where it may be thicket-forming; 50–1700 m
USES. None recorded on specimens from our area
CONSERVATION NOTES. Widespread in common habitat; least concern (LC)

SYN. *K. eupapposa* Cufod. in Stuttg. Beitr. Naturk. 218: 5, t. 5a (1970). Type: Ethiopia, Awash Valley, *Sebald* 2882 (WU!, holo., STU, iso.)
　　K. kleinioides sensu auct. mult., e.g. T.T.C.L.: 154 (1949), *non* (Sch. Bip.) M.R.F.Tayl.

NOTE. Said to resemble *Sarcostemma viminale* when sterile, except for the ridged stems.

7. **Kleinia odora** (*Forssk.*) *DC.*, Prodr. 6: 339 (1838); C. Jeffrey in K.B. 41: 925 (1986); P. Halliday, Noteworthy sp. Kleinia, in Hookers Ic. Pl. 29 (4): 125, t. 3899 (1988); U.K.W.F. ed. 2: 224 (1994). Type: Yemen, *Forsskål* s.n. (C, holo.)

Succulent suberect or scrambling shrub 0.6–3 m high, usually leafless, sometimes forming ragged thickets; stems often much branched, succulent, to 1 cm in diameter, cylindrical or ribbed, green with white or darker stripes. Leaves towards the tips of young shoots, quickly deciduous, succulent, lanceolate, becoming less fleshy when mature and up to 4.5 cm long and 1 cm wide, glabrous. Capitula discoid in terminal few-headed umbelliform cymes; involucre cylindrical, 10–15 mm long, 2.5–4 mm in diameter; bracts of calyculus 1–4, inconspicuous, up to 1 mm long; phyllaries 5, pale green, sometimes tipped with brown, 10–15 mm long. Ray florets absent; disc florets white, sometimes pale yellow or greenish-yellow, corolla 12–20 mm long, tube glabrous, expanded in upper part, lobes 1.3–2 mm long. Achenes ± 6.3 mm long, hairy; pappus 17–20 mm long in flower, 26–29 mm long in fruit. Fig. 139 (page 686).

UGANDA. Karamoja District: Lokitanyala, June 1953, *Dawkins* 803! & Lodoketeminit, July 1959, *Kerfoot* 1312! & km 200 on Kitale–Moroto road, Dec. 1959, *Napper* 1507!
KENYA. Northern Frontier District: 38 km N of Maralal on Baragoi road, June 1979, *Gilbert et al.* 5465!; Turkana District: Loya, Jan. 1965, *Newbould* 6808!; Teita District: NW of Voi, Mar. 1974, *Faden & Faden* 74/271!
TANZANIA. Masai District: Engaruka road, Feb. 1970, *Richards* 25493! & Olduvai Gorge, Aug. 1989, *Chuwa* 2832!; Lushoto District: Mkomazi, July 1955, *Semsei* 2148!
DISTR. **U** 1; **K** 1–3, 6, 7; **T** 2, 3; Ethiopia, Somalia; Yemen, Saudi Arabia
HAB. *Acacia* bushland, semi-desert scrub, often on rocky hillsides; may be thicket-forming, and possibly thrives in over-grazed areas; 10–1800 m
USES. 'Bark' a substitute for tobacco (*Newbould*); browsed by goats and game (*Wamukoya, Mumiukwa*)
CONSERVATION NOTES. Widespread in common habitat; least concern (LC)

SYN. *Cacalia odora* Forssk., Fl. Aegypt.-Arab.: 146 (1775)
   Senecio odorus (Forssk.) Defl., Voy. Yemen: 155 (1889)
   Kleinia longiflora sensu auct. mult., e.g. T.T.C.L.: 154 (1949); F.P.U.: 180 (1962), *non* DC.
   Senecio anteuphorbium (L.) Sch.Bip. var. *odorus* (Forssk.) G.D.Rowley in Nat. Cact. Succ. J. 13: 78 (1958) & in Jacobsen, Succ. Plants: 365 (1974)
   Kleinia sp. *A* of U.K.W.F. ed. 1: 483 (1974)

8. **Kleinia implexa** (*Bally*) *C.Jeffrey* in K.B. 41: 926 (1986). Type: from a plant cultivated at Nairobi from Kenya, Namanga, *Bally* 4671 (K!, holo.)

Succulent herb; stems succulent, prostrate and rooting at the nodes, the flowering branches to 30 cm high, densely shortly pubescent. Leaves succulent, oblanceolate or oblanceolate-spatulate, 2–4.4 cm long, 0.6–2 cm wide, attenuate into basally slightly expanded petioloid base, apex acute or subacute, apiculate, densely shortly pubescent. Capitula discoid, solitary, erect; peduncle 9–20 cm high, glabrous; involucre cylindrical, 15–21 mm long, 12–20 mm in diameter; bracts of calyculus 2–4, linear, lax, 3–6 mm long, shortly pubescent towards apex; phyllaries 13, 15–21 mm long, slightly striate. Ray florets absent; disc florets orange to red; corolla 14–18.5 mm long, tube glabrous, slightly expanded in upper half, lobes 1.2–1.5 mm long. Achenes 4.5 mm long, glabrous; pappus 12–19.5 mm long in flower, 35 mm long in fruit.

KENYA. Machakos District: SW foot of Mua Hills, Jan. 1965, *Gillett* 16619!; Masai District: Chyulu Plains, Nongiyiaa Big Kopje, July 1992, *Luke* 3206!; Kwale District: Mackinnon Road, Sep. 1953, *Drummond & Hemsley* 4092!
TANZANIA. Masai District: Merkerstein, Jan. 1936, *Greenway* 4351!; Handeni District: Ngobore, Sep. 1933, *Burtt* 4965!
DISTR. **K** 4, 6, 7; **T** 2, 3; not known elsewhere
HAB. *Acacia–Commiphora* bushland, in sandy or rocky sites; 150–1650 m
USES. None recorded on specimens from our area
CONSERVATION NOTES. Six Kenyan and two Tanzanian sites; locally common in two of these, but in a diminishing habitat; near threatened (NT)

SYN. *Senecio implexus* Bally in Candollea 18: 18, fig. 3 (1962)
   Notonia implexa (Bally) Agnew, U.K.W.F. ed. 1: 484 (1974); U.K.W.F. ed. 2: 224 (1994)
   Notoniopsis implexa (Bally) B.Nord. in Op. Bot. 44: 72 (1978)

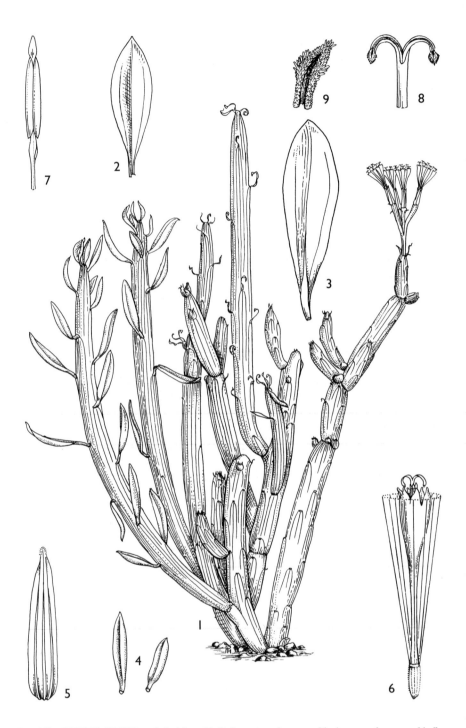

FIG. 139. *KLEINIA ODORA* — **1**, habit, × ²/₃; **2–3**, mature leaves, × ²/₃; **4**, young leaves, × ²/₃; **5**, phyllary, × 4; **6**, floret, × 4; **7**, anther, × 8; **8**, style, × 8; **9**, stigmatic appendage, × 32. 1 & 4–9 from *Collenette* s.n., 2 from *Smith & Henchie* 4423, 3 from *Collenette* 2720. Drawn by Pat Halliday, from Hooker's Icones Plantarum 39, 4.

9. **Kleinia schweinfurthii** (*Oliv. & Hiern*) *A.Berger* in Monatsschr. Kakt. 15: 11 (1905); Chiov., Result. Sci. Miss. Stef.-Paoli, Coll. Bot. 1: 106 (1916); F.P.S. 3: 37 (1956); C. Jeffrey in K.B. 41: 926 (1986). Type: Sudan, Dar Ferit, *Schweinfurth* 28 (K!, holo.)

Succulent herb, tuberous-rooted; flowering stems erect, 14–30 cm high, glabrous, leafy or bearing only reduced bractiform leaves. Leaves sessile, elliptic, 3–12.5 cm long, 1.2–6 cm wide, margins coarsely and irregularly sinuate-serrate or laciniate, apex obtuse to acute, minutely acuminate-apiculate; reduced bractiform leaves scale-like, reddish-brown, 1.1–2.5 × 0.5–0.7 cm wide. Capitula discoid, 1–3, terminal; involucre 19–21 mm long, ± 15 mm in diameter; bracts of calyculus 3–4, lanceolate, glabrous, 3–6 mm long; phyllaries 13, glabrous, 19–21 mm long. Ray florets absent; disc florets vivid scarlet, corolla 17–22 mm long, tube glabrous, expanded in upper half, lobes 1.8–2 mm long. Achenes 7.5 mm long, glabrous; pappus in flower 17 mm long, in fruit 28.5 mm long.

KENYA. Machakos District: Kibwezi Plains, Ithaba Swamp, Sep. 1938, *Teophilo* in *Bally* 7595!; Kwale District: Kinango–Gandini, Jan. 1996, *Luke & Luke* 4427!
TANZANIA. District unclear: near Lake Rukwa, Nov.–Dec. 1935, *Michelmore* 1431!; Kilosa District: Mikumi National Park, Sep. 1970, *Thulin & Mhoro* 1080!
DISTR. **K** 4, 7; **T** 4/7 (Lake Rukwa), 6; Nigeria, Sudan, Malawi
HAB. Dry bushland on shale soil; 200–900 m
USES. None recorded on specimens from our area
CONSERVATION NOTES. Wide distribution; least concern (LC)

SYN. *Notonia schweinfurthii* Oliv. & Hiern, F.T.A. 3: 407 (1877); U.K.W.F. ed. 2: 225 (1994)
    N. dalzielii Hutch. in F.W.T.A. 2: 149 (1931). Type: Nigeria, Gimi, Katagum, *Dalziel* 351 (K!, holo.)
    N. incisifolia Bally in Journ. E.Afr. Nat. Hist. Soc. 18: 127 (1946), *nom. non rite publ.*
    Senecio ballyi G.D.Rowley in Nat. Cact. Succ. Journ. 10: 31 (1955), *nom. non rite publ.*
    Notoniopsis schweinfurthii (Oliv. & Hiern) B.Nord. in Op. Bot. 44: 73 (1978)

10. **Kleinia oligodonta** *C.Jeffrey* in K.B. 41: 926 (1986). Type: Tanzania, Masai District, 24 km N of Loliondo, *Bally* 10638 (K!, holo.)

Perennial herb with fleshy horizontal rhizome and erect fleshy flowering stem 10–60 cm high, sometimes distally purplish-brown-tinged. Leaves succulent, sessile, oblanceolate, obovate or elliptic in outline, 4–13 cm long, 0.7–4 cm wide, margins entire or usually pinnately dentate or laciniate, glabrous, sometimes white-veined, often mottled or blotched with dull purple or red. Capitula discoid, 1(–3), terminal, nodding in bud; involucre cylindrical, 14–20 mm long, 13–18 mm in diameter; bracts of calyculus absent or obscure; phyllaries usually 8, rarely 5, 14–20 mm long. Ray florets absent; disc florets bright orange to rich red, corolla 14–20.5 mm long, tube glabrous, expanded in upper two-thirds, lobes 2–2.5 mm long. Achenes 4.5–7 mm long, ribbed, glabrous; pappus in flower and fruit 12–19 mm long.

KENYA. 'Uasin Gishu', Jan. 1932, *Harvey* 118!; Masai District: Morijo Loita, July 1961, *Glover et al.* 2259! & Narok town W of Longonot, Aug. 1971, *Kokwaro & Mathenge* 2763!
TANZANIA. Ngara District: Bugufi, Ntobeye, Dec. (no year), *Tanner* 5668!; Masai District: Serengeti, 13 km S of Lake Lagarja, May 1974, *Kreulen* 341!; Moshi District: Rongai Ranches, Apr. 1957, *Greenway* 9186!
DISTR. **K** 3, 6; **T** 1, 2; not known elsewhere
HAB. Grassland or bushed grassland, often among rocky outcrops; 900–2200 m
USES. None recorded on specimens from our area
CONSERVATION NOTES. Fairly widespread in common habitats, probably least concern (LC)

SYN. *Notonia oligodonta* (C.Jeffrey) Agnew in U.K.W.F. ed. 2: 225 (1994)

11. **Kleinia petraea** (*R.E.Fr.*) *C.Jeffrey* in K.B. 41: 926 (1986); Blundell, Wild Flow. E. Afr.: fig. 426 (1987); P. Halliday, Noteworthy sp. Kleinia, in Hookers Ic. Pl. 29 (4): 67, t. 3887 (1988). Type: Tanzania, Morogoro District, Uluguru, Ukami, Pembacito, *Busse* 291 (B†, holo, EA!, iso.)

Succulent perennial herb with fleshy creeping or procumbent stems to 50 cm long and 8 mm across, and erect flowering scapes to 40 cm high; stems thick, grey-green, glabrous. Leaves erect or spreading, thick, succulent, sessile, oblanceolate, obovate or obovate-spatulate, 2–6.6 cm long, 0.7–3.2 cm wide, cuneate or attenuate into a petioloid base, margins entire, apex rounded, minutely acuminate-apiculate, glabrous, often tinged purple, mauve or brown, sometimes glaucous. Capitula discoid, 1–5, terminal; involucre cylindrical, 15–24 mm long, 12–15 mm in diameter; bracts of calyculus 1–3 or absent, ovate or lanceolate, bract-like, 7–13 mm long; phyllaries 6–10, usually 8, thick, succulent, glaucous or pruinose, 15–24 mm long. Ray florets absent; disc florets orange, orange-red or orange-yellow, corolla 15.5–24 mm long, tube glabrous, slightly expanded above the middle, lobes 1.3–1.5 mm long. Achenes 5 mm long, glabrous, pappus 18–25 mm long.

KENYA. Northern Frontier District: Seya R. 30 km SE of Maralal, Feb. 1974, *Bally & Carter* 16557!; Eldama Ravine District: Eldama–Nakuru road, equator, Oct. 1981, *Gilbert & Mesfin* 6420!; Machakos District: Kyamutheke Hill, June 1992, *Luke* 3172!
TANZANIA. Musoma District: Serengeti, Engari Nanyuki, June 1962, *Greenway & Turner* 10696!; Moshi District: Lake Magadini, Dec. 1970, *Richards & Arasululu* 26524!; Kondoa District: near Kondoa, Feb. 1937, *B.D.Burtt* 5563!
DISTR. **K** 1, 3, 4, 6; **T** 1, 2, 5, 6; not known elsewhere
HAB. Grassland and bushed grassland, usually on stony ground or near rocky outcrops, may be locally common; 1300–2450 m
USES. Minor medicinal against cattle fleas and lice (*Glover*); as ornamental (*Patel & Mungai*)
CONSERVATION NOTES. Fairly widespread and not uncommon; least concern (LC)

SYN. *Senecio petraeus* Muschl. in E.J. 43: 70 (1909), *nom. illegit.*, *non* Boiss. & Reut. (1852) nec Klatt (1882)
      Notonia petraea R.E.Fr. in Act. Hort. Berg. 9: 148 (1928); U.K.W.F. ed. 2: 224, t. 94 (1994)
      Senecio jacobsenii G.D.Rowley in Nat. Cact. Succ. Journ. 10: 31 (1955). Type as for *Notonia petraea*
      Notoniopsis petraea (R.E.Fr.) B.Nord. in Op. Bot. 44: 72 (1978)

NOTE. A specimen from Uganda, 'native garden near Kampala', Mar. 1935, *Chandler* 1150! says 'ordinary cultivation near house' and so must be considered non-native.

12. **Kleinia amaniensis** (*Engl.*) *A.Berger*, Stapel. und Klein.: 384 (1910); Bally in Journ. E.A. Nat. Hist. Soc. 18: 127 (1946); C. Jeffrey in K.B. 41: 927 (1986); P. Halliday, Noteworthy sp. Kleinia, in Hookers Ic. Pl. 29 (4): 17, t. 3877 (1988). Type: a plant cultivated at Berlin, from Tanzania, Amani, *Engler* s.n. (B†, holo.); neotype: plant cult. at Nairobi, *Verdcourt* 3583 (K!, neo., chosen by Halliday)

Succulent perennial herb 75–140 cm high; stems often much branched, concolorous, glaucous when young. Leaves succulent, sessile, obovate to oblanceolate, 8–17 cm long, 2–6 cm wide, cuneate or attenuate into a petioloid base, margins entire, apex rounded, minutely obtusely apiculate, glabrous, glaucous at least when young. Capitula discoid, numerous in large lax spreading terminal paniculiform bracteolate glaucous cymes, nodding; involucre cylindrical, slightly contracted just below the apex, 11–18 mm long, 10–11 mm in diameter; bracts of calyculus few, lax, ovate or lanceolate, 7–8 mm long; phyllaries 7–9 or 13, usually 8, glaucous, tinged purple, 11–20 mm long. Ray florets absent; disc florets orange, dull orange-yellow or yellow, with strong musty odour, corolla 10–18 mm long, tube glabrous, expanded above the middle, lobes 1–1.3 mm long. Achenes glabrous, 6.5 mm long; pappus 7.5–11 mm long.

TANZANIA. Shinyanga District: Shinyanga, June 1931, *B.D. Burtt* 3443!; Lushoto District: W Usambara, Bungu–Ambangulu, Feb. 1916, *Peter* 54230!; Kondoa District: Kinyasi Scarp, Jan. 1928, *B.D. Burtt* 951!

DISTR. **T** 1, 3, 5, 6; not known elsewhere

HAB. Wooded grassland, rocky sites, rock faces; 1050–2200 m

USES. None recorded on specimens from our area

CONSERVATION NOTES. None of the collections from the wild is more recent than 1938, and much of its original distribution area has been put under sisal cultivation. The plant is established in cultivation among succulent growers. Data deficient (DD)

SYN. *Notonia amaniensis* Engl. in N.B.G.B. 4: 182, t. (1905)
Senecio amaniensis (Engl.) H.J.Jacobsen in Sukkulentenk. 4: 89 (1951) & Lex. Succ. Pl.: 365 (1974)

NOTE. This species has been cultivated at the National Museums of Kenya (1960, *Williams Sangai* 729! & 1963, *Verdcourt* 3583!)

13. **Kleinia mweroensis** (*Baker*) *C.Jeffrey* in K.B. 41: 927 (1986); P. Halliday, Noteworthy sp. Kleinia, in Hookers Ic. Pl. 29 (4): 63, t. 3886 (1988). Type: Zambia, Mwero, Kalongwizi R., *Carson* 15 (K!, holo.)

Perennial succulent herb, tuberous-rooted; stems semi-prostrate, succulent, 8–20 cm long, cylindrical, thick, tuberculate, longitudinally marked with darker green stripes, glabrous. Leaves narrowly elliptic, 4–11 cm long, 2–3 cm wide, base attenuate and petioloid, margins entire, apex rounded with a short acumen, thin, glabrous. Capitula discoid, solitary, terminal on up to 20 cm long pink-tinged glabrous scapes; involucre 15–23 mm long, 12–14 mm in diameter; bracts of calyculus absent or obscure, linear, 3 mm long; phyllaries 9–13, 15–23 mm long. Ray florets absent; disc florets orange-red, corolla 19–28 mm long, tube glabrous, slightly expanded in upper two-thirds, lobes 3.5–4 mm long. Achenes 7 mm long, glabrous; pappus 16–25 mm long.

TANZANIA. Kigoma District: 36 m S of Uvinza [Uvinsa], Aug. 1950, *Bullock* 3258!; Ufipa District: top of Kawa Falls, Oct. 1956, *Richards* 6344!

DISTR. **T** 4; Zambia

HAB. On or among rocks; 1200–1750 m

USES. None recorded on specimens from our area

CONSERVATION NOTES. Only two specimens from Tanzania; in Zambia restricted to a small area; at least vulnerable (VU-D2)

SYN. *Senecio mweroensis* Baker in K.B. 1895: 290 (1895)

14. **Kleinia picticaulis** (*Bally*) *C.Jeffrey* in K.B. 41: 927 (1986); P. Halliday, Noteworthy sp. Kleinia, in Hookers Ic. Pl. 29 (4): 33, t. 3880 (1988). Type: Kenya, Nanyuki District: Ngare Ndare, *Joy Bally* in *Bally* CM 11611 (EA!, holo.)

Perennial succulent herb; root tuberous and horizontal, ovoid, ± 2 cm long; stem erect, usually unbranched, cylindrical, 8–20 cm high, obscurely tuberculate, greenish with purple longitudinal stipes. Leaves subulate, 2–7(–12) cm long, up to 3 mm thick, glabrous. Capitula discoid, 1–2, terminal on 5–22 cm long scapes; involucre 20–23 mm long, 10–12 mm in diameter; bracts of calyculus obscure or absent; phyllaries 7–10, 20–23 mm long. Ray florets absent; disc florets bright red; corolla 24 mm long, tube glabrous, slightly expanded in upper half, lobes 3 mm long. Achenes not seen mature but ovary 4–6.5 mm long, hairy; pappus 19.5–22 mm long.

KENYA. Nanyuki District: Ngare Ndare, Nov. 1943, *Joy Bally* in *Bally* CM 11611!; Laikipia District: 45 km NE of Rumuruti, Feb. 1993, *Newton & Powys* 4245!; Meru District: S end of Lewa Downs Ranch, Jan. 1991, *Newton* 3802!

TANZANIA. Pare District: near Lembeni, June 1915, *Peter* 53975! & Same, May 1928, *Haarer* 1365! & 30 km N of Same, Jan. 1974, *Bally & Carter* 16375!
DISTR. **K** 4, 6; **T** 3; Sudan and Ethiopia
HAB. Grassland or open dry bushland; 800–1050 m
USES. None recorded on specimens from our area
CONSERVATION NOTES. Six specimens seen only; Halliday cites another five from Ethiopia and Sudan; data deficient (DD)

SYN. *Notonia subulata* Bally in J. E.A. Nat. Hist. Soc. 18: 127 (1946), *nom. non rite publ.*
Senecio subulatifolius G.D.Rowley in Nat. Cact. Succ. Journ. 10: 31 (1955), *nom. non rite publ.*
S. picticaulis Bally in Candollea 19: 163, fig. 17 (1964)
Notonia picticaulis (Bally) Cufod. in B.J.B.B. 37, suppl.: 1103 (1967) quoad basionym; U.K.W.F.: 484 (1974); U.K.W.F. ed. 2: 224 (1994)
Notoniopsis picticaulis (Bally) B.Nord. in Op. Bot. 44: 73 (1978)

15. **Kleinia patriciae** *C.Jeffrey* in K.B. 41: 927 (1986). Type: Plate drawn by Patricia Halliday from material cultivated from plants from Tanzania, Mbeya District, Poroto Mts S of Chimala, *Bally & Carter* 16464 (K!, holo.)

Perennial tuberous-rooted succulent herb; stem cylindrical, tuberculate, longitudinally marked with darker green, 15–23 cm long. Leaves succulent, cylindrical, 3–10 cm long, ± 1 mm in diameter when dry, glabrous. Capitula discoid, solitary, terminal on 12–21 cm long bracteate scapes; involucre 17–20 mm long, 12–13 mm in diameter; bracts of calyculus obscure or absent; phyllaries ± 12, 17–20 mm long, glaucous. Ray florets absent; disc florets red, corolla 17–19 mm long, glabrous, tube hardly expanded above the middle, lobes 2–3 mm long. Achenes unknown, but ovary 4 mm long, sparsely shortly hairy; pappus 15 mm long.

TANZANIA. Mbeya District: Kimani River falls at Nyengenge, June 1979, *Leedal* 5510! & idem, June 1990, *Congdon* 274! & Poroto Mts, 4 km up track to Matamba, June 1990, *Carter et al.* 2589!
DISTR. **T** 7; not known elsewhere
HAB. Woodland on rocky slopes; 1300–1850 m
USES. None recorded on specimens from our area
CONSERVATION NOTES. Known from a small area, four collections; at least vulnerable (VU-D2)

16. **Kleinia leptophylla** *C.Jeffrey* in K.B. 41: 927 (1986); P. Halliday, Noteworthy sp. Kleinia, in Hookers Ic. Pl. 29 (4): 27, t. 3879 (1988). Type: from a plant cultivated at RBG Kew from material from Kenya, Northern Frontier District, Huri Hills, Gabr Bori, *Bally* 12546 (K!, holo.)

Succulent perennial herb, tuberous-rooted; stems succulent, erect or procumbent, 11–30 cm long, 10–15 mm in diameter, tuberculate, deep green with 3 darker lines beneath each leaf-base and small paler dots between the lines. Leaves subulate, 2.5–6 cm long, 2–3(–6) mm wide and thick, clear green, not glaucous. Capitula discoid, solitary, terminal on 10–27 cm long bracteate scapes glaucous in upper half; involucre 14–20 mm long, 10–15 mm in diameter; bracts of calyculus obscure or absent; phyllaries 8–15, 14 mm long, glaucous. Ray florets absent; disc florets reddish-orange or magenta, corolla 14.7 mm long, glabrous, tube expanded above the middle, lobes 3 mm long. Achenes unknown, ovary 4 mm long, sparsely shortly hairy; pappus 9–10 mm long.

KENYA. Northern Frontier District: Huri Hills, Gabr Bori, Feb. 1963, *Bally* 12546! & Dec. 1987, *Newton* 3226!
DISTR. **K** 1; Ethiopia (Mt Mega), Saudi Arabia
HAB. Grassland; ± 1350 m
USES. None recorded on specimens from our area
CONSERVATION NOTES. Two Kenyan collections; also known from one site in Ethiopia and one in Saudi Arabia; vulnerable (VU-D2) but probably undercollected

17. **Kleinia schwartzii** *L.E.Newton* in Cact. Succ. Journ. (USA) 65, 5: 287 (1993). Type: from a plant flowering in cultivation based on material collected in Northern Frontier District, Marsabit, Burola Mt, *Schwartz & Powys* 73K (K!, holo.)

Succulent herb to 70 cm high; stem erect, succulent and up to 25 mm in diameter, green with 3 reddish lines below each node, the base enlarging and becoming tuberous and to 60 mm in diameter. Leaves densely set, semi-terete with adaxial groove, to 8 cm long, 3–4 mm in diameter, green, glabrous. Capitula solitary or up to 3, stalk of individual heads 9–42 cm long; bracts on inflorescence linear, 5–8 mm long; involucre 15–19 mm long, 15–20 mm in diameter; bracts of calyculus absent; phyllaries 19–20, 15–19 mm long, 2–5 mm wide. Florets red, the tube 15 mm long and yellow-green at base, lobes 3.5 mm long. Achenes unknown, but ovary 3 mm long, hirsute; pappus 10–14 mm long.

KENYA. Plant flowering in cultivation based on material collected in Northern Frontier District, Marsabit, Burola Mt, Oct. 1986, *Schwartz & Powys* 73K!
DISTR. **K** 1; possibly in Ethiopia
HAB. In soil pockets on rock slopes; altitude not given, ± 1000 m?
USES. None recorded on specimens from our area
CONSERVATION NOTES. Data deficient; known from only the type, but this area is undercollected; at least vulnerable (VU-D2)

SYN. *Senecio mweroensis* Baker forma *schwartzii* (L.E.Newton) G.D.Rowley in Bradleya 14: 83 (1996)

18. **Kleinia gregorii** (*S.Moore*) *C.Jeffrey* in K.B. 41: 927 (1986); Blundell, Wild Flow. E. Afr.: fig. 522 (1987); P. Halliday, Noteworthy sp. Kleinia, in Hookers Ic. Pl. 29 (4): 39, t. 3881 (1988). Type: Kenya, District unclear, Malewa R., *Gregory* s.n. (BM!, holo.)

Perennial succulent herb, rhizomatous, almost leafless; rhizomes subterranean, white with purple tips; stems erect, usually unbranched, succulent, 9.5–30 cm high, to 1.4 cm in diameter, subcylindrical, 5-ridged, tubercular, green or grey-green with the ribs usually darker green or reddish-brown. Leaves represented by narrow scales 2–6 mm long, shrivelling quickly. Capitula discoid, solitary or rarely 2 together, terminal on usually solitary 5–19 cm long scapes, green or dull purplish; bracts small, filiform; involucre cylindrical, 14–20 mm long, 10–12 mm in diameter; bracts of calyculus obscure or absent; phyllaries 7–10, reddish at least towards the apex, 14–20 mm long. Ray florets absent; disc florets rich bright red, corolla 16–23.5 mm long, tube glabrous, slightly expanded in upper two-thirds, lobes 2.5–4 mm long. Achenes 4–5 mm long, ribbed, glabrous or with sparse short hairs; pappus 16–22 mm long.

KENYA. Naivasha/Masai District: Mt Suswa lower slopes, Aug. 1952, *Verdcourt* 705!; Machakos District: 1 km N of Marimbeti Railway Station, Feb. 1972, *Mwangangi & Gillett* 1965!; Masai District: Siyabei Gorge, July 1962, *Glover & Samuel* 3158!
TANZANIA. Musoma District: Naabi Hill, Apr. 1961, *Greenway & Turner* 10024!; Masai District: Merkerstein, Jan. 1936, *Greenway* 4348!; Arusha District: Mt Meru slopes, Dec. 1970, *Richards & Arasululu* 26596!
DISTR. **K** 3, 4, 6; **T** 1, 2; not known elsewhere
HAB. *Acacia–Commiphora* bushland on stony ground, often at base of shrubs, thicket edges, rocky sites; (1100–)1500–2250 m
USES. Eaten by goats only (*Glover & Samuel*); roots eaten by snakes according to Masai (*Glover et al.*)
CONSERVATION NOTES. Not uncommon; probably least concern (LC)

SYN. *Notonia gregorii* S.Moore in J.L.S. 35: 353 (1907); U.K.W.F. ed. 2: 224, t. 94 (1994)
    Senecio gregorii (S.Moore) Jacobsen, Handb. Sukk. Pflanz. 2: 1025 (1954), *nom. illeg., non* F. Muell. (1859)

19. **Kleinia pendula** (*Forssk.*) *DC.*, Prodr. 6: 339 (1838); C. Jeffrey in K.B. 41: 927 (1986); P. Halliday, Noteworthy sp. Kleinia, in Hookers Ic. Pl. 29 (4): 47, t. 3883 (1988). Type: Yemen, *Forsskål* s.n. (C, holo.)

Perennial herb; stem 12–15 cm long, 1–1.5 cm in diameter, prostrate, creeping and rooting at the nodes, or pendulous from cliffs, succulent, green with white dots and darker lines, glabrous. Leaves scale-like, 3–9 mm long, shrivelling quickly. Capitula discoid, solitary or rarely 2 together, terminal on erect 4–21 cm long green scapes; bracts small, filiform; involucre 15–20 mm long, 12–14 mm in diameter; bracts of calyculus obscure or absent; phyllaries 8–13, green with reddish margins, 15–20 mm long. Ray florets absent; disc florets red, corolla 15–26 mm long, tube puberulous, slightly expanded in upper two-thirds, lobes ± 3 mm long. Achenes 2.5–5 mm long, ribbed, hairy; pappus 15–23 mm long.

KENYA. Northern Frontier District: Sul-Sul Mudde, Nov. 1990, *Newton & Powys* 3706!; 10 km S of Moyale, Nov. 1974, *Hale* 55
DISTR. **K** 1; Eritrea, Ethiopia, Somalia; Yemen, S Arabia
HAB. Grassland on rocky slope; 800–1250 m
USES. None recorded on specimens from our area
CONSERVATION NOTES. Fairly widespread; least concern (LC)

SYN. *Cacalia pendula* Forssk., Fl. Aegypt.-Arab.: 145 (1775)
　　Senecio pendulus (Forssk.) Sch. Bip. in Flora 28: 500 (1845)
　　Notonia trachycarpa Kotschy in Sitzber. Acad. Wien Mat.-Nat. Abt. 2: 370, t. 8 (1805); Oliv. & Hiern, F.T.A. 3: 408 (1877). Type: Eritrea, 'inter saxa montium in regno Boghos', *Hansal* 10 (W!, holo.)
　　Senecio gunnisii Baker in K.B. 1895: 217 (1895). Type: Somalia, Golis Range, *Cole* s.n. & *Lort Phillips* s.n. (K!, syn.)
　　Notonia pendula (Forssk.) Chiov. in Ann. Ist. Bot. Roma 8: 188 (1904)

20. **Kleinia vermicularis** *C.Jeffrey* in K.B. 41: 928 (1986). Type: Tanzania, Lushoto District, Pare Mts near Mkomazi, *B.D.Burtt* 6417 (K!, holo.)

Perennial stoloniferous leafless succulent herb; stems prostrate, creeping, rooting at the nodes, thick, cylindrical, succulent, dark green. Leaves represented by numerous narrow spine-like straw-coloured scales 9–12 mm long. Capitula discoid, solitary, terminal on solitary 11–15 cm long peduncles; bracts linear, narrow; involucre cylindrical, 19–21 mm long, 10–13 mm in diameter; bracts of calyculus obscure or absent; phyllaries 11, 19–21 mm long. Ray florets absent; disc florets bright red, corolla 20.5–22 mm long, tube hairy, especially in upper half, slightly expanded in upper two-thirds, lobes 2.5 mm long. Achenes 5.5 mm long, shortly hairy, pappus 19–20 mm long.

TANZANIA. Lushoto District: Mt Lasa, Sep. 1935, *Greenway* 4053! & Pare Mts near Mkomazi, Nov. 1937, *B.D.Burtt* 6417! & Mbalu Hill, Jan. 1948, *Bally* 5753!
DISTR. **T** 3; not known elsewhere
HAB. Rocky slope bushland; 650–700 m
USES. None recorded on specimens from our area
CONSERVATION NOTES. Small distribution area; known from 3 specimens; not collected since 1948. At least endangered (EN B2ab) if not extinct

21. **Kleinia grantii** (*Oliv. & Hiern*) *Hook.f.* in Bot. Mag. 125: t. 7691 (1899); C. Jeffrey in K.B. 41: 928 (1986); P. Halliday, Noteworthy sp. Kleinia, in Hookers Ic. Pl. 29 (4): 87, t. 3891 (1988); Lisowski, Aster. Fl. Afr. Centr. 2: 421 (1991). Type: Tanzania, ?District (see note), M'bumi, *Grant* s.n. (K!, holo.)

Perennial succulent herb with thick fleshy tubers or rhizomes; stems ± trailing, to 1 cm in diameter, flowering scapes erect, 9–50 cm high. Leaves sessile, recurved, succulent, ± glaucous, often tinged with purple, especially beneath, obovate,

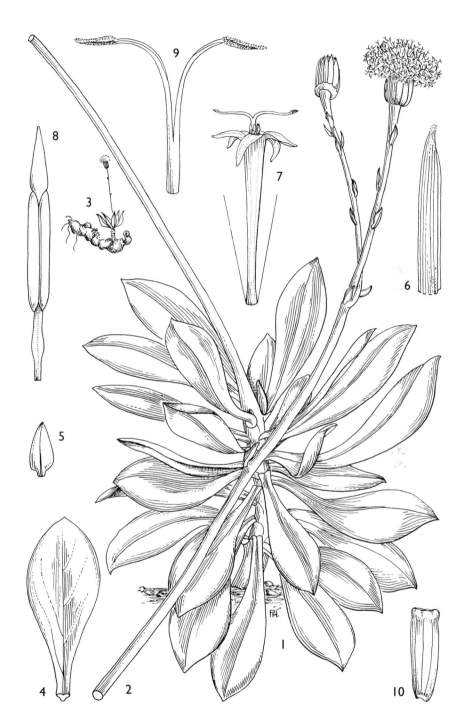

FIG. 140. *KLEINIA GRANTII* — **1–2**, habit, × ²/₃; **3**, whole plant with root, much reduced; **4**, leaf, × ²/₃; **5**, bract from scape, × 1; **6**, phyllary, × 2; **7**, floret, × 3; **8**, anther, × 12; **9**, style arms, × 8; **10**, achene, × 8. 1–2 & 4–5 from *Bailes* 225, 3 from *Bally* 8505, 6–9 from *Bailes* 115, 10 from *Bally* 11702. Drawn by Pat Halliday, from Hooker's Icones Plantarum 39, 4.

oblanceolate to linear-lanceolate, 3–11 cm long, 0.5–3 cm wide, cuneate or attenuate
into a comparatively wide petioloid base, margins entire, apex obtuse to acute,
apiculate, sometimes with recurving margins. Capitula 1–3, terminal, discoid;
involucre cylindrical, 12–21 mm long, 13–20 mm in diameter; bracts of calyculus
obscure or absent; phyllaries 8–15, red-tinged, 12–21 mm long. Ray florets absent;
disc florets bright red, corolla 12.5–19 mm long, tube glabrous, slightly expanded in
upper two-thirds, lobes 1.2–5 mm long. Achenes 3–4 mm long, glabrous; pappus
9–15 mm long in flower, to 50 mm long in fruit (fide *Lisowski*). Fig. 140 (page 693).

KENYA. Northern Frontier District: Mt Kulal, Narangani, June 1968, *Oteke* 68!; Laikipia District:
    45 km NE of Rumuruti, Feb. 1993, *P. Powys* in *Newton* 4239!; Kisumu, Feb. 1915, *Dummer* 1838!
TANZANIA. Musoma District: Serengeti, Seronera, Apr. 1961, *Greenway* 10010!; Arusha District:
    Sanya Juu, Jan. 1967, *Richards* 21875!; Njombe District: near Luilo, Sep. 1970, *Thulin & Mhoro*
    1179!
DISTR. **K** 1, 3–7; **T** 1–3, 5/6, 7; Guinea, Mali, Congo (Kinshasa), Rwanda, Burundi, Ethiopia,
    Somalia, Zambia
HAB. Grassland, scattered tree grassland, woodland, open bushland, often in stony sites;
    500–1900 m
USES. None recorded on specimens from our area
CONSERVATION NOTES. Least concern (LC)

SYN. *Notonia grantii* Oliv. & Hiern, F.T.A. 3: 407 (1877); F.P.U.: 180 (1962); U.K.W.F. ed. 2:
    225 (1994)
    N. coccinea Oliv. & Hiern, F.T.A. 3: 407 (1877). Type: Ethiopia, Somak Efat, *Roth* 36 (K!,
        holo.) – the collector is cited in F.T.A. as "Roth" but on the label of the type the name
        reads more like "Rohr"
    Senecio longipes Baker in K.B. 1895: 217 (1895). Type: Somalia, *Cole* s.n. (K!, holo.)
    Kleinia coccinea (Oliv. & Hiern) A.Berger in Monatsschr. Kakt. 15: 11 (1905)
    Senecio coccineus (Oliv. & Hiern) Muschl. in E.J. 43: 55 (1909), *nom. illegit., non* Klatt (1886)
    S. phellorhizus Muschl. in E.J. 43: 70 (1909). Type: Tanzania, Masai District, Lamuniane Mts,
        Jaeger 365a (B†, holo.)
    Notonia bequaertii De Wild., Pl. Bequaert. 5: 444 (1929). Type: Congo (Kinshasa), Kabare,
        Bequaert 5357 (BR!, holo.)
    Senecio coccineiflorus G.D.Rowley in Nat. Cact. Succ. J. 10: 31 (1955). Type as for *Notonia*
        coccinea
    Notoniopsis coccinea (Oliv. & Hiern) B. Nord. in Op. Bot. 44: 70 (1978)
    N. grantii (Oliv. & Hiern) B. Nord. in Op. Bot. 44: 70 (1978)

NOTE. The locality of the type (Grant: M'bumi) is unclear; the altitude is at 1750 (presumably
    feet); this is probably either **T** 5 or **T** 6.

   22. **Kleinia abyssinica** (*A.Rich.*) *A.Berger* in Monatsschr. Kakt. 15: 11 (1905); F.P.S.
3: 38 (1956); C. Jeffrey in K.B. 41: 928 (1986); Blundell, Wild Flow. E. Afr.: fig. 521
(1987); P. Halliday, Noteworthy sp. Kleinia, in Hookers Ic. Pl. 29 (4): 77, t. 3889
(1988); Lisowski, Aster. Fl. Afr. Centr. 2: 423, t. 89 (1991). Type: Ethiopia, 'Kouiaetha'
(I read Konasetha), *Dillon* s.n. (P!, holo.)

   Erect or prostrate perennial succulent herb 30–200(–300) cm high, glabrous, with
fleshy tuberous or rhizomatous rootstock; stems succulent, to 2.5 cm across near the
base, green, glabrous, sometimes glaucous. Leaves succulent, sessile, broadly ovate to
elliptic, obovate or oblanceolate, 6–26 cm long, 1–13 cm wide, cuneate to attenuate
into a petioloid base, margins entire, often recurving, apex acute to rounded, shortly
acuminate-apiculate, glabrous, often glaucous, pale green, usually variously
variegated with red or purple and/or with whitish veins. Capitula discoid, 1–10 in lax
terminal bracteate cymes, nodding in bud, erect in flower; bracts ovate-elliptic to
lanceolate; involucre cylindrical or urn-shaped, 15–31 mm long, up to 35 mm in
diameter; bracts of calyculus obscure or absent; phyllaries 7–15, pale green, often
glaucous, 15–31 mm long. Ray florets absent; disc florets bright red or orange-red,
corolla 12.5–28 mm long, glabrous, tube slightly expanded in upper two-thirds, lobes
2–8 mm long. Achenes 5–9 mm long, glabrous; pappus 8–16.5 mm long.

FIG. 141. *KLEINIA ABYSSINICA* VAR. *ABYSSINICA* — **1–2**, habit, × ²/₃; **3**, phyllary, × 2; **4**, floret, × 3; **5**, anther, × 12; **6**, style arms, × 6; **7**, achene, × 6. 1–2 & 4–7 from *Field & Powys* 51, 3 from *Gillett* 14128. Drawn by Pat Halliday, from Hooker's Icones Plantarum 39, 4.

1. Plant usually erect, 60–300 cm high; leaf base rather wide and
   winged; inflorescence with bracts usually comparatively
   wide and conspicuous .... var. *abyssinica*
   Plant often prostrate or arching, 15–50 cm high; leaf base
   narrow; inflorescence bracts usually comparatively small
   and narrow .... var.
   hildebrandtii

var. **abyssinica**

Plant erect, 60–240 cm high. Leaves decurrent onto a comparatively wide, winged, petioloid base, 10–26 cm long, 4–13 cm wide. Capitula 1–10; inflorescence with bracts usually comparatively wide and conspicuous; phyllaries usually 13, rarely 8 or 14, 15–31 mm long. Fig. 141 (page 695).

UGANDA. Karamoja District: foothills of Lochewar, Oct. 1952, *Verdcourt* 819!; Toro District: Buranga Pass, Oct. 1951, *Osmaston* 1361!; Mengo District: 2 km N of Kakoge, Dec. 1955, *Langdale-Brown* 1812!
KENYA. Northern Frontier District: Moyale, Nov. 1952, Gillett 14128! & Mathews Range, June 1959, *Kerfoot* 1111!; Elgon, Feb. 1948, *Hedberg* 45!
TANZANIA. Mbulu District: Lake Manyara National Park above Endabash, June 1965, *Greenway & Kanuri* 11852!; Dodoma District: 5 km on Kilimatinde–Dodoma road, Apr. 1988, *Bidgood et al.* 1160!; Iringa District: Irundi Hill, Apr. 1987, *Lovett et al.* 1911!
DISTR. U 1–4; K 1, 3, 4, 6, 7; T 1–3, 5, 7; Central African Republic, Congo (Kinshasa), Burundi, Sudan, Ethiopia, Zambia, Malawi
HAB. Grassland with scattered trees or bushes, thicket edges, wooded grassland, woodland, often in rocky sites; 500–2250 m
USES. Minor medicinal for wounds (*Jeffery*), against rheumatism (*Glover*)
CONSERVATION NOTES. Least concern (LC)

SYN. *Notonia abyssinica* A.Rich., Tent. Fl. Abyss. 1: 444, t. 59 (1848); Oliv. & Hiern, F.T.A. 3: 407 (1877); F.P.U.: 180 (1962); Cribb & Leedal, Mount. Flow. E. Africa: 163, t. 44c (1982); Maquet in Fl. Rwanda 3: 654, fig. 203/2 (1985); U.K.W.F. ed. 2: 225, t. 94 (1994)
Senecio nyikensis Baker in K.B. 1897: 271 (1897). Type: Malawi, Nyika, *Whyte* s.n. (K!, holo.)
Notonia optima S.Moore in J. B. 45: 329 (1907). Type: Uganda, Semliki Valley, *Bagshawe* 1276 (BM!, holo.)
Senecio superbus De Wild. & Muschl. in B.S.B.B. 49: 236 (1913). Type: Congo (Kinshasa), Lake Tanganyika shore, *Kassner* 3025a (K!, P!, iso.)
Notoniopsis abyssinica (A.Rich.) B.Nord. in Op. Bot. 44: 70 (1978)

var. **hildebrandtii** (*Vatke*) *C.Jeffrey* in K.B. 41: 928 (1986); Blundell, Wild Flow. E. Afr.: fig. 523 (1987); P. Halliday, Noteworthy sp. Kleinia, in Hookers Ic. Pl. 29 (4): 83, t. 3890 (1988). Type: Kenya, Mombasa, *Hildebrandt* 1946 (W, holo., K!, M!, P!, iso.)

Plant decumbent then ascending or erect, 15–120 cm high. Leaves cuneate to attenuate into narrow petioloid base, sometimes with whitish veins, often with recurving margins, 3–19 cm long, 0.5–5.5 cm wide. Capitula 1–3(–6); inflorescence bracts usually comparatively small and narrow; phyllaries 8–13, 15–24 mm long.

UGANDA. Karamoja District: Lodoketeminit, July 1959, *Kerfoot* 1306!; Ankole District: Mbarara Plantation, Ruti, Jan. 1953, *Osmaston* 2804!; Masaka District: Kyotera, Nov. 1945, *Purseglove* 1856!
KENYA. Turkana District: Ortum, Marich Pass, Oct. 1977, *Carter & Stannard* 50!; Masai District: top of Ngong Hills, July 1972, *Kibue* 181!; Kilifi District: Jaribuni, June 1973, *Musyoki & Hansen* 971!
TANZANIA. Mbulu/Masai District: Ngorongoro, Seneto Crater wall, Jan. 1989, *Pocs & Chuwa* 89/034v!; Tanga District: Mtotohovu, Dec. 1935, *Greenway* 4255!; Iringa District: Iringa, Mar. 1932, *Lynes* 198!
DISTR. U 1, 2, 4; K 2–4, 6, 7; T 2, 3, 7; Congo (Kinshasa), Rwanda, Burundi, Ethiopia
HAB. In rocky sites in dry grassland or bushland, woodland, open forest; 20–2700 m
USES. Warmed leaves are applied against rheumatism, or used for wounds
CONSERVATION NOTES. Least concern (LC)

SʏN. *Notonia hildebrandtii* Vatke in Oest. Bot. Zeitschr. 27: 197 (1877); Maquet in Fl. Rwanda 3: 654, fig. 203/1 (1985)
Crassocephalum notonoides S.Moore in J. B. 40: 341 (1902). Type: Kenya, Machakos District: Kibwezi, *Kassner* 718, 719 (B!, syn.)
Gynura notonoides (S.Moore) S.Moore in J. B. 50: 212 (1912)
Notonia urundiensis De Wild., Pl. Bequaert. 5: 443 (1929). Type: Burundi, Kitete, *Elskens* 25 (BR!, holo.)

100. GYNURA

Cᴀss. in Dict. Sci. Nat. 34: 391 (1825), *nom. cons.*; F.G. Davies in K.B. 33: 335–342 (1978); C. Jeffrey in K.B. 41: 929–930 (1986)

Perennial herbs, erect or climbing. Leaves alternate, simple, dentate or lobed. Capitula rarely solitary, usually in terminal corymbose cymes, discoid. Involucre with calyculus; receptacle epaleate. Florets with the upper part campanulate or infundibuliform. Anthers obtuse or slightly sagittate at base, with ovate or lanceolate apical appendage. Prominent style-arms long, exserted, gradually tapered, sometimes coloured. Achenes ribbed; pappus of many fine bristles.

About 40 species, mostly in SE Asia but also in mainland Asia and Africa, Madagascar.

1. Climbing or scrambling herb, 1.5–12 m long . . . .   1. *G. scandens*
   Erect or decumbent herb 2
2. At least some leaves deeply lobed to pinnatisect 3
   Leaves dentate or denticulate, not lobed, or with only a few
   lobes near base 5
3. Capitula 2–3; leaves glandular; plant of black cotton soils
   and bushed grassland . . . .   4. *G. pseudochina*
   Capitula 4 or more together; leaves without glands (?);
   plants usually in forest or habitats derived from forest 4
4. Herb 0.6–3 m high; leaves 8–20 cm wide . . . .   2. *G. valeriana*
   Herb 0.1–0.6 m high; leaves 3–4 cm wide . . . .   5. *G. colorata*
5. Leaves pubescent, to 20 × 4 cm; florets orange or red . . . .   3. *G. amplexicaulis*
   Leaves glabrous, 4–5 × 2–2.5 cm; florets yellow . . . .   6. *G. campanulata*

1. **Gynura scandens** *O.Hoffm.* in P.O.A. C: 416 (1895); U.K.W.F. ed. 1: 471 (1974); F.G. Davies in K.B. 33: 337 (1978); Maquet in Fl. Rwanda 3: 655 fig. 204/1 (1985); C. Jeffrey in K.B. 41: 929 (1986); Blundell, Wild Flow. E. Afr.: fig. 425 (1987); Halliday & Hind in Kew Mag. 7, 3: 109, plate 151 (1990); Lisowski, Aster. Fl. Afr. Centr. 2: 427, t. 90 (1991); U.K.W.F. ed. 2: 225, t. 95 (1994). Type: Tanzania, Lushoto District: Usambara, Lutindi, *Holst* 3315 (K!, lecto., BM!, M!, W, isolecto., chosen by Davies)

Climbing herb, 1.5–12 m long, slightly fleshy and with unpleasant smell. Leaves ovate or triangular, 2.5–12 cm long, 1.5–8 cm wide, cuneate to truncate or cordate into a petioloid base 1.5–6 cm long, auriculate or not, margins denticulate to dentate or shallowly lobed (lyrate), apex obtuse or acute, puberulous; 3–5-veined from base. Capitula several to many in terminal corymbs; involucre campanulate, 8–11 mm long; bracts of calyculus linear-lanceolate, 3–5 mm long; phyllaries 10–13, often with darker tips, 8–11 mm long, sparsely pubescent. Florets orange or less often yellow, 10–16 mm long, expanding in upper part, lobes 1.2–2 mm long, hooded at tip. Achenes 4.5 mm long, hairy in the grooves; pappus white, 8–11 mm long. Fig. 142 (page 698).

UɢᴀɴᴅA. Ankole District: Rugongo, Jan. 1971, *Rwaburindore* 510!; Busoga District: Bukizibu, June 1953, *G.H.S. Wood* 765!; Mengo District: 16 km Wabusana–Luwero, July 1956, *Langdale-Brown* 2253!

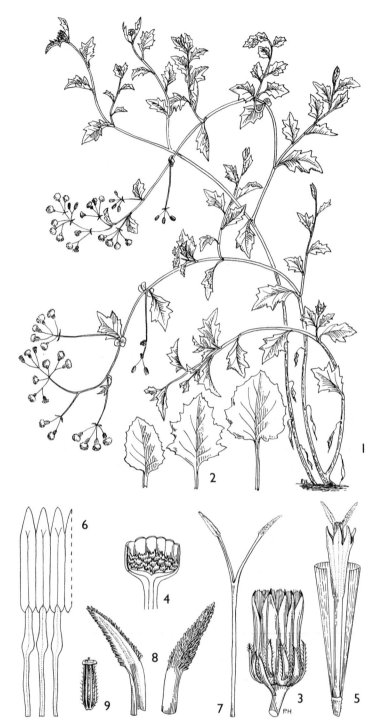

FIG. 142. *GYNURA SCANDENS* — **1**, habit; **2**, range of leaves; **3**, young capitulum, × 2; **4**, receptacle, × 2; **5**, floret, × 3; **6**, stamens, × 9; **7**, style, × 3; **8**, style arm appendage, adaxial and abaxial view, × 9; **9**, achene, × 4 $^{1}/_{2}$. Drawn by Pat Halliday from living material. Reproduced from Kew Mag. 7, 3.

Kenya. Nakuru District: 1 km N of Lake Nakuru, Aug. 1967, *Mwangangi* 86!; Masai District: Migori Bridge, June 1961, *Glover et al.* 1807!; Kilifi District: Mangea Hill, Mar. 1989, *Luke & Robertson* 1789!

Tanzania. Lushoto District: Kwamkoro Forest Reserve, May 1987, *Iversen* 87/372!; Ufipa District: Nsanga Forest, Aug. 1960, *Richards* 13009!; Morogoro District: Nguru Mts, Maskati Mission, Mar. 1988, *Bidgood et al.* 476!

Distr. **U** 2–4; **K** 3–7; **T** 1–4, 6–8; Congo (Kinshasa), Rwanda, Burundi, Angola, Zambia, Malawi

Hab. Moist forest, especially in forest margins and clearings, also in secondary vegetation, riverine forest, *Acacia xanthophloea* woodland; 1–2550 m

Uses. Minor medicinal against headache (*Magogo*); eaten by goats (*Glover*)

Conservation notes. Least concern (LC)

Syn. *Crassocephalum scandens* (O.Hoffm.) Hiern, Cat. Welw. Pl.: 595 (1898)
> *C. ruwenzoriensis* S.Moore in J.L.S. 35: 352 (1902). Type: Uganda, Ruwenzori, *Scott Elliot* 7777 (BM!, holo.)
> *C. auriformis* S.Moore in J.L.S. 37: 171 (1904). Type: Uganda, Lake Victoria, *Bagshawe* 657 (BM!, holo.)
> *Gynura taylorii* S.Moore in J.Bot. 43: 23 (1906). Type: Kenya, Kilifi District: Rabai, *Taylor* s.n. (BM!, holo.)
> *Senecio seretii* De Wild. in Ann. Mus. Congo 5, 3: 315 (1910); T.T.C.L.: 159 (1949). Type: Congo (Kinshasa), R. Arabi-Gonbari, *Seret* 707 (BR!, holo.)
> *S. agathionanthes* Muschl. in Z.A.E. 2: 385 (1911). Type: Rwanda, Rugege, Rukarara, *Mildbraed* 901 (B†, holo.)
> *Gynura auriformis* (S.Moore) S.Moore in J.B. 50: 212 (1912)
> *G. ruwenzoriensis* (S.Moore) S.Moore in J.B. 50: 213 (1912)
> *G. brownii* S.Moore in J.B. 54: 281 (1916). Type: Uganda, Mengo District: 100 km NW of Kampala, *E. Brown* in *Dummer* 2723 (BM!, holo.)
> *Senecio rutshuruensis* De Wild., Pl. Bequaert. 5: 130 (1929). Type: Congo (Kinshasa), Rutshuru, *Bequaert* 6054 (BR!, holo.)
> *S. seretii* De Wild. var. *divergentiramus* De Wild., Pl. Bequaert. 5: 132 (1929). Type: Congo (Kinshasa), Fendula, Nyakolonge, *Scaetta* 666 (BR!, holo.)
> *S. subalatipetiolatus* De Wild., Pl. Bequaert. 5: 134 (1929). Type: Congo (Kinshasa), Blukwa, *Claessens* 1488 (BR!, holo.)
> *S. variostipellatus* De Wild., Pl. Bequaert. 5: 135 (1929). Type: Congo (Kinshasa), between Mboga and Lesse, *Bequaert* 2980 (BR!, holo.)

Note. Looks quite like *Solanecio nandensis* but distinct in the flower colour (white to pale yellow in *S. nandensis*) and the longer pappus (5–8 mm in *S. nandensis*). The *Solanecio* also has 8 phyllaries and lacks the unpleasant smell.

2. **Gynura valeriana** *Oliv.* in Hook. Ic. 16: t. 1507 (1886) & in Trans. Linn. Soc. 11, 2: 339 (1887); U.K.W.F. ed. 1: 471 (1974); F.G. Davies in K.B. 33: 338 (1978); C. Jeffrey in K.B. 41: 929 (1986); Lisowski, Aster. Fl. Afr. Centr. 2: 433 (1991); U.K.W.F. ed. 2: 225 (1994). Type: Tanzania, Kilimanjaro, *Johnston* 103 (K!, holo.)

Perennial herb 0.6–1.8(–3) m high, with unpleasant smell; stem fleshy, erect or procumbent, leafy throughout or only in lower part, with scattered hairs. Leaves thinly fleshy, dark green or grey-green, purplish beneath, ovate to elliptic, 15–50 cm long (including petiolate base), 8–20 cm wide, base decurrent and auriculate, margins deeply lobed to pinnatisect with 1–10 pairs of oblong lobes, margins of lobes denticulate or dentate, apex acute, glabrous or with scattered hairs. Capitula in large inflorescences of lateral and terminal crowded groups of 5–20 capitula; stalks of individual capitula 1–3 cm long; involucre obconical, 6–12 mm long; bracts of calyculus 1–6 mm long; phyllaries 10–15, 6–12 mm long, 1 mm wide, with darker tip, glabrous or with scattered hairs. Florets yellow or orange, less often orange-red, scented, 10–12 mm long, expanded in upper third, lobes 1–1.2 mm long; anthers conspicuously yellow. Achenes 2 mm long, glabrous; pappus white, 7–10 mm long.

Kenya. Masai District: Chyulu Hills, June 1938, *Bally* 8176! & idem, Dec. 1993, *Luke & Luke* 3919!

TANZANIA. Arusha District: Mt Meru, Nov. 1969, *Richards* 24612!; Lushoto District: Kitivo N
Forest Reserve, Sep. 1966, *Semsei* 4105!; Kilosa District: Ukaguru Mts between Mandege &
Ihanga rock, June 1978, *Thulin & Mhoro* 2880!
DISTR. **K** 4/6, 6; **T** 2, 3, 6, 7; Congo (Kinshasa), Malawi
HAB. Moist sites in forest, on streamside rocks, and sometimes persisting after forest clearing,
rarely as epiphyte in trees; 700–2150 m
USES. Leaves eaten as a vegetable (*Wallace*)
CONSERVATION NOTES. More than 20 Tanzanian specimens; presumably least concern (LC)

3. **Gynura amplexicaulis** *Oliv. & Hiern*, F.T.A. 3: 403 (1877); Hoffmann in P.O.A.
C: 416 (1895); F.P.S. 3: 34 (1956); U.K.W.F.: 471 (1974); F.G. Davies in K.B. 33: 338
(1978); C. Jeffrey in K.B. 41: 930 (1986); Lisowski, Aster. Fl. Afr. Centr. 2: 431 (1991);
U.K.W.F. ed. 2: 225 (1994). Type: Sudan, Niam-niam land, Bodumo, *Schweinfurth*
3770 (K!, holo., BM!, W, iso.)

Perennial herb, rhizomatous, 0.3–1.2 m high; stems erect, sparsely branched,
tinged with red, pubescent; leaves mostly (but not all) arranged near base of stem.
Leaves fleshy, narrowly rhomboid to linear, 4–20 cm long, 0.4–4 cm wide, attenuate
to a rounded base, sagittate or amplexicaul, margins shallowly dentate, the teeth
fewer towards the apex, some teeth pointing backwards, apex attenuate, pubescent.
Capitula 3–9, in terminal corymbs, with many florets; involucre campanulate or
infundibuliform, (8–)11–16 mm long; bracts of calyculus 3–8 mm long; phyllaries
13–14, reddish green, (8–)11–16 mm long, pubescent. Florets orange to red,
(9–)11–15 mm long, expanded in upper part, lobes 1–1.2 mm long. Achenes 2.5–3 mm
long, glabrous; pappus 8–13 mm long.

UGANDA. Ankole District: Bunyaruguru, Feb. 1939, *Purseglove* 563!; Teso District: Serere, Mar.
1932, *Chandler* 592!; Masaka District: 2 km W of Lyantonde, May 1972, *Lye* 6884!
KENYA. Elgon, 1930, *Lugard* 240!; Trans Nzoia District: Moi's Bridge, July 1965, *Tweedie* 3071!;
Kericho District: Kibajet Estate, Oct. 1948, *Bally* 6477!
TANZANIA. Bukoba District: Kakindu, Oct. 1931, *Haarer* 2327!; Biharamulo District: Lusahunga,
Oct. 1960, *Tanner* 5359!
DISTR. **U** 2, 4; **K** 3, 5; **T** 1; Congo (Kinshasa), Rwanda, Burundi, Sudan
HAB. Swampy sites in grassland, grassland, secondary vegetation, a weed in cultivation;
1050–1900(?–2400) m
USES. None recorded on specimens from our area
CONSERVATION NOTES. Least concern (LC)

SYN. *Crassocephalum amplexicaule* (Oliv. & Hiern) S.Moore in J.B. 50: 211 (1912)
Gynura claessensii De Wild., Pl. Bequaert. 5: 91 (1929). Type: Congo (Kinshasa), Kilo,
Claessens 1295 (BR!, holo.)
Senecio claessensii (De Wild.) Humbert & Staner in B.J.B.B. 14: 105 (1936)

NOTE. *G. amplexicaulis* sensu F.W.T.A. ed. 1 is *G. pseudochina* L.

4. **Gynura pseudochina** (*L.*) *DC.*, Prodr. 6: 299 (1838); F.G. Davies in K.B. 33: 638
(1979); C. Jeffrey in K.B. 41: 930 (1986); Blundell, Wild Flow. E. Afr.: fig. 366 (1987);
Lisowski, Aster. Fl. Afr. Centr. 2: 432 (1991); U.K.W.F. ed. 2: 225 (1994). Type: Indonesia,
herb. *Royen* 164 (L, syn.) & fig. 335, t. 258 in Dill., Hort. Elth.: 345 (1732), syn.

Perennial herb 0.4–1 m high, with unpleasant musky smell; tuberous rootstock, the
tubers to 10 cm long and up to 5 cm across; stems ± fleshy, with leaves in rosette or
leafy to half their length, ± pubescent. Leaves slightly fleshy, green or purplish, the
basal leaves ovate or spatulate, 4–22 cm long, 2.5–11 cm wide, attenuate into a
petioloid base, margins ± entire, apex obtuse; middle and upper leaves narrower,
elliptic, obovate or narrowly obovate, lobed to pinnatisect, 6–25 cm long, 2–7 cm
wide, base clasping the stem in the uppermost leaves, margins 1–6-lobed, the lobes
toothed or lobed; all leaves pubescent or glabrous, glandular. Capitula terminal,
2–3 together, erect; involucre campanulate, 8–13 mm long, 13–15 mm in diameter;

bracts of calyculus 1–3 mm long; phyllaries 10–16, (6–)8–13 mm long, pubescent and sometimes glandular. Florets orange or yellow, corolla 9–12.5 mm long, expanded in upper part, lobes 0.9–1.7 mm long. Achenes 3 mm long, pubescent or glabrous; pappus 7–11 mm long.

KENYA. Meru District: Isiolo, Apr. 1971, *Kimani* 281!; Masai District: Loitokitok–Emali road, May 1974, *Kokwaro et al.* 3520!; Kwale/Kilifi District: Mayi ya Chumvi, Mar. 1902, *Kassner* 475!
TANZANIA. Moshi District: Engare Nairobi, June 1944, *Greenway* 6877!; Lushoto District: 2 km SW of Gare turnoff on Mombo road, June 1953, *Drummond & Hemsley* 2932!; Morogoro District: Morogoro, Nov. 1955, *Drummond & Hemsley* 7367!
DISTR. **K** 1, 4, 6, 7; **T** 2, 3, 4, 6, 7; Sierra Leone to Congo (Kinshasa), Burundi, Ethiopia, Somalia, Angola, Zambia, Malawi; tropical Asia
HAB. Boggy grassland, black cotton soil, often with *Acacia drepanolobium*, less often in bushed grassland or secondary woodland; 0–1800 m
USES. None recorded on specimens from our area
CONSERVATION NOTES. Least concern (LC)

SYN. *Senecio pseudochina* L., Sp. Pl.: 867 (1753)
    Gynura miniata Welw., Apont.: 586 (1859); Oliv. & Hiern, F.T.A. 3: 403 (1877); Bally in J. E.A. N.H. Soc. 18: 124, t. 20/4 (1944); F.W.T.A. ed. 2, 2: 243 (1963); U.K.W.F. ed. 1: 471 (1974); F.G. Davies in K.B. 33: 339 (1978). Type: Angola, Pungo Andongo, Caghuy, *Welwitsch* 3595 (BM!, holo., BR!, iso.)
    G. miniata Welw. var. *orientalis* O.Hoffm. in P.O.A. C: 416 (1895); Chiov., Fl. Som. 11: 270, fig. 155 (1932). Type: Tanzania, Uzaramo, *Stuhlmann* 7776 (not seen)
    G. rusisiensis R.E.Fr. in Wiss. Ergebn. Schwed. Rhod.–Kongo Exped. 1: 342 (1911). Type: Tanzania, Mpanda District: Rusisi Valley between Mpanda and Mecherenge, *Fries* 1435 (UPS, holo.)
    Senecio somalensis Chiov., Result. Sc. Miss. Stef.-Paoli Somal. Ital. 1: 106 (1916). Type: Somalia, Baidoa, *Paoli* 1110 (FT!, holo.)
    Gynura eximia S.Moore in J.B. 56: 225 (1918). Type: Angola, Kaconda, *Gossweiler* 3638 (BM!, K!, iso.)
    G. variifolia De Wild., Pl. Bequaert. 5: 93 (1929). Type: Congo (Kinshasa), Rutshuru, *Bequaert* 5627 (BR!, holo.)
    [*G. amplexicaulis* sensu F.W.T.A. ed. 1, 2: 148 (1931), *non* Oliv. & Hiern]
    G. somalensis (Chiov.) Cuf. in Nuov. Giorn. Bot. Hal. n.a. 1: 112 (1943)

NOTE. Gossweiler states the type of *Gynura eximia* is an annual.

5. **Gynura colorata** *F.G.Davies* in K.B. 33: 340 (1978); C. Jeffrey in K.B. 41: 929 (1986). Type: Tanzania, Lushoto District, Mt Mlinga, *Greenway* 6063 (K!, holo., EA, iso.)

Perennial herb 10–60 cm high, decumbent at base; rhizome spreading, or small tuber present; stem with a rosette of leaves at or a little above ground level, pubescent. Leaves quite variable, the lowest ovate or suborbicular, 4–6 cm long, 3–4 cm wide, cuneate into a 2.5–5 cm long petioloid base, margins entire or remotely lobed-dentate, apex obtuse; middle leaves ovate, elliptic or lyrate, 11–23 cm long, lobed; upper leaves lyrate to pinnatisect, 6–25 cm long, 3–5 cm wide, with 1–4 pairs of lobes, the terminal lobe largest, base cuneate into a petioloid base or auriculate; all leaves green above, purplish beneath, glabrous or sparsely hairy. Capitula 4–30 on a long peduncle with one to few bracts; stalks of individual capitula 1–2 cm long; involucre turbinate, 8.5–10 mm long; bracts of calyculus 1–4 mm long; phyllaries 12–14, 8.5–10 mm long, glabrous. Florets yellow, 7–10 mm long, expanded in upper part, lobes 0.7–1.4 mm long. Achenes 4 mm long when immature, glabrous; pappus 6–10 mm long.

KENYA. Kilifi District: Kauma, along small tributary of Ndzovuni R., June 1973, *Musyoki & Hansen* 974!
TANZANIA. Lushoto District: E Usambaras, Sigi Valley above Amani, May 1915, *Peter* 53876! & Kisiwani, May 1950, *Williams Sangai* 20!; Tanga District: Kange Gorge, June 1956, Faulkner 1872!
DISTR. **K** 7; **T** 3; not known elsewhere
HAB. Moist forest, nearly always among rocks; 0–850 m

Uses. None recorded on specimens from our area
Conservation notes. One specimen from Kenya, nine from Tanzania, the most recent
collection there from 1968; diminishing habitat; vulnerable (VU-B1ab)

6. **Gynura campanulata** *C.Jeffrey* in K.B. 41: 929 (1986); U.K.W.F. ed. 2: 225 (1994).
Type: Kenya, Kiambu, *T.H. Jackson* 15 (BM!, holo.)

Subsucculent herb ± 45 cm high; stems glabrous. Basal leaves petiolate, blade elliptic
or ovate-elliptic, 4–5 cm long, 2–2.5 cm wide, cuneate into petiole at the base, margins
remotely dentate, apex obtuse to subacute, shortly apiculate, glabrous; petiole 1.5–3 cm
long, bearing a few hairs; upper leaves sessile, oblanceolate, coarsely serrate, with a few
narrow pinnately arranged lobes at the base. Capitula discoid, 10–12 in terminal
corymbs; stalks of individual capitula glandular-hairy, 1.5–4.5 cm long; bracts linear-
lanceolate, glandular; involucre 12–13 mm long, 7 mm in diameter; bracts of calyculus
8–10, oblanceolate, ciliate, lax, 5–6 mm long; phyllaries c. 13, glabrous, 12–13 mm
long. Ray florets absent; disc florets yellow; corolla ± 13 mm long, tube ± 12 mm long,
lobes ± 1 mm long; Achenes immature, 1.5 mm long; pappus 11–13 mm long.

Kenya. Kiambu, Dec. 1935, *T.H. Jackson* 15!
Distr. **K** 4; known from only the type
Hab. Wet scrub; ± 1615 m
Uses. None recorded on specimens from our area
Conservation notes. Known from only a single specimen; at least vulnerable (VU-D2) if
not extinct

INSUFFICIENTLY KNOWN SPECIES

G. fischeri O.Hoffm. in P.O.A. C: 416 (1895). Type: Tanzania, N Mara District:
Ukira, *Fischer* 362 (B†, holo.)
The type has been destroyed, but from the description this is probably a synonym
of *Solanecio angulatus*.

G. meyeri-johannis O.Hoffm. in P.O.A. C: 416 (1895). Type: Tanzania, Kilimanjaro,
Meyer 374 (B†, holo.)
The type has been destroyed, but from the description this is possibly a synonym
of *Senecio syringifolius*.

HELIANTHEAE*

Cass. in J. Phys. Chim. Hist. Nat. 88: 189 (1819) & in Dict. Sci. Nat. 20:
354–385 (1821)

Herbs, shrubs, or sometimes small trees, rarely climbers. Leaves usually opposite,
entire or lobed, often 3-veined from base. Capitula solitary or in corymbose to
paniculate inflorescences, radiate or discoid, rarely disciform. Phyllaries usually in
1–3 series, if uniseriate often with oil glands. Receptacle paleate or epaleate, when
paleate the paleae sometimes enveloping the florets and achenes. Ray florets, when
present, nearly always female, often yellow; disc florets nearly always bisexual,
generally 5-lobed, rarely the outer somewhat zygomorphic, often yellow. Anthers
often with apical appendage, usually ovate with a constricted base, and often with
blackened thecae. Style with linear branches, usually with sterile appendages of fused
hairs; stigmatic areas usually in two lines. Achenes usually black or brown, those of
ray florets often 3-cornered, those of disc florets terete or angular. Pappus of scales
or awns, less often absent or of bristles.

* by H.J. Beentje & D.J.N. Hind

300 genera and ± 3300 species. Main distribution in Americas, but with many species in Africa and Asia; many species have become world-wide weeds.

There has been disagreement over the tribe, and Karis & Ryding, in Asteraceae: Cladistics and Classification (1994) split the tribe into Heliantheae s.s. and Helenieae; but state this is a provisional unit. For the Flora we retain the broad concept.

1. Most leaves alternate (some lower ones may be
   opposite) 2
   All leaves opposite (though some within the
   branching of the inflorescence may be alternate) 14
2. Heads are glomerules (heads of heads), the
   constituent capitula with 2 filiform female and
   several male florets; leaves simple 3
   Capitula free from each other; leaves simple or
   compound 4
3. Pappus of 6–19 scales .... 101. **Blepharispermum** (p. 709)
   Pappus a ring of hairs, or absent .... 102. **Athroisma** (p. 714)
4. Ray florets absent; capitula unisexual; pappus
   absent 5
   Ray florets usually present; capitula of mixed
   female and male or hermaphrodite florets 6
5. Fresh leaves aromatic; achenes enclosed in
   involucre with tiny straight spines; stem pilose 130. **Ambrosia** (p. 813)
   Fresh leaves not aromatic; achenes enclosed in
   involucre with hooked bristles; stem scabrid .. 132. **Xanthium** (p. 817)
6. Leaves entire, simple or lobed 7
   Leaves pinnatisect to bipinnate 9
7. Ray floret 3–5 mm long; phyllaries 1-seriate;
   pappus absent .... 117. **Sclerocarpus** (p. 757)
   Ray floret > 25 mm long; phyllaries 2- or more-
   seriate; pappus of bristles or scales 8
8. Leaves entire; ray florets ± 14; pappus of 2
   caducous bristles .... (cultivated) *Helianthus*
   Leaves usually lobed; ray florets 8–12; pappus of
   scales with awns .... 118. **Tithonia** (p. 759)
9. Pappus absent; rays 1–2.5 mm long .... 128. **Chrysanthellum** (p. 810)
   Pappus present 10
10. Pappus of 8–10 scales 11
    Pappus of 1–2 setae (occasionally mixed with
    3–4 scales) 12
11. Leaves with submarginal oil glands; ray floret
    3–4 mm long; each pappus scale with 2
    aristae; involucre 1-seriate .... (cultivated) *Thymophylla*
    Leaves without oil glands, though normal glands
    are present; ray floret, if present, 1–2 mm
    long; pappus scales acute .... 104. **Schkuria** (p. 722)
12. Pappus of 2 recurved awns enclosed in paleae . 131. **Parthenium** (p. 815)
    Pappus erect 13
13. Phyllaries connate, with oil glands; pappus of
    mixed small and large setae .... 106. **Tagetes** (p. 725)
    Phyllaries free, without oil glands; pappus of 2
    aristae .... 129. **Glossocardia** (p. 813)
14. Ray florets absent 15
    Ray florets present (but may be very small, e.g. in
    Enydra) 20
15. Pappus of bristles or setae 16
    Pappus absent, or of reduced hairs < 0.2 mm long 18

16. Florets orange to red; pappus often of hooked
    bristles .... 103. **Hypericophyllum** (p. 719)
    Florets yellow or white; pappus of a few bristles or
    setae 17
17. Annual; achenes without corky margin .... 108. **Acmella** (p. 730)
    Perennial; achenes with corky margins .... 109. **Spilanthes** (p. 735)
18. Capitula in congested glomerules, these axillary
    and sessile .... 105. **Flaveria** (p. 725)
    Capitula free or in stalked glomerules 19
19. Achene surrounded by connate phyllaries .... 115. **Lagascea** (p. 754)
    Phyllaries free .... 108. **Acmella** (p. 730)
20. Ray florets red, mauve, pink or purple 21
    Ray florets white, yellow or yellow-orange 24
21. Leaves pinnatisect with linear segments; achenes
    with tapering beak .... 127. **Cosmos** (p. 808)
    Leaves entire or pinnatisect with wide segments;
    achenes not beaked 22
22. Capitula axillary and subsessile; rays 6–9 mm long 114. **Aspilia** (p. 746)
    Capitula terminal and stalked; rays much longer 23
23. Plants with tuberous roots .... (cultivated) *Dahlia*
    Roots not tuberous .... (cultivated) *Zinnia*
24. Achenes of outer florets covered in spiny phyllaries 121. **Acanthospermum** (p. 766)
    Achenes free, not covered in spiny phyllaries or
    paleae 25
25. Achenes winged (at least the outer ones) 26
    Achenes not winged (but may have a narrow margin) 27
26. Pappus of two tiny teeth; rays large .... (cultivated) *Coreopsis*
    Pappus of 2–3 awns 2–4.5 mm long; rays 1.5–3 mm
    long .... 113. **Synedrella** (p. 744)
27. Pappus absent, or of a minute rim only 28
    Pappus present: of scales, bristles or aristae, or a
    small crown 0.5–1 mm high 38
28. Leaves divided .... 126. **Bidens** (p. 778)
    Leaves simple or lobed, not divided 29
29. Ray < 3 mm long 30
    Ray > 3 mm long 36
30. Capitula 1–4-flowered, in glomerules; leaf-bases
    connate .... 105. **Flaveria** (p. 725)
    Capitula not in glomerules 31
31. Phyllaries 4, in 2 pairs; leaves glandular-punctate 122. **Enydra** (p. 768)
    Phyllaries more than 4; leaves not glandular-
    punctate (*Micractis* has glands on lower leaf
    surface) 32
32. Phyllaries with stalked glands .... 125. **Sigesbeckia** (p. 776)
    Phyllaries without stalked glands 33
33. Ray florets white (disc florets may be yellow) 34
    Ray florets yellow 35
34. Rays > 5; achenes subglabrous; pappus absent or
    of a tiny rim .... 107. **Eclipta** (p. 728)
    Rays 4–5(–8); achenes hairy (pappus usually
    present, of narrow scales to 1.5 mm, but
    sometimes absent) .... 119. **Galinsoga** (p. 761)
35. Leaves glabrous or sparsely pilose; receptacle
    conical, the whole capitulum becoming
    conical in fruit .... 108. **Acmella** (p. 730)
    Leaves pubescent and glandular beneath;
    receptacle flat .... 124. **Micractis** (p. 776)

36. Leaves often lobed; phyllaries 1-seriate; florets
      white . . . .   116. **Montanoa** (p. 757)
      Leaves unlobed; phyllaries 2–3-seriate; florets
      yellow to orange 37
37. Paleae forming a tube around the florets or
      achene, very visible, acuminate . . . .   111. **Melanthera** (p. 737)
      Paleae ± flat . . . .   123. **Guizotia** (p. 770)
38. Pappus of mixed setae and scales; phyllaries with
      oil glands . . . .   106. **Tagetes** (p. 725)
      Pappus of either scales or bristles or aristae, not
      mixed, or of a small cup-like crown; phyllaries
      without oil glands 39
39. Pappus of 10–20 scales 40
      Pappus of bristles or aristae, or a small crown 41
40. Shrub; ray cream to yellow, 3–6 mm long; scales
      2.5–4.3 mm long . . . .   (cultivated) *Calea*
      Annual herbs; ray white, < 2 mm; scales < 1.5 mm
      long . . . .   119. **Galinsoga** (p. 761)
41. Pappus of 2–4 aristae, often barbed 42
      Pappus of bristles, or a small crown 43
      Note – aristae are needle-like and quite thick, bristles are more hair-like

Bidens: aristae   Melanthera bristles   Tridax bristles   Aspilia: crown

42. Achenes without a beak 47
      Achenes with a beak 5–15 mm long . . . .   127. **Cosmos** (p. 808)
43. Pappus of many plumose bristles . . . .   120. **Tridax** (p. 764)
      Pappus of bristles, often few, never plumose, or a
      small crown 44
44. Pappus a laciniate crown with 1–2 longer fine
      setae or bristles . . . .   114. **Aspilia** (p. 746)
      Pappus of free bristles 45
45. Receptacle conical; capitulum becoming conical
      in fruit . . . .   108. **Acmella** (p. 730)
      Receptacle ± flat 46
46. Ray floret white, 1–1.5 mm long . . . .   110. **Blainvillea** (p. 735)
      Ray floret yellow, 11–20 mm long . . . .   111. **Melanthera** (p. 737)
47. Achenes with ciliate margins . . . .   *Acmella uliginosa*
      Achenes without ciliate margins 48
48. Stems usually round; aristae on achene neither
      flat nor caducous . . . .   126. **Bidens** (p. 778)
      Stem ± tetragonal; aristae flat and caducous . . .   *Melanthera abyssinica*

Cultivated species:

Calea urticifolia (*Mill.*) *DC.*, Prodr. 5: 674 (1836); Lisowski, Asterac. Fl. Afr. Centr.
1: 184, t. 40 (1991). Syn. *Solidago urticifolia* Mill., Gard. Dict. ed. 8: Solidago 30 (1768)
      Originally from Central America; cultivated in gardens and sometimes an escape,
but does not seem to naturalize. A shrub 1–3.5 m high; leaves opposite, ovate, 3–13
× 1–8 cm, 3-veined from base. Capitula in small terminal cymes; ray florets 4–7, cream

to yellow, 3–6 mm long; disc florets yellow, 4.5–6 mm long. Achenes blackish, 2–2.7 mm long, the pappus of 10–20 subulate scales 2.5–4.3 mm long. Kenya, Nairobi Arboretum, Feb. 1952, *Williams Sangai* 349! & July 1955, *T.C.Brown* in EA 10878! & Feb. 1980, *Gillett* 22757!. Lisowski states this taxon occurs in Tanzania, but we have not seen specimens from there.

Coreopsis *L.*
Achenes flattened; pappus of 2 small tufts of bristles < 1 mm long.
Coreopsis grandiflora *Nutt.*, Fl. S. U.S. 235 (1860); U.O.P.Z.: 213 (1949).
A cultivated herb with large yellow ray florets. Kenya, Nairobi, July 1965, *Patel* in EA 13189!; Chiromo, Jan. 1970, *Mathenge* 528!; Tanzania, Lushoto, Dec. 1972, *Kyessi* 12!
Coreopsis lanceolata *L.*, Sp. Pl.: 908 (1753); D.J.N. Hind in Fl. Masc. 109: 196 (1993).
A cultivated carpet-forming herb with large yellow ray florets. Uganda, Masaka, July 1971, *Lye & Katende* 6506!.

Dahlia *Cav.* – note the names of these popular ornamentals are not authoritative – there are many cultivated Dahlia species and cultivars and much confusion in identification!
Dahlia excelsa *Benth.* in Maund, Botanist, 2: t. 88 (1838).
Perennial herb with tuberous roots; opposite pinnate leaves; capitula with large mauve-pink ray florets. Kenya – Kiambu District: Muguga, July 1964, *Greenway* 11759!; Tanzania – Lushoto District: Lushoto, June 1964, *Semsei* 3866! & Feb. 1970, *Shabani* 525! & Amani, Oct. 1969, *Ngoundai* 412!
Dahlia imperialis *Roezl.* in Regel, Gartenfl.: 243, t. 407 (1863).
A perennial herb to 3 m high with large mauve ray florets. Kenya – Nairobi, Feb. 1953, *Jex-Blake* H51/53! & Aug. 1965, *Kokwaro* 264!; Tanzania – Iringa District: Mufindi, Lugoda Tea Estate, May 1968, *Renvoize & Abdallah* 2079!

Gaillardia *Foug.*
U.O.P.Z.: 270 (1949) mentions "spp. and vars." of this American genus cultivated on Zanzibar and Pemba, but does not specify taxa. We have seen no specimens.

Helianthus *L.*
Helianthus annuus *L.*, Sp. Pl.: 904 (1753); U.O.P.Z.: 293 (1949); Maquet in Fl. Rwanda 3: 630 (1985); D.J.N. Hind in Fl. Masc. 109: 190 (1993).
Sunflower, originally from America, is cultivated in many places in East Africa for its seeds, as an oil crop. Occasionally it seems to escape and become naturalised, as in **T** 6, Uzaramo District: Dar es Salaam on airport road, Apr. 1977, *Wingfield* 3392a! and **T** 8, Songea District: Mkaku River SW of Kitai, Mar. 1956, *Milne-Redhead & Taylor* 9075! These escapes are small annual herbs, 20–120 cm high. Leaves alternate, ovate, to 8 × 3.5 cm and scabridulous to the touch, petiolate. Capitula terminal and solitary; ray florets ± 14, yellow, 35 × 10 mm; disc florets very many, yellow orange. In cultivation the plant and all its parts may become much larger. Both the cited specimens came from road-sides. Fig. 143 (page 707). Also Kenya, Athi River, *Luke & Luke* 10291.
U.O.P.Z.: 293–294 (1949) also mentions *H. debilis* and *H. tuberosus* (Jerusalem artichoke) as being grown in Zanzibar and Pemba. We have seen no specimens of either species; but we know it is growing in Hort. Luke in Karen and Tigoni.

Thymophylla *Lag.*
Thymophylla tenuiloba (*DC.*) *Small*, Fl. SE U.S.: 1295 (1903); D.J.N. Hind in Fl. Masc. 109: 225 (1993). Syn. *Hymenatherum tenuilobum* DC., Prodr. 5: 642 (1836); U.O.P.Z.: 301 (1949), as *tenuifolium*.
Annual or perennial herb 10–30 cm high; stems puberulous. Leaves usually alternate, pinnatifid, 1.5–3 cm long, the 9–13 segments filiform, aromatic. Capitula terminal and solitary; peduncle 3–5 cm long; involucre 5–6 mm long, 1-seriate. Ray florets yellow, ray 3–4 mm long; disc florets yellow, ± 2.5 mm long. Achenes ± 3 mm long; pappus of 10 oblong scales, each with 3 aristae at apex. Originally from Mexico

FIG. 143. *HELIANTHUS ANNUUS* — **1**, habit, × ²/₃; **2**, mature leaf, × ²/₃; **3**, floret, × 6. 1–3 from *Hilliard* 52/73. Drawn by Juliet Williamson.

FIG. 144. *THYMOPHYLLA TENUILOBA* — **1**, habit, × ²/₃; **2**, capitulum, × 6. 1–2 from *Pole-Evans & Erens* 5006. Drawn by Juliet Williamson.

and Texas; two cultivated specimens are: 10 km S of Mombasa, Dec. 1975, *Bally* 16953!; Kwale District: Ras Kikadini, June 1983, *Robertson* 3599!; and also cultivated on Zanzibar and possibly Pemba (U.O.P.Z.), sometimes escaped and naturalised, as in Mombasa, Sep. 1953, *Jex-Blake* EA 10359! Fig. 144.

Zinnia *L.*

Zinnia violacea *Cav.*, Ic. 1: t. 57, p. 81 (1791). Syn. *Z. elegans* Jacq., Ic. Pl. Rar. 3: 15, t. 589 (1793) & Collect.: 152 (1796); U.O.P.Z.: 492 (1949); Wild in Kirkia 6: 54 (1967); D.J.N. Hind in Fl. Masc. 109: 177 (1993).

Originally from Mexico, now widely cultivated; occasionally it may become an escape but it does not seem to naturalize. It resembles the next species but has larger leaves and larger ray florets, up to 25 × 12 mm. Nairobi: City Park, Dec. 1970, *Mathenge* 733!

Zinnia peruviana (*L.*) *L.*, Syst. Nat. ed. 10: 1221 (1759); Wild in Kirkia 6: 54 (1967); D.J.N. Hind in Fl. Masc. 109: 179, t. 59 (1993).

Annual herb, 30–60 cm high. Leaves (sub-)sessile, ovate or oblong, to 4.5 × 2 cm, scabridulous. Capitula terminal and solitary; ray florets red, pink or purple, to 15 × 11 mm; disc florets brownish. Pappus absent. Originally from Central and South America, now a widely cultivated plant that sometimes becomes naturalized and has been described as a weed: **K** 4, Nairobi and Kiambu, July 1948, *Bally* 6321! & Nairobi, Kabete, without date, *Gardner* 3596! **T** 1, Mwanza District: near Mwanza, Oct. 1932, *Geilinger* 3205! The two Kenyan specimens came from coffee plantations; the Tanzanian specimen has no indication of habitat.

U.O.P.Z.: 492 (1949) also mentions *Z. haageana*, but we have seen no specimens.

101. BLEPHARISPERMUM

DC. in Wight, Contrib. Bot. Ind. 11 (1834); T. Eriksson in Plant Syst. Evol. 182: 149–227 (1992)

Shrubs or trees, sometimes scrambling. Leaves alternate, sometimes fasciculate on short-shoots, simple. Capitula in dense heads of heads, the **glomerules**, these solitary or grouped in cymes, each glomerule formed of many capitula and usually subtended by bracts, capitula disciform. Phyllaries 3 per capitulum (primary head); receptacle paleate. Florets 2–6, 2 filiform female and 2–4 functionally male disc florets. Anthers caudate. Achenes 3-angled, often with twin-hairs (hairs of two parallel rows of cells); pappus of scales or absent.

15 species, Africa to India and Sri Lanka.

1. Leaves with dense glands beneath 2
   Leaves without glands 3
2. Leaves with entire margin; glomerules in much branched
   inflorescences, the branches arching .... 1. *B. arcuatum*
   Leaves with dentate margins; glomerules in little-
   branched inflorescences, the branches straight ...... 2. *B. pubescens*
3. Short-shoots present 4
   Short-shoots absent 5
4. Leaves lanceolate or oblanceolate, the apex obtuse to
   acuminate .... 3. *B. ellenbeckii*
   Leaves spatulate, the apex retuse .... 4. *B. minus*
5. Bracts of glomerule and individual capitula truncate ... 9. *B. villosum*
   Bracts of glomerule and individual capitula acute to
   acuminate 6
6. Leaves densely pubescent or tomentose, the hairs to
   1 mm long .... 6. *B. canescens*
   Leaves glabrous or sparsely pubescent with hairs < 0.5 mm 7
7. Glomerules obovoid, with two series of common bracts at
   base .... 7. *B. brachycarphum*
   Glomerules globose, without or with just one series of
   common bracts 8
8. Stalks of glomerules glabrous; all bracts subtend capitula . 5. *B. xerothamnum*
   Stalks of glomerules pubescent just below the glomerule;
   usually a few to one row of common bracts, without
   capitula .... 8. *B. zanguebaricum*

1. **Blepharispermum arcuatum** *T.Eriksson* in Plant Syst. Evol. 182: 189, fig. 21 (1992). Type: Tanzania, Lushoto District: Mkuzi, *Drummond & Hemsley* 2069 (S, holo., B, BR!, EA, K!, LISC, iso.)

Evergreen shrub, scrambling or climbing, to several m high; young branches zig-zag, pubescent. Leaves ovate to lanceolate, 6–11 cm long, 2–5.5 cm wide, base truncate to attenuate, margins entire, apex shortly acuminate, puberulous above, densely pubescent and densely glandular beneath; petiole 1–3 cm long. Glomerules in stalked leafy cymes near the ends of branches, the individual glomerules globose and 10–17 mm across, with up to 30 capitula; capitula with 2 female and 2 male florets. Phyllaries 3.5–6 mm long; paleae 5–7 mm long. Female florets 1.5–2.5 mm long, glandular; male florets 3.5–5 mm long, glandular. Achenes of female florets narrowly obovoid, 3–4 mm long, with twin-hairs; pappus of 8–15 flat scales 0.5–2 mm long. Achenes of male florets 1–2 mm long, with pappus of 10–19 scales 1–3 mm long.

TANZANIA. Arusha District: 30 km on Kibaya–Zoissa road, July 1965, *Leippert* 6034; Lushoto
    District: Mkuzi, June 1990, *Eriksson et al.* 561; Iringa District: Kihesa Hill, July 1984, *Mhoro* 4347
DISTR. **T** 2 (fide EA), 3, 7; Madagascar
HAB. Woodland or forest; 1500–1650 m
USES. None recorded on specimens from our area
CONSERVATION NOTES. Four specimens known from Tanzania, one from Madagascar; probably
    vulnerable (VU-D2)

2. **Blepharispermum pubescens** *S.Moore* in J.L.S. 37: 168 (1905); Maquet in Fl.
Rwanda 3: 618, fig. 190/2 (1985); Lisowski, F.A.C. Compositae (Inuleae): 213 (1989);
T. Eriksson in Plant Syst. Evol. 182: 192, fig. 23–24 (1992). Type: Uganda, Ankole
District: near Mulema, *Bagshawe* 225 (BM!, lecto., chosen by Eriksson)

Evergreen scrambling shrub 1.5–4.5 m long; young branches pubescent. Leaves
triangular, ovate or lanceolate, 2–10 cm long, 1–6.4 cm wide, base truncate to
attenuate, margins entire or more usually sparsely dentate, especially near the
base, apex acute to acuminate, pubescent and glandular on both surfaces; petiole
5–22 mm long. Glomerules 1–3 together near the ends of branches, the individual
glomerules globose and 10–18 mm across, with up to 50 capitula; capitula with
2(–3) female and 2 male/bisexual florets; phyllaries 4.5–6 mm long; paleae 5–7 mm
long. Flowers white to cream; female florets 2–3 mm long, glandular; male florets
3.5–4.5 mm long, glandular. Achenes of female florets narrowly obovoid, 3–4 mm
long, with twin-hairs; pappus of 8–12 flat scales 0.5–3 mm long. Achenes of male
florets 1–2 mm long, with pappus of ± 10 scales 0.5–3 mm long.

UGANDA. W Nile District: Uleppi, Mar. 1935, *Eggeling* 1696!; Busoga District: S of Buswale,
    Siavona Hill, Mar. 1952, *Wood* 660!; Masaka District: 1–2 km E of Kikoma, Oct. 1969, *Lye &*
    Rwaburindore 4416!
KENYA. Masai District: NE of Lolgorien, Dec. 1979, *Msafiri* 996; Masai Mara, Kerigodwa, July
    1978, *Kuchar* 9294 & Kurao Plains, June 1979, *Kuchar* 10627
TANZANIA. Biharamulo District: Biharamulo Game Reserve, Kigombe, Oct. 1973, *Ludanga*
    1604; Mwanza District: Mwanza, Nov. 1930, *Burtt* 2905! & Bwiru, Apr. 1952, *Tanner* 665!
DISTR. **U** 1–4; **K** 5, 6; **T** 1; Congo (Kinshasa), Rwanda
HAB. Lake-side or riverine thicket and secondary bush, also forest and forest margins near lake;
    1100–1800 m
USES. None recorded on specimens from our area
CONSERVATION NOTES. Not uncommon in a widespread habitat; least concern (LC)

3. **Blepharispermum ellenbeckii** *Cufod.* in Nuov. Giorn. Bot. Ital. 50: 107 (1943);
T. Eriksson in Plant Syst. Evol. 182: 195, fig. 25 (1992). Type: Ethiopia, Dande fort,
Corradi 2129 (FT, lecto., W, isolecto., chosen by Eriksson)

Shrub 1–3 m high, deciduous, sparsely branched; young branches glabrous. Leaves
subsessile, lanceolate or oblanceolate, 1–5.5 cm long, 0.3–1.2 cm wide, base cuneate,
margins entire and revolute, apex obtuse to acuminate, glabrous except for some
scattered hairs on the margins and sometimes on the veins beneath. Glomerules
solitary and terminal, globose and 10–16 mm across, with up to 60 capitula; capitula
with 2 female and 2 male florets; phyllaries 2.5–3.5 mm long; paleae 3.5–5.5 mm
long. Flowers white; female florets 2–3 mm long, glabrous; male florets 3.5–5 mm
long, glabrous. Achenes of female florets narrowly obovoid, 2–3.5 mm long, with
twin-hairs; pappus of 2 lateral scales 1–2.5 mm long and 4–10 shorter scales on the
abaxial side. Achenes of male florets 1–1.5 mm long, with pappus of 10–16 flat scales
2.5–3.5 mm long.

KENYA. Northern Frontier District: Moile Hill, 11 km S of Laisamis, Nov. 1977, *Carter &*
    Stannard 737! & Wajir, Mission compound, Apr. 1978, *Gilbert & Thulin* 1138!; Tana R. District:
    Kora, Dec. 1984, *Mungai & Rucina* 404/84
DISTR. **K** 1, 4, 7; Sudan, Ethiopia, Somalia

HAB. Dry deciduous bushland, often in the more open parts; 150–1050 m
USES. None recorded on specimens from our area
CONSERVATION NOTES. Widespread in a common habitat; least concern (LC)

4. **Blepharispermum minus** *S.Moore* in J.B. 40: 340 (1902); T. Eriksson in Plant Syst. Evol. 182: 199, fig. 26 (1992). Type: Kenya, Kwale District, Taru [Taro], *Kässner* 521 (BM!, holo., K!, W, Z, iso.)

Deciduous shrub 1–4 m high, sparsely branched; branchlets glabrous or nearly so. Leaves coriaceous, greyish green, spatulate, 0.5–2.5 cm long, 0.5–1.5 cm wide, base cuneate, margins entire, apex retuse (rarely acute), glabrous and without glands; petiole 1–6 mm long. Glomerules solitary and terminal on short side shoots or spurs, globose and 7–11 mm across, with up to 35 capitula; capitula with 2 female and 2 male florets; phyllaries 2.5–3.5 mm long; paleae 3.5–5 mm long. Flowers ?white; female florets 1.5–2 mm long, glabrous, the lobes glandular; male florets 2–3 mm long, glandular. Achenes of female florets narrowly obovoid, 1.5–2.5 mm long, with twin-hairs; pappus of 2 lateral scales 1–1.5 mm long and 2–5 shorter scales on the abaxial side. Achenes of male florets 0.5–2 mm long, with pappus of 7–16 flat scales 0.5–2 mm long.

KENYA. Northern Frontier District: 50 km NE from Garissa, May 1977, *Kuchar et al.* 5847; Tana River District: Tana R. National Primate Reserve, Mar. 1990, *Luke et al.* TPR 650!; Kilifi District: Galana Ranch near Kadakatha, June 1975, *Bally* 16886!
DISTR. **K** 1, 7; not known elsewhere
HAB. Open bushland or bushed grassland, scattered tree grassland; 30–450 m
USES. None recorded on specimens from our area
CONSERVATION NOTES. Habitat not uncommon, but the species is restricted in distribution. Least concern verging to near threatened? (LC–NT)

5. **Blepharispermum xerothamnum** *Mattf.* in N.B.G.B. 13: 295 (1936); T. Eriksson in Plant Syst. Evol. 182: 203, fig. 29 (1992). Type: Tanzania, Dodoma District: Mt Mlimwa, *Troll* 5588 (B, lecto., chosen by Eriksson; W, iso.)

Deciduous or evergreen scrambling shrub 1–5 m high; stem much branched, branches virgate and arching; branchlets glabrous or rarely pubescent. Leaves leathery, yellow-green, ovate to elliptic, 3–12 cm long, 1–5.5 cm wide, base attenuate, margins entire to obscurely dentate or serrate, apex acute to slightly acuminate, glabrous and without glands or with scattered hairs above and scabridulous beneath; petiole 5–15 mm long. Glomerules 2–20 together in cymes or rarely solitary, terminal; individual glomerules globose and 9–16 mm across, with up to 130 capitula; capitula with 2 female and 2 male florets; phyllaries 3–4 mm long; paleae 4–6 mm long. Flowers greenish turning white; female florets 1.5–2.5 mm long, glabrous or with a few glands; male florets 3–4 mm long, glabrous or rarely glandular. Achenes of female florets broadly obovoid, 2.5–3.5 mm long, with twin-hairs; pappus of 2 lateral scales 1–2 mm long and 1–6 shorter scales on the abaxial side. Achenes of male florets 0.5–1.5 mm long, with pappus of 7–14 flat scales 0.5–3 mm long.

TANZANIA. Dodoma District: Chenene Forest Reserve, May 1978, *Ruffo* 1332!; Mpwapwa District: Mpwapwa, on Gulwe road, Apr. 1932, *Burtt* 3916! & 5 km S of Mpwapwa, Apr. 1988, *Bidgood et al.* 945!
DISTR. **T** 5; not known elsewhere
HAB. Bushland or woodland, may be common on termite hills; 600–1250 m
USES. None recorded on specimens from our area
CONSERVATION NOTES. Status of habitat unknown; 11 specimens seen. Data deficient (DD)

6. **Blepharispermum canescens** *T.Eriksson* in Plant Syst. Evol. 182: 206, fig. 31 (1992). Type: Tanzania, Iringa District: Kidatu, *Mhoro* 808 (UPS, holo., DSM, iso.)

Fig. 145. *BLEPHARISPERMUM ZANGUEBARICUM* — **1**, habit, × ²/₃; **2**, glomerule, × 2; **3**, capitulum, × 10; **4**, female floret, × 12; **5**, male floret, × 12. 1–5 from *Abdallah & Vollesen* 95/10. Drawn by Juliet Williamson.

Shrub, probably deciduous, up to 1 m high; young branches densely pubescent. Leaves elliptic to slightly ovate, 6.5–10 cm long, 2.5–4.5 cm wide, base attenuate, margins entire to obscurely crenate, apex acute, pubescent to tomentose on both surfaces, eglandular; petiole 7–12 mm long. Glomerules 3–5 together in cymes, terminal; individual glomerules globose and 9–10 mm across, with up to 100 capitula; capitula with 2 female and 2 male florets; phyllaries 2–2.5 mm long; paleae 3–4 mm long. Flowers of unknown colour; female florets 1.5 mm long, glabrous; male florets 2.5 mm long, glabrous. Achenes of female florets broadly obovoid, 2–2.5 mm long, with twin-hairs; pappus of 2 lateral scales 1 mm long and 0–1 shorter scales on the abaxial side. Achenes of male florets 0.5–1 mm long, with pappus of 6–11 flat scales 0.5–1.5 mm long.

TANZANIA. Iringa District: Kidatu, Mar. 1971, *Mhoro* 808 (type)
DISTR. **T** 7; known from only the type
HAB. Riverine forest; altitude?
USES. None recorded on specimens from our area
CONSERVATION NOTES. The type locality has now been destroyed due to the building of a hydro-electric dam; data deficient (DD) but possibly extinct

7. **Blepharispermum brachycarphum** *Mattf.* in N.B.G.B. 13: 293 (1936); T. Eriksson in Plant Syst. Evol. 182: 208, fig. 32 (1992). Type: Tanzania, Lindi District: Lake Lutamba, *Schlieben* 6215 (M, lecto., B, BM!, BR!, HBG, S, Z, isolecto., chosen by Eriksson)

Scrambling shrub, probably evergreen, 1–2 m high; much branched, the young branches pubescent. Leaves ovate, lanceolate or elliptic, 4–12 cm long, 1.5–6.6 cm wide, base truncate to attenuate, margins entire or obscurely crenate, apex acute to acuminate, glabrous to sparsely pubescent and eglandular on both surfaces; petiole 5–15 mm. Glomerules 2–80 together in cymes, terminal; individual glomerules obovoid and 9–13 mm across, with up to 80 capitula; capitula with 2 female and 2 male florets; phyllaries 3 mm long; paleae 4.5–5.5 mm long. Flowers of unknown colour; female florets 1.5–2.5 mm long, glabrous; male florets 2.5–3.5 mm long, glabrous. Achenes of female florets obovoid, 2–3 mm long, with twin-hairs; pappus of 2 lateral scales 1–2 mm long and 1–4 shorter scales on the abaxial side. Achenes of male florets 0.5–1 mm long, with pappus of 6–15 flat scales 0.5–3 mm long.

TANZANIA. Lindi District: Mt Mtandamula, May 1903, *Busse* 2724 & Lake Lutamba, Milola stream valley, Sep. 1934, *Schlieben* 5275! & Chitoa Forest Reserve, June 1995, *G.P.Clarke* 66!
DISTR. **T** 8; N Mozambique
HAB. Woodland near lakes or rivers, dry forest; 150–450 m
USES. None recorded on specimens from our area
CONSERVATION NOTES. Data deficient (DD)

8. **Blepharispermum zanguebaricum** *Oliv. & Hiern*, F.T.A. 4: 336 (1877); T. Eriksson in Plant Syst. Evol. 182: 211, fig. 33 (1992). Type: Kenya, Mombasa, *Kirk* s.n. (K!, holo.)

Evergreen shrub, 1–3 m high or scrambling to 8 m long; stem near base up to 5 cm in diameter; young branches densely pubescent. Leaves ovate, lanceolate or elliptic, 3–12(–19) cm long, 1–7(–10) cm wide, base truncate to obtuse or cuneate, margins entire to dentate, apex acute to acuminate, glabrous and sparsely pubescent, especially on veins beneath, and eglandular; petiole 5–12 mm. Glomerules 3–40 together in cymes, terminal; individual glomerules globose and 11–15 mm across, with up to 70 capitula; capitula with 2 female and 2 male florets; phyllaries 2–4 mm long; paleae 4–5.5 mm long. Flowers greenish white or white, fragrant; female florets 2–2.5 mm long, glabrous; male florets 3–4 mm long, glabrous. Achenes of female florets obovoid, 2–3.5 mm long, with twin-hairs; pappus of 2 lateral scales 1–1.5 mm long and 1–4 shorter scales on the abaxial side. Achenes of male florets 0.5–1.5 mm long, with pappus of 7–13 flat scales 0.5–3 mm long. Fig. 145 (page 712).

KENYA. Kitui District: Mutha Hill, Jan. 1942, *Bally* 1645!; Masai District: Laitokitok [Loitokitok], July 1962, *Ibrahim* 667!; Teita District: Tsavo National Park, Mzinga Hill, Apr. 1966, *Gillett* 17225!

TANZANIA. Same District: Mkomazi Game Reserve, Ibaya Hill, Apr. 1995, *Abdallah & Vollesen* 95/10!; Korogwe District: Ndolwa Forest Reserve, Dec. 1960, *Semsei* 3111!; Bagamoyo District: Pongwe, Apr. 1970, *Harris et al.* 4334!

DISTR. **K** 4, 6, 7; **T** 2, 3, 6; not known elsewhere

HAB. Dry forest or forest margins, coastal woodland, thicket and secondary bushland; 1–1300 m

USES. Minor medicine for gonorrhoea and stomach-ache

CONSERVATION NOTES. Fairly widespread; least concern (LC)

NOTE. The altitude of 1825 m for Laitokitok is very much out of the normal range of this species.

9. **Blepharispermum villosum** *O.Hoffm.* in E.J. 38: 202 (1906); T. Eriksson in Plant Syst. Evol. 182: 219, fig. 39 (1992). Type: Ethiopia, Arussi-Galla, Mana, *Ellenbeck* 2001 (W, fragm., lecto., chosen by Eriksson)

Deciduous shrub 1–3 m high, much branched, once described as scrambling; young branches pubescent or glabrous. Leaves ovate to elliptic, 2–9.5 cm long, 1–6 cm wide, base subcordate to attenuate, margins entire or nearly so, apex acute, densely pubescent to almost glabrous on both surfaces and eglandular; petiole 4–14 mm. Glomerules solitary, terminal; individual glomerules globose and 10–15 mm across, with up to 75 capitula; capitula with 2 female and 2 male florets; phyllaries 2–3 mm long; paleae 3–5 mm long. Flowers greenish white, white or pale yellow; female florets 1.5–2.5 mm long, glabrous; male florets 2–3 mm long, glabrous. Achenes of female florets obovoid, 1.5–2.5 mm long, with twin-hairs; pappus of 2 lateral scales 0.5–1 mm long and 2–6 shorter scales on the abaxial side. Achenes of male florets 0.5–1.5 mm long, with pappus of 7–17 flat scales 0.5–2.5 mm long.

KENYA. Northern Frontier District: Ajas near Buna, Jan. 1949, *Dale* 706! & base of Kulal, July 1958, *Verdcourt* 2267! & 18 km on Ramu–Mandera road, May 1978, *Gilbert & Thulin* 1393!

DISTR. **K** 1, 4; Ethiopia, Somalia

HAB. Dry bushland, especially *Acacia–Commiphora*, where it may be locally common; 150–1000(–1350) m

USES. None recorded on specimens from our area

CONSERVATION NOTES. Least concern (LC)

SYN. *B. fruticosum* Klatt c. *lapathifolium* Chiov., Fl. Somal.: 200 (1929). Type: Ethiopia, El Ure, *Paoli* 1071 (FT, holo.) – the 'c.' might stand for form, variety or subspecies.

EXCLUDED SPECIES

Blepharispermum fruticosum Klatt, a name often used for East African specimens, is a taxon restricted to Ethiopia and Somalia.

102. **ATHROISMA**

DC. in Arch. Bot. 2: 516 (1833); T. Eriksson in Bot. J. Linn. Soc. 119: 101–184 (1995)

Perennial herbs or subshrubs. Leaves alternate, simple or dissected. Capitula in dense heads of heads, the **glomerules**, disciform or discoid, the glomerules arranged in synflorescences or solitary; phyllaries 0–2 per capitulum; receptacle paleate. Capitula with 2 filiform female florets and few to many hermaphrodite disc florets. Anthers caudate. Achenes slightly flattened; pappus absent or occasionally a ring of connected twin-hairs.

11 species in Africa, Madagascar and E Asia.

1. Leaves gland-dotted on both surfaces or at least above 2
   Leaves without gland-dots 4
2. Leaves sessile . . . . 4. *A. gracile* ×
boranense
   Leaves with petiole 10–25 mm long 3
3. Young branches densely pubescent; leaves 2–5 cm wide   1. *A. hastifolium*
   Young branches ± glabrous; leaves 0.7–1.5 cm wide . . . 2. *A. pusillum*
4. Glomerules sessile at the nodes . . . . 5. *A. stuhlmannii*
   Glomerules terminal, stalked 5
5. Perennial herbs, rarely annual; branching sympodial;
   K 1–7; **T** 1–3 . . . . 3. *A. gracile*
   Annual herbs; branchng monopodial; **U** 1–3; **T** 4 . . . 6. *A. inevitabile*

1. **Athroisma hastifolium** *Mattf.* in N.B.G.B. 13: 298 (1936); T. Eriksson in J.L.S. 119: 151, t. 29 (1995). Type: Tanzania, Mbulu District: between Oldeani and Lake Manyara, *Troll* 5605 (B, lecto., chosen by Eriksson)

Annual or perennial herb, 15–60 cm high, the base slightly woody; branches densely pubescent, usually glandular. Leaves triangular to ovate, 3–6.5 cm long, 2–5 cm wide, base truncate to almost hastate to almost attenuate, margins serrate, apex acute to acuminate, pubescent, glandular, 3-veined from base; petiole 10–25 mm. Glomerules solitary but branches close together may give an impression of 2–6 together, each 13–20 mm in diameter, with 20–40 capitula; phyllaries 3–4 mm long; paleae 4–5 mm long. Florets white or slightly pink, female florets 2, 2–2.5 mm long, glandular; disc florets 10–15, 2.5–3 mm long, glandular. Achenes ovoid to ellipsoid, 1–1.5 mm long, slightly hairy, without glands; pappus absent.

KENYA. Masai District: 48 km Magadi Road, Jan. 1933, *van Someren* 2468! & Ngong Hills near Kekonyokie, June 1963, *Verdcourt* 3664! & Kajiado to Namanga road, May 1974, *Kokwaro* 3430!
TANZANIA. Mbulu District: Lake Manyara National Park, May 1961, *Karani* 111! & S of Msasa R., Dec. 1963, *Greenway & Kirrika* 11173!; Moshi District: Ngare Nairobi, July 1943, *Greenway* 6719!
DISTR. **K** 6; **T** 2; not known elsewhere
HAB. Dry bushland, bushed grassland, scattered tree grassland, abandoned cultivations; possibly a pioneer, often on clay; 950–1950 m
USES. None recorded on specimens from our area
CONSERVATION NOTES. Fourteen specimens from Tanzania, five from Kenya, but from a range of habitats; probably least concern (LC)

2. **Athroisma pusillum** *T.Eriksson* in J.L.S. 119: 157 (1995). Type: Kenya, Kwale District: between Samburu & Mackinnon road, *Drummond & Hemsley* 4073 (K!, holo., B, BR!, EA, FT, K!, iso.)

Annual herbs 25–50 cm high; young branches almost glabrous. Leaves narrowly triangular, 2–6 cm long, 0.7–1.5 cm wide, base shallowly cordate, margins dentate near base, apex acuminate, glandular, 5-veined from base; petiole 10–20 mm long. Glomerules solitary on short lateral branches, 5–9 mm in diameter, with 35–45 capitula; phyllaries absent; paleae 2 mm long. Florets pale mauve, female florets absent; disc florets 9–15, 1.5 mm long, glandular. Achenes ellipsoid, 1 mm long, slightly hairy, without glands; pappus absent.

KENYA. Teita District: Ndara to Dida Harea km 3, Jan. 1972, *Faden & Faden* 72/120!; Kwale District: between Samburu & Mackinnon road, Aug. 1953, *Drummond & Hemsley* 4073!
DISTR. **K** 7; not known elsewhere
HAB. In seasonally swampy areas within *Acacia-Commiphora* bushland; 350–500 m
USES. None recorded on specimens from our area
CONSERVATION NOTES. Known from only two specimens; at least vulnerable (VU-D2)

3. **Athrosima gracile** (*Oliv.*) *Mattf.* in N.B.G.B. 13: 302 (1936); T. Eriksson in J.L.S. 119: 165, t. 36 (1995). Type: Tanzania/Kenya, Kapté country, *Thompson* s.n. (K!, holo.)

Perennial or rarely annual herb 0.1–1 m high, usually much branched; base of stem woody; young branches glabrous. Leaves linear to elliptic or obovate, 2–12 cm long, 0.1–2.5 cm wide, base attenuate, margins entire to dentate or serrate, apex acute to acuminate, glabrous; basal veins 1–3. Glomerules solitary and terminal, 9–16 mm in diameter, with 20–50 capitula; phyllaries absent; paleae 2.5–4 mm long. Florets white, pink or purple, female florets absent; disc florets 8–22, 1.5–2.5 mm long, glandular. Achenes ellipsoid, obovoid or almost globose, 1–1.5 mm long, slightly hairy, without glands; pappus absent or of short hairs. Fig. 146 (page 717).

NOTE. Eriksson states the capitula occur in groups of 1–5(–7), but I believe that is a matter of interpretation. I see them as solitary on small leafy branches, and because the plant is densely branched the heads look close together. Eriksson sees them as branched, leafy synflorescences. Of course, the whole upper part of the plant acts as a large inflorescence!

subsp. **gracile**; T.Eriksson in J.L.S. 119: 168, t. 36 (1995)

Perennial herb 0.1–0.6 m high. Leaves sessile, linear to narrowly elliptic or narrowly obovate, 2–7.5 cm long, 0.1–0.5 cm wide, margins entire or with a few teeth. Paleae 2.5–3.5 mm long. Florets white, pink or pale mauve. Achenes ellipsoid to globose, 1–1.5 mm long, 0.5 mm across, smooth; pappus absent or of small hairs.

KENYA. Northern Frontier District: Marsabit, July 1942, *J. Bally* 1855!; Machakos District: Kilungu, Kithembe Hill, June 1982, *Mwangangi* 2333!; Masai District: Ndara quarantine area, July 1957, *Trapnell* 2343!
TANZANIA. Musoma District: Togoro Plains, Feb. 1968, *Greenway & Kanuri* 13192!; Maswa District: Serengeti, Banagi, Mar. 1965, *Leippert* 5604!; Mbulu District: Kitingi, Mar. 1965, *Hukui* 22!
DISTR. **K** 1, 3, 4, 6; **T** 1, 2; not known elsewhere
HAB. Grassland, bushed grassland, open places in dry bushland, roadsides, dry forest margins; 1200–2100 m
USES. Cattle stomach medicine (*Glover*)
CONSERVATION NOTES. Due to wide range of habitats, probably least concern (LC)

SYN. *Sphaeranthus gracilis* Oliv. in J.L.S. 21: 400 (1885)
    Polycline gracilis (Oliv.) Oliv. in Hook., Ic. Pl. 23: t. 2293 (1894)

subsp. **psylloides** (*Oliv.*) *T.Eriksson* in J.L.S. 119: 170, t. 36 (1995). Type: Tanzania, Kilimanjaro, *Smith* s.n. (K!, holo.)

Annual or perennial herb 0.2–1 m high. Leaves sessile or subsessile, linear to elliptic, 2.5–12 cm long, 0.1–2.5 cm wide, margins entire or dentate to serrate. Paleae 3–4 mm long. Flowers white (pink in bud). Achenes obovoid, 1–1.5 mm long, 0.5–1 mm across, tuberculate or rugose; pappus absent or of hairs.

UGANDA. Busoga District: 6 km W of Kaliso, Sep. 1952, *Wood* 38 (fide EA)
KENYA. Machakos District: Katumani Farm near Machakos, June 1958, *Bogdan* 4604!; Masai District: S of Kajiado–Namanga road, June 1957, *Greenway* 9197!; Teita District: Bura Station, July 1958, *Semlingwa* 21!
TANZANIA. Arusha District: by Ngare Nanyuki road, Dec. 1968, *Richards* 23359!; Pare District: Mkomazi Game Reserve, below Ibaya Camp, May 1995, *Abdallah & Vollesen* 95/157!; Korogwe District: Mombo, Aug. 1966, *Semsei* 4086!
DISTR. **U** 3; **K** 4, 6, 7; **T** 2, 3; Rwanda
HAB. Grassland, bushed grassland, roadsides, in moist depressions; 0–1950 m
USES. None recorded on specimens from our area
CONSERVATION NOTES. Due to wide range of habitats, probably least concern (LC)

SYN. *Polycline psyllioides* Oliv. in Hook., Ic. Pl. 23: t. 2293 (1894)
    P. haareri Dandy in K.B. 1926: 437 (1926). Type: Tanzania, Arusha District: Doinyo Sambu, *Haarer* B150 (K!, holo.)

FIG. 146. *ATHROISMA GRACILE* — **1**, habit, × ²/₃; **2**, glomerule, × 2; **3**, capitulum, × 6; **4**, floret, × 14; **5**, floret cut open, × 10; **6**, anther, × 14; **7**, ripening achene, × 14. 1 from *Luke & Luke* 6173, 2–7 from *Verdcourt* 685. Drawn by Juliet Williamson.

Athroisma psyllioides (Oliv.) Mattf. in N.B.G.B. 13: 302 (1936); Maquet in Fl. Rwanda 3: 616, fig. 189/1 (1985); Blundell, Wild Fl. E. Afr.: pl. 89 (1987); Lisowski, F.A.C. Compositae (Inuleae): 210, t. 44 (1989)

A. haareri (Dandy) Mattf. in N.B.G.B. 13: 302 (1936)

4. **Athroisma gracile** *Cufod.* × **boranense**; T. Eriksson in J.L.S. 119: 150 (1995)

Annual or perennial herb or subshrub 20–40 cm high; young branches densely pubescent. Leaves sessile, narrowly ovate, elliptic or obovate, 2–8 cm long, 0.6–3 cm wide, base attenuate, margins serrate or subentire, apex acute to acuminate, sparsely hairy, glandular, 3-veined from base. Glomerules solitary and terminal on small leafy branchlets, 11–20 mm in diameter; phyllaries 3–3.5 mm long; paleae 3.5–4 mm long. Florets white or slightly pink because of pink anthers, female florets 2 or absent, 1.5–2 mm long; disc florets 2 mm long. Achenes obovoid to ellipsoid, 1–1.5 mm long, slightly hairy and with or without glands; pappus absent.

KENYA. Northern Frontier District: Dida Galgala, Feb. 1953, *Gillett* 15093! & 38 km N of Sabule Airstrip, Nov. 1978, *Brenan et al.* 14841!; Nairobi, Muthaiga, Apr. 1932, *Mainwaring in Napier* 1863!
TANZANIA. Masai District: Serengeti, July 1973, *Mwasumbi* 11172; Mbulu District: Mt Hanang, Feb. 1946, *Greenway* 7567! & Mar. 1965, *Richards* 20002!
DISTR. **K** 1, 4, 7; **T** 2; not known elsewhere
HAB. Grassland, dry bushland, semi-desert scrub, usually in sites where water is present seasonally; 50–1950 m
USES. None recorded on specimens from our area
CONSERVATION NOTES. (not given for hybrids)

NOTE. The Tanzanian specimens lack the female florets, but are different from *A. gracile* in the gland-dotted leaves. *A. boranense* occurs in Ethiopia.

5. **Athroisma stuhlmannii** *O.Hoffm.* in E.J. 20: 233 (1894); T. Eriksson in J.L.S. 119: 174, fig. 40 (1995). Type: ?Tanzania, Shinyanga District: Seke [Sseke], *Stuhlmann* 4194 (B†, holo.); neotype: Moshi District: 5 km NE of Magugu, *Eriksson, Kalema & Leliyo* 538 (B, neo., K!, NHT, S, isoneo.)

Annual herb 15–90 cm high; young branches glabrous. Leaves narrowly ovate to narrowly obovate, 3.5–8.5 cm long, 0.5–2.5 cm wide, base attenuate, margins serrate, apex obtuse to acute, glabrous; 3 main veins from base; petiole absent or up to 17 mm long. Glomerules solitary at the nodes, sessile, 9–18 mm in diameter; phyllaries absent; paleae 2.5–4 mm long. Florets white or slightly pink because of pink anthers, female florets absent; disc florets 1.5–2.5 mm long. Achenes obovoid, 1–1.5 mm long, slightly hairy and glandular near the apex; pappus a ring of basally connected hairs 0.1 mm long.

KENYA. Kisumu District: Nyandala–Kisumu, Dec. 1968, *Kokwaro* 1653 & Kibos, Dec. 1976, *Ochieng* 2 & Nanga, Aug. 1934, *Turner* 6790
TANZANIA. Musoma District: Majimoto, Apr. 1959, *Tanner* 4081!; Mbulu District: 6 km N of Magugu, Apr. 1964, *Welch* 569!; Dodoma District: 7 m on Bahi–Kilimatinde road, Apr. 1988, *Bidgood et al.* 1202!
DISTR. **K** 5; **T** 1, 2, 5; Zambia, Zimbabwe
HAB. Seasonally flooded hardpan soils in the bushed grassland or scattered tree grassland zone; 850–1300 m
USES. None recorded on specimens from our area
CONSERVATION NOTES. Least concern (LC)

6. **Athroisma inevitabile** *T.Eriksson* in J.L.S. 119: 178, fig. 41 (1995). Type: Rwanda, Kibungu, Rwinkavu, *Lewalle* 2611 (BR!, holo., EA, K!, iso.)

Annual herb 20–80 cm high; young branches glabrous. Leaves linear, elliptic, narrowly ovate or narrowly obovate, 3.5–10 cm long, 0.3–3 cm wide, base attenuate, margins serrate to subentire, apex acuminate to acute, glabrous; 3 main veins from base; petiole absent or up to 15 mm long. Glomerules solitary and terminal, 9–15 mm in diameter; phyllaries absent; paleae 3 mm long. Florets white, rarely pale mauve, female florets absent; disc florets 1.5–2 mm long. Achenes obovoid or ellipsoid, 1 mm long, slightly hairy and glandular near the base and apex; pappus absent or of a ring of a few hairs 0.1 mm long.

UGANDA. Acholi District: Achua R., no date, *Liebenberg* 246!; Teso District: Serere, Aug. 1932, *Chandler* 866!; Busoga District: Kaliro Plantation, May 1953, *G.H.S.Wood* 741!
TANZANIA. Kigoma District: Uvinza to Saline, Feb. 1926, *Peter* 36233! & Uvinza, Feb. 1958, *Friend* 36!
DISTR. **U** 1–3; **T** 4; Rwanda
HAB. Swampy or seasonally swampy sites, or grassland near river; 1050–1350 m
USES. None recorded on specimens from our area
CONSERVATION NOTES. Six specimens from Uganda, two from Tanzania, and six from Rwanda; probably least concern (LC)

NOTE. Close to *A. stuhlmannii.*

103. **HYPERICOPHYLLUM**

Steetz in Peters, Reise Mossamb. Bot. 2: 498 (1863); G.V. Pope in Kirkia 10: 104–111 (1975).

Perennial herbs; rhizome woody. Leaves opposite, entire, glandular, margins often scabrid. Capitula usually solitary on long stalks, discoid; phyllaries in several rows; receptacle epaleate. Corolla often orange to red, deeply 5-lobed; anther appendage narrowly ovate. Achenes ribbed; pappus of several bristles or setae, often hooked at apex.

7 species, restricted to Africa.

1. Leaves scabrid or hispidulous . . . .   1. *H. angolense*
   Leaves glabrous or nearly so, often the margins and midrib scabrid 2
2. Pappus setae 5–10 mm long, winged at base, occasionally hooked . . . .   2. *H. compositarum*
   Pappus setae 2–5 mm long, usually unwinged, always hooked   3. *H. elatum*

1. **Hypericophyllum angolense** (*O.Hoffm.*) *N.E.Br.* in J.L.S. 35: 122 (1902); G.V. Pope in Kirkia 10: 106, t. 2 (1975). Type: Angola, Huila, Chela, *Newton* s.n.; Huila, *Antunes* s.n.; Malange, *Teucsz* in *von Mechow* 470 (all B†, syn.)

Perennial herb 0.2–1 m high; rootstock creeping, short and stout with long brown roots; stem ribbed, scabrid. Leaves opposite, the lower appressed to the ground, oblanceolate to narrowly oblanceolate, less often ovate or elliptic, 3–20 cm long, 0.5–10 cm wide, base attenuate and connate with that of the opposite leaf, margins entire, apex obtuse, scabrid or hispid-pilose on both surfaces, glandular; with 3 main veins. Capitula solitary on long stalks, sometimes apparently branched inflorescences with very long stalks to individual heads; involucre 7–15 mm long; phyllaries 4–15 mm long, pubescent or glabrous. Florets many; corolla orange or reddish-orange, 9–10 mm long, pilose. Achenes ellipsoid or narrowly obovoid, 4–9 mm long, ribbed, hispidulous, glandular; pappus of bristles 3–7 mm long, winged near base, hooked at apex.

TANZANIA. Mpanda District: Mpanda–Uvinza road, May 2000, *Bidgood et al.* 4453!; Njombe
   District: 13 km past fork on Mbeya–Njombe road, May 1962, *Boaler* 549! & Songea District:
   73 km on Songea–Njombe road, Mar. 1991, *Bidgood et al.* 2098!
DISTR. **T** 4, 5, 7, 8; Congo (Kinshasa), Angola, Zambia, Malawi, Mozambique
HAB. *Brachystegia* or *Brachystegia–Uapaca* woodland or secondary woodland; 950–1950 m
USES. None recorded on specimens from our area
CONSERVATION NOTES. Least concern (LC)

SYN. *Jaumea angolensis* O.Hoffm. in Bol. Soc. Brot. 10: 178 (1892)
   Hypericophyllum scabridum N.E.Br. in J.L.S. 35: 122, t. 6 (1902). Types: Malawi, between
      Kondowe and Karonga, *Whyte* s.n. & Manganja Hills, *Kirk* s.n. & Shire Highlands near
      Blantyre, *Buchanan* 73 & 439 (all K!, syn.)

   2. **Hypericophyllum compositarum** *Steetz* in Peters, Reise Mossamb. Bot. 2: 498, t.
50 (1863); Verdcourt in K.B. 6: 363 (1952); F.P.S. 3: 37 (1956); G.V. Pope in Kirkia
10: 107 (1975). Type: Mozambique, Boror, Rios de Sena, *Peters* s.n. (B†, holo.)

Perennial herb to 0.3–1.5(–2.4) m high; rhizome short, creeping, with spreading
fleshy roots; young stem sparsely pubescent, with opposite branches in upper part
or simple. Leaves sessile or nearly so, narrowly ovate, 1–18 cm long, 1–6 cm wide,
the lowermost and uppermost smaller, base cordate to sub-auriculate and semi-
amplexicaul with the leaf-bases connate, margins entire or undulate, scabrid, apex
obtuse or acute, glabrous or sparsely puberulous, punctate. Capitula 3–21 in
terminal cymes, hemispherical; involucre 10–13 mm long; phyllaries spreading,
pale green and purplish, ovate, elliptic or lanceolate, 4–13 mm long, glabrous but
for the ciliate margins. Florets bright orange-red or red; corolla 10–12 mm long,
glabrous or with pubescent base. Achenes narrowly turbinate, (4.5–)6–7 mm long,
densely hairy; pappus of setae 5–10 mm long, winged and hairy at base, apex
straight or hooked.

TANZANIA. Ufipa District: Tatanda Mission, Apr. 1997, *Bidgood et al.* 3441!; Iringa District: 30
   km W of Mafinga, May 1986, *Lovett et al.* 732!; Kilwa District: Madaba, July 1983, *Kalambo* in
   EA 16886!
DISTR. **T** 3 (fide EA), 4, 6–8; Zambia, Malawi, Mozambique, Zimbabwe
HAB. *Brachystegia* or *Brachystegia–Uapaca* woodland or secondary grassland; 750–1800 m
USES. None recorded on specimens from our area
CONSERVATION NOTES. Least concern (LC)

SYN. *Jaumea compositarum* (Steetz) Benth. in G.P. 2: 397 (1873); Oliv. & Hiern, F.T.A. 3: 395 (1877)
   J. johnstonii Baker in K.B. 1898: 153 (1898). Types: Malawi, Nyika Plateau, *Whyte* 228 &
      Masuku Plateau, *Whyte* s.n., and between Mpata and Nyika Plateau, *Whyte* s.n. (all K!, syn.)

NOTE. *Milne-Redhead & Taylor* 9859a from **T** 8 is this species, but with a pappus that is only
   3–3.5 mm long. The pappus setae are hairy at the base, though, and are slightly winged; the
   leaves are also much more like *H. compositarum* than of *H. elatum*. *Burtt* 3641 from **T** 5,
   Kazikazi, has the pappus slightly longer (5 mm) but is otherwise similar.

   3. **Hypericophyllum elatum** (*O.Hoffm.*) *N.E.Br.* in J.L.S. 35: 122 (1902); Verdcourt
in K.B. 6: 363 (1952); G.V. Pope in Kirkia 10: 108 (1975). Type: Tanzania, Iringa
District: Uhehe, Rungembe, *Goetze* 723 (B†, holo.)

Perennial herb 60–210 cm high; rootstock fleshy; stem solitary or several branched
in upper parts. Leaves opposite, sessile or shortly petiolate, elliptic or broadly elliptic,
3–24 cm long, 1–11 cm wide, base subcordate or slightly cuneate, margins entire or
undulate, apex obtuse; glabrous or thinly glandular-puberulous, the margins nearly
always scabrid; 3–5-veined from base; petiole up to 5 mm long but more often absent.
Capitula solitary on long stalks or the upper part of the plant branched so several
capitula are close together with small leaves on their branches, resembling a cyme;

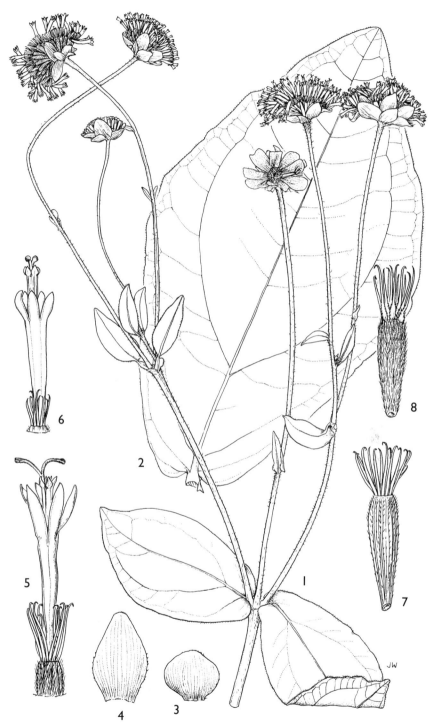

FIG. 147. *HYPERICOPHYLLUM ELATUM* — **1**, habit, × ²/₃; **2**, large leaf, × ²/₃; **3**, outer phyllary, × ²/₃; **4**, inner phyllary, × 1 ¹/₂; **5–6**, florets from different plants, × 4; **7–8**, achenes from different plants, × 4. 1 & 6 & 8 from *Abdallah* 1133, 2 from *Pielou* 170, 3–5 from *Bullock & Burnett* 49/90, 7 from *Sigara* 210. Drawn by Juliet Williamson.

involucre cup-shaped, 8–18 mm long; phyllaries green, the inner yellow-green, 4–18 mm long, glabrous but ciliate. Florets bright orange, many; corolla 9–15 mm long, glabrous or with pubescent base. Achenes narrowly turbinate, 5.5–8 mm long, densely hairy; pappus of setae 2–5 mm long, slightly flattened, neither hairy nor winged (rarely a few slightly winged at base), hooked at apex. Fig. 147 (page 721).

Tanzania. Tanga District: 32 km from Handeni on Turiani road, July 1982, *Abdallah* 1133!; Kigoma District: Mt Livandabe, June 1997, *Bidgood et al.* 4294!; Dodoma District: 24 m W of Itigi Station, Apr. 1964, *Greenway & Polhill* 11574!
Distr. T 3–8; Congo (Kinshasa), Zambia, Malawi, Mozambique, Zimbabwe, South Africa
Hab. Woodland and wooded grassland, ?often near termite hills, sometimes in recent fallow land; (200–)950–1950 m
Uses. Minor medicinal for malaria & post-childbirth (*Harwood*)
Conservation notes. Least concern (LC)

Syn. *Jaumea elata* O.Hoffm. in E.J. 28: 506 (1900)

104. SCHKUHRIA

Roth, Catalecta 1: 116 (1797), *nom. cons.*; Heiser in Ann. Missouri Bot. Gard. 32: 265–278 (1945); G.V. Pope in Kirkia 10: 119 (1975)

Annual herbs. Leaves alternate or sometimes opposite, pinnatisect with filiform lobes, rarely simple, glandular. Capitula in a leafy panicle, heterogamous; phyllaries in 1 to few rows; receptacle epaleate. Corolla with female ray florets and hermaphrodite disc florets; anther appendage narrowly ovate. Achenes black, narrowly obpyramidal; pappus of hyaline scales.

Six species in the Americas.

Schkuhria pinnata (*Lam.*) *Thell.* in F.R. 11: 308 (1912); G.V. Pope in Kirkia 10: 119, t. 5 (1975); U.K.W.F. ed. 2: 218 (1994). Type: a plant cultivated in the Jardin Botanique de Paris, possibly from Peru (P-LA, holo.)

Annual herb 5–50 cm high; much branched, the branches glandular and pilose. Leaves alternate, pinnatisect or the upper simple and filiform, in outline 1–6 cm long, 0.1–2 cm wide, with 3–7 lobes, these filiform or themselves divided, pilose or glabrous, glandular. Capitula turbinate or ovoid, on stalks to 3.5 cm long; involucre 4.5–6 mm long; phyllaries 2–6 mm long with yellow apex, glandular. Ray floret 1, sometimes absent, yellow, 1–2 mm long, glandular; disc florets 4–8, 2 mm long with glandular lobes. Achenes black, 3–4.5 mm long, ribbed, hairy; pappus scales 8, pale brown with purple marks or purplish, lanceolate, 1–2.5 mm long, hairy. Fig. 148 (page 723).

Uganda. Karamoja District: Moroto, June 1972, *Wilson* 2146A (fide EA)
Kenya. Nakuru District: 48 km N of Nakuru on Baringo road, Nov. 2000, *Smith et al.* 75!; Nairobi, Eastleigh Community Centre, May 1971, *Mwangangi & Mukenya* 1588!; South Kavirondo District: Mbita Point, Nov. 1981, *Gachathi & Opon* 21/81!
Tanzania. Moshi District: Moshi township, Aug. 1951, *Greenway* 8572!
Distr. U 1; K 3–6; T 2; Ethiopia, Mozambique, Zimbabwe, Botswana; originally from America but now a widespread weed
Hab. Ruderal sites, gardens, arable land, occasionally in dry woodland, grassland or bushland; may form almost pure stands; 950–1950 m
Uses. Minor medicinal for chest and stomach pain (*Okaka*)
Conservation notes. Least concern (LC)

Syn. *Pectis pinnata* Lam. in Journ. Hist. Nat. Paris 2: 150, t. 31 (1792)
      Rothia pinnata (Lam.) Kuntze, Rev. Gen. Pl. 3, 2: 170 (1898)

Note. The first East African collection is from 1922 at Kisumu.

Fig. 148. *SCHKUHRIA PINNATA* — **1**, habit, × ²/₃; **2**, capitulum, × 6; **3**, ray floret, × 6; **4**, ray floret, × 16; **5**, disc floret, × 16; **6**, mature achene, × 6. 1 from *Verdcourt* 415, 2–7 from *Hepper & Jaeger* 6692. Drawn by Juliet Williamson.

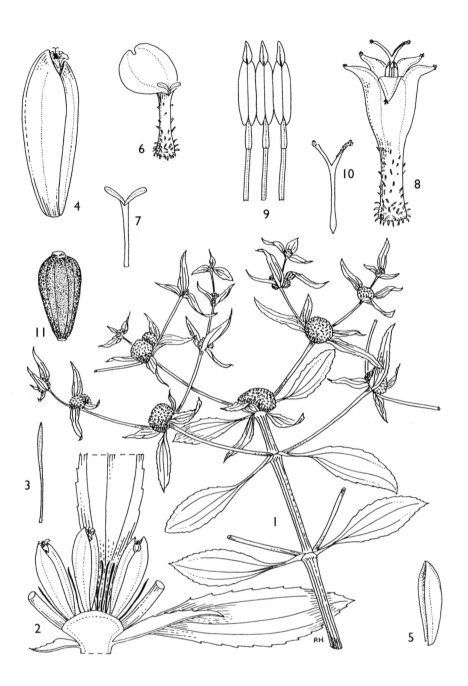

Fig. 149. *FLAVERIA TRINERVIA* — **1**, habit, × ²/₃; **2**, glomerule section, × 9; **3**, palea of glomerule receptacle, × 12; **4**, capitulum, × 12; **5**, phyllary, × 6; **6**, ray floret, × 18; **7**, style of ray floret, × 18; **8**, disc floret, × 18; **9**, stamens, × 24; **10**, style of disc floret, × 18; **11**, achene, × 12. 1–11 from *Lallmahomed* in MAU 15766. Drawn by Pat Halliday, from Flore des Mascareignes.

105. FLAVERIA

Juss., Gen. Pl.: 186 (1789); G.V. Pope in Kirkia 10: 111 (1975); Powell in Ann. Missouri Bot. Gard. 65, 2: 590–636 (1978)

Annual or perennial herbs or shrubs. Leaves opposite, with connate leaf bases, 3-veined from base. Capitula in terminal or axillary dense cymes or glomerules, heterogamous or much reduced to a single floret; phyllaries in several rows; receptacle epaleate. Corolla with female ray floret, or only one ray on the outside of the outer capitulum, and hermaphrodite disc florets; anther appendage narrowly ovate. Achenes black, narrowly obovoid; pappus absent or rarely of a few scales.

21 species from the Americas and Australia; two now pantropical weeds.

Flaveria trinervia (*Spreng.*) *C.Mohr* in Contrib. U.S. Nat. Herb. 6: 810 (1901); F.P.S. 3: 29 (1956); G.V. Pope in Kirkia 10: 111, t. 3a (1975); D.J.N. Hind in Fl. Masc. 109: 220, t. 77 (1993); U.K.W.F. ed. 2: 218 (1994). Type: cultivated plant in Halle Bot. Gard. (P, holo.)

Annual herb 10–75(–200) cm high, erect or procumbent, much branched, the branches opposite and decussate, green or red to pinkish, pilose or more often glabrous. Leaves green or yellowgreen, narrowly elliptic to oblanceolate, 1–7.5 cm long, 0.3–2 cm wide, base narrowed into a pseudopetiole but the very base widening and connate with the opposite leaf, margins serrulate to serrate, 3-veined from base, glabrous or puberulous. Capitula many in congested axillary and terminal cymes or glomerules; individual capitula oblong, heterogamous, some with a single female floret, some with 1 female and 1 bisexual floret, some with 2–4 bisexual florets only; involucre 4–5 mm long; phyllaries few, 2-seriate, 4–5 mm long. Female florets with ray yellow, suborbicular, 0.5–1 mm long, the tube green and ± 1 mm long; disc florets yellow with green base, 2–2.5 mm long, with pilose base. Achenes black, narrowly obovoid, 1.7–2.6 mm long, glabrous; pappus absent. Fig. 149 (page 724).

KENYA. Machakos District: Kiboko, Feb. 1949, *Bogdan* 2230!; Central Kavirondo District: Kisumu, May 1962, *Tweedie* 2358!; Mombasa Island, Jan. 1953, *Drummond & Hemsley* 1047!
TANZANIA. Pangani District: Bweni, Sep. 1955, *Semsei* 2316! & Pangani town, Dec. 1969, *Botany students* DSM 1267!; Rufiji District: Mafia, Kilindoni, Sep. 1937, *Greenway* 5250!; Zanzibar, 1930, *Taylor* 389!
DISTR. **K** 3–6, 7; **T** 3, 6; **Z**; pantropical weed, originally from the Americas
HAB. Waste ground, swamp, lake margins, a weed of cultivation; 0–1150 m
USES. None recorded on specimens from our area
CONSERVATION NOTES. Least concern (LC)

SYN. *Oedera trinervia* Spreng., Bot. Gard. Halle: 63 (1800)

NOTE. Much of our material had been mis-named as *F. australasica* Hook.
    The record cited for Zanzibar seems to be the earliest from East Africa.

106. TAGETES

L., Sp. Pl.: 887 (1753); G.V. Pope in Kirkia 10: 115 (1975)

Annual or perennial herbs, strongly aromatic when crushed. Leaves opposite or alternate, usually pinnate, sometimes simple, gland-dotted. Capitula solitary or in dense corymbs, terminal, heterogamous; phyllaries in a single row (rarely in 2 rows), the margins connate, glandular; receptacle epaleate. Ray florets few, in a single series, female; style branches filiform; disc florets few to many, bisexual; anthers with an acute appendix at apex, style branches subacute and pilose. Achenes black, narrowly cylindrical; pappus of 1–3 setae and 7–9 much shorter scales with ciliate margins.

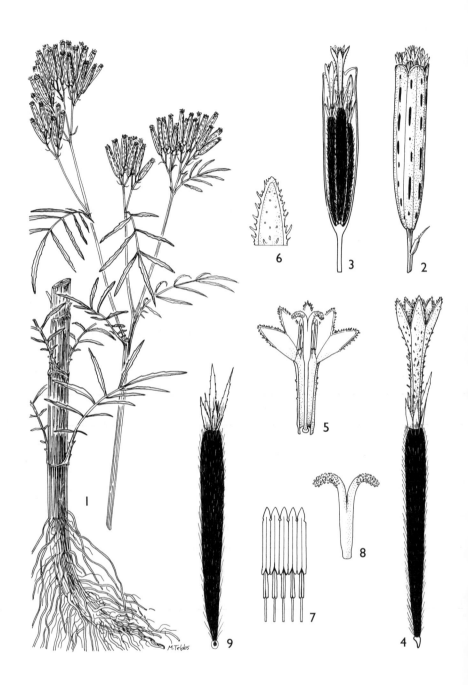

FIG. 150. *TAGETES MINUTA* — **1**, flowering stem with stem base and root, × ²/₃; **2**, capitulum, × 4; **3**, capitulum section, × 4; **4**, floret, × 9; **5**, floret section, × 9; **6**, floret lobe, × 20; **7**, stamens, × 20; **8**, style arms, × 20; **9**, achene, × 9. All from *Hind et al.* 3339. Drawn by Margaret Tebbs, from Flora de Bahia.

About 50 species from the Americas; two now widespread weeds.

1. Involucre narrowly cylindric, < 12 mm long .... 1. *T. minuta*
   Involucre ovoid, > 12 mm long 2
2. Flowers yellow; capitula usually solitary .... 2. *T. erecta*
   Flowers reddish or orange; capitula often in corymbs .... 3. *T. patula*

1. **Tagetes minuta** *L.*, Sp. Pl.: 887 (1753); F.P.U.: 176 (1962); G.V. Pope in Kirkia 10: 115, t. 4a (1975); Maquet in Fl. Rwanda 3: 642, fig. 199/3 (1985); Blundell, Wild Fl. E. Afr.: pl. 233 (1987); D.J.N. Hind in Fl. Masc. 109: 224 (1993); U.K.W.F. ed. 2: 218 (1994). Type: Dillenius, Hort. Eltham. t. 280 f. 362 (1732), lecto., chosen by Delgado-Montaño in Jarvis & Turland, Taxon 47: 368 (1998)

Annual herb, 10–250 cm high; whole plant aromatic; stems much branched in larger plants, almost woody in larger plants, ribbed, glabrous, glandular. Leaves mostly opposite but often alternate in upper part of plant, dark green, pinnatisect, elliptic in outline, 3–30 cm long, 0.7–8 cm wide, the rachis thinly winged, lobes up to 17, linear-oblong, to 11 cm long and 1 cm wide, with orange glands. Capitula many in dense terminal corymbs, narrowly cylindric; involucre 8–12 mm long; phyllaries 3–4, yellow-green, fused, glabrous, with brown or orange linear glands. Ray florets 2–3, pale yellow to cream, the ray 2–3.5 mm long; disc florets 4–7, yellow to dark yellow, 4–5 mm long. Achenes black, narrowly ellipsoid, 6–7 mm long, pilose; pappus of 1–2 setae to 3 mm long and 3–4 scales to 1 mm long, these with ciliate apex. Fig. 150 (page 726).

UGANDA. West Nile District: Paida, Aug. 1953, *Chancellor* 179!; Ankole District: Kamatalisi, Oct. 1951, *Jarrett* 202!; Teso District: Kasilo, Dec. 1931, *Chandler* 302!
KENYA. Uasin Gishu District: Eldoret area, Oct. 1951, *Williams Sangai* 292!; Fort Hall District: Thika, July 1967, *Faden* 67/481!; Teita District: Bura Girls High School, July 1972, *Cheseny* 104!
TANZANIA. Ngara District: Bugufi, Nyamiaga, Sep. 1960, *Tanner* 5086!; Pare District: Mpinji–Mamba, June 1969, *Mshigeni* 1013!; Mbeya District: Mbeya Peak Forest Reserve, Apr. 1959, *Myembe* 144!
DISTR. **U** 1–3; **K** 1–7; **T** 1–3, 5, 7, 8; originally from South America but now a widespread weed in Africa, S Europe, S Asia and Australia
HAB. Weed of cultivation and post-cultivation, in ruderal sites; may occur in dense patches; (850–)1300–2750 m
USES. Minor medicinal for ear-ache (*Koritschoner*), inflamed sinuses (*Tanner*); used to repel siafu by the Elgon Masai (*Cheseny*); poisonous (*Muladi*)
CONSERVATION NOTES. Least concern (LC)

NOTE. Said to have been imported with animal forage during the First World War. As the plant can go from seedling to flowering in a few days, it is very difficult to eradicate.

2. **Tagetes erecta** *L.*, Sp. Pl.: 887 (1753); Maquet in Fl. Rwanda 3: 642 (1985); D.J.N. Hind in Fl. Masc. 109: 224, t. 78 (1993). Type: Herb. Linn. No. 1009.3 (LINN, lecto. designated by Howard in Fl. Lesser Antilles 6: 601(1989))

Annual herb 0.4–1.8 m high. Leaves opposite, the upper alternate, elliptic in outline, pinnatisect with narrowly winged midrib, to 20 cm long, glabrous and glandular. Capitula solitary, on long stalks which are thickened near the apex; involucre 12–20 mm long; phyllaries 5–8, connate, glabrous except for apex, glandular. Ray florets 5–8, golden yellow, the ray 10–25 mm long; disc florets many, 10–16 mm long. Achenes black, 7–10 mm long; pappus of 1–2 setae 6–12 mm long and 3–4 scales 3–5 mm long.

TANZANIA. Lushoto District: Lushoto, July 1970, *Mshana* 76!; Kigoma District: Kasogi, Sep. 1959, *Harley* 9517! & 9518!; Morogoro District: Uluguru Mts, Bunduki, Mar. 1969, *Batty* 438!
DISTR. **T** 3, 4, 6; 'African marigold', widely cultivated and sometimes naturalized; originally from the Americas

HAB. Lake shore or by streams; 750–2100 m
USES. None recorded on specimens from our area
CONSERVATION NOTES. Least concern (LC)

3. **Tagetes patula** *L.*, Sp. Pl.: 887 (1753); G.V. Pope in Kirkia 10: 117, t. 4b (1975); Maquet in Fl. Rwanda 3: 642, fig. 199/4 (1985). Type: Herb. Linn. No. 1009.1 (LINN, lecto. designated by Hind in Jarvis et al., Regnum Veg. 127: 92 (1993))

Annual to 80 cm high; whole plant aromatic. Leaves opposite, the upper alternate, elliptic in outline, pinnatisect with narrowly winged midrib, to 20 cm long, glabrous and glandular. Capitula loosely corymbose, on long stalks which are thickened near the apex; involucre 15–20 mm long; phyllaries 5–7, connate, glabrous except for apex, glandular. Ray florets yellow, orange or reddish, the ray 5–15 mm long; disc florets many, 10–16 mm long. Achenes black, 7–10 mm long; pappus of 1–2 setae 6–10 mm long and 3–4 scales 3–5 mm long.

KENYA. Naivasha District: NW Lake Naivasha, June 1982, *Mwangangi* 2335 (fide EA)
TANZANIA. Lushoto District: Lushoto, July 1970, *Mshana* 78!; Kigoma District: Selambula, Sep. 1958, *Newbould & Jefford* 2419!; Mpwapwa District: Mpwapwa, Dec. 1969, *Kaufmann* 1!
DISTR. **K** 3, 4; **T** 3–5; 'French marigold'; originally from Central America, widely cultivated and sometimes naturalized
HAB. Ruderal sites; 700–?1500 m
USES. None recorded on specimens from our area
CONSERVATION NOTES. Least concern (LC)

107. **ECLIPTA**

L., Mant. Pl. Alt.: 286 (1771), *nom. conserv.*

Annual herbs. Leaves opposite. Capitula axillary or terminal, solitary or in pairs or in panicles, heterogamous, radiate; involucre 2-seriate; receptacle paleate. Ray florets female or sterile, with a small ray; disc florets hermaphrodite, tubular to campanulate, 4–5-lobed; anthers with rounded terminal appendage; style-arms obtuse. Achenes ± compressed, rugose to tuberculate; pappus absent or of 2 short awns.

3–4 species; one of which is a pantropical weed.

Eclipta prostrata (*L.*) *L.*, Mant. Pl. Alt.: 286 (1771); F.P.S. 3: 26, t. 4 (1956); Wild in Kirkia 6: 59 (1967); Maquet in Fl. Rwanda 3: 622, fig. 191/3 (1985); Lisowski, Asterac. Fl. Afr. Centr. 1: 223 (1991); D.J.N. Hind in Fl. Masc. 109: 181, t. 60 (1993); U.K.W.F. ed. 2: 215 (1994). Typotype: Herb. Sloane 94: 175 (BM) [several specimens]

Annual or short-lived perennial herb, erect or decumbent, sometimes scrambling, 5–90 cm high and long, sometimes mat-forming; stem fleshy or sappy, sometimes tinged with red; branches appressed scabrid-pubescent. Leaves narrowly ovate, narrowly elliptic or elliptic, 2–12 cm long, 0.3–4 cm wide, base cuneate or attenuate, margins serrate or serrate-crenate, apex acute or subacute, appressed scabrid-pubescent on both surfaces; petiole to 3 mm long. Capitula hemispheric, stalked for up to 1 cm; involucre to 6 mm long; phyllaries 8–11, appressed-pubescent; paleae not very visible. Ray florets in several series, inconspicuous, white, the ray 1–2 mm long; disc florets white, rarely yellow, 1–2 mm long, with puberulous lobes, but especially distinctive when young when they are flat and mid- to dark green. Achenes brown or black with pale margins, oblong, 2–2.5 mm long, flat or 3-gonous, subglabrous; pappus absent or of a small puberulous rim. Fig. 151 (page 729).

UGANDA. West Nile District: Arua, Nov. 1940, *Purseglove* 1068!; Ankole District: Kyibego, Nyabusozi, Apr. 1970, *Katende* 136!; Mengo District: km 78 Kampala–Masindi, Feb. 1963, *Tallantire* 613!

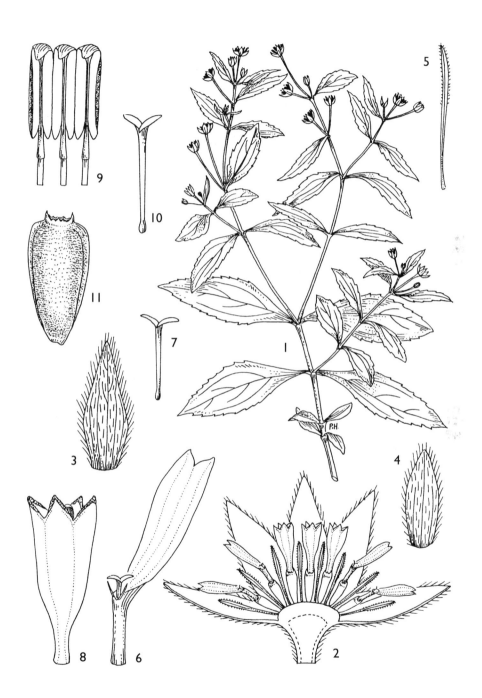

FIG. 151. *ECLIPTA PROSTRATA* — **1**, habit, × ²/₃; **2**, capitulum section, × 12; **3–4**, phyllaries, × 6; **5**, palea, × 18; **6**, ray floret, × 18; **7**, style of ray floret, × 18; **8**, disc floret, × 24; **9**, stamens, × 18; **10**, style of disc floret, × 12; **11**, achene, × 9. 1–7 & 9–11 from *Morin* s.n., 8 from *Guého* in MAU 18823. Drawn by Pat Halliday, from Flore des Mascareignes.

Kenya. Turkana District: Lokori, Mar. 1965, *Newbould* 7331!; Baringo District: Baringo, Feb.
  1962, *Tweedie* 2295!; Tana River District: Mchelelo, Mar. 1990, *Tana River Primate Res. Exp.* 103!
Tanzania. Mwanza District: Nyamanoro, May 1987, *Kisena* 387!; Ufipa District: Kipili, Feb.
  1950, *Bullock* 2385!; Uzaramo District: Dar es Salaam University, Mar. 1969, *Mwasumbi* 10488!;
  Zanzibar: Mnazi Moja, May 1964, *Faulkner* 3381!
Distr. U 1–4; K 1–7; T 1–7; Z; a pantropic weed
Hab. Moist sites such as swamp edges, river or lake banks, edge of rice-fields, black cotton soil;
  0–1250 m
Uses. None recorded on specimens from our area
Conservation notes. Least concern (LC)

Syn. *Verbesina prostrata* L., Sp. Pl.: 902 (1753)
  V. alba L., Sp. Pl.: 902 (1753). Type: Herb. Linn. No. 1020 (LINN, lecto. designated by
    D'Arcy in Woodson & Schery; Ann. Miss. Bot. Gard. 62: 1102 (1975))
  Cotula alba (L.) L., Syst. Nat. ed. 12, 2: 564 (1767)
  Eclipta erecta L., Mant. Pl. Alt.: 286 (1771), *nom. illeg., typ. cons.* of *Eclipta*, based on *Verbesina
    alba* L.
  E. alba (L.) Hassk., Pl. Jav. Rar.: 528 (1848); Oliv. & Hiern, F.T.A. 3: 373 (1877)

108. ACMELLA

Pers., Syn. Pl. 2: 472 (1807); R.K. Jansen in Syst. Bot. Monogr. 8: 1–115 (1985)

Annual or perennial herbs. Leaves opposite and decussate, 3-veined from base.
Capitula axillary or terminal, solitary or grouped, ovoid, radiate or discoid; involucre
1–3-seriate; receptacle conical, paleate. Radiate florets absent or up to 22,
functionally male; style with obtuse branches. Disc florets many, hermaphrodite, 4–5-
merous; style with obtuse branches. Achenes of ray florets triangular in cross-section,
black; pappus absent or of up to 10 setae; achenes of disc florets compressed, black;
pappus absent or of up to 10 setae.

30 species, mostly from the Americas but several from the Old World; one widespread weed.

1. Heads radiate 2
   Heads discoid 3
2. Achenes glabrous and epappose .... 1. *A. caulirhiza*
   Achenes ciliate, with pappus of a few bristles .... 2. *A. uliginosa*
3. Corolla white 4
   Corolla yellow or yellow-orange; capitula 11–17 mm in
     diameter .... 3. *A. oleracea*
4. Capitula 5–9 mm in diameter; pappus of 2–3 soft bristles .. 4. *A. radicans*
   Capitula 7–13 mm in diameter; pappus of 2(–4) stiff awns .. *Spilanthes costata*

1. **Acmella caulirhiza** *Del.*, Voy. Meroe 4: 335, t. 64, fig. 7 (1827); R.K. Jansen in Syst.
Bot. Monogr. 8: 37, map (1985); Lisowski, Asterac. Fl. Afr. Centr. 1: 228, t. 50 (1991);
D.J.N. Hind in Fl. Masc. 109: 207, t. 71 (1993); U.K.W.F. ed. 2: 216, t. 88 (1994), as
calirhiza. Type: Sudan, Sennar, *Caillaud* s.n. (MPU, holo., not found)

Perennial or annual herb, decumbent or scrambling, often rooting at the nodes,
up to 15 cm high or 60 cm long; stems sappy, often reddish or purplish, glabrous or
pilose. Leaves ovate, 1–7 cm long, 0.8–5 cm wide, base attenuate, margins dentate,
apex acute or obtuse, glabrous or pilose; 3-veined from base; petiole 1–15 mm long,
narrowly winged. Capitula solitary and hemispherical, becoming conical, radiate,
4–14 mm long, 4–10 mm in diameter, on stalks 2–7 cm long; involucre 3.5–6 mm
long, of 2 series of phyllaries. Ray florets 10–15, yellow, 2.2–3.3 mm long, the ray
1.4–2.3 mm long. Disc florets many, yellow or slightly orange, 1.1–1.9 mm long.
Achenes 1.3–2.4 mm long, glabrous; pappus usually absent but sometimes some setae
present on achenes of outer florets. Fig. 152 (page 731).

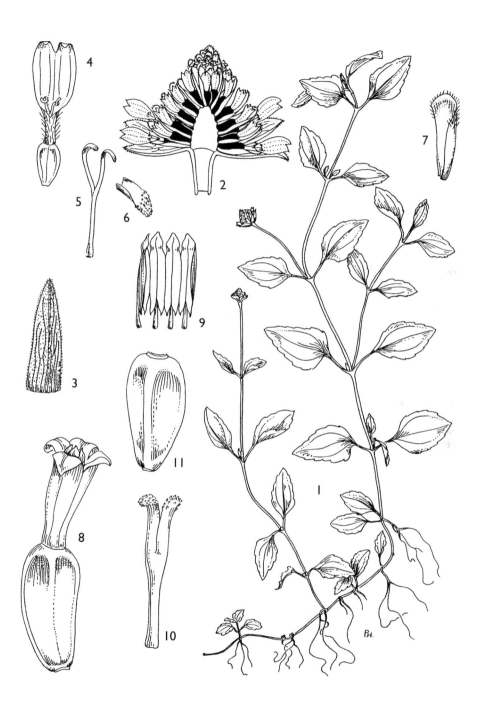

FIG. 152. *ACMELLA CAULIRHIZA* — **1**, habit, × ²/₃; **2**, capitulum section, × 4; **3**, phyllary, × 6; **4**, ray floret, × 8; **5**, style of ray floret, × 30; **6**, style branch, much enlarged; **7**, palea, × 8; **8**, disc floret, × 18; **9**, stamens, × 36; **10**, style of disc floret, × 36; **11**, achene, × 18. 1 from *Orian* 104, 2–11 from *Guého* in MAU 19900. Drawn by Pat Halliday, from Flore des Mascareignes.

UGANDA. West Nile District: Chobi, Oct. 1967, *Buzigye* 10!; Ankole District: Bunyaruguru, Feb. 1939, *Purseglove* 565!; Mengo District: Kituza, June 1957, *Griffiths* 45!

KENYA. Northern Frontier District: Isiolo, July 1951, *Kirrika* 9!; Trans Nzoia District: Mt Elgon 0.5 km W of Suam Saw Mills, Dec. 1967, *Mwangangi* 475!; Machakos District: 1 km E of Ithemboni, Aug. 1971, *Mwangangi* 1714!

TANZANIA. Arusha District: Ngurdoto Crater, Oct. 1965, *Greenway & Kanuri* 11953!; Ufipa District: 4 km on Safu road from Sumbawanga–Mbala road, Nov. 1992, *Gereau et al.* 5009!; Njombe District: Livingstone Mts, near source of Ngolo R., Feb. 1991, *Gereau & Kayombo* 3917!

DISTR. U 1–4; **K** 1–7; **T** 1–8; widespread in tropical and subtropical Africa and Madagascar

HAB. Swampy or seasonally wet sites, river-banks, cultivated areas, forest margins; 750–3100 m

USES. Used widely for fever, mouth and throat infections, chewed for toothache

CONSERVATION NOTES. Least concern (LC)

SYN. *Spilanthes africana* DC., Prodr. 5: 623 (1836). Type: South Africa, between Omtata & Omsamwubo, *Drège* 5086 (G-DC, holo., K!, iso.)
  S. caulirhiza (Del.) DC., Prodr. 5: 623 (1836); F.P.S. 3: 54 (1956)
  S. abyssinica A.Rich., Tent. Fl. Abyss. 1: 415 (1848). Type: Ethiopia, near Adoa, *Schimper* (no number given; Jansen says 145 and P, lecto., BR, G, GOET, HBG, K, M, MPU, NY, OXF, P, S, W, iso.; this number not found at BM, BR, K or P, though in all these herbaria there is a *Schimper* 245 with this locality)
  S. acmella auct. mult., *non* L.
  Acmella mauritiana auct. mult., *non* Rich.
  Spilanthes mauritiana auct. mult., *non* (Rich.) DC. e.g. F.P.U.: 172 (1962); Blundell, Wild Fl. E. Afr.: pl. 385 (1987)

NOTE. A pioneer that may colonize bare ground, and a very common plant in moist ruderal locations. "Creeping but rising if the grass is long" (Tweedie in litt.).

2. **Acmella uliginosa** (*Swartz*) *Cass.*, Dict. Sci, Nat. 24: 331 (1822); R.K.Jansen in Syst. Bot. Monogr. 8: 55, map (1985); Lisowski, Asterac. Fl. Afr. Centr. 1: 232, t. 51 (1991). Type: Jamaica, *Swartz* s.n. (BM!, holo.)

Annual herb 10–100 cm high; stems erect, scrambling or sometimes decumbent, green or reddish, glabrous or pilose. Leaves lanceolate, narrowly ovate or ovate, 1–8.5 cm long, 0.3–3 cm wide, base attenuate, margins sinuate to dentate, apex acute to acuminate, glabrous to sparsely pilose; petiole 2–32 mm long, narrowly winged. Capitula solitary, radiate, 4–6 mm in diameter, 6–8 mm long, on stalks 1–6 cm long; involucre 2–4 mm long, of 1 series of green phyllaries. Ray florets 4–7, pale yellow, 1.8–3.6 mm long, the ray 0.9–2.2 mm long. Disc florets many, yellow to orange-yellow, 1–1.6 mm long. Achenes 1.2–1.8 mm long, ciliate; pappus of 2–4 bristles 0.1–0.7 mm long.

KENYA. Teita District: Taita Hills, Mnyuchi, Dec. 1997, *Mwachala et al.* 570!; Kwale District: Shimba Hills, Mwalunganji, Oct. 1999, *Luke & Luke* 6032!

TANZANIA. Tanga District: Mnyusi, Manta, Aug. 1964, *Semsei* 3888!; Mpanda District: 11 km on Mpanda–Uruwira road, Mar. 1994, *Bidgood et al.* 2745!; Songea District: Mbinga, Mitomoni, May 1991, *Ruffo & Kisena* 3341!

DISTR. **K** 7; **T** 3, 4, 6–8; pantropical weed

HAB. Damp ground in the moist forest zone, swampy sites, grassland near rivers, a weed of rice fields; 0–1200 m

USES. None recorded on specimens from our area

CONSERVATION NOTES. Least concern (LC)

SYN. *Spilanthes uliginosa* Swartz, Nov. Gen. Sp.: 110 (1788)

NOTE. *Bidgood et al.* from **T** 7, km 68 on Tunduma–Sumbawanga road, lacks rays, but otherwise is this species.

3. **Acmella oleracea** (*L.*) *R.K.Jansen* in Syst. Bot. Monogr. 8: 65 (1985); Lisowski, Asterac. Fl. Afr. Centr. 1: 234, t. 52 (1991); D.J.N. Hind in Fl. Masc. 109: 209 (1993); U.K.W.F. ed. 2: 216 (1994); D.J.N. Hind & Biggs in Curtis Bot. Mag. 20: 31–39 (2003). Type: Herb. Linn. No. 974.5 (LINN, lecto., designated by Jansen in Syst. Bot. Monogr. 8: 65 (1985))

Annual herb, the stems decumbent or erect; stems often reddish, glabrous. Leaves broadly ovate to deltate, 5–11 cm long, 4–8 cm wide, base truncate or with a short attenuate part, margins dentate, apex acute, usually glabrous. Capitula solitary, elongated-conical, discoid, 10–24 mm long, 11–17 mm in diameter; phyllaries in 3 series, 5–7 mm long. Disc florets many, yellow to orange, 2.7–3.3 mm long. Achenes black, 2–2.5 mm long, ciliate; pappus of 2 bristles 0.3–1.5 mm long.

UGANDA. Masaka District: Sesse Is., Bugalla Is., Oct. 1958, *Symes* 483!
KENYA. Machakos District: Donyo Sabuk, cultivated in Farm managers's garden, July 1954, *Bally* 9837!
TANZANIA. Mwanza District: Mwanza, Oct. 1932, *Geilinger* 3245!; Zanzibar: Ziwani, June 1930, *Vaughan* 1375!
DISTR. **U** 4; **K** 4; **Z**; known only from cultivation and escapes from cultivation
HAB. Unclear; the Zanzibari specimen might be from a clove shamba
USES. A well-known salad vegetable
CONSERVATION NOTES. Least concern (LC)

SYN. *Spilanthes oleracea* L., Syst. Nat. 2: 534 (1767)

4. **Acmella radicans** (*Jacq.*) *R.K.Jansen* in Syst. Bot. Monogr. 8: 69 (1985). Type: Venezuela; Jansen does not indicate a type but says plate 584 in Jacq., Icon. Pl. Rar. 3 (1792) 'fixes the application' (lectotype?)

Annual herb 6–30(–100?) cm high; stem erect to ascending, rarely rooting at the nodes, green to purple, glabrous to pilose. Leaves ovate to narrowly ovate, 1–8 cm long, 1–5 cm wide, base attenuate, margins denticulate to dentate, apex acute, glabrous to sparsely pilose and with hispid margins; petiole 4–37 mm long. Capitula solitary or in groups of 2–3, long-stalked, elongate-conical, discoid, 6–11 mm long, 5–9 mm in diameter; phyllaries 6–9, 2-seriate, 4–7 mm long. Disc florets many, greenish white or white, 1.6–2 mm long. Achenes black, 2–2.7 mm long, ciliate; pappus of 2–3 subequal bristles 0.4–1.6 mm long.

var. **radicans**; R.K.Jansen in Syst. Bot. Monogr. 8: 70 (1985)

TANZANIA. Mbeya District: Iyunga Secondary School, Sep. 1968, *Wingfield* 600/d! & between Mbozi and Lake Rukwa, May 1975, *Hepper et al.* 5467! & 19 km on Mbozi–Tunduma road, *Magogo* 2418!
DISTR. **T** 7; central and South America, now a fairly widespread weed
HAB. Waste ground and roadsides, river-banks; 1200–1700 m
USES. None recorded on specimens from our area
CONSERVATION NOTES. Least concern (LC)

SYN. *Spilanthes radicans* Jacq., Collectanea 3: 229 (1791)

NOTE. The three cited specimens are the only ones from our area seen by us.

FIG. 153. *SPILANTHES COSTATA* — **1**, habit, × ²/₃; **2**, capitulum, × 3; **3**, floret with achene within palea, × 6; **4**, floret, × 16; **5**, achene, × 8. 1 & 5 from *Chandler* 1276, 2 from *Purseglove* 537, 3–4 from *Symes* 455. Drawn by Juliet Williamson.

109. **SPILANTHES**

Jacq., Enum. Pl. Carib.: 8 (1760); R.K. Jansen in Syst. Bot. 6, 3: 231–257 (1981)

Perennial herbs. Leaves opposite and decussate, 3–5-veined from base. Capitula axillary or terminal, solitary, ovoid, discoid; phyllaries 2–3-seriate, subequal; receptacle paleate. Disc florets many, bisexual, 5-lobed; style with bulbous base. Achenes compressed, with a cork-like margin; pappus of 1–4 awns.

Six pantropical species.

Spilanthes costata *Benth.* in Hook., Niger Fl.: 436 (1849); R.K. Jansen in Syst. Bot. 6, 3: 247 (1981); Maquet in Fl. Rwanda 3: 631, fig. 194/2 (1985); Lisowski, Asterac. Fl. Afr. Centr. 1: 225, t. 49 (1991). Type: Ivory Coast, Cape Palmas, *Vogel* 41 (K!, lecto., chosen by Jansen)

Perennial herb, creeping, decumbent or sometimes straggling and then to 1.2 m high; stems semi-succulent, glabrous to pilose. Leaves ovate, 1–9 cm long, 1–5 cm wide, base rounded to truncate, margins denticulate, apex acute, 3-veined from base, usually scabridulous. Capitula axillary, solitary, on stalks to 6 cm long, ovoid, discoid; involucre 6–12 mm long; phyllaries in 2 unequal series, the outer 6–12 mm long, the inner 4–10 mm long, glabrous or ciliate, occasionally pilose. Florets many, white, 2–2.7 mm long. Achenes black, obovoid, 2.7–3.5 mm long, at maturity with cork-like margins, ciliate; pappus of 2(–4) subequal awns 0.9–2.1 mm long. Fig. 153 (page 734).

UGANDA. Kigezi District: Lake Edward, Rwenshama, Mar. 1951, *Purseglove* 3597!; Masaka District: 0.5 km E of Port Bell pier, Jan. 1969, *Lye & Rwaburindore* 1188!; Mengo District: Kajansi Forest, July 1935, *Chandler* 1276!
KENYA. Central Kavirondo District: Kisumu, Feb. 1915, *Dummer* 1794!
TANZANIA. Arusha District: Arusha Town, Aug. 1997, *Phillipson* 4894!; Mpanda District: Edith Bay, no date, *Cunnington* 40!
DISTR. **U** 2, 4; **K** 5, 6; **T** 2, 4; W Africa from Ivory Coast to Ethiopia and Angola
HAB. Shores or banks of lakes and streams, less often in forest margins; 900–1350 m
USES. None recorded on specimens from our area
CONSERVATION NOTES. Least concern (LC)

110. **BLAINVILLEA**

Cass. in Dict. Sc. Nat. 29: 493 (1823)

Annual or perennial herbs; branching often trichotomous. Leaves opposite or subopposite, scabrid. Capitula in lax cymes or solitary at branch forks, heterogamous, radiate or disciform; involucre bracts subequal; receptacle paleate, the paleae clasping the florets. Ray florets white, in 1–2 rows with short ray, fertile; disc florets distally campanulate. Achenes 3-gonous, often compressed; pappus of 2–5 barbellate bristles and often of unequal, shortly connate scales.

10 species; pantropical.

Blainvillea acmella (*L.*) *Philipson* in Blumea 6: 350 (1950). Type: Sri Lanka, 'habitat in Zeylona', collector not cited, Herb. Hermann 2: 10, No. 309 (BM, lecto. designated by Koster & Philipson in Blumea 6: 349, f. 1 (1950))

Annual herb, erect, 0.3–1.2 m high; stems hardly to much branched, scabrid-pilose. Leaves elliptic to ovate, to 12 cm long, to 8 cm wide, base cuneate, margins serrate-crenate, apex acuminate, 3-veined from base, scabridulous. Capitula slightly elongate-

FIG. 154. *BLAINVILLEA ACMELLA* — **1**, habit, × ²/₃; **2**, capitulum, × 6; **3**, floret within palea, × 5; **4**, ray floret, × 16; **5**, disc floret, × 4; **6**, achene, × 5. 1 & 3–5 from *Abdallah & Vollesen* 95/12, 2 from *Gillett* 13045, 6 from *Gilbert* 6274. Drawn by Juliet Williamson.

campanulate; involucre 10 mm long; phyllaries pale green with darker veins, oblong to elliptic, pilose and glandular. Ray florets whitish, 3–5, ray 1–1.6 mm long, glabrous; disc florets whitish, 3.6 mm long. Achenes narrowly obovoid, 3–5 mm long; pappus of 1–3 weak bristles to 1 mm long, on the edge of a shallow cup. Fig. 154 (p. 736)

KENYA. Northern Frontier District: Dandu, May 1952, *Gillett* 13045!; Machakos District: Seven Forks, Kindaruma Dam, Apr. 1967, *Gillett & Faden* 18113!; Masai District: 40 km from Nairobi on Magadi road, June 1981, *Gilbert* 6274!
TANZANIA. Pare District: Mkomazi Game Reserve, Ibaya Hill, Apr. 1995, *Abdallah & Vollesen* 95/12!; Dodoma District: 8 km on Kilimatinde–Dodoma road, Apr. 1988, *Bidgood et al.* 1168!; Mbeya District: Igawa, Apr. 1962, *Polhill & Paulo* 1972!; Pemba: Mkoani, Aug. 1929, *Vaughan* 475!
DISTR. **K** 1, 3, 4, 6, 7; **T** 1–3, 5, 7; **P**; widely distributed in tropical Africa and Asia
HAB. Shady sites in dry bushland or woodland, especially where disturbed, also in riverine fringe within this zone; 450–1350 m
USES. None recorded on specimens from our area
CONSERVATION NOTES. Least concern (LC)

SYN. *Verbesina acmella* L., Sp. Pl.: 901 (1753)
   Acmella mauritiana L.C.Rich. in Pers., Syn. Pl. 2: 472 (1807), *nom. superfl.* based on *V. acmella*
   Blainvillea gayana Cass., Dict. Sc. Nat. 47: 90 (1827); Oliv. & Hiern, F.T.A. 3: 375 (1877); Hepper in F.W.T.A. ed. 2, 2: 237 (1963); Wild in Kirkia 6: 38 (1967); U.K.W.F. ed. 2: 215 (1994). Type: from a plant cultivated from seed from Senegal (P, holo.). Fig. 154 (page 736)
   Spilanthes mauritiana (L.C.Rich.) DC., Prodr. 5: 625 (1836)
   Blainvillea rhomboidea sensu Oliv. & Hiern, F.T.A. 3: 375 (1877); F.P.S. 3: 13 (1956), *non B. rhomboidea* Jacq.
   Wedelia triseta Peter in Abh. Ges. Wiss. Gottingen, n.f. 13, 2: 94 in clavi (1928). Type: Tanzania, Lushoto/Pare District: Buiko to Hedaru, *Peter* 11066 and Lushoto District: Pangani R. 6 km N of Buiko, *Peter* 10436 (both B†, syn.)

111. **MELANTHERA**

Rohr in Skrift. Nat. Selsk. Kobenh. 2: 213 (1792); Wild in Kirkia 5: 1–18 (1965); Lisowski, Asterac. Fl. Afr. Centr. 1: 204–218 (1991); Wagner & Robinson in Brittonia 53, 4: 539–561 (2001)

Wollastonia Decne. in Nouv. Ann. Mus. Par. 3: 414 (1834); Fosberg & Sachet in Smithsonian Contrib. Botany 45: 1–40 (1980)

Annual or perennial herbs, sometimes scandent. Leaves opposite, usually scabrid. Capitula solitary or in lax cymes, shallowly hemispheric, sometimes becoming globose in fruit (in *M. scandens*), radiate (in Africa); involucre 2–3-seriate, the inner scarious and sheathing the achenes; receptacle paleate, the paleae enveloping the disc florets and usually obovate with ciliate upper margins. Ray florets conspicuous, style in these florets present or absent; disc florets tubular or nearly so, narrowed proximally; style branches narrowing to an acute apex or thickened just below the apex. Achenes obovoid or obconical, compressed to 3-angled; pappus of 0–12 often unequal bristles, ?deciduous or fragile.

35 species in Africa, the Pacific Islands and Central America.

1. Leaves sessile, narrowly ovate or ovate, with rounded leaf
    base, margins serrate; style absent in ray florets . . . . 1. *M. abyssinica*
   Leaves petiolate or if (sub-)sessile not as above 2
2. Style present in ray florets 3
   Style absent in the neuter ray florets 5

3. Decumbent herb of coastal dunes; leaf base cuneate . . . .   5. *M. biflora*
   Erect herb or climber, inland plants; leaf base cuneate in
   M. richardsiae 4
4. Pappus absent; leaf margins serrate-crenate; phyllaries 8–12,
   ovate . . . .   3. *M. richardsiae*
   Pappus setae 10–15; leaf margins often irregularly toothed;
   phyllaries many, usually narrowly ovate . . . .   2. *M. scandens*
5. Annual herb . . . .   *M. robinsonii**
   Perennial herbs . . . .   4. *M. pungens*

1. **Melanthera abyssinica** (*A.Rich.*) *Oliv. & Hiern*, F.T.A. 3: 382 (1877); Wild in Kirkia
5: 13 (1965); C.D.Adams in F.W.T.A. ed. 2, 2: 240 (1963); E.P.A.: 1131 (1967). Type:
Ethiopia, Tigray, Mt. Scholoda [Selleuda], *Schimper* 334 (P!, lectotype, chosen by Mesfin)

Perennial herb 1–2 m high; stem ± tetragonal, sulcate. Leaves sessile, ovate to
ovate-lanceolate, 2–12 cm long, 1–5 cm wide, base cordate, margins serrate, apex
acute or attenuate, 3-veined from base, scabrid on both surfaces. Involucre
hemispheric becoming obconical in fruit, 6–8 mm high at anthesis, up to 15 mm
long in fruit; phyllaries 3–4-seriate, green, the outer 6–8 mm long and strigose, the
inner yellowish white, 5–8 mm long, hispid outside. Capitula terminal and solitary or
in lax corymbs; peduncle 2–12 cm long, scabrid; paleae yellowish white, acute at
apex, hispid outside. Ray florets orange-yellow, up to 15 mm long, 5 mm wide, ray
pilose beneath; disc florets yellow, 3–5 mm long. Achenes dark brown to black,
slightly quadrangular, 3–3.5 mm long, glabrous, surfaces convex; pappus of 2
caducous flat aristae 3.5 mm long and minute bristly scales.

TANZANIA. Kigoma District: Gombe National Park, Mar. 1964, *Pirozynski* 567!
DISTR. **T** 4; Sierra Leone, Guinea, Cameroon, Central African Republic, Congo (Kinshasa),
   Sudan, Ethiopia, Angola, Zambia, Zimbabwe; also in Yemen
HAB. Margins of riverine forest; altitude unclear
USES. None recorded on specimens from our area
CONSERVATION NOTES. Least concern (LC)

SYN. *Wurschmittia abyssinica* A.Rich., Tent. Fl. Abyss. 1: 412 (1848)

2. **Melanthera scandens** (*Schum. & Thonn.*) *Roberty* in Bull. I.F.A.N. 16: 68 (1968);
F.P.S. 3: 40 (1956, with Brenan as comb. author); F.P.U.: 171, fig. 110 (1962);
C.D.Adams in F.W.T.A. ed. 2, 2: 240 (1963); Wild in Kirkia 5: 5 (1965) & in Kirkia 6:
50 (1967); Maquet in Fl. Rwanda 3: 626, fig. 193/1 (1985); Lisowski, Asterac. Fl. Afr.
Centr. 1: 205 (1991); U.K.W.F. ed. 2: 216, t. 89 (1994). Type: Ghana, *Thonning* s.n. (C,
holo., K, photo!)

Perennial herb to 1 m high, often scandent and then to 4 m high, branched or ±
unbranched; branches ± quadrangular, scabrid. Leaves deltate, narrowly to broadly
ovate or oblong, 2.5–13 cm long, 0.6–7.5 cm wide, base truncate, cordate or abruptly
cuneate and often lobed at basal angles, margins serrate-crenate to almost entire,
apex obtuse to acuminate, scabrid on both surfaces and sometimes almost tomentose
beneath; strongly 3-veined from base; petiole 0.5–5 cm long, scabrid. Capitula
terminal, solitary or in lax corymbs, flat at first becoming globose with age; stalks of
individual capitula to 18 cm long, scabrid; involucre 4–8 mm long, 2–3-seriate, the
phyllaries ovate, 4–8 mm long; paleae narrowly obovate and usually acuminate. Ray
florets 6–20, orange-yellow or yellow, the ray oblong or elliptic, 11–20 mm long,
3–6 mm wide, sparsely pilose abaxially, tube 1.5–3.5 mm long, style present. Disc
florets to 6 mm long, pilose on the lobes. Achenes obovoid, 2–2.5 mm long, slightly
3-angled, pilose; setae 10–15, 0.5–2 mm long. Fig. 155 (page 739).

* The single East African *M. robinsonii* specimen cited by Wild, Lazarus Thomas 26, is *M. pungens*

Fig. 155. *MELANTHERA SCANDENS* — **1**, habit, × ²⁄₃; **2**, inflorescence, × ²⁄₃; **3**, capitulum, × 2; **4**, ray floret, × 4; **5**, ray floret, abaxial surface, × 4; **6**, disc floret, × 8; **7**, anthers, × 12; **8**, achene with seta detail, × 12. 1–2 & 4–7 from *Richards* 27238, 3 from *Bidgood et al.* 4641, 8 from *Tanner* 1047. Drawn by Juliet Williamson.

Syn. *Buphthalmum scandens* Schum. & Thonn., Beskr. Guin. Pl.: 392 (1827)
    Lipotriche brownii DC., Prodr. 5: 544 (1836), as *brownei.* Type: Congo, Congo R., ca. 1816,
        Smith s.n. in herb. *Brown* (G-DC, holo., BM, K!, iso.)
    Melanthera brownii (DC.) Sch.Bip. in Flora 27: 673 (1844); S. Moore in J.L.S. 40: 116 (1911)

Leaves narrowly oblong-ovate, often with basal lobes . . . .   subsp. *madagascariensis*
Leaves ovate to broadly ovate, without basal lobes . . . . . .   subsp. *subsimplicifolia*

NOTE. Subsp. *scandens* occurs in West Africa, and subsp. *dregei* in southern Africa.

subsp. **madagascariensis** (*Baker*) *Wild* in Kirkia 5: 7 (1965); Lisowski, Asterac. Fl. Afr. Centr. 1: 208, t. 45 (1991). Type: Madagascar, Central Madagascar, *Baron* 2344 & 2534, *Humblot* 410 (all K!, syn.)

Leaves usually narrowly oblong-ovate, ± lobed at the basal angles and therefore sometimes hastate, base usually cordate, margins irregularly crenate-dentate, upper surface often bullate, with prominent reticulate venation beneath.

UGANDA. Acholi District: Pailyech village near Pakiri, May 1971, *Okello-Degaouchii* 15!; Toro District: Mungilo, Bwamba Forest reserve, July 1960, *Paulo* 601!; Mengo District: Kyadondo, 0.5 km E of Port Bell pier, Jan. 1969, *Lye & Rwaburindore* 1187!
KENYA. Nairobi, Museum Bridge on Nairobi R., Sep. 1971, *Mwangangi* 1776!; South Nyanza District: Mbita Point, Nov. 1981, *Gachathi & Opon* 27/81!; Masai District: Nguruman, 9 km NE of Entasekera, Oct. 1977, *Fayad* 252!
TANZANIA. Lushoto District: between Derema and Mambo Estates, Nov. 1986, *Borhidi & Steiner* 86/406!; Kigoma District: Mahale Mts, Kasoje, Aug. 1984, *Takahata* 427!; Iringa District: 30 km from Mafinga [Sao Hill] on Mbeya road, Mar. 1988, *Bidgood et al.* 804!
DISTR. **U** 1–4; **K** 1, 3–7; **T** 1–4, 6, 7; Congo (Kinshasa), Ethiopia, Angola, Zambia, Botswana, Namibia
HAB. Swamps and swamp margins, riverine and lake margins, occasionally in forest margins; 250–2100 m
USES. Minor medicinal for sores (*Shillito*); stem used for toothbrush (*Tanner*)
CONSERVATION NOTES. Least concern (LC)

Syn. *Melanthera madagascariensis* Baker in J.L.S. 21: 418 (1885); Humbert, Fl. Madag. Composées 3: 658, t. 120/1–4 (1963)
    M. swynnertonii S.Moore in J.L.S. 40: 116 (1911). Type: Mozambique, Lower Buzi R., *Swynnerton* 1882 (BM!, holo.)

subsp. **subsimplicifolia** *Wild* in Kirkia 5: 6 (1965) & in Kirkia 6: 51 (1967); Lisowski, Asterac. Fl. Afr. Centr. 1: 207 (1991). Type: Cameroon, Likomba, *Mildbraed* 10752 (K, holo., not found)

Leaves drying dark green, ovate to broadly ovate, not lobed near base, base itself cordate or abruptly cuneate. Ray florets orange-yellow or less often yellow.

UGANDA. Kigezi District: S Maramagambo Forest Reserve, May 1970, *Lock* 70/42!; Ankole District: Kibale National Park, near Kanyawara, Oct. 1995, *Poulsen et al.* 1019!; Mengo District: 27 km from Kampala on Fort Portal road, Feb. 1959, *Lind* 2368!
KENYA. North Kavirondo District: Kakamega Forest Station, Sep. 1949, *Maas Geesteranus* 6239!
TANZANIA. see Note
DISTR. **U** 2, 4; **K** 5; **T** (see Note); from West Africa to our area and southwards to Zimbabwe and Mozambique
HAB. Moist forest margins; 1050–1600(–1850) m
USES. None recorded on specimens from our area
CONSERVATION NOTES. Least concern (LC)

NOTE. This seems to be more like a form than a subspecies. All material from Tanzania which has the look of this subspecies also displays characteristics of subsp. *madagascariensis*, e.g. *H.M. Lloyd* 4 from Magusiro Island in **T** 1 and *Kahurananga* 2627 from **T** 4, Mwese Hills, seem to have both types of leaves on the same plant. *Sangster* 11 from Budongo has incipient basal lobes on some leaves. Intermediate specimens include *Semsei* 3510 from **T** 3, Kwamkoro; *Bidgood et al.* 2645 from **T** 4, Nkansi; *Pocs* 88/178d from **T** 6, Nguru Mts; and *Luke & Luke* 4870 from **T** 7, Udzungwa Mts. All of these come from forest margins between 450 and 1200 m.

3. **Melanthera richardsiae** *Wild* in Kirkia 5: 9 (1965) & in Kirkia 6: 49 (1967), as *richardsae*, Maquet in Fl. Rwanda 3: 626, fig. 173/2 (1985); Lisowski, Asterac. Fl. Afr. Centr. 1: 210, t. 46 (1991). Type: Zambia, Mbala, Chilongowelo, *Richards* 1491 (K!, holo., BR, E, EA, K!, iso.)

Climbing herb to 2 m high; stems appressed-pilose, glabrescent. Leaves ovate, 2.5–13 cm long, 1.5–9 cm wide, base abruptly cuneate, margins serrate-crenate, apex acute to acuminate, scabrid on both surfaces, 3-veined from base; petiole to 3 cm long. Capitula terminal, solitary or 2–3 together on stalks up to 10 cm long; involucre 8–13 mm long, phyllaries 2–3-seriate, ovate to elliptic, white-pilose; paleae oblong, acuminate. Ray florets ± 10, yellow, tube 2.7 mm long, glabrous, ray ± 14 mm long and minutely pilose on the veins abaxially, style 3.8–5.5 mm long; disc florets yellow, tube 4.2–6 mm long, glabrous, lobes ± 1 mm long, pilose. Achenes grey-black, obovoid, 2.7–3.5 mm long, slightly 3-angled, puberulous; pappus absent.

UGANDA. Toro District: Ndali, Apr. 1932, *Hazel* 259!; Kigezi District: Kachwekano Farm, May 1949, *Purseglove* 2824!
TANZANIA. Kigoma District: Kasoje, July 1959, *Newbould & Harley* 4399! & Mt Livandabe, June 1997, *Bidgood et al.* 4319!; Mpanda District: Ntakatta Forest, June 2000, *Bidgood et al.* 4642!
DISTR. U 2; T 4; Congo (Kinshasa), Rwanda, Zambia (Mbala area only)
HAB. Forest or thicket margin, often in forest margins by streams or lakes; 750–2100 m
USES. None recorded on specimens from our area
CONSERVATION NOTES. 2 Ugandan and 4 Tanzanian specimens; despite the fairly common habitat type a rare plant; data deficient (DD)

4. **Melanthera pungens** *Oliv. & Hiern*, F.T.A. 3: 382 (1877); F.P.S. 3: 43 (1956); Wild in Kirkia 5: 14 (1965); Lisowski, Asterac. Fl. Afr. Centr. 1: 216 (1991). Type: Sudan, Djur, Seriba Ghattas, *Schweinfurth* 1990 (K!, holo., BM!, iso.)

Perennial herb 0.3–1.2 m high; often bushy, in southern populations with woody rootstock, the branches often purplish, quadrangular, appressed-scabrid. Leaves narrowly to broadly ovate, oblong or elliptic, sometimes ovate, 1.5–14 cm long, 0.6–7 cm wide, base subcordate to cuneate, margins coarsely serrate or dentate to remotely serrate, apex acute or acuminate, scabrid on both surfaces, 3-veined from base; petiole absent or up to 10 mm long. Capitula terminal, solitary or in few-headed cymes, shallowly hemispherical but becoming globose with age, radiate; stalks of individual capitula 1–12 cm long; involucre 4–8.5 mm long, 2–3-seriate, the phyllaries ovate and pale green, the apices darker, pilose and often pectinate; paleae 6–9 mm long, acuminate to long-attenuate to elongate-setose, often yellow, often eroding and becoming fibrose with age. Ray florets 6–20, yellow, orange-yellow or orange, tube ± 1 mm long, glabrous, style absent, ray 3–11(–13) mm long, 1–4.5 mm wide, densely pilose abaxially; tube florets darker yellow to orange-yellow, tube 3–4 mm long, lobes 0.5–0.9 mm long with stiff hairs. Achenes ellipsoid, 1–2.8 mm long, 3–4-gonous, pilose; setae absent or 2–4, quickly deciduous, 0.5–3.3 mm long.

var. **pungens**

Usually single-stemmed herb. Paleae with elongate-setose apex. Rays 3–8 mm long.

UGANDA. Bunyoro District: without precise location, Nov. 1907, *Brown* 409!; Teso District: Serere, Dec. 1931, *Chandler* 146!; Mengo District: 14 m S of Nakasongola, Feb. 1956, *Langdale-Brown* 1954!
KENYA. West Suk District: Kapenguria, May 1932, *Napier* 1929!; Trans-Nzoia District: S Mt Elgon, June 1939, *Tweedie* 459!; North Kavirondo District: Kakamega, Mar. 1944, *Carson* H19!
TANZANIA. Tabora District: Kakoma, Dec. 1985, *Lloyd* 52!; Mpanda District: Mahale Mts, Bilenge, Feb. 1984, *Uehara* 8401a!
DISTR. U 2–4; K 2, 3, 5; T 4, 7 (see Note); Senegal, Mali, Congo (Kinshasa), Sudan
HAB. Grassland, bushland, old cultivation; 1000–2100 m

Uses. None recorded on specimens from our area
Conservation notes. Least concern (LC)

Syn. *M. acuminata* S.Moore in J.L.S. 25: 344 (1902). Type: Kenya, 'Kavirondo', *Scott Elliot* 7052
    (BM!, holo., K!, iso.)
  M. ugandensis S.Moore in J.L.S. 54: 256 (1916). Type: Uganda, District unclear, Uamananzi,
    Dummer 2564 (BM!, holo., K!, iso.)
  M. albinervia O.Hoffm. subsp. *acuminata* (S.Moore) Wild in Kirkia 5: 12 (1965); Lisowski,
    Asterac. Fl. Afr. Centr. 1: 215 (1991)

var. **albinervia** (*O.Hoffm.*) *Beentje* **comb. et stat. nov.** Type: Angola, Benguela, Quimbundo,
Ancieta 119 (COI, holo., LISU, iso.)

Often many-stemmed herb, the stems often striped or tinged with purple, from a woody
rootstock. Paleae with acuminate apex and often yellow. Rays 6–11(–13) mm long.

Uganda. (see Note)
Tanzania. Kondoa District: Kikore [Kikori], Jan. 1930, *Burtt* 2730!; Iringa District: E Mt Ngolo
    near summit, Feb. 1991, *Gereau & Kayombo* 3902!; Songea District: 32 km E of Songea, R.
    Murira, Jan. 1956, *Milne-Redhead & Taylor* 8277!
Distr. **U** (see Note); **T** 4–8; Congo (Kinshasa), Angola, Zambia, Malawi, Mozambique, Zimbabwe,
    Namibia (Caprivi), South Africa
Hab. *Brachystegia* or *Berlinia* woodland, less often in grassland; (250–)850–1800 m
Uses. None recorded on specimens from our area
Conservation notes. Least concern (LC)

Syn. *Melanthera albinervia* O.Hoffm. in Bol. Soc. Brot. 13: 30 (1896); Wild in Kirkia 5: 11 (1965)
    & in Kirkia 6: 52 (1967); Lisowski, Asterac. Fl. Afr. Centr. 1: 212, t. 47 (1991)
  Aspilia zombensis Baker in K.B. 1898: 152 (1898). Type: Malawi, Mt Zomba, *Whyte* s.n. (K!,
    holo., BM!, iso.)
  A. zombensis Baker var. *longifolia* S.Moore in J.L.S. 25: 345 (1902). Types: Malawi, Shire
    Highlands, *Scott Elliot* 8555 (BM!, K!, syn.) & *Buchanan* 24 (BM, syn., not found)

Note. Wild distinguished his *W. albinervia* subsp. *acuminata* from subsp. *albinervia* as "a more
robust plant.. with larger leaves whose venation is more prominent beneath. The leaves also
differ in drying grey-green. It seems to be somewhat intermediate between *M. albinervia* and
M. pungens. More investigation in the field is required." I have failed to find consistent
differences in leaf size, venation prominence or leaf colour between the subspecies of Wild.
The type of subsp. *acuminata* agrees with all our *pungens* material, including the type, and I
believe this is the same taxon. *M. albinervia* subsp. *albinervia* differs only in the apex of the
paleae, which is acuminate rather than elongate-setose, and the rays of the outer florets
which are often (but not always) longer. In *M. albinervia* sensu stricto the paleae often turn
the same yellow as the florets, which happens only rarely in *M. pungens*. *M. pungens* is a
northern taxon, occurring in Uganda and Kenya with only a few Tanzanian specimens; *M.
albinervia* sensu stricto is more southern, occurring from central Tanzania southwards to
Angola and South Africa. Nevertheless, these taxa are very close, and intermediates occur
(e.g. *Purseglove* 1232 from **U** 2 which has acuminate paleae, or *St.Clair Thompson* 491 from **T**
7 which has setose paleae). The epithet *pungens* has priority, and I believe the best solution
is to distinguish the two taxa at varietal level as their distribution areas overlap, and
differences are minimal.
    Subspecies *caudata* Wild has leaves less than 4 cm long and the paleae caudate; it occurs in
Zambia and southern Congo (Kinshasa).

5. **Melanthera biflora** (*L.*) *Wild* in Kirkia 5: 4 (1965) & in Kirkia 6: 48 (1967);
Wagner & Robinson in Brittonia 53, 4: 551 (2001). Type: India (LINN 1021.4, lecto.
designated by Wild in Kirkia 5: 4 (1965))

Perennial herb, woody herb or weak shrub, erect to 1 m high or decumbent with
spreading branches sometimes rooting at the nodes, sometimes scandent; branches
ribbed, glabrous or nearly so. Leaves thick, ovate, 2–13 cm long, 1–10 cm wide, base
cuneate to almost rounded, margins dentate, apex acuminate, scabridulous-pilose on
both surfaces; petiole 7–44 mm long. Capitula solitary or in few-headed corymbs,

Fig. 156. *MELANTHERA BIFLORA* — **1**, habit, × ²/₃; **2**, capitulum, × 2; **3**, ray floret, × 8; **4**, disc floret, × 8; **5**, achene, × 8. 1 & 3–4 from *Drummond & Hemsley* 3987, 2 from *Jex-Blake* 2279, 5 from *Napier* 6233. Drawn by Juliet Williamson.

radiate; stalks of individual capitula 2–6(–8.5) cm long; involucre bell-shaped to obconical, 4–9 mm long; phyllaries ovate to lanceolate, with thickened apex, pilose; paleae acuminate, keeled, pilose and ciliate. Ray florets yellow, tube 1.9–2 mm long, ray 5–9.5 mm long, 1.7–5 mm wide, hairy on 2 veins, style 3.8–4.5 mm long; disc florets yellow, tube 3.5–4 mm long, lobes 1 mm long, with a few hairs. Achenes obconical, 3–4 mm long and 1.8 mm across when mature, with pubescent apex; pappus absent (in ours). Fig. 156 (p. 743).

KENYA. Kilifi District: Vipingo, Dec. 1953, *Verdcourt* 1072! & Watamu, Apr. 1974, *Ng'weno* 22!; Kwale District: Diani, Dec. 1975, *Kokwaro* 3903!
TANZANIA. Tanga District: Boma Peninsula near Bomondani, Aug. 1953, *Drummond & Hemsley* 3616!; Bagamoyo District: Kaole, Feb. 1968, *Harris* 1400!; Uzaramo District: Dar es Salaam, Oyster Bay, June 1969, *Mwasumbi* 10526!; Pemba: Tundoni, July 1901, *Lyne* 134!
DISTR. **K** 7; **T** 3, 6; **Z**, **P**; from East Africa including Mozambique to the Pacific Islands and Australia
HAB. Sandy shore above the high water mark, also in bush on coral rag; 0–15 m
USES. None recorded on specimens from our area
CONSERVATION NOTES. Least concern (LC)

SYN. *Verbesina biflora* L., Sp. Pl. ed. 2: 1272 (1762)
    Wollastonia biflora (L.) DC., Prodr. 5: 546 (1836). Fig. 156 (page 743)
    Wollastonia zanzibarensis DC., Prodr. 5: 547 (1836). Type: Tanzania, Zanzibar, *Bojer* s.n. (G-DC, holo., K!, iso.)
    Wedelia biflora (L.) Wight, Contrib. Bot. Ind.: 18 (1837); Oliv. & Hiern, F.T.A. 3: 376 (1877); U.O.P.Z.: 486 (1949)

113. **SYNEDRELLA**

Gaertn., Fruct. 2: 456, t. 171 (1791), *nom. cons.*; Turner in Phytologia 76: 39–51 (1994)

Annual herbs, erect or procumbent. Leaves opposite, simple, 3-veined from base. Capitula axillary or terminal, small, usually in glomerules but sometimes solitary, heterogamous; involucre with few phyllaries, the outer foliaceous, the inner resembling the paleae and enveloping the marginal achenes; receptacular paleae subtending the internal florets. Ray florets 3–8, fertile; disc florets 5–10, hermaphrodite, fertile, with 4 corolla lobes and 4 stamens, the anthers sagittate and with obtuse or truncate apical appendage. Achenes of marginal florets ellipsoid, compressed, with winged margins; pappus of 2–3 short awns. Achenes of inner florets compressed, wings present or absent, tuberculate on one surface; pappus of 2(–4) awns.

Two species from tropical America; one has become a pantropic weed.

Synedrella nodiflora (*L.*) *Gaertn.*, Fruct. 2: 456, t. 171 (1791); Maquet in Fl. Rwanda 3: 638, fig. 196/1 (1985); Lisowski, Asterac. Fl. Afr. Centr. 1: 118 (1991); D.J.N. Hind in Fl. Masc. 109: 183, t. 61 (1993). Type: Tropical America, 'habitat in Caribaeis' (LINN 1021.7, lecto., designated by Howard in Fl. Lesser Antilles 6: 600 (1989))

Annual herb, erect, 15–120 cm high, hardly branched to much branched, the stems scabridulous but glabrescent. Leaves ovate or elliptic, 2–12 cm long, 1–5 cm wide, base cuneate and decurrent, margins crenulate-serrate or rarely subentire, apex acute or obtuse, 3-veined from base, scabridulous on both surfaces; petiole winged, to 3 cm long. Capitula axillary and terminal, in dense sessile glomerules; involucre 7–12 mm long; outer phyllaries green, the inner phyllaries scarious; paleae with eroded or ciliate margins. Ray florets 3–8, yellow, the ray 1.5–3 mm long; disc florets 4–11, yellow, 2–4 mm long. Achenes of ray florets 4–5 mm long, winged, with 2 short awns; achenes of disc florets 4–6 mm long, the outer winged, the inner without wings, pappus of 2–3 awns 2–4.5 mm long. Fig. 157 (page 745).

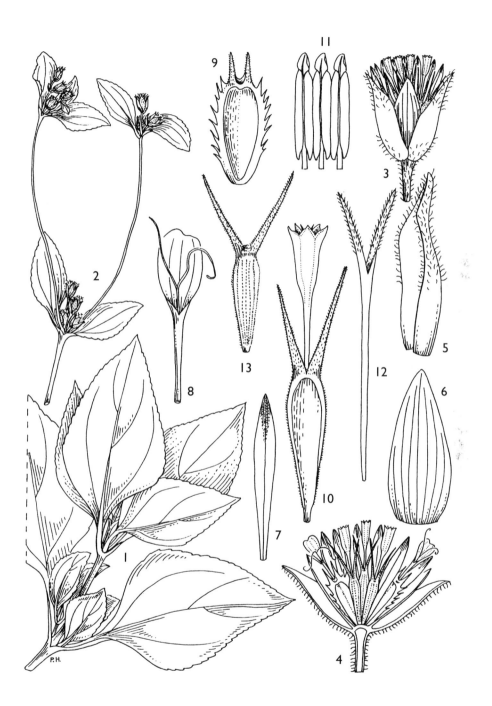

FIG. 157. *SYNEDRELLA NODIFLORA* — **1**, habit, × ²/₃; **2**, flowering branch, × ²/₃; **3**, young capitulum, × 4; **4**, capitulum section, × 4; **5–6**, outer and inner phyllaries, × 6; **7**, palea, × 6; **8**, ray floret, × 12; **9**, ray floret achene, × 6; **10**, disc floret, × 12; **11**, anthers, × 32; **12**, disc floret style, × 24; **13**, disc floret achene, × 6. 1–2 from *Cadet* 1739, 3–13 from *Cadet* 1675. Drawn by Pat Halliday, from Flore des Mascareignes.

UGANDA. Busoga District: 5 km W of Busesa, Oct. 1962, *Lewis* 6030!; Mengo District: Kiwala, 1917, *Dummer* 3264! & Kituza, June 1957, *Griffiths* 46!
KENYA. Kilifi District: Kaya Pangani, Nov. 1978, *Brenan et al.* 14568!; Kwale District: Mrima Hill, Sep. 1957, *Verdcourt* 1919! & Msambweni, Aug. 1982, *Robertson* 3319!
TANZANIA. Tanga District: Longuza–Tongwe, Sep. 1936, *Greenway* 4636!; Kigoma District: Kibirizi, Apr. 1994, *Bidgood & Vollesen* 3169!; Uzaramo District: Pugu Forest reserve, July 1969, *Mwasumbi* 10568!; Zanzibar, Nungwi, July 1950, *Oxtoby* 27!
DISTR. **U** 3, 4; **K** 4, 7; **T** 3, 4, 6, 7; **Z**; originally from the Caribbean, now a pantropical weed
HAB. Waste places, cultivated land, secondary vegetation; a common weed forming large colonies; 0–900(–1200) m
USES. None recorded on specimens from our area
CONSERVATION NOTES. Least concern (LC)

SYN. *Verbesina nodiflora* L., Cent. I. Pl.: 28 (1755)
    Wedelia cryptocephala Peter in Abh. Ges. Wiss. Gottingen, n.f. 13, 2: 94 in clavi (1928). Type: Lushoto District: Usambara, various localities, *Peter* 23449, 19602, 19832, 19919, 25220, 24655 (B†, syn.)

NOTE. Not many specimens from, and probably uncommon in, Uganda and coastal Kenya (we have cited nearly all sheets from each country!), but a common weed in coastal Tanzania.

114. **ASPILIA**

Thouars, Gen. Nov. Madag.: 12 (1806); Wild in Kirkia 5: 197–228 (1966); Lisowski, Asterac. Fl. Afr. Centr. 1: 184–204 (1991)

Annual or perennial herbs, less often shrubs. Leaves opposite, scabrid or hispid. Capitula terminal, solitary or in lax cymes, or axillary and in congested corymbs, radiate; phyllaries 2–3-seriate, the outer herbaceous, the inner scarious; receptacle with conduplicate paleae. Ray florets with the ray 2-lobed at apex, with or without style, not producing mature achenes; disc florets tubular; anther bases obtuse or shortly sagittate, thecae and appendages often black; style branches gradually narrowing, pubescent at margins. Achenes obovoid, compressed, pubescent; pappus a laciniate cupule, often with a few setae.

21 species in tropical Africa.
A difficult genus, and one of the most unsatisfactory treatments HB has produced; though better, we believe, than what came before.
Wild, a taxonomist we are usually happy to follow, did not produce a key we can work with. We are not surprised about this: variation seems to be all over the place. Intermediates occur between the key characters used below, especially for the last couplet; still, the taxa are ± distinct.
Treated as a generic synonym of *Wedelia* Jacq. by authors such as Bremer (Asteraceae, cladistics and classification), Strother, Robinson, and Turner. We believe this is incorrect.

1. Plants annual; capitula usually ± subsessile among the
   leaves; styles absent in ray florets 2
  Plants perennial; capitula usually stalked and above the
   leaves; styles absent or present in ray florets 3
2. Rays white or violet to purple; anther appendages often
   purple; widespread . . . .   1. *A. kotschyi*
  Rays yellow; anther appendages black; **U** 2, **T** 4 . . . .   2. *A. ciliata*
3. Leaves ovate to trullate; plant of coastal dunes; **K** 7 . . . .   5. *A. macrorrhiza*
  Leaves elliptic to ovate; widespread 4
4. Anther appendages black or darkened; often some outer
   phyllaries shorter than median and inner ones; paleae
   acuminate from wider 'shoulders'; style present in ray
   florets . . . .   4. *A. africana*
  Anther appendages yellow or pale; outer phyllaries as
   long as, or longer than median and inner ones 5

5. Paleae acuminate to almost caudate; plant trailing, or with
   many short erect stems; leaves usually 2–5 × 0.8–2 cm;
   capitula shorter and broader than in *A. mossambicensis*;
   styles absent from ray florets . . . . 3. *A. pluriseta*
   Paleae obtuse to acute; plant usually erect; leaves usually
   larger; styles absent or present in ray florets . . . . 6. *A. mossambicensis*

1. **Aspilia kotschyi** (*Hochst.*) *Oliv.* in Trans. Linn. Soc. 29: 98 (1873); Oliv. & Hiern,
F.T.A. 3: 381 (1873); Muschler in E.J. 50, Suppl.: 342 (1914); F.W.T.A. 2, 1: 146
(1931); F.P.S. 3: 10 (1956); F.P.U.: 171 (1962); C.D.Adams in F.W.T.A. ed. 2, 2: 239
(1963); Wild in Kirkia 5: 202 (1966) & in Kirkia 6: 40 (1967); Maquet in Fl. Rwanda
3: 623, fig. 192/4 (1985); Lisowski, Asterac. Fl. Afr. Centr. 1: 187 (1991); U.K.W.F. ed.
2: 216 (1994). Type: Ethiopia, Arash-Cool, *Kotschy* 103 (BM!, BR!, K!, LD, M, iso.)

Annual herb 0.3–1.8 m high, erect, little branched to much branched; stem
scabrid-hispid, often with purple spots at the base of the hairs. Leaves (sub-)sessile,
narrowly ovate to narrowly elliptic, 3–16(–20) cm long, 0.6–3(–5) cm wide, base
obtuse, truncate or semi-amplexicaul, sometimes cuneate (Lisowski), margins
subentire or shallowly crenate, apex acute to attenuate, hispid on both surfaces.
Capitula axillary, solitary and subsessile among the upper leaves; involucre ovoid,
3-seriate, 8–14(–30) mm long; phyllaries green with scarious brown base, hispid, the
inner often with darker tips; paleae 6–10 mm long with obtuse or acute apex, often
with reddish tips. Ray florets dark red, purple or almost black, less often white,
2–5(–11), without styles, the ray almost circular, 6–9(–15 fide Lisowski) mm long,
tube 3.5–8 mm long, glabrous; disc florets coloured as the rays, 5–8 mm long,
glabrous but the lobes puberulous; anther appendages often purple. Achenes
obovoid, 5–6 mm long, pubescent; pappus a lacerate cupule to 1 mm high and
usually with 2 setae 1–3.5 mm long.

var. **kotschyi**; Wild in Kirkia 5: 203 (1966) & in Kirkia 6: 41 (1967); Lisowski, Asterac. Fl. Afr.
Centr. 1: 187, t. 41 (1991)

Corolla dark red, purple or almost black, the ray ± circular.

UGANDA. Acholi District: Queen Elizabeth National Park, Chobi, Oct. 1967, *Buzigye* 11!; Teso
   District: Serere, Dec. 1931, *Chandler* 240!; Mengo District: 18 km NW of Nakasongola, Sep.
   1955, *Langdale-Brown* 1509!
KENYA. W Suk District: Marich Pass, Ortum, Oct. 1977, *Carter & Stannard* 52!; Kisumu District:
   Kisumu, Feb. 1915, *Dummer* 1816!; Mombasa, Bamburi Farm, 1973, *Haller* 3!
TANZANIA. Kigoma District: Gombe National Park, Kakombe, Apr. 1992, *Mbago* 1044!; Kilwa
   District: Selous Game Reserve, Nahomba, May 1970, *Rodgers* 1058!; Zanzibar: Kisimbani, Mar.
   1961, *Faulkner* 2791!
DISTR. **U** 1–4; **K** 2, 3, 5, 7; **T** 1, 3, 4, 6–8, **Z**; West Africa from Senegal to Sudan, Ethiopia, and S
   to Angola, Zimbabwe and Mozambique
HAB. Grassland, moist grassland, waste land and weed of cultivation, less often in woodland,
   bushland, on rocky outcrops; may be locally common; 0–1500(–1950) m
USES. Minor medicinal for sore eyes and cuts (*Tanner*)
CONSERVATION NOTES. Least concern (LC)

SYN. *Dipterotheca kotschyi* Sch.Bip. in Flora 25: 435 (1842)

var. **alba** *Berhaut* in Bull. Soc. Bot. France 101: 375 (1954); C.D. Adams in F.W.T.A. ed. 2, 2:
239 (1963); Wild in Kirkia 5: 204 (1966) & in Kirkia 6: 41 (1967); Lisowski, Asterac. Fl. Afr.
Centr. 1: 190 (1991). Type: Senegal, Bargny, *Berhaut* 1726 (P!, holo.)

Corolla white, the ray spatulate, less wide than in var. *kotschyi*.

UGANDA. Bunyoro District: Magoma, Oct. 1970, *Katende* 641!; Busoga District: 14 km W of
   Kamuli, Apr. 1953, *G.Wood* 723!; Masaka District: Sese, Bukasa, June 1932, *A.S.Thomas* 134!

TANZANIA. Bukoba District: Bukoba, June 1954, *van Someren* s.n.!; Mpanda District: Rukwa, Tumba, Feb. 1952, *Siame* 145!; Zanzibar: Kisimbani, Mar. 1961, *Faulkner* 2792!
DISTR. U 1–4; T 1, 4, 6; Z; overall range as in var. *kotschyi*
HAB. Grassland, moist grassland, waste land and weed of cultivation; 90–1500 m
USES. Minor medicinal for gonorrhoea (*Maitland*)
CONSERVATION NOTES. Least concern (LC)

SYN. *A. polycephala* S.Moore in J.Bot. 45: 45 (1907). Type: Uganda, Toro District: Fort Portal, *Bagshawe* 993 (BM, holo.)

2. **Aspilia ciliata** (*Schumach.*) *Wild* in Kirkia 6: 41 (1967); Maquet in Fl. Rwanda 3: 624, fig. 192/2 (1985); Lisowski, Asterac. Fl. Afr. Centr. 1: 191, t. 42 (1991). Type: Ghana, without locality, *Thonning* s.n. (C, holo., K, photo.!)

Annual herb 0.3–1.3 m high, usually erect; branches sulcate, scabrid-pilose. Leaves (sub-)sessile or petiolate, ovate or narrowly ovate, 3–15 cm long, 1–5 cm wide, base cuneate or rounded and abruptly cuneate, margins subentire or shallowly serrate, apex acute, scabrid on both surfaces; 3-veined from base; petiole absent or to 18 mm long. Capitula in axillary and terminal short corymbs; stalks of individual capitula to 3.5 cm long but usually much less; involucre 3-seriate, 7–18 mm long; outer phyllaries herbaceous, inner scarious, scabrid-pilose; paleae 6–9 mm long with the upper part almost appendiculate, acute or abruptly acuminate, ciliate. Ray florets 5–15, yellow or pale yellow, style absent, ray 4–10 mm long, tube 1.7–3 mm, glabrous; disc florets yellow, 3.5–5.3 mm long, glabrous or nearly so; anther-appendages dark to black. Achenes obovoid, ± compressed, 4.7–6 mm long, pilose; pappus a lacerate cupule to 0.5 mm high and usually without setae, sometimes with 2 setae to 2.5 mm long.

UGANDA. Kigezi District: Kayonza, Apr. 1948, *Purseglove* 2646! & Mitano Gorge, Nov. 1950, *Purseglove* 3492!
TANZANIA. Kigoma District: Gombe National Park, Apr. 1961, *Siwezi* 131! & Kibirizi, Apr. 1994, *Bidgood & Vollesen* 3168!; Mpanda District: Mahali Mts, Utahya, 1958, *Newbould & Jefford* 2362!
DISTR. U 2; T 4; widespread in Africa from Senegal to Ethiopia and S to Zambia
HAB. On sandy soils near lakes or rivers, weed of cultivation; 650–1500 m
USES. Roots as minor medicinal for stomach trouble (*Newman*)
CONSERVATION NOTES. Least concern (LC)

SYN. *Verbesina ciliata* Schumach. in Schumach. & Thonn., Beskr. Guin. Pl.: 391 (1827)
　　Blainvillea prieuriana DC., Prodr. 5: 492 (1836); Oliv. & Hiern, F.T.A. 3: 375 (1877); F.P.S. 3: 13 (1956). Type: Senegal, *Leprieur* s.n. (G-DC, holo.)
　　Wirtgenia abyssinica Sch.Bip. in Walp., Repert. 6: 146 (1846). Type: Ethiopia, near Gapdiam, *Schimper* 819 (BM, BR!, K!, M, iso.)
　　W. schimperi A.Rich., Tent. Fl. Abyss. 1: 412 (1848). Type: Ethiopia, Tacaze, Djeladjeranne, *Schimper* 1684 (BM, K!, M, P, iso.)
　　Aspilia abyssinica (Sch.Bip.) Vatke in Linnaea 39: 495 (1875); Oliv. & Hiern, F.T.A. 3: 379 (1877); Muschler in E.J. 50, suppl.: 341 (1914)
　　A. schimperi (A.Rich.) Oliv. & Hiern, F.T.A. 3: 379 (1877); F.P.S. 3: 9 (1956)
　　A. dewevrei O.Hoffm. in B.S.B.B. 39, 3: 32 (1901); Muschler in E.J. 50, suppl.: 335 (1914). Types: Congo (Kinshasa), no locality given, *Dewevre* 898 & 920 (both BR!, syn.)
　　Wedelia ringoetii De Wild. in F.R. 13: 210 (1914). Type: Congo (Kinshasa), Shinsenda, *Ringoet* in Homblé 503 (BR!, syn.) & Kapiri Valley, *Homblé* 1103 (BR!, syn.)
　　Aspilia helianthoides (Schumach. & Thonn.) Oliv. & Hiern subsp. *ciliata* (Schumach.) C.D.Adams in Webbia 12: 245 (1956); F.W.T.A. ed. 2, 2: 239 (1963); Wild in Kirkia 5: 205 (1965)
　　A. helianthoides (Schumach. & Thonn.) Oliv. & Hiern subsp. *prieuriana* (DC.) C.D.Adams in Webbia 12: 246 (1956); F.W.T.A. ed. 2, 2: 239 (1963); Wild in Kirkia 5: 208 (1965)

NOTE. This taxon has white, cream or purple corollas in West Africa, and may have white flowers in Ethiopia and Sudan. So far, all our material has yellow flowers. DJNH wonders whether it really is all the same taxon, but Charles Jeffrey (pers. comm.) considers it is.
　　Lisowski has *Aspilia brachystephana* O.Hoffm. as a synonym of this taxon, but I have been unable to trace this taxon. The type is said to be *Gillet* s.n. from Kisantu, collected in 1900. I have not seen this 'type'.

3. **Aspilia pluriseta** *Schweinf.* in L.Höhn., Rudolf–Stephanie See: 862 (1892); Muschler in E.J. 50, suppl.: 341 (1914); Wild in Kirkia 5: 210 (1965) & in Kirkia 6: 43 (1967); Maquet in Fl. Rwanda 3: 624, fig. 192/1 (1985); Lisowski, Asterac. Fl. Afr. Centr. 1: 196 (1991); U.K.W.F. ed. 2: 216 (1994). Type: Kenya, "Massaihochland, Ndoro", *von Höhnel* 66 (B†, holo.); neotype: Kenya, Ukamba, *Hildebrandt* 2712 (B†, K!, neo., BM!, M, isoneo., chosen by Wild)

Perennial herb or subshrub, with multiple branches from woody rootstock, usually trailing with distal part erect and to 50 cm high, or with many short erect stems up to 50 cm high, sometimes (?in shade?) erect and to 1.5 m high; stem scabrid. Leaves (sub-)sessile, ovate to narrowly ovate, 2–5(–8) cm long, 0.8–2(–3.5) cm wide, base rounded, subcordate or broadly cuneate, margins serrate, apex obtuse or acute, scabrid on both surfaces; 3-veined from base; petiole 0–3(–5) mm long. Capitula terminal and solitary or up to 3 together, almost sessile among the upper leaves or on stalks 0.5–6 cm long; involucre ovoid, 3-seriate, 7–12(–16) mm long; phyllaries green, scabrid-pilose, the outer often long and reflexed; paleae 7–11 mm long with caudate-acuminate apex, especially in fruit. Ray florets yellow or orange-yellow, 9–16, without styles, the ray 10–15 mm long, tube 2.3–3.2 mm long, glabrous or pilose; disc florets yellow or brownish, 4.7–9 mm long, glabrous or puberulous; anther appendages yellow. Achenes obovoid, 2.5–4.5 mm long, pubescent; pappus a lacerate cupule to 0.7 mm high and sometimes with 1–2(–5) setae to 2.5 mm long.

UGANDA. Kigezi District: Shumbe Hill, Aug. 1949, *Purseglove* 3083!; Mbale District: Tororo, Apr. 1948, *Bally* 6194!; Mengo District: Bugiri near Kisubi, Aug. 1969, *Lye & Hamilton* 3619!
KENYA. Nakuru, Sep. 1933, *Napier* 5385!; N Kavirondo District: Kakamega, Mukulu Forest, Mar. 1976, *Gonget* 140!; Machakos/Masai District: Chyulu Hills, Saddle, Jan. 1997, *Luke & Luke* 4593!
TANZANIA. Mbulu District: Mbulumbul, June 1945, *Greenway* 7470!; Ufipa District: Wansambo Hill slopes near Wipanga, June 1987, *Mwasumbi et al.* 13183!; Rungwe District: Mwankinja, Jan. 1954, *Semsei* 1593!
DISTR. U 2–4; K 2–7; T 1–5, 7, 8; Congo (Kinshasa), Rwanda, Burundi, Zambia, Malawi, Mozambique, Zimbabwe, South Africa
HAB. Grassland, woodland, also a pioneer of cultivation and ruderal sites; may be locally common; (45–)1000–2200(–2600) m
USES. Minor medicinal against worms (*Jarrett*); leaves used as sandpaper (*Glover*); used in witchcraft, and for cleaning teeth (*Harwood*)
CONSERVATION NOTES. Least concern (LC)

SYN. *A. asperifolia* O.Hoffm. in P.O.A. C: 413 (1895); Muschler in E.J. 50, suppl.: 335 (1914). Syntypes: Tanzania, Kilimanjaro, *Volkens* 367 (B†, BM!, syn.), 541 (B†, syn.); Lushoto District: Usambara, Msinga, *Holst* 9127 (B†, COI, K!, M, syn.)
   A. gondensis O.Hoffm. in P.O.A. C: 413 (1895); Muschler in E.J. 50, suppl.: 336 (1914). Type: Tanzania, Tabora District: Igonda, *Böhm* 41 (B†, holo.); neotype: Tanzania, *Busse* 1366 (B†, EA, neo., chosen by Wild)
   ?*A. involucrata* O.Hoffm. in P.O.A. C: 413 (1895); Muschler in E.J. 50, suppl.: 341 (1914). Type: Kenya, "Massaihochland, Wadiboma", *Fischer* 330 (B†, holo.)
   A. vulgaris N.E.Br. in K.B. 1906: 164 (1906). Type: Zimbabwe, between Umtali and Harare [Salisbury], *Cecil* 43 (K!, holo.)
   A. brachyphylla S.Moore in J.L.S. 40: 115 (1911). Syntypes: Zimbabwe, Chirinda, *Swynnerton* 292 (K!, syn.) & 495 (ubi?, syn., not at BM)
   A. pluriseta Schweinf. subsp. *gondensis* (O.Hoffm.) Wild in Kirkia 5: 210 (1965) & in Kirkia 6: 44 (1967)

NOTE. Wild distinguishes subsp. *pluriseta* and subsp. *gondensis*, on "the narrowly oblong-ovate leaves and somewhat longer peduncles" plus the yellow anther-appendages of the latter. This seems a matter of degree not worthy of distinction (certainly not at subspecific level) and I unite the subspecies.
   Kuchar 23333 from T 5, Singida, has up to 5 setae on the achenes.
   Aspilia mendoncae Wild in Kirkia 5: 212 (1965) & in Kirkia 6: 42 (1967). Type: Malawi, Lower Kirk Range, *E.M.& W.* 945 (SRGH, holo., BM, LISC, iso.) is close: Wild says it is close to *asperifolia*, but erect; leaves narrowly elliptic; capitula 2–3 together; peduncles slender. He cites *Milne-Redhead & Taylor* 8409 from Songea as the only EA specimen, but this is a specimen of *A. mossambicensis*.

4. **Aspilia africana** (*Pers.*) *C.D.Adams* in Webbia 12: 236 (1956) & in F.W.T.A. ed. 2, 2: 238 (1963); F.P.U.: 171, fig. 109 (1962); Wild in Kirkia 5: 218 (1965); Maquet in Fl. Rwanda 3: 624, fig. 192/3 (1985). Type: Nigeria, Warri [Oware], *Beauvois* s.n. (?P, not found)

Perennial herb 25–130 cm high, erect; stem scabrid. Leaves lanceolate, 4–12 cm long, 0.7–3.7(–5) cm wide, base cuneate, margins serrate, apex attenuate to acuminate, scabrid on both surfaces and often also pubescent beneath; 3-veined from base; petiole 2–4 mm long, winged. Capitula in lax terminal corymbs; stalks of individual capitula 1–5 cm long; involucre ovoid, 3-seriate, 6–14 mm long; phyllaries recurved distally, some of the outer occasionally much shorter, scabrid; paleae ± 7 mm long, acuminate from a wider part (almost appendiculate). Ray florets 10–12, pale to rich yellow, ray 8–12 mm long, sometimes glandular abaxially, tube 2 mm long, style present; disc florets 4.8–6 mm long, glabrous or nearly so; anthers with black appendage. Achenes ovoid, compressed, 3–3.5 mm long, pubescent; pappus of a laciniate cupule to 1 mm long, without setae.

UGANDA. W Nile District: Maracha Rest Camp, July 1953, *Chancellor* 59!; Bunyoro District: Budongo Forest, July 1972, *Synnott* 1086!; Mubende District: Kasambya, May 1957, *Griffiths* 20!
TANZANIA. Mwanza District: Mbarika, Apr. 1953, *Tanner* 1377!; Kigoma District: Gombe National Park, Msekela Hill, Mar. 1996, *Kayombo et al.* 1227!; Chunya District: Mbangala, Feb. 1994, *Bidgood et al.* 2234!
DISTR. **U** 1–4; **T** 1, 4, 5, 7; widespread in tropical Africa
HAB. Grassland, woodland, forest margins, (abandoned) cultivation; may be locally common; 800–1800(–2100) m
USES. Minor medicinal for infected cuts and against bilharzia (*Tanner*) and against fever (*Lovett*)
CONSERVATION NOTES. Least concern (LC)

SYN. *Wedelia africana* Pers., Syn. Pl. 2: 490 (1807); P.Beauv., Fl. Owar. Benin 2: 19 , t. 69 (1810); DC., Prodr. 5: 439 (1836); Oliv. & Hiern, F.T.A. 3: 376 (1877); F.P.S. 3: 62 (1956)
    Aspilia latifolia Oliv. & Hiern, F.T.A. 3: 379 (1877); F.P.S. 3: 9 (1956). Type: Ghana, Accra, *Vogel* s.n. (K!, syn.); Nigeria, Niger R., *Baikie* s.n. (K!, syn.); Nigeria, Old Calabar, *Mann* s.n. (K, syn., not found) & *Monteiro* s.n. (K, syn., not found); Sudan, Djur land, Seriba Ghattas, *Schweinfurth* 2011 (K!, P!, syn.) & Niam-niam land, *Schweinfurth* 3737 (K!, syn.)
    Wedelia magnifica Chiov. in Ann. Bot. Rom. 9: 74 (1911). Type: Ethiopia, Gondar, Dembia, *Chiovenda* 1582 (FT, holo.)
    Aspilia congoensis S.Moore in J.Bot. 58: 45 (1920); Wild in Kirkia 5: 213 (1965); Lisowski, Asterac. Fl. Afr. Centr. 1: 194, t. 43 (1991). Type: Congo (Kinshasa), upper Welle Province, *Lacombley* 67 (BM!, holo., BR!, iso.), **syn. nov.**
    A. africana (Pers.) C.D.Adams subsp. *magnifica* (Chiov.) Wild in Kirkia 5: 219 (1965)

NOTE. Wild commented that he thought of treating *Aspilia congoensis* as a subspecies of *A. africana* but that the differences (which seem to boil down to "caudate-acuminate paleae like in *A. pluriseta*, not acute as in *A. africana*") were enough to keep it distinct. After having studied the type of *A. congoensis* and material of *A. africana* (sadly, I was unable to trace the type of *A. africana*) HB feels that the two are synonymous; paleae can vary considerably and within the range of *A. africana* sensu stricto attenuate paleae occur with regularity. The name *africana* is much older and has priority.

5. **Aspilia macrorrhiza** *Chiov.*, Result. Sci. Miss. Stef.-Paoli, Coll. Bot.: 105 (1916); Wild in Kirkia 5: 216 (1965). Type: Somalia, Mogadishu and Amarr Gegeb, *Paoli* 23 (FT, holo.)

Perennial herb, prostrate and creeping, mat-forming; stem scabrid. Leaves ovate or trullate, 1.5–8 cm long, 0.7–4 cm wide, base rounded and abruptly cuneate to attenuate into a winged pseudopetiole, margins serrate, apex acute, scabrid on both surfaces. Capitula terminal and solitary or in pairs; stalks of individual capitula 1–7.5 cm long; involucre ovoid, 2–3-seriate, 6–11 mm long; phyllaries green, scabrid-pilose; paleae with acute or apiculate apex. Ray florets yellow to orange, 8–9, usually with styles, the ray 7.5–8 mm long, tube 2.3–2.7 mm long; disc florets yellow or orange, 4.5–5 mm long; anther appendages yellow. Achenes obovoid with 4 fat ribs reaching as high as the cupule, 3.7–5.5 mm long, pubescent; pappus a lacerate cupule to 1 mm high and with or without 1 seta to 1.8 mm long.

KENYA. Lamu District: Osine, Oct. 1957, *Greenway & Rawlins* 9290! & Kiunga, Aug. 1961, *Gillespie* 181! & Kiwayu, Chole Is., Jan. 2000, *Luke & Luke* 6148!
DISTR. **K** 7; Somalia (Kismayu area)
HAB. Near the sea on dunes, coral or sandstone; 0–10 m
USES. None recorded on specimens from our area
CONSERVATION NOTES. Restricted to a narrow strip between Kismayu and Lamu, but as far as we know, habitat not under threat; least concern (LC) or near threatened (NT)

SYN. *Wedelia macrorrhiza* (Chiov.) Chiov., Fl. Somal. 2: 264, fig. 153 (1932)

6. **Aspilia mossambicensis** (*Oliv.*) *Wild* in Kirkia 5: 221 (1965) & in Kirkia 6: 46 (1967); Blundell, Wild Fl. E. Afr.: pl. 351 (1987); Lisowski, Asterac. Fl. Afr. Centr. 1: 203 (1991); U.K.W.F. ed. 2: 216, t. 88 (1994). Type: Tanzania, 6°S, 3800 feet, *Grant* s.n. (K!, holo.)

Perennial herb or shrub 10–250 cm high, with a single or many stems from short rootstock with numerous fibrous roots and stiff branching; stems often reddish or purple near base; branches scabrid-pubescent, sometimes also glandular. Leaves sessile or with petiole to 1 cm long, ovate, narrowly elliptic or lanceolate, 2.5–20 cm long, 1–8.5 cm wide, base rounded to cuneate, margins serrate or subentire, apex attenuate or acuminate, less often obtuse, very scabrid on both surfaces; 3-veined from base. Capitula terminal and solitary or in few-headed lax racemes; stalks of individual capitula to 14 cm long; involucre ovoid, 2–3-seriate, 5–17 mm long; outer phyllaries foliaceous, green, scabrid-pilose and sometimes glandular; paleae 6.5–12 mm long with acute to attenuate apex, keel often purple. Ray florets cream, yellow to orange, (5–)7–17, with or without styles, the ray 7–18.5 mm long, to 6.5 mm wide, tube 1.5–4.5 mm long; disc florets cream, yellow or orange, sometimes with purple line down from the lobes, 5.5–9 mm long, the lower part of tube and lobes puberulous; anther appendages yellow. Achenes narrowly obovoid or sub-cylindrical, 2.5–5.5 mm long, pubescent to almost glabrous; pappus a lacerate cupule to 1 mm high and with or without 1–2 setae to 3 mm long. Fig. 158 (page 752).

UGANDA. Karamoja: Napak Mts, June 1957, *J.Wilson* 354!; Toro District: Matiri, Jan. 1996, *Freidberg & Yarom* 43a!; Mengo District: Entebbe, July 1971, *Willemse* 3!
KENYA. Turkana District: Oropoi, Feb. 1965, *Newbould* 6892!; Naivasha District: Lake Naivasha SW, Apr. 1968, *Mwangangi* 772!; Kwale District: Shimba Hills, Longo Mwagandi, Mar. 1968, *Magogo & Glover* 325!
TANZANIA. Lushoto District: W Usambara Mts 3 km N of Mashewa, Apr. 1987, *Borhidi et al.* 87/013!; Ufipa District: Mbizi Forest reserve, Oct. 1987, *Ruffo & Kisena* 2846!; Iringa District: Livingstone Mts 9 km N of Mbilwa, Mar. 1991, *Gereau & Kayombo* 4400!
DISTR. **U** 1–4; **K** 1–7; **T** 1–8; Congo (Kinshasa), Burundi, Ethiopia, Somalia, and southward to South Africa
HAB. Ruderal sites, seasonal swamps, along rivers and lakes, in forest margins, woodland, wooded grassland, bushland and grassland; may be locally common; 45–2300 m
USES. Minor medicinal for sore eyes and sore gums (*Tanner*) and against lumbago (*Ruffo*); roots used for snuff (*Tanner*); possibly used as medicine by chimpanzees (*Turner*); leaves fed to animals to cure ulcers (*Murugu*)
CONSERVATION NOTES. Least concern (LC)

SYN. *Menotriche strigosa* Steetz in Peters, Reise Mossamb. Bot. 2: 475 (1863). Type: Mozambique, Rios de Sena, *Peters* s.n. (B†, holo.) [non *Aspilia strigosa* (Hook. & Arn.) W.B.Hemsl., 1881)
Wedelia mossambicensis Oliv. in Trans. Linn. Soc. 29: 97 (1873); Oliv. & Hiern, F.T.A. 3: 377 (1877)
W. menotriche Oliv. & Hiern, F.T.A. 3: 377 (1877). Type as for *Menotriche strigosa*

FIG. 158. *ASPILIA MOSSAMBICENSIS* — **1**, habit, × ²/₃; **2**, leaf detail, upper surface, × 1 ½; **3**, capitulum, × 1; **4**, capitulum, early seed stage, × 2; **5**, ray floret, × 3; **6**, disc floret, style showing, × 5; **7**, disc floret, anthers showing, × 5; **8**, achene, × 5. 1–2 from *Drummond & Hemsley* 2188, 3–7 from *Constabel* 4/7/86, 8 from *Bax* 242. Drawn by Juliet Williamson.

Aspilia wedeliiformis Vatke in Osterr. Bot. Zeit. 27: 197 (1877), as *wedeliaeformis*; Oliv. &
Hiern, F.T.A. 3: 461 (1877); Muschler in E.J. 50, suppl.: 338 (1914). Type: Kenya, Lamu,
Hildebrandt 1908 (B†, holo., W, fragm.)

A. monocephala Baker in K.B. 1898: 152 (1898); Muschler in E.J. 50, suppl.: 337 (1914).
Type: Malawi, Zomba, *Whyte & McClounie* s.n. (K!, syn.)

A. holstii Engl., P.O.A. C: 413 (1895); Muschler in E.J. 50, suppl.: 340 (1914). Type: Kenya,
Teita, *Hildebrandt* 2380 (B†, syn.), Tanzania, Lushoto District: Usambara, *Holst* 148 (B†,
syn.) & *Holst* 4334 (BM!, COI, EA, K!, M, syn.)

Coreopsis aspiliodes Baker in K.B. 1898: 153 (1898). Type: Malawi, Zomba, *Whyte* s.n. (K!, holo.)

Wedelia instar S.Moore in J.L.S. 35: 343 (1902). Type: Malawi, *Buchanan* 67 (BM!, holo.)

Aspilia chrysops S.Moore in J.L.S. 38: 459 (1900); Muschler in E.J. 50, suppl.: 338 (1914).
Type: Somalia, Laskaroto, *Donaldson Smith* s.n. (BM!, holo.)

A. aspilioides (Baker) S.Moore in J.L.S. 40: 115 (1911); Lisowski, Asterac. Fl. Afr. Centr. 1:
202 (1991)

Wedelia affinis De Wild. in F.R. 13: 209 (1914). Types: Congo (Kinshasa), Shaba [Katanga],
Corbisier in herb. *Homblé* 606 (BR!, syn.), 612 (BR, syn.)

W. katangensis De Wild. in F.R. 13: 209 (1914). Types: Congo (Kinshasa), Shaba [Katanga],
Welgelegen, *Homblé* 794 (BR!, syn.), Esschen, *Homblé* 887 (BR!, syn.), 894 (BR!, syn.)

Aspilia ritellii Chiov., Fl. Somal. 2: 266 (1932). Type: Somalia, Isola di Alexandra, *Tozzi* 280
(FT, holo.)

A. tanganyikensis Lawalrée in B.J.B.B. 19: 222 (1949); Lisowski, Asterac. Fl. Afr. Centr. 1: 202
(1991). Type: Congo (Kinshasa), Mtoto, *Van Meel* 1169 (BR!, holo.)

A. vernayi Brenan in Mem. N.Y. Bot. Gard. 8, 5: 479 (1954). Type: Malawi, Zomba Mt, *Brass*
16215 (K!, holo., PRE, SRGH, iso.)

A. natalensis sensu Wild in Kirkia 5: 213 (1965), except South African material, & Wild in
Kirkia 6: 44 (1967) & Lisowski, Asterac. Fl. Afr. Centr. 1: 197, t. 44 (1991), non *Wedelia
natalensis* Sond. sensu stricto

A. gillettii Wild in Kirkia 5: 215 (1965), **syn. nov.** Type: Ethiopia, Mt Mega, *Gillett* 14355 (K,
holo., not found)

NOTE. "One of the most characteristic plants of East Africa" (Burtt, 1932, **T** 5)
A very polymorphic species. Wild (1967): "Extremely polymorphic, at times approaches *A.
africana* but differs in cuneately based leaves and yellow rather than dark or black anther
appendages. The purple markings on paleae and disc florets appear apparently at random
throughout the range; intermediates are frequent".

HB has struggled long over the systematics of *A. monocephala/mossambicensis* and the host
of synonyms. I believe the East and Central African material is all one polymorphic taxon
(Charles Jeffrey agrees, pers. comm.), and the oldest available name is *mossambicensis*. It is
quite possible the South African *natalensis* is the same as well but I feel it best to await a
critical revision. All the characters used to keep the constituent taxa separate were gradual
rather than absolute, and seem to boil down to the number of capitula (one or 2–3) and the
phyllary apices (obtuse or rounded/subacute or obtuse). The name *gillettii* is brought into
synonymy here, as no differences can be upheld.

Richards 21242 from **T** 7, Ruaha National Park, is described as an annual, but with yellow
florets with yellow anther appendages and the other characters apparent can only be this
species. The type of *Wedelia katangensis* is described as being annual shoots from a perennial
base in a fire-induced vegetation.

Wild considered *Wedelia ringoetii* to be a synonym of *A. mossambicensis*. HB agrees with
Lisowski that it is a synonym of *A. ciliata* instead.

INSUFFICIENTLY KNOWN SPECIES

Aspilia fischeri O.Hoffm. in P.O.A. C: 413 (1895); Muschler in E.J. 50, Suppl.: 340
(1914). Type: Tanzania, Umgamwesi, *Fischer* 370 (B†, holo.)

Treated by Wild in Kirkia 5: 225 (1966) as 'insufficiently known species'. Said by
Hoffmann to be related to *A. multiflora*, a synonym of *A. ciliata*. Fischer's locality is
not on any of my maps, or in the Gazetteer; it is probably the same as Unyamwezi, a
rather vague area in **T** 4.

Aspilia mildbraedii Muschler in Z.A.E.: 379 (1911); Muschler in E.J. 50, Suppl.: 336
(1914). Type: Tanzania, Bukoba District: Itara, *Mildbraed* 83 (B†, holo.).

Wild suggests this might just be an earlier name for *A. congensis*, now *A. africana*.

Aspilia subpandurata O.Hoffm. in P.O.A. C: 413 (1895); Muschler in E.J. 50, Suppl.: 338 (1914). Type: Tanzania, Bukoba, *Stuhlmann* 3668 & 3861 (both B†, syn.)

Wild says the form of the leaves suggests *A. kotschyi* but the plant is described as perennial.

115. **LAGASCEA**

Cav. in An. Cien. Nat. 6: 331 (1803), *nom. conserv.*; Stuessey in Fieldiana, Bot.: 38: 75–133 (1978)

Annual or perennial herbs. Leaves opposite. Capitula in dense glomerules surrounded by foliose bracts, these glomerules in turn in cymose or racemose inflorescences; phyllaries connate; capitula usually 1-flowered. Flowers bisexual; style branches attenuate, pilose on abaxial side; anther appendages narrowly ovate. Pappus of few scales or bristles.

8 species from Central America; one of these a pantropical weed.

Lagascea mollis *Cav.* in An. Cien. Nat. 6: 332, t. 44 (1803); D.J.N. Hind in Fl. Masc. 109: 189, t. 64 (1993); U.K.W.F. ed. 2: 214 (1994). Type: cultivated specimen from Madrid Botanical Garden, originating from Cuba, *Espinosa & Peralta* s.n. (MA, lecto., chosen by D.J.N. Hind)

Annual herb to 1 m high, rarely perennial, creeping or straggling, to 90 cm high or long, sometimes rooting at the nodes, sometimes mat-forming; stem sometimes purplish, with stipitate glandular hairs on young parts. Leaves opposite, narrowly ovate or ovate, 1–7 cm long, 0.5–4 cm wide, base obtuse to attenuate, margins subentire to serrate, apex acute to acuminate, short-pubescent to slightly scabridulous; 3–5-veined from base; petiole 5–27 mm long. Glomerules solitary and terminal, campanulate, 8–13 mm long and 8–30 mm in diameter, with 8–25 1-flowered capitula; subtending bracts lanceolate to obovate, 5–15 mm long, 1–6 mm wide; involucre of capitula 4–5 mm long, 1 mm in diameter, with elongate glands. Florets white or with blue tinge, 4–5 mm long. Achenes brown or black, 3 mm long, hairy near apex; pappus a minute crown, pubescent; achene usually surrounded by connate phyllaries topped by scales 2–2.5 mm long. Fig. 159 (page 755).

UGANDA. Mbale District: 3.7 km NW of Kamankoli, Aug. 2001, *Lye & Namaganda* 25241!; Busoga District: Buwaiswa near Kamuli, Feb. 1952, *G.H.S. Wood* 436!; Mengo District: Sonde, E of Namungongo, June 1994, *Rwaburindore* 3736!
KENYA. South Nyanza District: Mbita Point, Nov. 1981, *Gachathi & Opon* 20/81!; Kilifi District: Magarini, Jan. 1983, *Robertson* 3473!; Mombasa, Miritini area, Sep. 1993, *Luke* 3842!
TANZANIA. Moshi District: Moshi town, Aug. 1971, *Shabani* 753!; Mpanda District: Mpanda old mine, June 1968, *Sanane* 196!; Morogoro District: 24 km on Morogoro–Kimboza track, Apr. 1988, *Bidgood et al.* 1257!
DISTR. **U** 3, 4; **K** 4, 5, 7; **T** 2, 4, 6; originally from Central America, now a pantropical weed
HAB. Ruderal sites, cultivation, seasonally wet grassland; 50–1700 m
USES. None recorded on specimens from our area
CONSERVATION NOTES. Least concern (LC)

NOTE. The first East African record is from Kisumu, Kenya, in 1926.

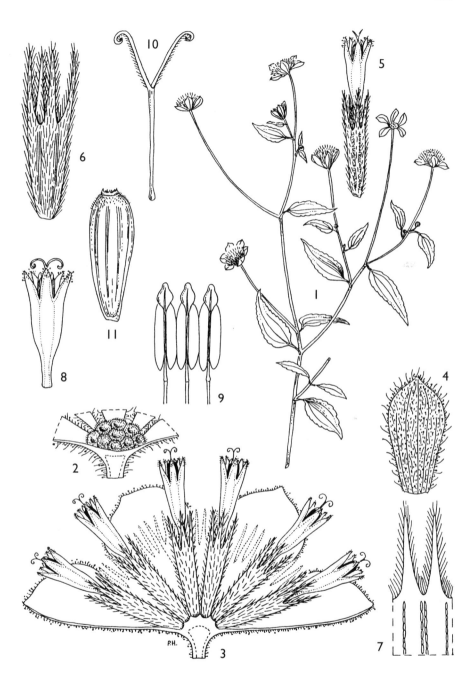

FIG. 159. *LAGASCEA MOLLIS* — **1**, habit, × ²/₃; **2**, common receptacle, × 8; **3**, glomerule section, × 8; **4**, glomerule bract, × 4; **5**, capitulum, × 6; **6**, involucre, × 8; **7**, detail of involucre, × 10; **8**, floret, × 6; **9**, stamens, × 12; **10**, style, × 8; **11**, achene, × 12. **Note**: leaves should be opposite! 1 & 9 from *Guého* in MAU 18037, 2–8 & 10–11 from *Tweedie* 2902. Drawn by Pat Halliday, from Flore des Mascareignes.

FIG. 160. *MONTANOA HIBISCIFOLIA* — **1**, habit, × ²/₃; **2**, capitulum, schematic section, × 2; **3**, phyllary, × 6; **4**, palea, × 10; **5**, ray floret, × 3; **6**, disc floret, × 8; **7**, stamens, × 18; **8**, style of disc floret, × 12; **9**, palea of achene, × 2; **10**, achene, × 8. 1–8 from *Duljeet* in MAU 10465, 9–10 from *Ogilvie* 15. Drawn by Pat Halliday, from Flore des Mascareignes.

116. MONTANOA

Cerv. in La Llave & Lexarza, Nov. Veg. Desc. 2: 11 (1825); Funk in Mem. N.Y. Bot. Gard. 36: 1–133 (1982)

Shrubs, lianas or trees. Leaves opposite, often lobed. Capitula many, in terminal panicles; involucre 1–2-seriate, the outer phyllaries leafy, the inner scarious; paleae scarious. Ray florets neuter, white; inner florets hermaphrodite, with auriculate anthers and recurved style branches. Achenes usually compressed, black, glabrous; pappus absent.

Twenty species from tropical America; two of these are ornamental, and one of these sometimes naturalises in our area.

Montanoa hibiscifolia *Benth.* in Oersted, Vidensk. Meddel. Naturhist. Foren. Kjobenhavn 1852, 5–7: 89 (1852); Lisowski, Asterac. Fl. Afr. Centr. 1: 222 (1991); D.J.N. Hind in Fl. Masc. 109: 185, t. 62 (1993). Type: Nicaragua, Nueva Segovia, *Oersted* 235 (K!, lecto., C, isolecto)

Shrub or woody herb 1–6 m high; branches pubescent to glabrous. Leaves ovate to pentagonal, 7–40 cm long, 2.5–30 cm wide, entire or 3–5-lobed, the lobes themselves with secondary lobes, margins crenate to serrate, apices attenuate, pubescent to scabridulous on both surfaces and glandular beneath; petiole 1–8 cm long, often with auricles at apex. Capitula in corymbose panicles, pendulous; stalks of individual capitula 2–6 cm long; involucre 1-seriate; phyllaries 4–5 mm long, reflexed in fruit; paleae 3–3.5 mm long, to 10 mm long in fruit. Ray florets 7–8, ray white, 15–17 mm long, tube yellow, < 1 mm long; disc florets many, yellow, 1–1.2 mm long. Achenes brown or red-brown, 3–3.5 mm long, rugulose, the apex glandular; pappus absent. Fig. 160 (page 756).

UGANDA. Masaka District: 1 km E of Kitovu, May 1972, *Lye* 6849!; Mengo District: 0.5 km N of Nagalama Hospital, Oct. 1996, *Lye & Katende* 21927!
TANZANIA. Arusha District: Tengeru, July 1954, *Matalu* 3155! & July 1988, *Pocs* 88166!; Lushoto District: Lushoto, June 1970, *Mshana* 30!
DISTR. **U** 4; **T** 2, 3; originally from Central America, now a widespread ornamental that sometimes naturalises
HAB. Forest margins, thicket edge near swamp; 1100–1200 m
USES. None recorded on specimens from our area
CONSERVATION NOTES. Least concern (LC)

NOTE. The name *Montanoa bipinnatifida* (Kunth.) C.Koch has been used on many East African specimens, but by mistake; this is quite a different species.

117. SCLEROCARPUS

Murr., Syst. Veg. ed. 14: 783 (1784)

Annual or perennial herbs. Leaves alternate or sometimes opposite near the base of the plant, simple. Capitula in leafy cymes, radiate; phyllaries in 1 series, few and leafy; receptacle paleae clasping the florets. Ray florets with a very small ray; disc florets 4–5-lobed; anthers with apical appendage. Achenes obovoid; pappus absent or of a small rim.

Eight species in Central America and Africa.

Sclerocarpus africanus *Murr.*, Syst. Veg. ed. 14: 783 (1784); Oliv. & Hiern, F.T.A. 3: 374 (1877); F.P.S. 3: 48 (1956); C.D.Adams in F.W.T.A. ed. 2, 2: 235 (1963); Wild in Kirkia 6: 37 (1967); Lisowski, Asterac. Fl. Afr. Centr. 1: 181 (1991); U.K.W.F. ed. 2: 215 (1994). Type: protologue mentions only Jacq., Icon. pl. rar. M (iconotype)

FIG. 161. *SCLEROCARPUS AFRICANUS* — **1**, habit, × ²/₃; **2**, capitulum, × 3; **3**, ray floret, abaxial surface, × 6; **4**, ray floret, adaxial surface, × 6; **5**, disc floret within palea, × 4; **6**, disc floret, × 6; **7**, achene within palea, × 5; **8**, achene with palea remnants, × 5. 1 & 7–8 from *Lye* 4770b, 2 from *Bally* 4714, 3–6 from *Pielou* 169. Drawn by Juliet Williamson.

Annual herb, erect, 15–120 cm high; stems hispid-pubescent. Leaves opposite on the lower part of the stem, alternate higher up, ovate, 3–12 cm long, 1–6.5 cm wide, base cuneate, margins serrate or crenate, apex acute, scabrid or hispid on both surfaces; petiole to 4 cm long. Capitula terminal; involucre with ± 5 spreading leafy green bracts, 0.7–3 cm long, appressed-pubescent; paleae tubular with a large bulge caused by the ovary/achene, 9–13 mm long, pubescent. Ray florets few or absent, yellow to deep yellow, with a narrow tube 2.5–6 mm long, ray 2.5–4.5 mm long; disc florets greenish yellow, 6.5–7.5 mm long, puberulous near apex. Achenes clasped by paleae, black, asymmetrically obovoid, 5–8 mm long, palea firmly attached and protruding for 2–3 mm; pappus absent. Fig. 161 (page 758).

KENYA. Northern Frontier District: Marsabit, June 1960, *Oteke* 34!; Kitui District: 11 km W of Ukazzi, Jan. 1972, *Gillett* 19470!; Tana River District: Kora National Park, Dec. 1982, *van Someren* 883!
TANZANIA. Kigoma District: Kabogo Mts, Feb. 1963, *Kyoto University Expedition* 340!; Kilosa District: 3 km N of Mbuyuni, Aug. 1970, *Thulin & Mhoro* 803!; Iringa District: Mtera on Ruaha R., Feb. 1962, *Polhill & Paulo* 1309!
DISTR. **K** 1, 4, 6, 7; **T** 4, 6, 7; from Senegal to Sudan and South to Namibia and Mozambique
HAB. Sandy river-banks and lake-shores, sandy sites in dry bushland and woodland; 500–1350(–2100) m
USES. None recorded on specimens from our area
CONSERVATION NOTES. Least concern (LC)

118. **TITHONIA**

Juss., Gen. Pl.: 189 (1789); La Duke in Rhodora 84: 453–522 (1982)

Annual or perennial herbs or shrubs. Leaves alternate or sometimes opposite near the base, 3-veined from base. Capitula solitary on long lateral peduncles which thicken near apex, heterogamous, radiate; phyllaries 2–5-seriate; paleae clasping disc florets. Ray florets sterile, 2–3-toothed; disc florets tubular, 5-lobed; anthers with sagittate base and apical appendage; style-arms hairy abaxially. Achenes oblong; pappus a crown of free scales, usually with 1–2 awns.

11 species in Northern and Central America.

Most leaves ± cordate at base; phyllaries in 2 series, the
    inner row with acute apices; ray florets bright
    orange-red . . . . *T. rotundifolia* (cultivated)
Leaves cuneate to attenuate at base; phyllaries in 4
    series, the inner row with rounded or obtuse
    apices; ray florets yellow or orange-yellow . . . . . *T. diversifolia*

Tithonia rotundifolia (*Mill.*) *S.F.Blake* in Contrib. Gray Herb., n.s. 52: 41 (1917); Wild in Kirkia 6: 55 (1967); Maquet in Fl. Rwanda 3: 628 (1985); Lisowski, Asterac. Fl. Afr. Centr. 1: 218 (1991). Type: Mexico, Veracruz (where?); basionym *Tagetes rotundifolia* Mill., Gard. Dict. ed. 8: Tagetes no. 4 (1768)

This annual has been cultivated in Kenya: Nairobi Arboretum, Feb. 1952, *Williams Sangai* 346! & Kikuyu, Njogu Inn, July 1953, *Verdcourt* 990! It does not seem to become naturalized as it does in the Flora Zambesiaca area, where it is a common weed of roadsides and disturbed ground.
    Possibly the *T. speciosa* mentioned in U.O.P.Z.: 473 (1949) is this taxon.

Fig. 162. *TITHONIA DIVERSIFOLIA* — **1**, habit, × ²/₃; **2**, capitulum, schematic section, × 2; **3**, inner phyllary, × 2; **4**, ray floret, × 1; **5**, achene of ray floret, × 4; **6**, palea, × 3; **7**, disc floret, × 6; **8**, anthers, × 6; **9**, style of disc floret, × 4; **10**, achene of disc floret, × 4. All from *Duljeet* in MAU 10420. Drawn by Pat Halliday, from Flore des Mascareignes.

Tithonia diversifolia (*Hemsl.*) *A.Gray* in Proc. Amer. Acad. Arts 19: 5 (1883); U.O.P.Z.: 473 (1949); T.T.C.L.: 160 (1949); Wild in Kirkia 6: 56 (1967); Maquet in Fl. Rwanda 3: 628, fig. 193/2 (1985); Blundell, Wild Fl. E. Afr.: pl. 387 (1987); Lisowski, Asterac. Fl. Afr. Centr. 1: 219, t. 48 (1991); D.J.N. Hind in Fl. Masc. 109: 192, t. 65 (1993); K.T.S.L.: 564, ill., map (1994); U.K.W.F. ed. 2: 216, t. 89 (1994). Type: Mexico, *Borgeau* 2319 (K!, lecto., BR, FT, G, P, S, US, iso.)

Annual or short-lived perennial herb or slightly scandent shrub 0.7–3 m, often branching; stems 4-angled, pilose but soon glabrescent. Leaves alternate, pseudo-petiolate, ovate to obovate, deeply 3–5-lobed or unlobed in some upper leaves, 6–23 cm long, 3–18 cm wide, base cuneate to attenuate, margins crenate to subentire, apex acuminate, scabrid above, pubescent to tomentose beneath; pseudo-petiole to 10 cm long, with 2 auricles 2–8 mm long at base (always?). Capitula terminal on side branches, solitary; stalks of individual capitula to 21 cm long, widening near apex and in extreme cases the widened part to 6 cm long and 1 cm across; involucre 4-seriate, 14–20 mm long; phyllaries 5–17 mm long; paleae ± 12 mm long, pubescent or glabrous. Ray florets 8–12, yellow or orange-yellow, tube 1.9–3 mm and pubescent, ray 40–75 mm long; disc florets cylindrical, 7.5–10 mm long, puberulous near base. Achenes oblong, 5–6 mm long, 4-angled, pubescent; pappus of scales to 2 mm with 2 setae to 4 mm long. Fig. 162 (page 760).

UGANDA. Mengo District: Kampala, Apr. 1917, *Dummer* 3161! & Makerere University, Mar. 1953, *Wigg* 1067! & Kajansi Fish Farm between Kampala & Entebbe, Sep. 1960, *Kendall & Richardson* 1b!
KENYA. Nairobi, July 1971, *E.Polhill* 366!; Machakos District: Kanzui, Aug. 1985, *J.Muasya* 575!; N Kavirondo District: Kakamega–Kisumu road, Nov. 1951, *Trapnell* 2194!
TANZANIA. Biharamulo District: Hamugongo, no date, *Tanner* 6055!; Lushoto District: Lawns Hotel, May 1969, *Ngoundai* 329!; Tanga District: Tanga–Moa road, June 1957, *Semsei* 2647!; Pemba: Chake Chake, July 1929, *Vaughan* 423!
DISTR. U 4; K 3–5, 7; T 1–3; Z; P; originally from Central America, now a weed in higher rainfall areas in tropical Africa
HAB. Roadsides, waste ground, valley grassland, also used as a hedge-plant; 0–1950 m
USES. A hedge-plant; minor medicine for childrens' stomachache (*Tanner*) and stomach upsets (*Okaka*) but too much can be poisonous (*Okaka*)
CONSERVATION NOTES. Least concern (LC)

SYN. *Mirasolia diversifolia* Hemsl., Biol. Centr. Amer. Bot. 2: 168, t. 47 (1881)

119. GALINSOGA

Ruiz & Pav., Prodr. Fl. Pér.: 110, t. 24 (1794); Canne in Rhodora 79: 319–389 (1977); Canne-Hilliker in Taxon 41: 661–666 (1992)

Annual herbs. Leaves opposite. Capitula terminal or in upper axils, stalked, heterogamous, radiate; involucre 1–3-seriate; paleae flat. Ray florets female, ray 3-lobed, disc florets hermaphrodite, cylindric, the basal part narrow, 5-toothed; anthers obtuse at base; style arms densely papillose-pubescent. Achenes obovoid, those of ray florets enclosed by a group of phyllaries and paleae; pappus of linear or narrow scales with or without terminal awns, but sometimes absent in ray florets.

Thirteen species from the Americas; two of these have become widespread weeds.

Peduncle hairs long and spreading; paleae 3-fid; pappus scales without terminal awns . . . . 1. *G. parviflora*
Peduncle hairs short and appressed; paleae undivided; pappus scales with terminal awns . . . . 2. *G. quadriradiata*

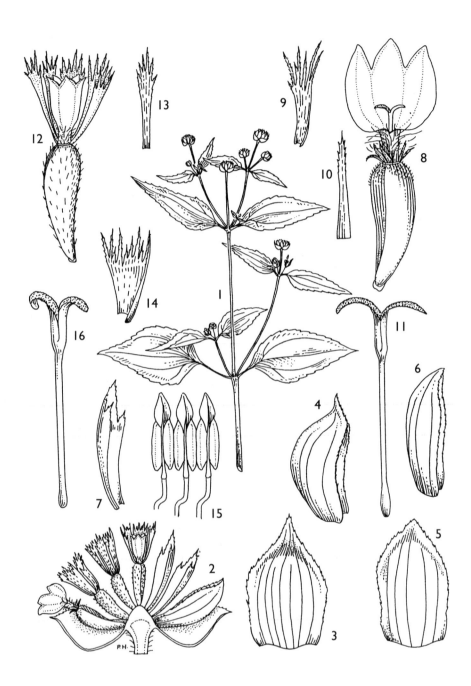

FIG. 163. *GALINSOGA PARVIFLORA* — **1**, habit, × ²/₃; **2**, capitulum section, × 8; **3–4**, phyllaries, × 12; **5–6**, outer paleae, × 12; **7**, inner palea, × 12; **8**, ray floret, × 18; **9–10**, pappus scales of ray floret, × 60; **11**, style of ray floret, × 30; **12**, disc floret, × 18; **13–14**, disc floret pappus scales, × 18; **15**, stamens, × 60; **16**, disc floret style, × 60. 1 from *Cadet* 3362, 2–16 from *Cadet* 5868. Drawn by Pat Halliday, from Flore des Mascareignes.

1. **Galinsoga parviflora** *Cav.*, Icon. 3: 41, t. 281 (1794); F.P.S. 3: 30 (1956); F.P.U.: 174, fig. 114 (1962); Wild in Kirkia 6: 9 (1967); Canne in Rhodora 79: 373 (1977); Maquet in Fl. Rwanda 3: 637, fig. 197/1 (1985); Lisowski, Asterac. Fl. Afr. Centr. 1: 121 (1991); D.J.N. Hind in Fl. Masc. 109: 206, t. 70 (1993); U.K.W.F. ed. 2: 218 (1994). Type: from a plant grown at Madrid originating from Peru (MA, holo.)

Annual herb 10–75(–100) cm high, erect, spreading or decumbent; branches hispidulous or pilose, glabrescent. Leaves petiolate, membranous, ovate, 1–11 cm long, 0.5–7 cm wide, base broadly cuneate, margins crenate to subentire, apex acute or acuminate, scabridulous-pilose or glabrous on both surfaces; 3-veined from base; petiole to 25 mm long. Capitula in few-headed cymes in upper leaf-axils, globose, on slender stalks to 2 cm long; involucre 2.5–3.5 mm long; phyllaries green, membranous, ovate, pilose or glabrous; paleae 3-fid, to 2.6 mm long, ciliolate. Ray florets white, 4–5(–8), the ray broadly ovate, 0.8–2 mm long, 3-lobed, tube 0.8–1 mm long, pilose; disc florets yellow or less often orange, 0.8–1.4 mm long, the tube puberulous. Achenes 1.2–2.5 mm long, puberulous; pappus absent or of 15–20 ovate laciniate scales, usually obtuse at apex, 1–1.5 mm long. Fig. 163 (page 762).

UGANDA. West Nile District: Paida, Aug. 1953, *Chancellor* 189!; Kigezi District: Kachwekano Farm, July 1949, *Purseglove* 2979!; Masaka District: Kisasa, May 1972, *Lye* 6947!
KENYA. Mt Elgon, 1930, *Lugard* 107!; Meru District: Nyambeni Hills, Dec. 1968, *Riling* 29!; Masai District: Rotian area, May 1961, *Glover et al.* 1298!
TANZANIA. Mbulu District: Lake Manyara National Park, Endabash R., Mar. 1964, *Greenway & Kanuri* 11314!; Pare District: Chome Shengena Forest, Sep. 1999, *Mlangwa & Masanyika* 532!; Iringa District: Ngowasi [Ngwazi], May 1989, *Kayombo* 624!
DISTR. U 1–4; K 2–7; T 1–3, 6–8; originally from Central and S America, now a tropical, subtropical and occasionally temperate weed
HAB. A very common weed of cultivation, roadsides, waste places, river-sides, forest margins; often the first coloniser on bare ground in higher rainfall areas, may form dense and pure stands; (300–)900–2300(–2750) m
USES. Boiled leaves used as a vegetable (*Meyerhoff, Mtali*); minor medicine for back-ache (*Tanner*)
CONSERVATION NOTES. Least concern (LC)

NOTE. Reported to have been introduced first in 1919 in Embu (fide *Sunman*)

2. **Galinsoga quadriradiata** *Ruiz & Pav.*, Syst. Veg.: 198 (1798); Canne in Rhodora 79: 355 (1977). Type: Peru, Lima, *Ruiz & Pavon* s.n. (MA, holo., P, iso.)

Annual herb 20–80 cm high; branches scabridulous. Leaves petiolate, ovate, 2.5–7 cm long, 1–5 cm wide, base cuneate, margins crenate-serrate, apex acute, scabridulous to glabrous on both surfaces; 3-veined from base; petiole to 35 mm long. Capitula in few-headed cymes, terminal or in upper axils; involucre 2.6–3.5 mm long; phyllaries pilose to glabrous; paleae ciliolate. Ray florets white, 4–5, the ray 1–1.6 mm long; disc florets greenish, 1.2–1.7 mm long, pubescent. Achenes black, 1.5–1.7 mm long, pilose; pappus of ± 18 narrow scales, each with apical awn to 1.2 mm long.

UGANDA. Mengo District: Makerere University Hill, Aug. 2001, *Lye* 25091!
KENYA. Trans Nzoia District: Kitale, Aug. 1964, *Tweedie* 2873!
TANZANIA. Lushoto District: Jaegertal, Apr. 1985, *Kisena* 242!
DISTR. U 4; K 1, 3, 4; T 3; probably a widespread weed in Africa but often confused with G. *parviflora*
HAB. Waste places, roadsides, gardens; 1200–1900 m
USES. Used in dyeing (*Hindmarsh*)
CONSERVATION NOTES. Least concern (LC)

SYN. *Wilburgia urticifolia* Kunth, Nov. Gen. Sp. Pl. 4: 257, t. 389 (1818). Type: Ecuador, between Mulalo and Pansache, *Humboldt & Bonpland* 3055 (P, holo.)
Adventina ciliata Raf., New Fl. Bot. N. Amer. 1: 67 (1836). Type: unknown (not at PH)

Galinsoga ciliata (Raf.) S.F.Blake in Rhodora 24: 35 (1922); C.D. Adams in F.W.T.A. ed. 2, 2: 230 (1963); Wild in Kirkia 6: 10 (1967); Maquet in Fl. Rwanda 3: 638, fig. 197/2 (1985); Lisowski, Asterac. Fl. Afr. Centr. 1: 123 (1991)

G. urticifolia (Kunth) Benth. in Oersted, Vidensk. Meddel. Naturhist. Foren. Kjobenhavn 1852, 5–7: 102 (1852); U.K.W.F. ed. 2: 218 (1994)

120. **TRIDAX**

L., Sp. Pl.: 900 (1753); Gen. Pl. ed. 5: 382 (1754)

Annual or perennial herbs. Leaves opposite (or alternate in non-weedy species). Caputula terminal, solitary on long stalks (or paniculate in non-weedy species), radiate or discoid; involucre 1–5-seriate; paleae linear and keeled. Ray florets female; disc florets hermaphrodite, cylindric; anthers with sagittate base and terminal appendage; style arms linear, pubescent, style-base with globose swelling. Achenes with pappus of many plumose bristles.

30 species from Central and South America; one of these has become a pantropical weed.

Tridax procumbens L., Sp. Pl.: 900 (1753); U.O.P.Z.: 477 (1949); F.P.S. 3: 55 (1956); F.P.U.: 174, fig. 115 (1962); C.D. Adams in F.W.T.A. ed. 2, 2: 230, t. 248 (1963); Wild in Kirkia 6: 10 (1967); Maquet in Fl. Rwanda 3: 636, fig. 196/2 (1985); Lisowski, Asterac. Fl. Afr. Centr. 1: 124, t. 30 (1991); D.J.N. Hind in Fl. Masc. 109: 211, t. 72 (1993); U.K.W.F. ed. 2: 218, t. 90 (1994). Type: Mexico, Vera Cruz, *Houston* s.n. (BM-HC 418, *Tridax* 1, lecto., designated by Powell in Brittonia 17: 80(1965))

Annual or perennial herb, sprawling or ascending but with erect inflorescences, 15–50 cm tall; stems scabrid-pilose, rooting at the nodes. Leaves petiolate, narrowly ovate or ovate, often slightly 3-lobed, 2–7(–12) cm long, 1–4(–6) cm wide, base cuneate, margins coarsely dentate to incised-dentate, apex acute to acuminate, scabrid on both surfaces; petiole 4–30 mm long. Capitula solitary and terminal, urn-shaped; stalks of individual capitula 7–25 cm long; involucre 7–8 mm long; phyllaries in 2–3 series, green, ovate and pilose; paleae linear, 6–8 mm long, pilose near apex. Ray florets 2–6, white, cream or pale yellow, tube 3.5 mm long, pilose near apex, ray ovate, 2.5–5 mm long, 3-dentate at apex; disc florets forming a cone, yellow, 5–7 mm long, puberulous near base. Achenes narrowly obovoid, black, 2–2.5 mm long, hairy; pappus of many unequal plumose bristles 2.5–3 mm long in marginal florets, 4–7.5 mm long in central florets. Fig. 164 (page 765).

UGANDA. West Nile District: Metu, Sep. 1953, *Chancellor* 293!; Ankole District: Mbarara, June 1939, *Purseglove* 828!; Mengo District: Entebbe, Dec. 1922, *Maitland* 529!
KENYA. North Kavirondo District: Webuye [Broderick Falls], May 1958, *Tweedie* 1540!; Tana River District: Kora National Park, Asako, Aug. 1983, *Mutangah* 77!; Kwale District: Shimba Hills, Marere, Apr. 1968, *Magogo & Glover* 723!
TANZANIA. Mpanda District: Kabungu, Aug. 1948, *Semsei* 99!; Uzaramo District: Dar es Salaam, University campus, May 1986, *Kisena* 266!; Iringa District: Lupingu, Jan. 1991, *Gereau & Kayombo* 3788!; Zanzibar, Pete, July 1972, *Robins* 33!
DISTR. **U** 1–4; **K** 1, 3–7; **T** 1–8; **Z**; originally from Central America, now a pantropical weed
HAB. A weed of cultivation, also on waste ground, an invader of bare soil; may form patches or mats; 0–1700(–2850) m
USES. Minor medicine against backache and infantile diarrhoea (*Tanner*), and against stomach-ache and malaria (*Glover*)
CONSERVATION NOTES. Least concern (LC)

NOTE. A common weed of cultivation; invades lawns like the European daisy. Earliest records from East Africa – Uganda, 1922 at Entebbe; Kenya, 1915 at Kisumu; Tanzania, 1926 near Pangani and in W Usambaras.
Burtt 2503 from **T** 1 states the plant is very attractive to butterflies.

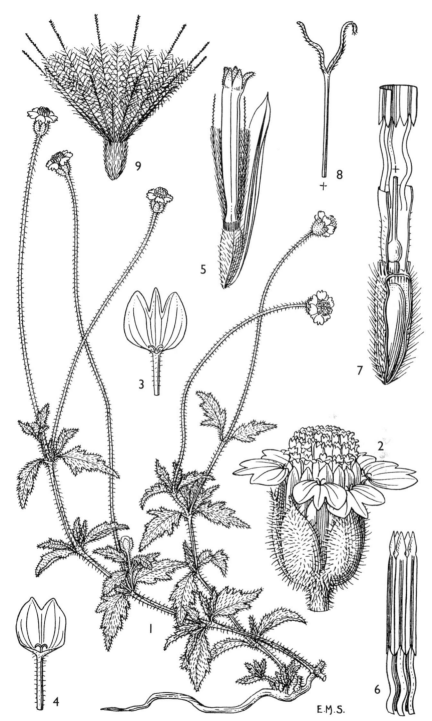

Fig. 164. *TRIDAX PROCUMBENS* — **1**, habit, × ²/₃; **2**, capitulum, × 4; **3**, ray floret, ovary removed, × 4; **4**, ray floret, ovary removed, × 4; **5**, disc floret, × 4; **6**, stamens, × 16; **7**, section of ray floret, × 16; **8**, style apex, × 16; **9**, achene, × 6. 1 & 4 & 9 from *Robinson* 811, 2–3 & 5–8 from *Adams* s.n. Drawn by Margaret Stones, from the Flora of West Tropical Africa.

121. **ACANTHOSPERMUM**

Schrank, Pl. Rar. Hort. Monac.: t. 53 (1819); D.J.N. Hind in Fl. Masc. 109: 170 (1993)

Annual or perennial herbs. Leaves opposite. Capitula solitary, axillary or in branch forks, sessile or shortly stalked, heterogamous, radiate, enlarging in fruit; involucre of two series, the outer herbaceous, the inner closely enveloping the outer florets; receptacle paleate. Outer florets female, radiate with small ray; style branches strongly divergent, linear and obtuse. Inner florets bisexual but functionally male, campanulate or infundibuliform; style branches connate, club-shaped; anthers with truncate bases and ovate apical appendix. Achenes of the female florets covered by the indurate and echinate phyllaries, the achene itself smooth; achenes in disc florets abortive; pappus absent.

Eight species from tropical and subtropical America, several of which have become pantropical weeds.

1. Plant prostrate, branching rather irregularly; capitula stalked
      for up to 1 cm; fruit with all spines to 2 mm long . . . .   1. *A. glabratum*
   Plant erect, regularly dichotomously branched; capitula ± sessile;
      fruit compressed, with the two terminal spines 4–5 mm long   2. *A. hispidum*

1. **Acanthospermum glabratum** (*DC.*) *Wild* in Kirkia 6: 6 (1967); Lisowski, Asterac. Fl. Afr. Centr. 1: 114 (1991); D.J.N. Hind in Fl. Masc. 109: 171 (1993); U.K.W.F. ed. 2: 214 (1994). Type: Brazil, Rio de Janeiro, *Martius* 1823 (G-DC, lecto., microfiche!; chosen by Wild)

Annual herb, prostrate, often mat-forming. Leaves elliptic, broadly ovate to almost round, 1–4.5 cm long, 0.8–4 cm wide, base truncate or shortly cuneate, margins subentire to dentate, apex acute or obtuse, pubescent but quickly glabresecnt, densely glandular; petiole to 1 cm long. Capitula axillary, on a short stalk; involucre 3 mm long, accrescent in fruit and then up to 10 mm long; phyllaries ciliate. Florets yellow or greenish cream, the rays 1.4 mm long and glandular; disc florets ± 2 mm long, glandular. Fruits usually 4–5, oblong, 8–10 mm long, 5–7-ribbed, the angles with 1–2 rows of hooked spines 1–2 mm long, apex with large pore, glandular.

KENYA. Kiambu District: Muguga, Apr. 1953, *Verdcourt* 920!; Nairobi, Loresho, Mar. 1974, *Gillett* 20452!; Machakos District: Kilungu, Jan. 1972, *Mwangangi* 1958!
TANZANIA. Iringa District: 1.7 km NE of John's Corner on Iringa–Mbeya road, Nov. 1973, *Spjut & Muchai* 3493! & Igowole, Mar. 1989, *Kayombo & Kayombo* 261!
DISTR. **K** 3, 4; **T** 7; originally from tropical America but now a spreading pantropical weed
HAB. Ruderal; 150–1850 m
USES. None recorded on specimens from our area
CONSERVATION NOTES. Least concern (LC)

SYN. *A. xanthioides* (Kunth) DC. var. *glabratum* DC., Prodr. 5: 522 (1836)

NOTE. The first record from East Africa is from Nairobi/Muguga in 1953.
     Often confused with *A. australe* (Loefling) O.Kuntze which has not yet been found for our area. *A. australe* looks similar to *A. glabratum*, but differs from it in the higher number of fruits per capitulum: (6–)8–9; the fruits being up to 7 mm long; and the denser and persistent pubescence on the leaves and stems.

2. **Acanthospermum hispidum** *DC.*, Prodr. 5: 522 (1836); Wild in Kirkia 6: 5 (1967); Blundell, Wild Fl. E. Afr.: pl. 19 (1987); Lisowski, Asterac. Fl. Afr. Centr. 1: 112. t. 28 (1991); D.J.N. Hind in Fl. Masc. 109: 171, t. 56 (1993); U.K.W.F. ed. 2: 214 (1994). Type: Brazil, Bahia, *Salzman* s.n. (G-DC, holo., K!, iso.)

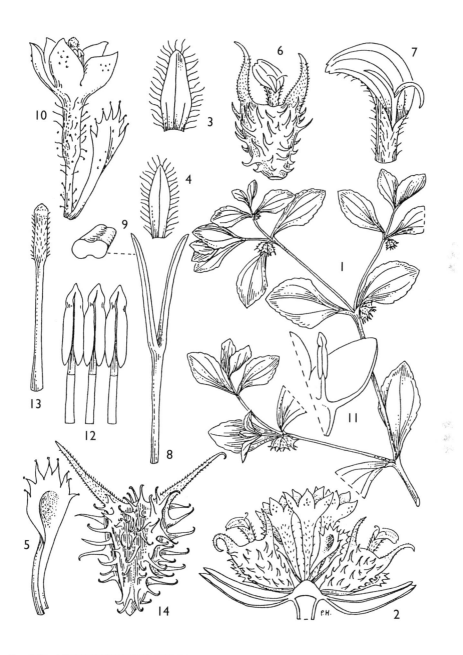

Fig. 165. *ACANTHOSPERMUM HISPIDUM* — **1**, habit, × ²⁄₃; **2**, capitulum section, × 8; **3–4**, outer and intermediate phyllaries, × 6; **5**, palea, × 24; **6**, ray floret in inner phyllary, × 12; **7**, ray floret corolla and style, × 24; **8**, ray floret style, × 72; **9**, style branch section, much enlarged; **10**, disc floret and palea, × 18; **11**, stamen insertion, much enlarged; **12**, stamens, × 24; **13**, disc floret style, × 24; **14**, achene and palea, × 6. 1 & 14 from *Cadet* 24, 2–13 from *Guého* in MAU 16433. Drawn by Pat Halliday, from Flore des Mascareignes.

Annual herb, 0.3–1.3 m high, erect; stem branching regularly and dichotomously, hispid. Leaves oblong or obovate, 1–12 cm long, 0.5–8 cm wide, base cuneate, margins subentire to coarsely dentate, apex obtuse to acute, hispid and gland-dotted on both surfaces; petiole absent or short. Capitula sessile in leaf axils or at the branching point of stems, enlarging in fruit; involucre ± 7 mm long; phyllaries ciliate, the inner with soft spiny processes enlarging in fruit. Ray florets pale yellow, elliptic, 1.5–2 mm long; disc florets darker yellow, 1.5–2.5 mm long, glandular-puberulous. Achenes enclosed in the enlarged inner phyllaries, the whole 5–6 mm long with two 4–5 mm long hooked spines at the apex, shorter hooked spines all over fruit. Fig. 165 (page 767).

Uganda. Karamoja District: Lokitonyala, Apr. 1960, *Wilson* 864!
Kenya. Kisumu-Londiani District: Kisumu Airport, Jan. 1969, *Kokwaro* 1800!; Baringo District: 57 km from Nakuru on Marigat road, Nov. 2000, *Smith et al.* 88!; Tana River District: Kora National Park, 5 km from Research Camp towards Adamsons Camp, Aug. 1983, *Mutangah* 39!
Tanzania. Lushoto District: Kikwajuni, Aug. 1971, *Magogo* 117!; Tabora District: Tabora town, Kiloleni, Dec. 1975, *Ruffo* 980!; Morogoro District: Morogoro, June 1952, *Semsei* 739!
Distr. **U** 1; **K** 1–7; **T** 2–8; **Z**; originally from South America but now a pantropical weed
Hab. Weed of cultivation, ruderal sites; 0–1700 m
Uses. None recorded on specimens from our area
Conservation notes. Least concern (LC)

Note. The first record from East Africa is from Dodoma in 1945.

122. **ENYDRA**

Lour., Fl. Cochinch. 2: 510 (1790); Lack in Willdenowia 10: 3–12 (1990)

Perennial herbs of wet habitats. Leaves opposite. Capitula axillary, sessile or subsessile, radiate or disciform; involucre with 4–6 phyllaries in opposite pairs, the outer leafy; paleae sheathing the florets. Ray florets in 1–several rows, female, the ray short, 3–4-dentate, or absent; disc florets hermaphrodite, fertile or the innermost sterile, tubular, 5-lobed; anthers obtuse at base, with terminal appendage; style arms linear, puberulous. Achenes surrounded by paleae, oblong, compressed; pappus absent.

Ten species, pantropical.

Enydra fluctuans *Lour.*, Fl. Cochinch. 2: 625 (1793); Oliv. & Hiern, F.T.A. 3: 372 (1877), as '*Enhydra*'; F.P.S. 3: 28 (1956); C.D. Adams in F.W.T.A. ed. 2, 2: 242 (1963); Wild in Kirkia 6: 35 (1967); Maquet in Fl. Rwanda 3: 622, fig. 191/4 (1985); Lisowski, Asterac. Fl. Afr. Centr. 1: 176 (1991); U.K.W.F. ed. 2: 215 (1994). Type: Cochinchina, *Loureiro* s.n. (BM, holo./iso.)

Perennial herb, creeping and rooting at the nodes, the flowering stems erect and to 20(–30) cm high; stems hollow, flushed red or purple, pubescent to glabrous. Leaves (sub-)sessile, narrowly oblong, 1–12 cm long, 0.3–1.5 cm wide, base sub-hastate to broadly cuneate, margins entire or serrate, apex acute to obtuse, sparsely pubescent to glabrous, glandular-punctate. Capitula axillary, solitary, urn-shaped; involucre 9–15 mm long; phyllaries 4, in 2 opposite pairs, the outer broadly ovate; paleae 3.2–5.5 mm long, sheathing the disc florets, keeled, ciliate and glandular near apex. Ray florets greenish yellow to lemon yellow or white, tube 1.2–1.8 mm long, rays 0.4–0.7 mm long; disc florets yellow, 2.4–3 mm long. Achenes brown, narrowly oblong, 2.4–3.7 mm long, gland-dotted; pappus absent. Fig. 166 (page 769).

Uganda. Kigezi District: Lake Bunyonyi, Apr. 1970, *Katende & Lye* K146!; Mengo District: Entebbe, Oct. 1935, *Chancellor* 1442! & Busiro, Kisi, Sep. 1969, *Lye* 3945!

FIG. 166. *ENYDRA FLUCTUANS* — **1**, habit, × ²/₃; **2**, capitulum, × 4; **3**, ray floret, palea removed, × 10; **4**, disc floret and palea, × 8; **5**, disc floret, × 10; **6**, disc floret opened, × 8. 1–2 from *Richards* 21619, 3–6 from *Purseglove* 3350. Drawn by Juliet Williamson.

KENYA. Naivasha District: Lake Naivasha, Crescent Is., Sep. 1983, *Hayes* 2002!; Central Kavirondo District: Lake Victoria between Kangao and Kerebi, Dec. 1964, *Gillett* 16353!; Masai District: Amboseli, Ol Tukai causeway, Oct. 1962, *Verdcourt* 3285!

TANZANIA. Masai District: Manyara National Park, Manali pa Nyati, Nov. 1969, *Richards* 24710!; Moshi District: Lake Nyumba ya Mungu, Korogwe, Aug. 1974, *Mhoro & Backeus* 2234!; Mpanda District: Lake Katavi, Jan. 1950, *Bullock* 2337!

DISTR. U 2–4; **K** 3–6; **T** 1, 2, 4; widespread in tropical Africa and SE Asia to Australia

HAB. Swamps or very muddy margins of lakes and rivers, from completely aquatic to half-terrestrial; 700–1950 m

USES. None recorded on specimens from our area

CONSERVATION NOTES. Least concern (LC)

123. GUIZOTIA

Cass. in Dict. Sci. Nat. 59: 237, 247, 248 (1829), *nom. cons.*; J.Baagøe in Bot. Tidsskr. 69: 1–39 (1974)

Annual or perennial herbs or shrubs. Leaves opposite or rarely ternate, the upper leaves often alternate, dotted with resin/oil drops on lower leaf surface. Capitula terminal, solitary or in cymes or panicles, radiate; involucre 2-seriate, the outer leafy, the inner grading into the paleae; paleae 3-veined, the inner keeled. Ray florets female with 3-dentate ray; disc florets cylindric, with pubescent tube, 5-lobed; anthers obtuse or sagittate at base, with apical appendage; style arms short and pilose. Achenes black or brown, obovoid to obconical, the ray-floret achenes 3-angled, the inner 4-angled; pappus absent.

Six species in tropical Africa.

1. Creeping herb, rooting at the nodes, to 30 cm long; heads
    solitary in upper axils . . . .   5. *G. jacksonii*
   Erect or decumbent herbs or shrubs, not rooting at nodes;
    heads in small groups 2
2. Shrub to 4.5 m high; leaves with petiole 15–55 mm long; **U** 1
    (Imatong Mts) . . . .   4. *G. arborescens*
   Herbs or much smaller shrubs; leaves sessile 3
3. Leaves pandurate, 3-veined from base; **K** 1 . . . .   3. *G. zavattarii*
   Leaves ovate, lanceolate or oblanceolate, not 3-veined 4
4. Achenes 3.5–5.7 mm long; paleae 5-veined, with sessile
    glands; cultivated plant sometimes naturalized . . . .   1. *G. abyssinica*
   Achenes 1.8–3 mm long; paleae 3-veined, with glandular
    hairs; common wild plant and weed . . . .   2. *G. scabra*

1. **Guizotia abyssinica** (*L.f.*) *Cass.* in Dict. Sci. Nat. 59: 237, 248 (1829); Oliv. & Hiern, F.T.A. 3: 384 (1877); E.P.A.: 1132 (1967); J.Baagøe in Bot. Tidsskr. 69: 20, fig. 16 (1974); Lisowski, Asterac. Fl. Afr. Centr. 1: 127 (1991); D.J.N. Hind in Fl. Masc. 109: 202, t. 69 (1993); U.K.W.F. ed. 2: 216 (1994). Type: Linnean herbarium no. 1033 (LINN, lecto., chosen ?by Baagøe)

Annual herb 1–2 m high, erect; stems often purplish, pilose to glabrous. Leaves sessile, subconnate-perfoliate, lanceolate to oblanceolate, 10–15 cm long, 2–6 cm wide, base truncate to cordate, margins entire to serrate, apex acute, scabrid on both surfaces, with sessile glands. Capitula in terminal few-headed cymes; stalks of individual capitula 2–12 cm long; involucre 7–10(–32) mm long, the outer phyllaries foliaceous, pilose, the inner scarious; paleae 5–9 mm long, 5-veined. Ray florets yellow, 6–8(–15), tube 1–2.8 mm long, ray 8–14(–21) mm long; disc florets many, yellow, 4–5.5 mm long. Achenes 3.5–5.7 mm long; pappus absent.

UGANDA. Mengo District: possibly Entebbe, no date, *Snowden* 2055!
KENYA. Trans-Nzoia District: Kimilil, Nov. 1984, *Hohl* 281 (fide EA)
DISTR. **U** 4; **K** 3; probably native in N Ethiopia, naturalized in India and Réunion
HAB. Ruderal; 1500–2000 m
USES. Formerly (?still) cultivated as an oil crop (e.g. **T** 2, Arusha, June 1948, *Bally* 6338!), and known as Niger Oil.
CONSERVATION NOTES. Least concern (LC)

SYN. *Polymnia abyssinica* L.f., Sp. Pl. suppl.: 383 (1781)
  Verbesina sativa Sims in Curtis Bot. Mag. 26: t. 1017 (1807). Type: cult. in Brompton, England by Salisbury (?not preserved, name typified by illustration)
  Ramtilla oleifera DC. var. *angustior* DC. in Wight, Contrib. Bot. Ind.: 18 (1834). Type: Wallich catalogue 3194 (G-DC, holo.)
  R. oleifera DC. var. *sativa* (Sims) DC. in Wight, Contrib. Bot. Ind.: 18 (1834)
  Guizotia abyssinica (L.f.) Cass. var. *sativa* (DC.) Oliv. & Hiern, F.T.A. 3: 385 (1877)
  G. abyssinica (L.f.) Cass. var. *angustior* (DC.) Oliv. & Hiern, F.T.A. 3: 385 (1877)

2. **Guizotia scabra** (*Vis.*) *Chiov.*, Annuario Reale Ist. Bot. Roma 8: 184 (1904); F.P.S. 3: 32 (1956); F.P.U.: 172, fig. 111 (1962); C.D. Adams in F.W.T.A. ed. 2, 2: 230 (1963); Wild in Kirkia 6: 12 (1967); E.P.A.: 1133 (1967); J.Baagøe in Bot. Tidsskr. 69: 25 (1974); Maquet in Fl. Rwanda 3: 639, fig. 198/1 (1985); Blundell, Wild Fl. E. Afr.: pl. 365 (1987); Lisowski, Asterac. Fl. Afr. Centr. 1: 128 (1991); U.K.W.F. ed. 2: 216, t. 89 (1994). Type: Sudan, Fazokel, Tumad, Kassan, *Kotschy* 501 (FT, holo., FT, G, GH, K!, M, S, W, iso.)

Perennial or annual herb, erect or decumbent, (0.1–)0.3–2(–3?) m high; stem hardly to moderately branched, yellow, brown or reddish, scabrid to glabrous. Leaves sessile, narrowly lanceolate to ovate, 3–17 cm long, (0.5–)1–4(–5) cm wide, base connate and ± perfoliate to auriculate to cordate, margins entire to serrate, scabrid-ciliate, apex acute, scabrid to glabrous. Capitula in corymbose cymes; stalks of individual capitula 0–4(–10) cm long; involucre 7–15(–28) mm long, the outer phyllaries green, herbaceous; paleae scarious, 4–8 mm long, ciliate. Ray florets 7–15, yellow or orange-yellow, tube 1.3–2.5 mm long, ray 11–16 mm long, strongly lobed, the outside occasionally glandular; disc florets many, yellow or orange-yellow, 3.4–5.5(–9) mm long. Achenes 1.8–3 mm long, 4-gonous, glabrous; pappus absent. Fig. 167 (page 772).

subsp. **scabra**; J.Baagøe in Bot. Tidsskr. 69: 25, fig. 19, 20 (1974)

UGANDA. Acholi/Karamoja District: Lonyili Peak, Dec. 1971, *Katende* 1392!; Mbale District: Elgon, Nyangaya, Dec. 1938, *A.S.Thomas* 2701!; Buganda District: Lubowa Estate S of Mutungo, Aug. 1974, *Katende* 2216!
KENYA. Northern Frontier District: Marsabit, Aug. 1957, *Verdcourt* 1810!; Nakuru District: Londiani, Nov. 1967, *Perdue & Kibuwa* 9040!; North Kavirondo District: 3 km W of Musanda, Oct. 1971, *Gillett* 19350!
TANZANIA. Msai/Mbulu District: Ngorongoro Crater rim, May 1989, *Chuwa* 2743!; Ufipa District: 2 km on Tatanda–Mbala road, Apr. 1997, *Bidgood et al.* 3384!; Iringa District: Lake Nkwazi, May 1973, *Shabani* 976!
DISTR. **U** 1–4; **K** 1–5; **T** 1, 2, 4, 5, 7, 8; Nigeria to Ethiopia and south to Congo (Kinshasa) and Mozambique
HAB. Grassland, swampy grassland, wooded grassland, a weed of cultivation and waste places, on Marsabit and Elgon in forest clearings; may be locally common; 800–3100 m
USES. Minor medicinal against syphilis (*Tanner*), against stomach-ache and gonorrhoea (*Maitland*)
CONSERVATION NOTES. Least concern (LC)

SYN. *Veslingia scabra* Vis., Nuovi Sagg. Acc. Sc. Padova, Mem. 1: 21 (1840)
  Guizotia schultzii Sch.Bip. in Walp. Repert. 6: 158 (1846), *nom. illeg.*; A. Rich., Tent. Fl. Abyss. 1: 407 (1848); Oliv. & Hiern, F.T.A. 3: 385 (1877). Type: Ethiopia, Scholoda Mts, *Schimper* I, 350 (TUB, lecto., BM!, BR!, FT, G, GH, K!, M, S, iso., selected by Baagøe who says it is a superfluous name)

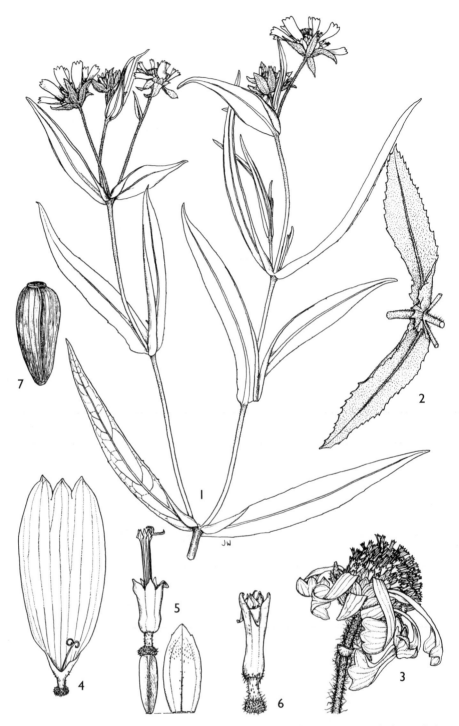

FIG. 167. *GUIZOTIA SCABRA* — **1**, habit, × ²/₃; **2**, leaf pair, × ²/₃; **3**, overmature capitulum, × 2; **4**, ray floret, × 4; **5**, disc floret with palea, × 6; **6**, disc floret, × 8; **7**, achene, × 10. 1 & 4–5 & 7 from *Richards* 8453, 2 from *Harley* 9206, 3 from *Jefford et al.* 322a, 6 from *Peter* 44223. Drawn by Juliet Williamson.

Guizotia nyikensis Baker in K.B. 1898: 153 (1898). Types: Zambia/Malawi, Nyika Plateau, *Whyte* 198 (K!, syn.) & Masuku Plateau, July 1896, *Whyte* s.n. (K!, syn.)

G. schultzii Sch.Bip. var. *sotikensis* S.Moore in J.L.S. 35: 344 (1902); R.E.Fr. in Acta Hort. Berg. 9, 6: 141 (1928). Type: Kenya, Kericho District: Sotik, 1889, *Jackson* s.n. (BM!, holo.)

G. collina S.Moore in J.L.S. 38: 262 (1908). Type: Uganda, Ruwenzori E, 26.3.1906, *Wollaston* s.n. (BM!, holo.)

G. eylesii S.Moore in J.Bot. 46: 43 (1908). Type: Zimbabwe, Mazoe, *Eyles* 349 (BM!, holo., K!, SRGH, iso.)

Wedelia oblonga Hutch., Gard. Chron. ser. 3, 40: 18 (1909). Type: Kenya, Ravine District: Eldama Ravine, 1898, *Whyte* s.n. (K!, syn.) & from seed collected in Kenya, 1908, *Diespecker* s.n. (K!, syn.)

Guizotia kassneri De Wild. in F.R. 13: 205 (1914). Type: Congo (Kinshasa), Lake Tanganyika, *Kassner* 3037 (BR!, holo., K!, iso.)

G. ringoetii De Wild. in F.R. 13: 204 (1914). Type: Congo (Kinshasa), Katanga, Shinsenda, *Ringoet* 6 (BR!, syn.) & Uvira, 1908, *Rouling* s.n. (BR!, syn.)

G. oblonga (Hutch.) Hutch. & Dalz. in K.B. 1933: 150 (1933)

G. scabra (Vis.) Chiov. var. *sotikensis* (S.Moore) Robyns, Fl. Sperm. Parc Nat. Albert 2: 526 (1947)

Coreopsis galericulata Sherff in Amer. J. Bot. 34: 156 (1947); Verdc. in K.B. 17: 498 (1964). Type: Kenya, Trans-Nzoia District: Kitale, *Webster* 8858 (EA, holo.)

NOTE. The other subspecies, subsp. *schimperi* (Sch.Bip.) J.Baagøe, is restricted to Ethiopia. It is distinct in the more annual habit and fewer outer phyllaries. There is one collection from Kitale (anno 1962), said to be of plants hanging on from being cultivated a few years before. This taxon is sometimes cultivated as 'wild Niger oil'. Charles Jeffrey (pers. comm.) feels this taxon is more closely related to *G. abyssinica* – possibly the wild ancestral form of the crop! - and should be transferred to that species.

3. **Guizotia zavattarii** *Lanza* in Miss. Biol. Borana, Racc. Bot.: 258, t. 83 (1939); E.P.A.: 1133 (1967); J.Baagøe in Bot. Tidsskr. 69: 31 (1974). Type: Ethiopia, Javello, *Cufodontis* 404 (FT, holo., iso.)

var. **zavattarii**; J.Baagøe in Bot. Tidsskr. 69: 31, fig. 23 (1974)

Perennial herb or shrub, erect, to 1 m high; stems lanate and glandular. Leaves sessile, pandurate, 4–10 cm long, 1.5–8 cm wide, base sub-connate-perfoliate to auricled, margins serrate-lacerate, apex acute, scabrid-pubescent, glandular; 3-veined from base. Capitula few in terminal corymbose cymes; stalks of individual capitula 1–6 cm long; involucre 5.5–10 mm long; outer 5 phyllaries leafy, the inner membranous, pilose and often glandular; paleae 5–6.5 mm long, ciliate. Ray florets yellow?, 8, tube 1.5–3 mm long, hairy, ray 7–12 mm long; disc florets 4–4.8 mm long. Achenes 2.8–3.5 mm long, flattened, 4-angled, glabrous; pappus absent.

KENYA. Northern Frontier District: Furroli, Sep. 1952, *Gillett* 13904! & Huri Hills, July 1957, *J.Adamson* 624!
DISTR. **K** 1; S Ethiopia (Mt Mega area)
HAB. *Olea-Juniperus* scrub on summit; 1500–2000 m
USES. None recorded on specimens from our area
CONSERVATION NOTES. Two Kenyan specimens; data deficient (DD)

SYN. *G. zavattarii* Lanza var. *opima* Lanza in Miss. Biol. Borana, Racc. Bot.: 2598, t. 84 (1939); E.P.A.: 1133 (1967). Type: Ethiopia, Javello, *Cufodontis* 518 (FT, holo., iso.)

G. zavattarii Lanza var. *hirsutissima* Cuf. in Nuovo Giorn. Bot. Ital. 50: 110 (1943); E.P.A.: 1134 (1967). Type: Ethiopia, Galla-Sidamo, Mega, *Corradi* 1750 (FT, syn., collections from 3 different dates)

NOTE. The other variety, var. *angustata* Cuf., is restricted to Ethiopia. It differs from the typical variety in the entire leaves, long peduncles, and short indumentum.

4. **Guizotia arborescens** *I.Friis* in Norwegian J. Bot. 18: 23, t. 1, map (1971); J. Baagøe in Bot. Tidsskr. 69: 22, fig. 17 (1974). Type: Ethiopia, Mt Maigudo, *Friis, Hounde & Jakobsen* 504 (C, holo., BR!, EA, ETH, FTI, K!, WAG, iso.)

Shrub to 4.5 m high, much branched; stems dark brown, the youngest parts puberulous, becoming glabrous. Leaves petiolate or the upper sessile, ovate to lanceolate, 7–19 cm long, 4–15 cm wide, base cordate to cuneate and connate with that of opposite leaf, margins entire, dentate or serrate, recurved, apex acute, veins and margins pilose otherwise glabrous or minutely hairy below, glandular on veins and margins; 3-veined from base; petiole 15–55 mm long. Capitula terminal several-headed cymes; stalks of individual capitula 4–10(–15) mm long; involucre 7–10(–15) mm long; outer 5 phyllaries larger than the inner; paleae similar to inner phyllaries. Ray florets 5–8, yellow, tube 1–2 mm long with hairy base, ray 6–11 mm long; disc florets yellow, tube 3.4–3.6 mm long, lobes 0.7–0.8 mm long. Achenes obovoid, 2.3–2.8 mm long; pappus absent.

UGANDA. Acholi District: SE Imatong Mts, Mt Lomwaga, Apr. 1945, *Greenway & Hummel* 7285!
DISTR. U 1; Sudan (Imatong), SE Ethiopia
HAB. Forest patch, where locally dominant; 2600 m
USES. None recorded on specimens from our area
CONSERVATION NOTES. One specimen known from Uganda; endemic to Imatong Mts and the Jima area of Ethiopia. Data deficient (DD)

5. **Guizotia jacksonii** (*S.Moore*) *J.Baagøe* in Bot. Tidsskr., 69 (1): 39 (1974). Type: Kenya, Kiambu District: Kikuyu, anno 1899, *Jackson* s.n. (BM!, holo.)

Perennial herb, creeping, sparsely branched, 3–30 cm long, rooting at the nodes and sometimes mat-forming; stems glabrous. Leaves of a pair unequal or equal, sessile, elliptic to oblanceolate, 1–6.2 cm long, 0.5–2.5 cm wide, base tapering to a connate part, margins remotely and minutely glandular-dentate and often revolute, apex acute or obtuse, glabrous or pilose to almost scabrid or pilose only on veins and margins. Capitula solitary in upper leaf axils; stalks of individual capitula 0.5–3 cm long; involucre 7–12 mm long; phyllaries ciliate; paleae yellowish, to 3 mm long. Ray florets 4–9, yellow, tube to 2 mm long, rays 7–12(–15) mm long; disc florets 7–13, 3.7–5 mm long. Achenes brown, 3.5 mm long; pappus absent.

UGANDA. Mt Elgon, Piswa trail, Oct. 1997, *Wesche* 1940!
KENYA. West Suk District: Kapseis, Aug. 1968, *Thulin & Tidigs* 68!; Naivasha District: Nyandarua/Aberdare Mts, Mutubio Gate, Oct. 2000, *Smith et al.* 52!; North Nyeri District: Mt Kenya, Sirimon Track, Dec. 1974, *Williams Sangai* 77!
DISTR. U 3; K 2–6; endemic to Elgon, Cherangani, Nyandarua/Aberdares, Mau and Mt Kenya
HAB. Moorland and heath zone, moist stream-sides, montane forest glades; 2350–3900 m
USES. None recorded on specimens from our area
CONSERVATION NOTES. Least concern (LC)

SYN. *Coreopsis jacksonii* S.Moore in J.L.S. 35: 347 (1902)
　　Guizotia reptans Hutch. in K.B. 1914: 17 (1914); R.E. Fr. in Acta Horti Berg. 9, 6: 141 (1928); Thulin in Bot. Not. 123: 489, 492 (1970); J. Baagøe in Bot. Tidsskr. 69: 23, fig. 18 (1974); U.K.W.F. ed. 2: 216 (1994). Type: Kenya, Nyandarua/Aberdare Mts, *Battiscombe* 530 (K!, holo.)
　　Bidens spathulata Sherff in Bot. Gaz. 76: 149, t. 13 (1923). Type: Kenya, Mt Kenya W slopes, *Mearns* 1291 (ubi?)
　　Bidens jacksonii (S.Moore) Sherff in Bot. Gaz. 81: 45 (1926)
　　Guizotia reptans Hutch. var. *keniensis* R.E.Fr. in Acta Horti Berg. 9, 6: 142, t. 2, fig. 5 (1928); Blundell, Wild Fl. E. Afr.: pl. 364 (1987). Types: Kenya, Mt Kenya W, Forest Station, *Fries & Fries* 756 (K!, syn.) & Mt Kenya, between Coles Mill & Forest Station, *Fries & Fries* 756a (S, syn.)
　　Coreopsis jacksonii S.Moore var. *arthrochaeta* Sherff in Bot. Gaz. 88: 302 (1929). Type: Kenya, Nyandarua/Aberdare Mts, Camp Gusisu, *Piemeisel & Kephart* 166 (US, holo.)

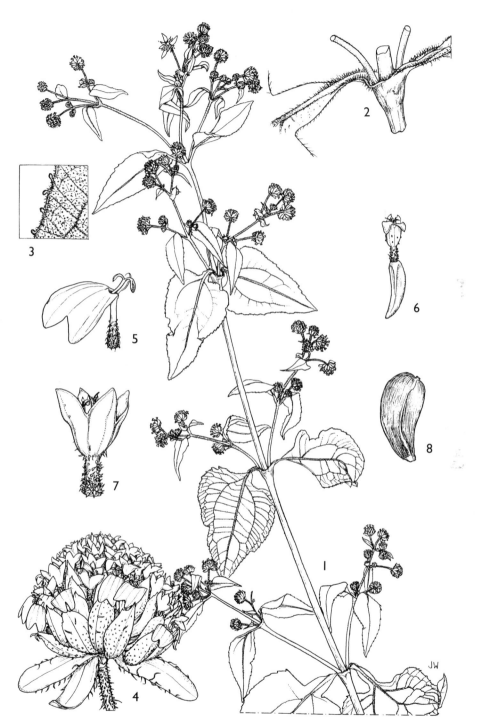

FIG. 168. *MICRACTIS BOJERI* — **1**, habit, × ²/₃; **2**, leaf axils, × 2; **3**, leaf surface detail – lower surface, × 4; **4**, capitulum, × 6; **5**, ray floret, × 16; **6**, disc floret, × 10; **7**, disc floret, achene removed, × 16; **8**, achene, × 10. 1–3 & 5 from *Verdcourt et al.* 3030, 4–5 & 7–8 from *Townsend* 2187a. Drawn by Juliet Williamson.

124. **MICRACTIS**

DC., Prodr. 5: 619 (1836)

Annual or perennial herbs. Leaves opposite, simple. Capitula in panicles, radiate; receptacle flat. Phyllaries 2-seriate; paleae enveloping the achenes and lower part of florets. Ray florets small, minutely 2-lobed; disc florets 4-lobed. Achenes ovoid-oblong; pappus absent.

Three species in Africa and Madagascar.

Micractis bojeri *DC.*, Prodr. 5: 620 (1836); Schulz in Gleditschia 18, 2: 214 (1990). Type: Madagascar, *Bojer* 59 (G-DC, holo., microfiche)

Annual or sometimes ?short-lived perennial herb 0.3–2.5 m high; stem often purple, branches scabridulous to glabrous. Leaves ovate or narrowly ovate, 2–22 cm long, 0.7–9.5 cm wide, base gradually narrowing, the very base connate with that of opposite leaf, margins finely serrate to serrate, apex acuminate or attenuate, scabrid or scabridulous above, thinly pubescent and glandular beneath; 3-veined from base. Capitula in terminal and axillary lax panicles; stalks of individual capitula 1–4 mm long; involucre 3–5.5 mm long (but sometimes closely subtended by leafy bracts to 12 mm long), 5–8 mm in diameter, the phyllaries pilose, ciliate; paleae 3 mm long, glandular. Ray florets many, in 2 rows but may look as if uniseriate, yellow, tube 0.5–0.8 mm long and glandular, ray 0.8–1 mm long; disc florets many, yellow, 1–1.8 mm long, the tube glandular. Achenes dark brown, obovoid, 1.8–2.5 mm long, 4-angled, glabrous; pappus absent. Fig. 168 (page 775).

UGANDA. Kigezi District: Kisoro, Bufumbura, Nov. 1946, *Purseglove* 2295!; Toro District: 5 km W of Kilembe, June 1970, *Lye & Katende* 5543!; Mbale District: Butandiga, Dec. 1938, *A.S.Thomas* 2562!
KENYA. Naivasha District: S Kinangop, Dec. 1960, *Verdcourt et al.* 3030!; Meru District: Kangeta school, Sep. 1967, *Hanid & Kiniaruh* 1025!; Kericho District: SW Mau, Sambret, *Kerfoot* 2902!
TANZANIA. Masai/Mbulu District: Ngorongoro Crater, Sep. 1932, *Burtt* 4333!; Lushoto District: Kungului, Mar. 1971, *Shabani* 675!; Mbeya District: above Kitakalo Mission, Dec. 1963, *Richards* 18562!
DISTR. **U** 2, 3; **K** 3–6; **T** 1–4, 7, 8; from Nigeria to Ethiopia and south to Congo (Kinshasa) and Malawi; Madagascar
HAB. Streamsides in grassland or forest, seasonally flooded grassland, swamp grassland; 1350–2750 m
USES. None recorded on specimens from our area
CONSERVATION NOTES. Least concern (LC)

SYN. *Limnogenneton abyssinicum* Sch.Bip. in Walp. Rep. 6: 147 (1846). Types: Ethiopia, near Adoa, *Schimper* 1099 (W, holo., FT, G, GH, K!, LG, LZ, MO, S, UPS, US, iso.)
    Cryphiospermum abyssinicum (Sch.Bip.) Schweinf., Beitr. Fl. Aeth.: 284 (1867)
    Sigesbeckia abyssinica (Sch.Bip.) Oliv. & Hiern, F.T.A. 3: 372 (1877); Wild in Kirkia 6: 37 (1967); Maquet in Fl. Rwanda 3: 620, fig. 191/1 (1985); Lisowski, Asterac. Fl. Afr. Centr. 1: 179, t. 39 (1991); U.K.W.F. ed. 2: 215, t. 88 (1994)
    Micractis abyssinica (Sch.Bip.) Chiov. in Ann. Bot. Roma 9: 73 (1911)

125. **SIGESBECKIA**

L., Sp. Pl.: 900 (1753); Schulz in Gleditschia 15: 205–210 (1987) & in Haussknechtia 3: 57–64 (1987)

Annual or perennial herbs. Leaves opposite. Capitula in lax panicles, heterogamous; involucre of 3–5 phyllaries with stalked glands; receptacle paleate. Ray florets 1-seriate, female, inconspicuous; disc florets hermaphrodite, 3–5-lobed; anthers with minute apical appendage; style branches short and flattened. Achenes usually incurved; pappus absent.

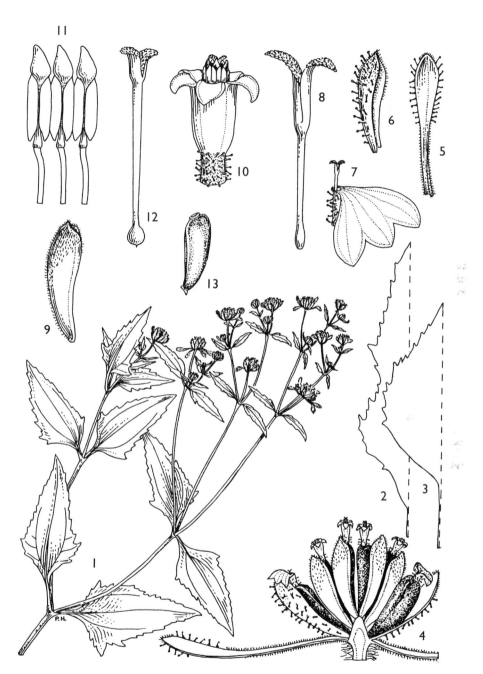

FIG. 169. *SIGESBECKIA ORIENTALIS* — **1**, habit, × ²/₃; **2–3**, lower leaves, schematic, × 8; **4**, capitulum section, × 6; **5**, part of outer phyllary, × 4; **6**, inner phyllary, × 6; **7**, ray floret, achene removed, × 20; **8**, ray floret style, × 60; **9**, palea, × 8; **10**, disc floret, achene removed, × 30; **11**, stamens, × 60; **12**, disc floret style, × 60; **13**, achene, × 6. 1 & 4–13 from *Guého* in MAU 15430, 2 from *Bosser* 11502, 3 from *Bijioux* 889. Drawn by Pat Halliday, from Flore des Mascareignes.

10 species, or possibly 3, from tropical Africa and Asia – or possibly the Americas as well. Taxonomy, again, is disputed.

Sigesbeckia orientalis *L.*, Sp. Pl.: 900 (1753); Oliv. & Hiern in F.T.A. 3: 372 (1877); F.P.U.: 172 (1962); Wild in Kirkia 6: 36 (1967); Maquet in Fl. Rwanda 3: 620, fig. 191/2 (1985); Lisowski, Asterac. Fl. Afr. Centr. 1: 178 (1991); D.J.N. Hind in Fl. Masc. 109: 176, t. 58 (1993); U.K.W.F. ed. 2: 215 (1994). Type: China, 'habitat in China media ad pagos', Sigesbeckia in Hort. Cliff., t. 23 (1738) (BM, lecto., designated by Stearn in Introd. Linnaeus' Sp. Pl. (Ray Soc. ed.): 47 (1957))

Annual herb 20–120 cm high; stems often red or purplish, pilose. Leaves ovate to triangular-hastate, 5–18 cm long, 3–11 cm wide, base cuneate to truncate, margins irregularly dentate or serrate, apex obtuse or acute, short-pubescent and often slightly scabridulous on both surfaces and glandular beneath; 3-veined from near base; petiole 1–5 cm long, often slightly winged. Capitula in lax leafy dichotomous panicles; involucre 5–7 mm long, green; outer phyllaries linear-spatulate, 5–15 mm long; inner phyllaries 5 mm long, all with stalked glands; paleae also often glandular. Ray florets usually 5, yellow, ray 1.3–1.5 mm long; disc florets ± 12, yellow, ± 1 mm long, glandular near base. Achenes dark brown or blackish, curved, 3–4 mm long, glandular-pubescent; pappus absent. Fig. 169 (page 777).

UGANDA. Ankole District: Bunyanguru, no date, *Purseglove* 466!; Mengo District: Old Entebbe, Jan. 1956, *Harker* 147! & Kampala, Makerere Hill, June 1952, *Lind* 101!
KENYA. Meru District: Meru town, Jan. 1933, *Napier* 2420!; North Kavirondo District: Webuye [Broderick Falls], Aug. 1965, *Tweedie* 3080!; Teita District: near Wusi Mission, Feb. 1953, *Bally* 8788!
TANZANIA. Moshi District: Lyamungu, Aug. 1932, *Greenway* 3051!; Lushoto District: Magamba Forest Reserve, no date, *Shabani* 949!; Morogoro District: Uluguru Mts above Morogoro town, May 1933, *Burtt* 4697!
DISTR. **U** 2, 4; **K** 3–5, 7; **T** 2, 3, 6; a widespread weed
HAB. Ruderal sites, weed of cultivation; 350–1750 m
USES. None recorded on specimens from our area
CONSERVATION NOTES. Least concern (LC)

NOTE. Spreading easily because of the sticky 'fruit'. The first record from East Africa is in 1880, from Uganda (unlocalized).
　　Sigesbeckia abyssinica is a synonym of *Micractis bojeri*.

126. **BIDENS**[*]

L., Sp. Pl.: 831 (1753) & Gen. Pl.: 362 (1754); Mesfin in K.B. 48: 437–516 (1993)

Annual or perennial herbs or shrubs. Leaves opposite, simple or variously compound with segments linear to broadly ovate. Capitula radiate, terminal, solitary or arranged in cymes; involucre of usually green outer phyllaries and variously coloured inner phyllaries; paleae membranous, usually oblong-linear, 2–many-striate. Ray florets neuter with aristate or exaristate ovary or pistillate and fertile with well developed achenes and aristae, rays yellow or orange, rarely white (*B. pilosa* L.) or lemon-yellow, striate; disc florets yellow to brownish, corolla 5-lobed. Achenes compressed or angled, striate-sulcate, ribbed, unmargined or margins narrowly winged or callose-thickened; at apex with or without aristae, aristae with or without antrorse and or retrorse barbs.

[*] by Mesfin Tadesse, Ohio State University, Department of Evolution, Ecology and Organismal Biology, 1753 Neil Ave., Columbus, OH 43210-1293, U.S.A.

340 species worldwide, most in the New World, but 63 species in Africa.

1. Most leaf teeth and apices setigerous, i.e., with
   hairs or bristle-like extensions (in *B. crocea*
   some leaf teeth may be sharply drawn out but
   not setigerous) 2
   Leaf teeth and apices not setigerous, merely acute,
   or leaf margins entire 5
2. Ray florets neuter; achene margins winged 3
   Ray florets pistillate and fertile; achene margins
   not winged 4
3. Involucre 3–5 mm long; capitulum 10–25 mm wide;
   achene margins often with corky thickenings;
   annual herb; **U** 1, 3 .... 24. *B. negriana* (p. 799)
   Involucre 5–9 mm long; capitulum 25–30 mm
   wide across the rays; achene margins with
   narrow flatwings; ?perennial herb; **K** 1 ...... 23. *B. nobiloides* (p. 799)
4. Leaves simple, basally lobed or bilobed, or 3-partite,
   lamina or segments elliptic or ovate-lanceolate ...... 12. *B. ternata* (p. 792)
   Leaves pinnately 3–9-lobed or -partite, the
   segments narrowly ovate to linear .... 13. *B. rueppellii* (p. 793)
5. Outer phyllaries dumb-bell shaped, dilated in
   upper half and also at base; all phyllaries with
   characteristic greyish-purple spots .... 17. *B. flagellata* (p. 796)
   Outer phyllaries not as above; phyllaries without
   such spots 6
6. Pappus of inner (disk) florets with retrorse
   (downward-pointing) barbs 7
   Pappus of inner florets absent, without barbs, or
   with antrorse (upward-pointing) barbs (or both
   retrorse and antrorse barbs) 21
7. Annual herbs 8
   Perennial herbs or shrubs 13
8. Leaf segments linear to filiform, 0.5–3 mm wide 9
   Leaf segments ovate to lanceolate, 5–30 mm wide
   or more 11
9. Involucre 7–15 mm long; achene apex convex or
   dome-shaped, aristae divaricate or, sometimes,
   at right angles with body .... 19. *B. lineariloba* (p. 797)
   Involucre 6–8 mm long (to 13 mm in fruit);
   achene apex concave or beaked, aristae erect or
   helically twisted 10
10. Ray florets 6–7 × 1–1.5 mm; capitulum 2–25-
    flowered, 2 cm or more tall in fruit; achenes
    beaked, with beak up to 24 cm long .... 28. *B. acuticaulis* (p. 802)
    Ray florets 15–20 × 3–4 mm; capitulum 5–15-
    flowered, up to 1.2 cm tall in fruit; achenes not
    beaked, 8–14 mm long .... 29. *B. diversa* (p. 803)
11. Leaves glandular-punctate (use × 10 lens); capitula
    2.5–4 cm wide; achenes oblong-elliptic, flat or
    compressed, bifacial .... 30. *B. schimperi* (p. 803)
    Leaves without visible glands; capitula, if radiate,
    0.5–2 cm wide; achenes linear-oblong, tetragonal 12

12. Ray florets yellow; outer phyllaries linear or oblanceolate, usually much longer than the inner phyllaries, marginal hairs up to 1 mm long; inner achenes much longer than the others, exceedingphyllaries and paleae by half when fully mature . . . .   31. *B. biternata* (p. 804)

Ray florets, when present, white or creamy-white; outer phyllaries spatulate, not exceeding inner phyllaries, marginal hairs up to 0.3 mm long; achenes ± uniform . . . .   32. *B. pilosa* (p. 806)

13. Leaves glandular-punctate (use × 10 lens) 14

Leaves without visible glands 17

14. Rays 30–35 × 12–17 mm; involucre 8–12 mm long; mature achenes 4–8 mm long . . . .   6. *B. holstii* (p. 787)

Rays less than 20 mm long; other characters not as combined above 15

15. Aristae of achenes 2–4(–5), barbed all over, inner achenes usually curved toward the apex; involucre 6–9 mm long; capitulum 3–4.5 cm wide   1. *B. hildebrandtii* (p. 782)

Aristae of achenes 2, with barbs mostly near top only, achenes straight; involucre 3–7 mm long; capitulum 1.5–3 cm wide 16

16. Stems procumbent or prostrate to scambling; leaves densely hispid-pilose; involucre 3–5 mm long . .   14. *B. whytei* (p. 794)

Stems erect or scrambling; leaves sparsely ciliate only on veins and margins; involucre 5–7 mm long . .   16. *B. taylorii* (p. 795)

17. Leaf segments linear to linear-elliptic, 2–8 mm wide 18

Leaf segments ovate to ovate lanceolate, 10–40 mm wide 19

18. Capitula 1–3 cm wide at anthesis; achenes 9–18 mm long, glabrous or sparsely pilose on margins . . .   9. *B. crocea* (p. 789)

Capitula (3–)3.5–7.5 cm wide at anthesis; achenes 4–10 mm long, strigose . . . .   11. *B. ugandensis* (p. 790)

19. Achene margins narrowly callose-thickened; pappus bristles with both antrorse and retrorse barbs; leaves simple to pinnatisect, segments ovate or broadly ovate-lanceolate, 3–10 × 1–4 cm; T 3, 5–7 . . . .   7. *B. magnifolia* (p. 788)

Achene margins not callose-thickened; pappus bristles retrorse; leaf characters not as combined above 20

20. Leaves bipinnatisect, membranous, dark green, with appressed white hairs on upper surface; capitula 4–6 cm wide at anthesis; achenes glabrous   18. *B. fischeri* (p. 796)

Leaves pinnate, thickish, grey-green, minutely spinulose- setose to scabrid hispid; capitula 3–3.5 cm wide atanthesis; achenes hairy . . . . .   15. *B. cinerea* (p. 794)

21. Disc florets and/or achenes without aristae, or aristae without barbs 22

Disc florets and/or achenes with aristae with antrorse barbs (note: may be absent on achenes of ray florets) 33

22. Disc florets and/or achenes without aristae 23

Disc florets and/or achenes with smooth aristae 24

23. Rays 23–30 mm long; leaves 6–19 × 6–19 cm, pinnatisect or pinnatipartite, without glands; **T** 3, 7, 8 . . . . *2. B. pinnatipartita* (p. 783)

Rays 8–13 mm long; leaves 2–5 × 1.5–4 cm, pedately tri-partite, punctate with glands; **U** 1, 3; **K** 3 . . *22. B. elgonensis* (p. 798)

24. Leaves glandular-punctate (use × 10 lens) 25

Leaves without clearly visible glands 26

25. Leaves bipinnatisect, with pilose veins; rays 30–35 mm long; **K** 3, 7; **T** 2, 3 . . . . *6. B. holstii* (p. 787)

Leaves simple or shallowly lobed, scabrid; rays 15–18 mm long; **T** 1, 4 . . . . *8. B. baumii* (p. 788)

26. Leaves simple to pinnately lobed 27

Leaves variously compound 31

27. Leaves narrowly linear-elliptic, ± 10 times as long as wide . . . . *11. B. ugandensis* (p. 790)

Leaves ovate or ovate-lanceolate, less than 10 times as long as wide 28

28. Capitula 6–8.5 cm wide at anthesis; involucre 10–18 mm high; outer phyllaries (5–)10–30 mm long 29

Capitula 3.5–6(–7) cm wide at anthesis; involucre 5–10 mm high (or 8–17 mm in *B. oblonga*); outer phyllaries 3–10(–15) mm long 30

29. Leaves ovate, opposite, usually hispid-tomentose, margins irregularly lobulate-dentate; outer and inner phyllaries 5–14 × 1–4.5 mm . . . . *3. B. kilimandscharica* (p. 783)

Leaves oblong-elliptic, opposite or whorled, margins regularly serrate, glabrous or hispid; outer phyllaries (9–)12–25 × 2–13 mm; inner phyllaries 9–16 × 3–5(–7) mm . . . . *4. B. buchneri* (p. 785)

30. Achenes 9–11 × 3–4.5 mm; outer phyllaries 7–11(–15) mm long; ray florets 8–13, 15–45 mm long . . . . *5. B. oblonga* (p. 786)

Achenes 6–11 × 1–1.7 mm; outer phyllaries 6–7 mm long; ray florets 6–8, 15–18 mm long . . . . *8. B. baumii* (p. 788)

31. Leaves tomentose or pubescent, the segments ± ovate; outer achenes without aristae, inner with 2 aristae each . . . . *3. B. kilimandscharica* (p. 783)

Leaves ± glabrous, only the margins scabrid or pubescent, the segments ± linear; all achenes bi-aristate 32

32. Mature achenes 2.5–3.5 mm wide, margins narrowly callose-thickened and bristled; **T** 4, 7 . . . . *10. B. ochracea* (p. 789)

Mature achenes 1–2(–2.3) mm wide, margins not thickened; widespread . . . . *11. B. ugandensis* (p. 790)

33. Leaves glandular-punctate (use × 10 lens) 34

Leaves without clearly visible glands 36

34. Ray florets pistillate and fertile; shrubby perennial herbs, 1–4.5 m high; **U** 2 . . . . *21. B. elliottii* (p. 798)

Ray florets neuter or female-sterile; annual herbs, 0.5–2.5 m high 35

35. Achenes 6–10 × 1–1.7 mm; leaves with long, shiny, silky hairs, or glabrous; capitula 5–8 cm wide at anthesis; **T** 1, 4, 5, 7, 8 . . . . *26. B. steppia* (p. 801)

Achenes 4–7 × 0.5–0.8 mm; leaves with short, pale hairs; capitula 2–5 cm wide at anthesis; **U** 2, 4; **K** 3; **T** 1, 2 . . . . *25. B. grantii* (p. 800)

1. **Bidens hildebrandtii** *O.Hoffm.* in E.J. 20: 234 (1895); Sherff in Field Mus. Nat. Hist., Bot. Ser. 16: 574, fig. 154 (1937); T.T.C.L.: 147 (1949); E.P.A.: 113 (1967); Mesfin in Symb. Bot. Upsal. 24: 97, fig. 48 (1984) & in K.B. 48: 449 (1993); K.T.S.L.: 554 (1994); U.K.W.F. ed. 2: 217 (1994). Type: Kenya, Teita District: Taita Hills, *Hildebrandt* 2432 (B†, holo., K!, lecto., BM!, M!, P!, W!, WU!, iso.; chosen by Mesfin)

Perennial herb or shrub 0.6–3 m high; stems several from rootstock, usually erect but sometimes becoming scandent, the young growth rarely pubescent, usually glabrous, often reddish or purple. Leaves pinnatisect, petiolate, triangular or ovate in outline, 3–14 cm long, 1–9 cm wide, base cuneate to attenuate, apex obtuse and apiculate, segments ovate, irregularly serrate, glabrous or sparsely pubescent on both surfaces, densely glandular-punctate; petiole 20–45 mm long. Capitula in lax corymbose cymes; stalks of individual capitula to 9 cm long; involucre 6–9 mm long; outer phyllaries reflexed at anthesis, glabrous or pubescent, the inner erect, pale orange, glabrous or pilose; paleae white with 10 brown lines, pilose. Ray florets 8, yellow, neuter, the tube 1.6–2 mm long, ray 16–20 × 5–7 mm; disc florets tubular, 5–6.5 mm long. Achenes black, shiny, oblong, 6–11.5 mm long, strigose; pappus of 2–4(–5) retrorsely barbed aristae 2–4 mm long.

KENYA. Turkana District: Murua Nysigar, Dec. 1988, *Beentje et al.* 3920!; Masai District: Chyulu plains, June 1991, *Luke* 2848!; Teita District: Taita Hills, Msau R. valley, May 1985, *Taita Hills Exped.* 649!
TANZANIA. Kilimanjaro, near old Moshi, Feb. 1914, *Peter* 53256!; Mbulu District: Kampi ya Nyoka, Aug. 1926, *Peter* 43500!
DISTR. **K** 1–4, 6, 7; **T** 2; Ethiopia
HAB. In rock crevices and on lava in the dry bushland zone, less often in scrub on hillsides; 600–1900 m
USES. For friction firesticks (*Gillett*)
CONSERVATION NOTES. Least concern (LC)

Syn. *B. lindblomii* Sherff in Field Mus. Nat. Hist., bot.ser. 16: 646 (1937). Type: Kenya, Machakos/Kitui District: Ukamba, *Lindblom* s.n. (S!, holo.)

 B. hildebrandtii O.Hoffm. var. *boranensis* Lanza in Miss. Biol. Borana 4: 263, fig. 85 (1939); E.P.A.: 113 (1967). Type: Ethiopia, Sidamo, Moyale, *Cufodontis* 628 (FT!, holo.)

 B. incumbens Sherff in Bot. Leafl. 5: 19 (1951); U.K.W.F.: 466 (1974). Type: Kenya, Teita District: Voi, *Napier* 1016 (EA!, holo., K!, iso.)

 B. incumbens Sherff var. *muthicola* Sherff in Bot. Leafl. 7: 18 (1952). Type: Kenya, Kitui District: Mutha Hill, *Bally* 1655 (EA!, holo., K!, iso.)

Note. The altitude of 1500 feet given by Bally for his collection from Mutha Hill is erroneous.

2. **Bidens pinnatipartita** (*O.Hoffm.*) *Wild* in Kirkia 6: 18 (1967); Mesfin in K.B. 48: 449 (1993). Type: Tanzania, Mbeya/Chunya District: Usafwa, *Goetze* 1041 (B†, holo., K!, isolecto., P!, iso., selected by Mesfin)

Perennial herb or soft-wooded shrub 0.5–3 m high (once described as a tree of 4.5 m); stems usually several from a thick rootstock, often tinged reddish, tomentose to glabrous. Leaves petiolate, pinnatisect to pinnatipartite, ovate in outline, 6–19 cm long, 6–19 cm wide, leaf segments narrowly ovate or ovate, margins serrate to deeply lobed, pubescent above, pubescent to tomentose beneath; petiole to 50 mm long. Capitula solitary or up to 5 together; stalks of individual capitula to 3 cm long; involucre cylindric, 7–15 mm long; outer phyllaries leafy, reflexed at anthesis, 7–15 mm long, pubescent to tomentose, inner phyllaries yellow, 7–16 mm long, spreading, pubescent; paleae 6–15 mm long, usually with 3–6 purple lines. Ray florets yellow, neuter, 6–13, tube 3–3.3 mm long, pubescent, rays 23–30 mm long, to 14 mm wide; disc florets yellow or orange-yellow, 5.5–8 mm long, sparsely puberulous in lower parts. Achenes shiny dark brown or black, obovoid, 4.5–7 mm long, glabrous; pappus usually absent, rarely with two entire or distally barbed yellow aristae to 2.5 mm long.

Tanzania. Lushoto District: Shume, Aug. 1982, *Kibuwa* 5544!; Mbeya District: Mbeya Peak Forest Reserve, Oct. 1958, *Gaetan Myembe* 90!; Songea District: Mpapa, Kiteza Hill, Oct. 1956, *Mgaza* 104!
Distr. **T** 3, 7, 8; Angola, Zambia, Malawi, Mozambique
Hab. Forest margins, bushland, bushed grassland, secondary bushland; 1500–2400 m
Uses. None recorded on specimens from our area
Conservation notes. Least concern (LC)

Syn. *Guizotia bidentoides* Oliv. & Hiern, F.T.A. 3: 386 (1877). Type: Malawi, Chikwawa [Shibisa], *Kirk* s.n. (K!, holo.), *non B. bidentoides* Britton (1893)
 Coreopsis pinnatipartita O.Hoffm. in E.J. 30: 432 (1901)
 Coreopsis lupulina O.Hoffm. in E.J. 30: 432 (1901). Type: Tanzania, Mbeya/Chunya District: Usafwa, *Goetze* 1069 (B†, holo., P!, isolecto., BM!, BR!, K!, iso., selected by Mesfin)
 C. whytei S.Moore in J.L.S. 35: 348 (1902). Type: Malawi, Mt Mlanje, *Whyte* 35 (BM!, holo.)

3. **Bidens kilimandscharica** (*O.Hoffm.*) *Sherff* in Bot. Gaz. 59: 309 (1915) & in Field Mus. Nat. Hist., Bot. Ser. 16: 606, fig. 174 (1937); T.T.C.L.: 148 (1949); Wild in Kirkia 6: 31 (1967); U.K.W.F.: 467 (1974); Blundell, Wild Fl. E. Afr.: pl. 354 (1987); Mesfin in K.B. 48: 451 (1993); K.T.S.L.: 554, fig. (1994); U.K.W.F. ed. 2: 218 (1994). Type: Tanzania, Kilimanjaro, anno 1890, *Abbott* s.n. (B†, lecto., chosen by Sherff; US, iso.)

Perennial herb 0.6–3 m high; stems slightly woody at base, little branched, pilose to thinly pubescent. Leaves variable, pinnatipartite to bipinnatisect or the uppermost and lowermost sometimes simple, 5–15 cm long, up to 10 cm wide, when lobed with ovate segments, when simple with serrate to lobed margins, the teeth mucronate to apiculate, apex acute to acuminate, tomentose or hispid-pubescent on both surfaces; petiole 10–40 mm long. Capitula terminal, solitary or in few-headed leafy cymes, erect; stalks of individual capitula 1–5(–21, fide EA) cm long; involucre 7–14 mm long; outer phyllaries 9–13, green, pubescent, inner phyllaries with scarious margins, pilose to pubescent; paleae 5.5–14 mm long, obtuse. Ray florets yellow or orange, neuter, 8–13, tube 3.5–4.5 mm long, pubescent, ray 20–40 mm long, 10–15 mm wide; disc florets

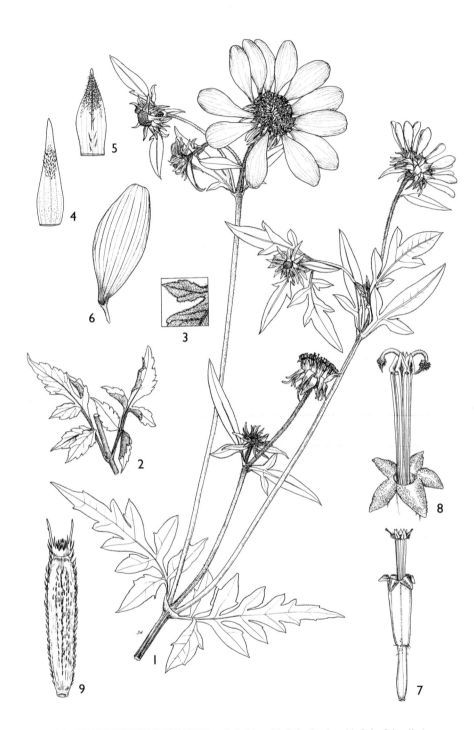

FIG. 170. *BIDENS KILIMANDSCHARICA* — **1**, habit, × ¹/₂; **2**, leaf pair, × ¹/₂; **3**, leaf detail; **4**, outer phyllary, inner surface, × 2; **5**, inner phyllary, outer surface, × 2; **6**, ray floret, × 1; **7**, disc floret, × 5; **8**, anthers and style, × 10; **9**, achene, × 5. 1 & 6–7 from *Mabberley & McCall* 7, 2–5 & 8 from *Verdcourt* 1615, 9 from *Dale* 2690. Drawn by Juliet Williamson.

yellow, 5–7 mm long. Achenes black or dark brown, 5.5–10 mm long, compressed, pilose; aristae absent in outer achenes, on inner 2, divergent, to 2.5 mm long, nude or with 1 or 2 antrorsely- or retrorsely-set barbs at apex or at base. Fig. 170 (page 784).

UGANDA. Acholi District: Chua, Agoro, Langia, Nov. 1945, *A.S.Thomas* 4381!
KENYA. Northern Frontier District: Mt Nyiro, July 1960, *Kerfoot* 2057!; Nakuru District: 12 km on Nakuru–Eldoret road, Dec. 1956, *Verdcourt* 1615!; Machakos/Masai District: N end of Chyulu Hills, May 1981, *Gilbert* 6185!
TANZANIA. Arusha District: Ngurdoto National Park, Tululusie Hill, Nov. 1965, *Greenway & Kanuri* 12313!; Lushoto District: W Usambara Mts, Mtumbi Forest Reserve, Feb. 1985, *Borhidi et al.* 85/646!; Mpanda District: Mwese Hill, May 1975, *Kahurananga et al.* 2637a!
DISTR. U 1; K 1–7; T 1–4, 7; Congo (Kinshasa), Burundi, Angola, Malawi, Mozambique, Zimbabwe
HAB. Grassland in rocky sites, scattered tree grassland, bushed or wooded grassland, bushland, forest margins; 1050–2800 m
USES. For staining wood (*van Someren*)
CONSERVATION NOTES. Least concern (LC)

SYN. *Coreopsis kilimandscharica* O.Hoffm. in E.J. 20: 234 (1894)
   Bidens volkensii O.Hoffm. in P.O.A. C: 415 (1895); Sherff in Field Mus. Nat. Hist., Bot. Ser. 16: 610, fig. 175 (1937). Type: Tanzania, Moshi District: Kware [Quari] R. below Machame, *Volkens* 1694 (B†, holo.)
   Coreopsis crataegifolia O.Hoffm. in E.J. 30: 431 (1901). Type: Tanzania, Njombe District: Yawuanda Mt, *Goetze* 851 (B†, holo., BM, lecto., chosen by Rayner in 1992)
   Bidens robustior S.Moore in J.L.S. 35: 350 (1902); Sherff in Field Mus. Nat. Hist., Bot. Ser. 16: 609, fig. 181 (1937); Wild in Kirkia 6: 31 (1967). Type: Kenya, Nakuru District: Elmenteita, *Scott Elliot* 6846 (BM!, holo., K!, iso.)
   B. ukambensis S.Moore in J.L.S. 35: 350 (1902); Sherff in Field Mus. Nat. Hist., Bot. Ser. 16: 609 (1937). Type: Kenya, Machakos/Kitui District: Ukamba, *Scott Elliot* 6462 (BM!, holo., K!, iso.)
   B. crataegifolia (O.Hoffm.) Sherff in Bot. Gaz. 76: 158 (1915) & in Field Mus. Nat. Hist., Bot. Ser. 16: 605, fig. 173 (1937)
   Coreopsis leptoglossa Sherff in Bot. Gaz. 76: 88 (1915) & in Field Mus. Nat. Hist., Bot. Ser. 11: 362 (1936). Type: Congo (Kinshasa), Lofuku R., *Kassner* 2871 (B†, holo., K!, lecto., chosen by Mesfin, BM!, P!, Z!, iso.)
   Bidens kilimandscharica (O.Hoffm.) Sherff var. *retrorsa* Sherff in Bot. Gaz. 92: 202 (1931); Sherff in Field Mus. Nat. Hist., Bot. Ser. 16: 607 (1937); T.T.C.L.: 148 (1949). Type: Tanzania, Moshi District: Baloti [Boloti], *Haarer* 1472 (K!, holo., EA!, iso.)
   B. insignis Sherff in Field Mus. Nat. Hist., Bot. Ser. 17: 591 (1939). Type: Kenya, Nakuru District: Menengai, *Brodhurst-Hill* 612 (K!, holo., EA!, iso.)
   B. kilimandscharica (O.Hoffm.) Sherff var. *oxymera* Sherff in Amer. J. Bot. 34: 155 (1947). Type: Kenya, Machakos District: Ol Doinyo Sapuk, *Napier* 6085 (EA!, holo., K!, iso.)
   B. meruensis Sherff in Bot. Leafl. 5: 18 (1951). Type: Tanzania, Arusha District: Mt Meru, *van Someren* in EA 3929 (EA!, holo.)
   B. cuspidata Sherff in Bot. Leafl. 9: 12 (1954). Type: Mozambique, Vila Gouveia, *Pole-Evans & Erens* 483 (SRGH!, holo., K!, L!, MO!, P!, iso.)
   B. crataegifolia (O.Hoffm.) Sherff var. *burttii* Sherff in Amer. J. Bot. 42: 561 (1955). Type: Tanzania, Arusha District: Mt Meru, *Burtt* 4117 (K!, holo., EA!, iso.)
   B. leptoglossa (Sherff) Lisowski, Asterac. Afr. Centr. 1: 146 (1991)

NOTE. Leaf variation is considerable as to dissection and indumentum: 3-lobed leaves are very common, with the terminal segments the largest, but up to 11 lobes per leaf do occur in the species.

4. **Bidens buchneri** (*Klatt*) *Sherff* in Bot. Gaz. 76: 158 (1923) & in Field Mus. Nat. Hist., Bot. Ser. 16: 594, fig. 166 (1937); E.P.A.: 1136 (1967); Mesfin in Symb. Bot. Upsal. 24, 1: 103, fig. 51 (1984); Asterac. Afr. Centr.: 166 (1991); Mesfin in K.B. 48: 454 (1993); U.K.W.F. ed. 2: 218, t. 90 (1994). Type: Angola, Malange, *Buchner* 31 (B†, holo.); Malange, between Sanzala Maquir & Quinje, *Raimundo & Matos* 1108 (BM!, neo., chosen by Mesfin)

Perennial herb or shrub, 0.6–2 m high; stem solitary or several, branched near apex, pubescent. Leaves opposite or rarely ternate, oblong-lanceolate to elliptic, 6–19 cm long, 1.5–5 cm wide, base cuneate to cordate, margins coarsely and irregularly serrate with thickened teeth, rarely lobed or bilobed at base, apex attenuate, glabrous or hispid-pubescent; sessile or with petiole to 15 mm long. Capitula terminal, solitary or few in corymbose cymes; stalks of individual capitula 5–50(–190, fide, EA) mm long; involucre 12–30 mm long; outer phyllaries 9–13, green, pubescent and inner phyllaries 8–10, pubescent or glabrous; paleae light brown, striped, 5–7.5 mm long. Ray florets yellow, neuter, 10–18, tube 4–6 mm long and pubescent, rays 30–50 mm long, 10–18 mm wide; disc florets yellow, 5–8 mm long, pubescent or glabrous in lower half. Achenes black, oblong, 6–13 mm long, pilose; aristae 2(–3), 0.5–3(–4) mm long, antrorsely barbed at base with 1 or 2 barbs in middle or 1 retrorse at apex.

UGANDA. Acholi District: Gulu, Keyo, Dec. 1938, *Eggeling* 3926!; Kigezi District: Nyakageme, May 1947, *Purseglove* 2414!; Mengo District: Mbuya Hill, Aug. 1969, *Rwaburindore* 83!
KENYA. West Suk District: Kapenguria, May 1932, *Napier* 1924!; Trans Nzoia District: Mt Elgon, Nov. 1957, *Symes* 219!; Kericho District: Ngoina Tea Estate, Dec. 1967, *Perdue & Kibuwa* 9369!;
TANZANIA. Morogoro District: Uluguru Mts above Morogoro, May 1933, *Burtt* 4696! & Apr. 1933, *Schlieben* 3837!; Nguru Mts near Maskati Missiona, June 1978, *Thulin & Mhoro* 3086!
DISTR. **U** 1–4; **K** 2, 3, 5; **T** 2, 4, 6, 7; Congo (Kinshasa), Rwanda, Burundi, Sudan, Angola
HAB. Hillside grassland, scattered tree grassland, bushland; 660–2500 m
USES. Minor medicinal against earache (*Bruce, Wallace*)
CONSERVATION NOTES. Least concern (LC)

SYN. *Coreopsis buchneri* Klatt in Leopoldina 25: 107 (1889)
    C. coriacea O.Hoffm. in P.O.A. C: 414 (1895). Type: ?Tanzania, Masai Plateau, *Fischer* 367 (B†, holo.)
    C. stuhlmannii O.Hoffm. in P.O.A. C: 415 (1895). Type: Tanzania, Bukoba District: Kitangule, *Stuhlmann* 1649 (B†, holo.)
    C. ruwenzoriensis S.Moore in J.L.S. 35: 345 (1902). Type: Kenya, Kavirondo, *Scott Elliot* 7410 (BM!, holo., K!, iso.)
    C. seretii De Wild. in Ann. Mus. Congo 2: 212 (1907). Type: Congo (Kinshasa), Mt Angbla, Uele R., *Seret* 306 (BR!, holo.)
    Bidens ruwenzoriensis (S.Moore) Sherff in Bot. Gaz. 59: 309 (1915)
    B. stuhlmannii (O.Hoffm.) Sherff in Bot. Gaz. 76: 158 (1923) & in Field Mus. Nat. Hist., Bot. Ser. 16: 598, fig. 169 (1937)
    B. seretii (De Wild.) Sherff in Bot. Gaz. 76: 162, fig. 14 (1923) & in Field Mus. Nat. Hist., Bot. Ser. 16: 596, fig. 168 (1937); F.P.U.: 174 (1962)
    B. coriacea (O.Hoffm.) Sherff in Bot. Gaz. 81: 52 (1923) & in Field Mus. Nat. Hist., Bot. Ser. 16: 601, fig. 172 (1937); U.K.W.F.: 467 (1967)
    B. bruceae Sherff in Bot. Gaz. 97: 606 (1923) & in Field Mus. Nat. Hist., Bot. Ser. 16: 603 (1937); T.T.C.L.: 147 (1949). Type: Tanzania, Morogoro District: Uluguru Mts, *Bruce* 26 (K!, holo.)
    B. bruceae Sherff var. *pubescentior* Sherff in Bot. Gaz. 97: 607 (1923) & in Field Mus. Nat. Hist., Bot. Ser. 16: 604 (1937); T.T.C.L.: 147 (1949). Type: Tanzania, Morogoro District, *Wallace* 294 (K!, holo.)
    B. bruceae Sherff var. *swynnertonii* Sherff in Field Mus. Nat. Hist., Bot. Ser. 16: 604 (1937); T.T.C.L.: 147 (1949). Type: Tanzania, Kilosa District: Hiwaga, *Swynnerton* 859 (BM!, holo., K!, iso.)

5. **Bidens oblonga** (*Sherff*) *Wild* in Kirkia 6: 30 (1967); Mesfin in K.B. 48: 455 (1993). Type: Zambia, Lake Tanganyika, *Carson* 106 (K!, holo.)

Perennial herb, 1–2 m high, with tuberous rootstock; stem 4-angled, pilose to glabrous. Leaves ovate in outline, simple to pinnately 3–5-lobed, 4–22 cm long, 3–15 cm wide, margins coarsely crenate to irregularly lobed, apex acute, hispid to glabrous; petiole to 50 mm long. Capitula terminal, solitary or in few-headed cymes; involucre 8–17 mm long; outer phyllaries 6–12, green, pubescent and inner phyllaries 8–10, pilose, appearing longer in fruit; paleae 8–10 mm long. Ray florets yellow or orange, neuter, 8–13, tube puberulous, rays 15–45 mm long, 5–10 mm wide; disc florets yellow. Achenes brown, oblong, 9–11 mm long, flat, setulose; aristae 2, 1–3 mm long, usually with a few antrorse barbs near the base, otherwise nude.

TANZANIA. Ufipa District: Chapota, Mar. 1957, *Richards* 8494! & 8495!

DISTR. **T** 4; Congo (Kinshasa), Angola, Zambia, Malawi, Zimbabwe
HAB. Rough grassland; ± 1650 m
USES. None recorded on specimens from our area
CONSERVATION NOTES. Least concern (LC)

SYN. *Coreopsis oblonga* Sherff in Bot. Gaz. 76: 80 (1923) & in Field Mus. Nat. Hist., Bot. Ser. 11: 374, fig. 2 (1936)
    Bidens nyikensis Sherff in Bot. Gaz. 81: 50 (1926) & in Field Mus. Nat. Hist., Bot. Ser. 16: 613, fig. 178 (1937). Type: Malawi, Nyika Plateau, *Whyte* 191 (K!, holo.)
    B. rhodesiana Sherff in Bot. Gaz. 92: 301 (1931) & in Field Mus. Nat. Hist., Bot. Ser. 16: 608, fig. 170/j–p (1937). Type: Zimbabwe, Odzani R. Valley, *Teague* 226 (K!, holo.)
    Coreopsis exilis Sherff in Bull. Jard. Bot. Etat 13: 290 (1935) & in Field Mus. Nat. Hist., Bot. Ser. 11: 385 (1936). Type: Congo (Kinshasa), Katuba, *Quarré* 419 (BR!, holo.)
    C. goffardii Sherff in Bull. Jard. Bot. Etat 13: 290 (1935) & in Field Mus. Nat. Hist., Bot. Ser. 11: 390 (1936). Type: Congo (Kinshasa), Kipilia–Lubumbashi, *Quarré* 1616 (BR!, holo.)
    Bidens dielsii Sherff var. *intermedia* Sherff in Amer. J. Bot. 42, 6: 562 (1955). Type: Zambia, Mbala, *Bullock* 2740 (K!, holo.)
    B. richardsiae Sherff in Amer. J. Bot. 42, 6: 562 (1955). Type: Zambia, Mbala, Chilongwelo, *Richards* 1546 (K!, holo.)
    B. exilis (Sherff) Lisowski, Asterac. Afr. Centr.: 140 (1991)

6. **Bidens holstii** (*O.Hoffm.*) *Sherff* in Bot. Gaz. 76: 79 (1923) & in Field Mus. Nat. Hist., Bot. Ser. 16: 536, fig. 134 (1937); Mesfin in K.B. 48: 456 (1993). Type: Tanzania, Lushoto District: Usambara Mts, *Holst* 76 (B†, holo.)

Perennial herb or shrub, 0.5–3.5 m high; stems several from base, branched near apex, often reddish, glabrous or sparsely pilose. Leaves ovate in outline, bipinnatisect, 3–19 cm long, 1.5–14 cm wide, segments lanceolate to ovate, margins dentate, apex acute to attenuate, pilose above, pilose on veins beneath, glandular-punctate and aromatic; petiole 1–7 cm long. Capitula terminal, solitary or few in ± corymbose cymes; stalks of individual capitula 1–10 cm long; involucre 8–12 mm long; outer phyllaries green, 10–16, reflexed in fruit, pilose, inner phyllaries greenish orange, ± 8, sometimes partly connate, pilose; paleae 6–8 mm long. Ray florets yellow, rarely orange, neuter, 8–10, rays 30–35 mm long, 12–17 mm wide; disc florets yellow. Achenes black or dark brown, oblong, flat, 4–8 mm long; aristae 2–3, to 2 mm long, nude or with a few barbs near the base or at middle, rarely with a single retrorse barb, rarely falling early.

KENYA. Naivasha District: Mt Longonot, Jan. 1982, *Gilbert & White* 6836!; Teita District: Vuria Peak, Apr. 1960, *Verdcourt & Polhill* 2718!; Kwale District: Kilibasi Hill, Nov. 1989, *Luke & Robertson* 2050!
TANZANIA. Masai District: Ngorongoro Crater rim, Oct. 1977, *Raynal* 19566!; Kilimanjaro N, Jan. 1990, *Pocs et al.* 90/006c!; Lushoto District: Chambogo Forest Reserve, July 1987, *Kisena* 42!
DISTR. **K** 3, 7; **T** 2, 3; not known elsewhere
HAB. Forest margins, montane bushland and heathland, grassland; (700–)1250–2800 m
USES. None recorded on specimens from our area
CONSERVATION NOTES. Least concern (LC)

SYN. *Coreopsis holstii* O.Hoffm. in P.O.A. C: 415 (1895)
    Bidens rupestris Sherff in Bot. Gaz. 76: 144 (1923). Type: Tanzania, Mt Meru, *Uhlig* 750 (B†, holo., EA, lecto., chosen by Rayner in 1992)
    B. holstii (O.Hoffm.) Sherff var. *rupestris* (Sherff) Sherff in Bot. Gaz. 90: 393 (1930) & in Field Mus. Nat. Hist., Bot. Ser. 16: 537, fig. 135 (1937)
    B. taitensis Sherff in Bot. Gaz. 90: 396 (1930) & in Field Mus. Nat. Hist., Bot. Ser. 16: 552, fig. 141 (1937); Blundell, Wild Fl. E. Afr.: pl. 356 (1987). Type: Kenya, Teita District: Taita Hills, *Hildebrandt* 2432a (B†, holo.)
    B. nobilis Sherff in Field Mus. Nat. Hist., Bot. Ser. 17: 593 (1939); T.T.C.L.: 148 (1949). Type: Tanzania, N Kilimanjaro above Rongai, *Rogers* 133 (K!, holo., BM!, EA!, iso.)
    B. napierae Sherff in Bot. Leafl. 5: 17 (1951). Type: Kenya, Naivasha District: Mt Longonot, *Napier* 218 (EA!, holo., K!, iso.)
    B. taitensis Sherff var. *aciculata* Sherff in Ann. Mag. Nat. Hist. 10: 44 (1957). Type: Kenya, Teita District: Ngaongao [Ngangao], *Lynes* 282 (BM!, holo.)

7. **Bidens magnifolia** *Sherff* in Bot. Gaz. 90: 390, fig. 4 (1930) & in Field Mus. Nat. Hist., Bot. Ser. 16: 575, fig. 160 (1937); Mesfin in Symb. Bot. Upsal. 26, 2: 199, fig. 2 (1986) & in K.B. 48: 458 (1993). Type: Tanzania, Lushoto District: Derema, Kwa Kiniari, *Holst* 2252 (B†, holo., K!, lecto., chosen by Mesfin)

Perennial herb or shrub, 0.6–3 m high; stem to 3 cm in diameter, 'ringed' with leaf scars, much branched, sparsely pubescent to glabrous. Leaves ovate in outline, simple (near stem apex) to pinnatisect, 6–20(–26) cm long, 5–16 cm wide, when pinnatisect with 3–9 ovate to lanceolate segments 3–10 cm long and 1–4 cm wide, the terminal segment largest, margins serrate, apex attenuate, sparsely scabridulous on both surfaces; petiole 1–7 cm long. Capitula in terminal lax corymbose cymes; stalks of individual capitula 2–20 cm long; involucre 6–12 mm long; outer phyllaries 6–12, green, often reflexed, glabrous or pilose, inner phyllaries 8–10, free or partly connate; paleae to 8 mm long. Ray florets yellow, neuter, 8, tube 2–3 mm long, pilose, rays 20–35 mm long, 6–15 mm wide; disc florets yellow to brownish yellow, 7–8 mm long, glabrous. Achenes dark brown, oblong to obovoid, 4.5–9.5 mm long, strigose, margins narrowly callose-thickened; pappus of 2(–4) aristae 1–3.5 mm long, with few barbs near apex, both antrorse and retrorse on the same arista.

TANZANIA. Tanga District: Mlinga Peak, Nov. 1986, *Borhidi et al.* 86/483!; Morogoro District: N Uluguru Reserve above Morningside, June 1953, *Semsei* 1251!; Iringa District: Mufindi, Rufuna Forest Reserve, Nov. 1982, *Macha* 136!
DISTR. **T** 3, 5–7; not known elsewhere
HAB. Forest clearings and margins, secondary vegetation in forest zone; 600–2300 m
USES. Used for string; leaves edible (*Wallace*); minor medicine against pneumonia & cough (*Koritschoner*)
CONSERVATION NOTES. Least concern (LC)

SYN. *Coreopsis frondosa* O.Hoffm. in E.J. 20: 414 (1895), *non Bidens frondosa* L. (1753)
   Bidens dolosa Sherff in Field Mus. Nat. Hist., Bot. Ser. 16: 596 (1937). Type: Tanzania, Iringa District: E Mufindi, *Greenway* 3475 (K!, holo., EA!, iso.)
   B. phelloptera Sherff in Bot. Gaz. 92: 204 (1931) & in Field Mus. Nat. Hist., Bot. Ser. 16: 631, fig. 189/j–s (1937). Type: Tanzania, Lushoto District: Usambara Mts, Mt Gonja, *Busse* 2257 (B†, holo., EA!, lecto., chosen by Mesfin)
   B. lynesii Sherff in Field Mus. Nat. Hist., Bot. Ser. 17: 590 (1939); T.T.C.L.: 148 (1949). Type: Tanzania, Iringa District: Dabaga, *Lynes* 4 (K!, holo.)
   B. magnifolia Sherff var. *versuta* Sherff in Amer. J. Bot. 34: 147 (1947). Type: Tanzania, Mpwapwa, *Hornby* 896 (EA!, holo., BR!, K!, iso.)

8. **Bidens baumii** (*O.Hoffm.*) *Sherff* in Bot Gaz. 59: 309 (1915) & in Field Mus. Nat. Hist., Bot. Ser. 16: 589, fig. 158 (1937); Lisowski, Asterac. Fl. Afr. Centr. 1: 163, fig. 37 (1991); Mesfin in K.B. 48: 460 (1993). Type: Angola, Mambunda, *Baum* 883 (B†, holo., W!, lecto., BM!,G!, K!, M!, selected by Rayner in 1992)

Perennial herb, 0.6–1.3 m high, erect to decumbent; stem single from a woody tuberous rootstock, little branched, glabrous. Leaves lanceolate to ovate, 5–15 cm long, 1–3 cm wide, base long attenuate but wider near stem, margins coarsely dentate to shallowly pinnately lobed, apex acute to attenuate, glandular-punctate, scabrid on both surfaces; pseudopetiole to 35 mm long. Capitula terminal, solitary or few in lax cymes; stalks of individual capitula to 16 cm long; involucre 8–11 mm long; outer phyllaries 8, green, pilose, the inner 8, pilose; paleae 7–8.5 mm long. Ray florets yellow, neuter, 6–8, tube to 3 mm long, rays 15–18 × 7–8 mm; disc florets yellow, 5.5–6 mm long, pilose. Achenes black, 6–11 mm long, pubescent; aristae 2, 1–2.5 mm long, without barbs or with barbs connate to arista, rarely absent.

TANZANIA. Bukoba District: Bugufi, Jan. 1936, *Chambers* K 52! & 55!; Ufipa District: Mbizi Forest, June 1980, *Hooper & Townsend* 1821! & Mbizi Mts, Apr. 1997, *Bidgood et al.* 3572!
DISTR. **T** 1, 4; Congo (Kinshasa), Rwanda, Burundi, Angola, Zambia, Malawi
HAB. Grassland, bushed grassland; 1800–2200 m

CONSERVATION NOTES. Least concern (LC)

SYN. *Coreopsis baumii* O.Hoffm. in Warb., Kunene–Sambesi Exped.: 419 (1903)
  C. scabrifolia Sherff in Bot. Gaz. 76: 86 (1923) & in Field Mus. Nat. Hist., Bot. Ser. 11: 397,
    fig. 158 (1936). Type: Congo (Kinshasa), Kundelungu, *Kassner* 2776 (B†, holo., BM!,
    lecto., BM!, K!, P!, Z!, chosen by Rayner in 1992)
  Bidens ruandensis Sherff in B.J.B.B. 13, 4: 285 (1935) & in Field Mus. Nat. Hist., Bot. Ser.
    16: 590 (1937). Type: Rwanda, Ruhengeri, *Scaetta* 427 (BR!, holo.)
  B. somaliensis Sherff var. *bukobensis* Sherff in Field Mus. Nat. Hist., Bot. Ser. 17: 584 (1939).
    Type: Tanzania, Ngara District: Bugufi, *Chambers* K21 (K!, holo.)

9. **Bidens crocea** *O.Hoffm.* in Bol. Soc. Brot. 10: 177 (1892), as *croceus*; Sherff in
Field Mus. Nat. Hist., Bot. Ser. 16: 585, fig. 161/a–g (1937); Lisowski, Asterac. Fl. Afr.
Centr. 1: 160 (1991); Mesfin in K.B. 48: 463 (1993). Type: Angola, Benguela,
Welwitsch 3964 (B†, holo., BM!, lecto., C!, BR!, G!, LISU!, K!, M!, P!, chosen by
Rayner, 1992)

Perennial herb 0.5–1.2 m high, rarely annual; stem single or several, glabrous, from
thickened woody rootstock. Leaves pinnatifid with 3–5 segments or bipinnatifid with
few side lobes, 5–15 cm long, segments linear, 1–6 cm long, 0.5–2 mm wide, glabrous;
petiole 0.5–7 cm long. Capitula terminal and solitary; stalks of individual capitula to
12 cm long; involucre 8–10 mm long; outer phyllaries 6–8, green, linear, glabrous or
sparsely pubescent, the inner phyllaries 8, oblong, glabrous but for the apex; paleae
linear. Ray florets golden yellow, neuter, 5–8, tube glabrous, rays 10–20 × 3–5 mm; disc
florets yellow, 3.8–5 mm long, glabrous or sparsely puberulous. Achenes dark brown
or black, flat, 9–18 mm long, glabrous or pilose on margin, curved near the apex
when fully mature; aristae 2(–3), to 4.3 mm long, retrorsely barbed.

TANZANIA. Iringa District: between Matanana and Malangali, Mar. 1962, *Polhill & Paulo* 1890!
  & 5 km from Mafinga on Madibira road, Mar 1988, *Bidgood et al.* 578!; Ufipa District:
  Sumbawanga, Chapota, Mar. 1957, *Richards* 8516!
DISTR. **T** 3, 4, 7; Congo (Kinshasa), Angola, Zambia
HAB. Open woodland; 1650–1900 m
USES. None recorded on specimens from our area
CONSERVATION NOTES. Least concern (LC)

SYN. *B. bequaertii* De Wild., Repert. Sp. Nov. 13: 204 (1914); Sherff in Field Mus. Nat. Hist., Bot.
    Ser. 16: 582, fig. 148/h–m (1937); Lisowski, Asterac. Fl. Afr. Centr. 1: 153 (1991). Type:
    Congo (Kinshasa), Lubumbashi, *Bequaert* 270 (BR!, holo.)
  B. palustris Sherff in Bot. Gaz. 76: 148 (1923) & in Field Mus. Nat. Hist., Bot. Ser. 16: 568,
    fig. 150/a–g (1937); Lisowski, Asterac. Fl. Afr. Centr. 1: 152 (1991). Type: Congo
    (Kinshasa), Kundelungu, *Kassner* 2599 (B†, holo., K!, lecto., BR!, HBK!, P!, Z!, iso.,
    chosen by Rayner, 1992)
  B. phalangiphylla Sherff in Bot. Gaz. 76: 152 (1923) & in Field Mus. Nat. Hist., Bot. Ser. 16:
    579, fig. 156/a–h (1937). Type: Tanzania, Lushoto District: Usambara, Doda, *Holst* 2967
    (B†, holo.)
  B. kasaiensis Lisowski in B.J.B.B. 58: 259 (1988) & Asterac. Fl. Afr. Centr. 1: 159 (1991).
    Type: Congo (Kinshasa), Kwango, Tshilualua R., *Devred* 3523 (BR!, holo.)

NOTE. A specimen from **T** 8: Songea District: Mbinga, Ndondo Hill, May 1991, *Ruffo & Kisena*
3282! is close to *B. crocea*, and is an annual plant. More material would be desirable to
determine the status of this taxon.

10. **Bidens ochracea** (*O.Hoffm.*) *Sherff* in Bot. Gaz. 76: 158, fig. 3 (1923); Mesfin in
K.B. 48: 465 (1993). Type: Tanzania, Iringa District: Bueni [Bweni], *Goetze* 731 (B†,
holo., BM!, lecto., chosen by Mesfin)

Perennial herb, up to 1.5 m high; stems several from the thickened or tuberous
rootstock, hardly branched, glabrous. Leaves ovate in outline, deeply pinnatipartite
with 3–7 segments or bipinnatipartite, simple in upper leaves, 4–10(–18) cm long,

2–7 cm wide, leaf segments linear, thick, 1.5–7 mm wide, margins scabrid or pubescent and slightly revolute, apex acute or mucronate; petiole to 5 cm long. Capitula 1–3 at apex of branches; stalks of individual capitula to 15 cm long; involucre 8–13 mm long; outer phyllaries 8–10, green, pilose at base, inner phyllaries 8–9, yellow with a red stripe, glabrous but for the apex; paleae to 10 mm long. Ray florets orange or golden yellow, neuter, 8–9, tube ± 3 mm long, pubescent, rays 20–35 × 10–14 mm; disc florets pale cream yellow, sparsely pubescent near base. Achenes black, narrowly obovoid, 6–7 mm long, narrowly callose-thickened, pilose; aristae 2, to 2 mm long, nude or sparsely antrorsely barbed.

TANZANIA. Ufipa District: 10 km on Tatanda–Mbala road, Apr. 1997, *Bidgood et al.* 3419!; Chunya District: Kepembawe, path to Muzibini, Mar. 1965, *Richards* 19830!; Iringa District: Dabaga Highlands, Kibengu, Feb. 1962, *Polhill & Paulo* 1483!
DISTR. **T** 4, 7; Zambia, Malawi
HAB. Woodland, grassland; 1050–2550 m
USES. None recorded on specimens from our area
CONSERVATION NOTES. Not uncommon within a common habitat: least concern (LC)

SYN. *Coreopsis ochracea* O.Hoffm. in E.J. 30 (431) (1901); Sherff in Bot. Gaz. 80: 375, fig. 19 (1925) & in Field Mus. Nat. Hist., Bot. Ser. 11: 380, fig. 3 (1936)
    C. cosmophylla Sherff in Bot. Gaz. 76: 90, fig. 9/h–n (1923). Type: Tanzania, Ufipa District, *Muenzner* 159 (B†, holo.)
    C. ochraceoides Sherff in Amer. J. Bot. 42: 564 (1955). Type: Tanzania, Iringa District: Ukinga & Ubena areas, *Ward* U31 (K!, holo., EA!, iso.)

11. **Bidens ugandensis** (*S.Moore*) *Sherff* in Bot. Gaz. 59: 309 (1915) & in Field Mus. Nat. Hist., Bot. Ser. 16: 583, fig. 159 (1937); U.K.W.F.: 467 (1974); Mesfin in Symb. Bot. Upsal. 24, 1: 98, fig. 49, 50 (1984); Lisowski, Asterac. Fl. Afr. Centr. 1: 157 (1991); Mesfin in K.B. 48: 466 (1993); U.K.W.F. ed. 2: 218, t. 90 (1994). Type: Uganda, Masaka District: Buddu, *Scott Elliot* 7520 (BM!, holo., K!, iso.)

Perennial or rarely annual herb 30–135 cm high, usually with several erect stems from the tuberous rootstock, stems little branched, sparsely pilose. Leaves simple and narrowly elliptic to linear, or pinnately 3–7-lobed to bipinnatifid with linear segments, 3–15 cm long, the leaf or segments 1–6(–15) mm wide, sessile, apex acute, glabrous except for the revolute margins which have 2 rows of short scabrid hairs, rarely leaf scabrid; pseudo-petiole to 7.5 cm long. Capitula terminal and few in lax cymes; stalks of individual capitula to 12 cm long; involucre 6–13 mm long; outer phyllaries green, 8–16, glabrous to pubescent, the inner 8–13 phyllaries with scarious margins, glabrous or nearly so; paleae thick, yellowish, 6–9 mm long. Ray florets yellow to orange, neuter, 8–16, tube 2–2.5 mm long, rays (10–)20–30 × (3–)8–12 mm; disc florets yellow to golden yellow, 6–8 mm long. Achenes black, ellipsoid, 4.5–11.5 mm long, strigose; aristae 2, 0.5–2 mm long, divergent, nude or with a few retrorse barbs near the apex. Fig. 171 (page 791).

UGANDA. West Nile District: Mt Otze, Oct. 1959, *Scott* in EA 11774!; Teso District: Serere, Dec. 1931, *Chandler* 338!; Masaka District: 7 km W of Kakuto, May 1971, *Lye & Katende* 6069!
KENYA. Trans-Nzoia District: Kitale, Mar. 1953, *Bogdan* 3694!; Uasin Gishu District: Kipkarren area, Mar. 1932, *Brodhurst-Hill* in *Napier* 703!; Kisumu-Londiani District: Muhoroni, 1907, *Battiscombe* 83!
TANZANIA. Buha District: Keza Mission 20 km NE of Kibondo, May 1994, *Bidgood & Vollesen* 3253!
DISTR. **U** 1–4; **K** 2, 3, 5; **T** 4; Cameroon, Central African Republic, Congo (Kinshasa), Burundi, Sudan
HAB. Grassland or wooded grassland, woodland; 1050–1950(–2400) m
USES. Formerly used as substitute for salt (*Chancellor*)
CONSERVATION NOTES. Least concern (LC)

SYN. *Coreopsis linearifolia* Oliv. & Hiern, F.T.A. 3: 390 (1877). Type: Sudan, Djur-land near Agada, *Schweinfurth* 27 (K!, holo., BM!, G!, W!, iso.), non *B. linearifolia* Sch.Bip. (1856)
    C. ugandensis S.Moore in J.L.S. 35: 347 (1902)

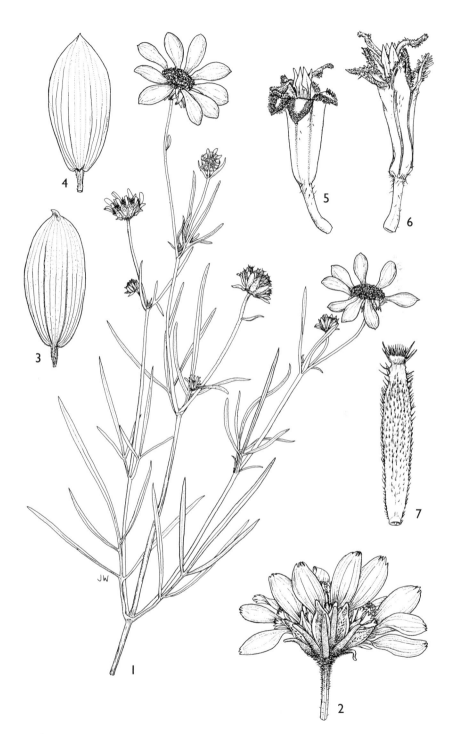

FIG. 171. *BIDENS UGANDENSIS* — **1**, habit, × ²/₃; **2**, capitulum, × 1.5; **3**, ray floret, abaxial, × 2; **4**, ray floret, adaxial, × 2; **5**, disc floret, × 8; **6**, disc floret, × 8; **7**, achene, × 5. 1 & 2–5 from *Scott* 11774, 2 & 7 from *Chancellor* 338, 6 from *Dawe* 914. Drawn by Juliet Williamson.

Bidens linearifolia (Oliv. & Hiern) Sherff in Bot. Gaz. 70: 109 (1920), *nom. illegit., non B. linearifolia* Sch.Bip. (1856)

B. *schweinfurthii* Sherff in Bot. Gaz. 59: 309 (1915) & in Field Mus. Nat. Hist., Bot. Ser. 16: 612, fig. 177 (1937); F.P.S. 3: 13 (1956); Maquet in Fl. Rwanda 3: 634 (1985). Type as for *C. linearifolia* Oliv. & Hiern

B. *rogersii* Sherff in Bot. Gaz. 81: 52 (1926) & in Field Mus. Nat. Hist., Bot. Ser. 16: 614, fig. 179 (1937). Type: Congo (Kinshasa), Sakania, *Rogers* 10046 (K!, holo.)

B. *chandleri* Sherff in Field Mus. Nat. Hist., Bot. Ser. 17: 597 (1939). Type: Uganda, Teso District: Serere, *Chandler* 939 (K!, holo.)

Coreopsis ochracea O.Hoffm. var. *lugardii* Sherff in Field Mus. Nat. Hist., Bot. Ser. 17: 608 (1939). Type: Kenya, West Suk District: Kapenguria, *Lugard* K4 (K!, holo.)

Bidens ugandensis (S.Moore) Sherff var. *longisquama* Sherff in Bot. Leafl. 5: 16 (1951). Type: Kenya, West Suk District: Kapenguria, *Jack* 324 (EA!, holo., K!, iso.)

B. *ugandensis* (S.Moore) Sherff var. *schweinfurthii* (Sherff) Lisowski, Asterac. Fl. Afr. Centr. 1: 158 (1991). Type as for *Coreopsis linearifolia* Oliv. & Hiern

B. *ugandensis* (S.Moore) Sherff var. *rogersii* (Sherff) Lisowski, Asterac. Fl. Afr. Centr. 1: 158 (1991)

12. **Bidens ternata** (*Chiov.*) *Sherff* in Bot. Gaz. 90: 391 (1930) & in Field Mus. Nat. Hist., Bot. Ser. 16: 626, fig. 187 (1937); Mesfin in Symb. Bot. Upsal. 24, 1: 72, fig. 36, 37 (1984); Lisowski, Asterac. Fl. Afr. Centr. 1: 168 (1991); Mesfin in K.B. 48: 467 (1993); U.K.W.F. ed. 2: 217, t. 90 (1994). Type: Ethiopia, Gondar, above Asoso, *Chiovenda* 2581 (FT!, holo.)

Perennial herb, 0.6–2 m high, stems erect, several from a woody rootstock, ± glabrous. Leaves opposite or ternate, sessile and simple, sometimes deeply lobed or pinnately 3-lobed, 3–22 cm long, 0.6–4 cm wide, base cuneate to cordate, margins crenate, serrate, deeply toothed or lobed, apex acute, with 2 lines of scabrid hairs on margins, otherwise scabrid to glabrous. Capitula few in lax terminal cymes; stalks of individual capitula to 8 cm long; involucre 5–22 mm long; outer phyllaries 11–30, green, inner phyllaries 13–15, orange, all pubescent outside; paleae yellow, 4–9 mm long. Ray florets yellow to orange-yellow, pistillate and fertile, 12–15, tube 1.5–2.3 mm long, rays 14–30 mm long; disc florets yellow, 3–5 mm long. Achenes black or brown, ellipsoid, 3–6.4 mm long; aristae 2, 0.5–4 mm long, antrorsely barbed.

UGANDA. Karamoja District: Mt Moroto, June 1970, *Lye et al.* 5627!; Toro District: Itwara Forest, Sep. 1941, *A.S.Thomas* 3937!; Ankole District: Igara, Mar. 1939, *Purseglove* 613!
KENYA. S Elgon, June 1937, *Tweedie* 394!; N Kavirondo District: Kakamega Forest, Oct. 1953, *Drummond & Hemsley* 4753!; Kisumu-Londiani District: Tinderet, Nov. 1933, *Mainwaring* 6086!
TANZANIA. Bukoba District: Ihangiro and Karagwe, date unknown (pre-1923), *Meyer* 532
DISTR. **U** 1, 2; **K** 3, 5; **T** 1; Cameroon, Congo (Kinshasa), Sudan, Ethiopia
HAB. Grassland, bushland, forest clearings and margins, ruderal ; 1500–2700 m
USES. None recorded on specimens from our area
CONSERVATION NOTES. Least concern (LC)
VARIATION. Mesfin (1984:75) distinguishes three varieties, based on length of petiole and shape of the leaf subtending the inflorescence. Var. **angustifolia** (Sherff) Mesfin [*Bidens superba* Sherff var. *angustifolia* Sherff in Bot. Leafl. 7: 20 (1952); Type: Uganda, Kigezi District: Ruhinda, *Purseglove* 3550 (K!, holo., EA!, MO!, iso.)] is distinct by narrower leaves subtending the inflorescence, and is known from only the type. All other FTEA material belongs to var. **ternata**. A third variety (var. *vatkei*) is only known from Ethiopia.

Basally lobed leaves are usually found in grassland or wooded grassland, while the 3-lobed leaves are found in secondary habitats and woodland or riverine margins.

SYN. *Coreopsis ternata* Chiov., Ann. Bot. Roma 9: 74 (1911)
C. *chrysantha* Vatke var. *simplicifolia* Vatke in Linnaea 39: 500 (1875); Oliv. & Hiern, F.T.A. 3: 388 (1877). Type: Ethiopia, Gondar, Dewari, *Schimper* s.n. (B†, holo., BM!, lecto., K!, iso., chosen by Mesfin)
C. *simplicifolia* (Vatke) Engl. in Abh. Konigl. Akad. Wiss.: 435 (1892)
C. *neumannii* Sherff in Bot. Gaz. 76: 85 (1923) & in Field Mus. Nat. Hist., Bot. Ser. 11, 6: 378 (1936). Type: Ethiopia, Gamo Gofa, Gardulla, *Neuman* 134 (B†, holo.)

C. bracteosa Sherff in Bot. Gaz. 76: 88 (1923) & in Field Mus. Nat. Hist., Bot. Ser. 11, 6: 444 (1936). Type: Tanzania, Bukoba District: Ihangiro and Karagwe, *Meyer* 532 (B†, holo.)

Bidens dielsii Sherff in Bot. Gaz. 90: 388 (1930) & in Field Mus. Nat. Hist., Bot. Ser. 16: 622, fig. 185 (1937) & in Amer. J. Bot. 34: 153 (1947); E.P.A.: 1137 (1967). Type as for *Coreopsis chrysantha* Vatke var. *simplicifolia* Vatke

B. neumannii Sherff in Bot. Gaz. 90: 394 (1930) & in Field Mus. Nat. Hist., Bot. Ser. 16: 625, fig. 186 (1937). Type: Ethiopia, Gamo Gofa, Gardulla, *Neuman* 135 (B†, holo.)

B. dielsii Sherff var. *medusoides* Sherff in Bot. Gaz. 91: 311 (1931) & in Field Mus. Nat. Hist., Bot. Ser. 16: 623 (1937) & in Amer. J. Bot. 34: 153 (1947). Type: Uganda, Ruwenzori, Wasa, *Maitland* 1004 (K!, holo.)

Coreopsis rueppellii Sch.Bip. var. *simplicifolia* (Vatke) Chiov. in Savoia-Aoste, Esplor. Uabi–Uebi Scebeli: 416 (1923)

Bidens superba Sherff in Amer. J. Bot. 22: 707 (1935) & in Field Mus. Nat. Hist., Bot. Ser. 16: 624 (1937) & in Amer. J. Bot. 34: 153 (1947); U.K.W.F.: 466 (1974). Type: Congo (Kinshasa), Nioka, *Jurion* 41 (BR!, holo.)

B. superba Sherff var. *brachycarpa* Sherff in Field Mus. Nat. Hist., Bot. Ser. 17: 599 (1939) & in Amer. J. Bot. 34: 153 (1947). Type: Uganda, Kigezi District: Mabungo Hill, *Snowden* 1594 (K!, holo.)

B. dielsii Sherff var. *incisior* Sherff in Field Mus. Nat. Hist., Bot. Ser. 17: 599 (1939). Type: Uganda, Karamoja District: Mt Moroto, *Liebenberg* 367 (K!, holo.)

B. gardullensis Cufod., E.P.A.: 1137 (1967). Type as for *Bidens neumannii* Sherff

13. **Bidens rueppellii** (*Walp.*) *Sherff* in Bot. Gaz. 90: 389 (1930) & in Field Mus. Nat. Hist., Bot. Ser. 16: 616, fig. 181 (1937); E.P.A.: 1141 (1967); U.K.W.F.: 466 (1974); Mesfin in Symb. Bot. Upsal. 24, 1: 81, fig. 42 (1984) & in K.B. 48: 469 (1993); U.K.W.F. ed. 2: 218 (1994). Type: Ethiopia, Gondar, Simen, *Rüppell* s.n. (P!, holo., FR!, iso.)

Perennial herb or subshrub 1–2(–3) m high, erect, with several stems from a thickened rootstock; stems sparsely hispid. Leaves ovate in outline, pinnatipartite to rarely bipinnatisect, up to 14 cm long, up to 9 cm wide, the primary segments ovate to narrowly lanceolate, margins serrate, hispid all over, rarely glabrous but for the margin. Capitula in few-headed lax terminal cymes; stalks of individual capitula to 12 cm long; involucre 5–11 mm long; outer phyllaries 9–13, green, pubescent, inner phyllaries 9–15, light brown, puberulous; paleae 5–8 mm long. Ray florets yellow, pistillate and fertile, 11–13, tube 1–2.4 mm long, rays 13–27 mm long; disc florets 3.8–5.5 mm long. Achenes black or grey, 3.4–6 mm long, strigose; aristae 2, 0.3–2.1 mm long, antrorsely barbed, sometimes falling early.

KENYA. Kericho District: SW Mau, Sambret, Oct. 1961, *Kerfoot* 2988!; Nakuru District: Nyahururu Falls, *Pierce* in EA 2568; Naivasha District: Kinangop Plateau, *Battiscombe* 945!
DISTR. **K** 1, 3, 5; Ethiopia
HAB. Not given for East African specimens, in Ethiopia on rocky mountain slopes, forest margins, river-banks; 2250–2800 m
USES. None recorded on specimens from our area
CONSERVATION NOTES. Least concern (LC)

SYN. *Coreopsis rueppellii* Walp. in Repert. Bot. Syst. 6: 163 (1846)
Verbesina rueppellii (Walp.) A.Rich., Tent. Fl. Abyss. 1: 410 (1848)
Coreopsis abyssinica Sch.Bip. var. *latisecta* Vatke in Linnaea 39: 499 (1875), *nom. nud.* pro parte
C. glaucescens Oliv. & Hiern, F.T.A. 3: 389 (1877). Type: Ethiopia, Tigray, Hedscha, *Schimper* 329 (K!, holo., BM!, C!, S!, US!, W!, WU!, Z!, iso.)
C. bella Hutch. in K.B. 1907: 364 (1907). Type: cultivated material from seed collected in Kenya by *Diespecker* (K!, holo.)
C. feruloides Sherff in Bot. Gaz. 80: 389 (1925) & in Field Mus. Nat. Hist., Bot. Ser. 11, 6: 379 (1936). Type: Kenya, Naivasha District: Kinangop Plateau, *Battiscombe* 945 (K!, holo., EA!, P!, iso.)
Bidens setigera (Sch.Bip.) Sherff var. *lobata* Sherff in Bot. Gaz. 92: 311 (1931) & in Field Mus. Nat. Hist., Bot. Ser. 16: 628 (1937). Type: cultivated material from seed collected in Kenya by *Diespecker* (K!, holo.)

Coreopsis rueppellii Walp. forma *angustisecta* Chiov. in Savoia-Aoste, Esplorazione dell Uabi–Uebi Scebeli: 416 (1932). Type: Ethiopia, Arssi NE of Mt Laggio, *Basile* 223 (TO, holo.)

Bidens articulata Sherff in Bot. Gaz. 94: 591 (1933) & in Field Mus. Nat. Hist., Bot. Ser. 16: 620 (1936). Type as for *C. glaucescens* Oliv. & Hiern

Coreopsis rueppellii Sch.Bip. var. *incisa* Sherff in Bot. Leafl. 5: 16 (1951). Type: Kenya, Nakuru District: Nyahururu Falls, *Pierce* in EA 2568 (EA!, holo.)

Bidens rueppeloides Sherff in Bot. Leafl. 5: 20 (1951), *nom.illegit. superfl.* Type as for *C. feruloides* Sherff

14. **Bidens whytei** *Sherff* in Bot. Gaz. 76: 145 (1923) & in Field Mus. Nat. Hist., Bot. Ser. 16: 564, fig. 148/a–g (1937); Mesfin in K.B. 48: 470 (1993). Type: Kenya, near Nairobi, *Whyte* s.n. (K!, holo.)

Perennial herb (sometimes annual?), procumbent or prostrate or scrambling with ascending shoots, 25–120 cm high; stems glabrous or pilose. Leaves ovate in outline, pinnately or sub-bipinnately compound, 1.5–8 cm long, 1–5.5 cm wide, segments ± linear and 1–3.5 mm wide, acute, glabrous or pilose, aromatic-glandular; pseudopetiole to 13 mm long. Capitula few in lax terminal cymes; stalks of individual capitula to 15 cm long; involucre 3–5 mm long; outer phyllaries 5–8, widened and hairy at apex, green with dark red centre, inner phyllaries 6–8, oblong, orange-brown with scarious margins; paleae orange-brown, 4–4.5 mm long. Ray florets bright yellow, neuter, 8, rays 10–12 mm long; disc florets orange-yellow, 3.5–4 mm long. Achenes dark brown or black, compressed, graded monomorphic, setose, 3–4 mm long (outer) or 7–10 mm long (inner); aristae 2, 1–2 mm long, retrorsely barbed near the apex.

KENYA. Laikipia District: Nyahururu [Thomsons Falls], Apr. 1956, *Verdcourt* 1479!; Machakos District: Kilungu, Kauti, Aug. 1971, *Mwangangi* 1718!; Masai District: Oltarakwai, *Glover et al.* 2517!

TANZANIA. Moshi District: Mengwe area, Kwa Kinabo, *Volkens* 365!

DISTR. **K** 1, 3, 4, 6, 7; **T** 2; not known elsewhere

HAB. Grassland, riverine, forest margins, secondary bushland (?always in dry forest zone?); (300–)1700–2400 m

USES. None recorded on specimens from our area

CONSERVATION NOTES. Least concern (LC)

SYN. *B. hoffmannii* Sherff in Bot. Gaz. 76: 146, fig. 12/h–n (1923) & in Field Mus. Nat. Hist., Bot. Ser. 16: 560, fig. 151/j–p (1937); T.T.C.L.: 148 (1949). Type: Tanzania, Moshi District: Mengwe areaa, Kwa Kinabo, *Volkens* 365 (G!, holo., BM!, iso.)

B. hoffmannii Sherff var. *angustata* Sherff in Field Mus. Nat. Hist., Bot. Ser. 17: 595 (1939). Type: Kenya, Uasin Gishu District: Turbo-Soy road, Brodhurst-Hill 215 (K!, holo.)

B. angustata (Sherff) Sherff in Amer. J. Bot. 34: 154 (1947). Type as for *B. hoffmannii* Sherff var. *angustata* Sherff

B. palustris Sherff var. *nematomera* Sherff in Bot. Leafl. 5: 16 (1951). Type: Kenya, Uasin Gishu District: Kipkarren, *Brodhurst-Hill* 215 (EA!, holo.)

NOTE. It is unclear whether the types of *B. hoffmannii* Sherff var. *angustata* and *B. palustris* Sherff var. *nematomera* are the same, or different taxa collected under the same number.

15. **Bidens cinerea** *Sherff* in Bot. Gaz. 59: 302 (1915) & in Field Mus. Nat. Hist., Bot. Ser. 16: 584, fig. 160 (1937); Mesfin in K.B. 48: 471 (1993); U.K.W.F. ed. 2: 217 (1994). Type: Tanzania, Kilimanjaro, *Smith* s.n. (K!, holo.)

Perennial or rarely annual herb with erect or ascending stems 30–100 cm long or high from woody rootstock; stem erect or decumbent, puberulous. Leaves grey-green, ovate in outline, imparipinnate, 2–7 cm long, 1–4 cm wide, leaf segments dentate to pinnatilobed, minutely spinulose-setose to scabrid to almost glabrous. Capitula terminal and solitary; stalks of individual capitula to 18 cm long; involucre

6–8 mm long; outer phyllaries 8, green, hispid, inner phyllaries with yellow margins; paleae yellow with maroon lines, 4–5 mm long. Ray florets yellow, neuter, 8, rays 15–17 × 7–8 mm, striate; disc florets yellow or brown, 3–3.5 mm long. Achenes black, linear, 7–10 mm long, setulose; pappus of 2(–3) retrorsely barbed aristae 1–2 mm long.

KENYA. Northern Frontier District: Isiolo–Mado Gashi road, N of Lochongale Mt, June 1977, *Gillett* 21349!; Kiambu District: Muguga, July 1961, *Verdcourt* 3198!; Masai District: Chyulu Hills, above Ol Doinyo Wuas Lodge, Oct. 1989, *Luke* 1966!
TANZANIA. Mbulu District: Mbulumbul, June 1945, *Greenway* 7468!; Arusha District: Arusha National Park, Londuka, Feb. 1969, *Richards* 23953!; Kilimanjaro, Ol Molog, June 1946, *Greenway* 7822!
DISTR. **K** 1, 4, 6; **T** 2; not known elsewhere
HAB. Dry forest, forest glades, dry bushland, grassland or almost bare rock; may be locally common; 900–1950 m
USES. None recorded on specimens from our area
CONSERVATION NOTES. Least concern (LC)

SYN. *B. lineata* Sherff in Bot. Gaz. 76: 84 (1923). Type: Tanzania, Kilimanjaro, *Endlich* 117 (B†, holo.)
   B. cinerea Sherff var. *tricuspidata* Sherff in Amer. J. Bot. 22: 705 (1935). Type: Tanzania, Arusha District: N Meru, *Uhlig* 2038 (B†, holo.)
   B. lineata Sherff var. *tenuipes* Sherff in Bot. Leafl. 5: 16 (1951). Type: Kenya, Machakos District: Ithaba to Chyulu Hills, *Bally* 141 (EA!, holo.)
   B. cinereoides Sherff in Bot. Leafl. 5: 21 (1951). Type: Kenya, Kiambu District: Kikuyu, *van Someren* 1166 (EA!, holo., K!, iso.)

NOTE. Seems to grow only on lava or lava-derived soils – except the population at Muguga (**K** 4) which is in dry forest soil but probably not on lava.

16. **Bidens taylorii** (*S.Moore*) *Sherff* in Bot. Gaz. 59: 309 (1915) & in Field Mus. Nat. Hist., Bot. Ser. 16: 576 (1937), as *taylori*; Mesfin in K.B. 48: 472 (1993). Type: Kenya, Kilifi District: Rabai, *Taylor* s.n. (BM!, holo.)

Perennial herb, 40–150 cm high, erect or scrambling; stem with reddish streaks. Leaves ovate in outline, 1–2-pinnatipartite, 3–14 cm long, 1–8 cm wide, pinnae lanceolate, apex indurate-apiculate, with sparsely ciliate margins and puberulous on veins, densely glandular-punctate. Capitula solitary, terminal, peduncle 4–16 cm long; involucre 5–7 mm long; outer phyllaries 6–8, green, glabrous or sparsely hairy, inner phyllaries orange; paleae 4.5 mm long, yellow, striped. Ray florets yellow, neuter, 8, the rays 10–15 × 4–5 mm; disc florets yellow, tubular, 3–4 mm long. Achenes black, cylindrical, compressed, 4.5–7.5 mm long, setose; aristae yellow, to 2 mm long, retrorsely barbed near the apex.

KENYA. Kwale District: Cha Simba, Feb. 1953, *Drummond & Hemsley* 1071! & Shimba Hills, Jan. 1964, *Verdcourt* 3927! & Longomwagandi, Feb. 1968, *Magogo & Glover* 19!
TANZANIA. Mwanza District: Kome Is., Apr. 1961, *Carmichael* 823!; Tanga District: 34 km N of Tanga, July 1955, *Faulkner* 1679!; Uzaramo District: S of Kibaha, Aug. 1969, *Harris & Harris* 3185!
DISTR. **K** 7; **T** 1, 3, 6–8; not known elsewhere
HAB. Bushed grassland, grassland, secondary bushland, forest margins; 0–850 m
USES. Used as a vegetable (*Graham, Magogo*)
CONSERVATION NOTES. Least concern (LC)

SYN. *Coreopsis exaristata* O.Hoffm. var. *gracilior* O.Hoffm. in P.O.A. C: 414 (1895). Type: Tanzania, Uzaramo District: Marambo, Mkambo [Mgambo], *Stuhlmann* 6403 (B†, holo.)
   C. taylori S.Moore in J.B. 44: 22 (1906)
   Bidens gracilior (O.Hoffm.) Sherff in Bot. Gaz. 76: 84 (1923) & in Bot. Gaz. 81: 29 (1926) & in Field Mus. Nat. Hist., Bot. Ser. 16: 565 (1937)
   B. gracilior (O.Hoffm.) Sherff var. *ukerewensis* Sherff in Bot. Gaz. 92: 203 (1931) & in Field Mus. Nat. Hist., Bot. Ser. 16: 566 (1937). Type: Tanzania, Mwanza District: Ukerewe, *Uhlig* 19 (B†, holo.)

B. praecox Sherff in Bot. Gaz. 92: 450 (1931) & in Field Mus. Nat. Hist., Bot. Ser. 16: 632 (1937); Lisowski, Asterac. Afr. Centr.: 169 (1991). Type: Tanzania, Lindi District: Mayanga, *Busse* 2523 (B, lecto., EA, iso., selected by Rayner, 1992)

B. schimperi Sch.Bip. var. *brachyceroides* Sherff in Bot. Leafl. 5: 17 (1951). Type: Kenya, Kwale District: Shimba Hills, *van Someren* 43 (EA!, holo.)

17. **Bidens flagellata** (*Sherff*) *Mesfin* in Symb. Bot. Upsal. 24: 94, fig. 47 (1984); Lisowski, Asterac. Afr. Centr.: 162 (1991); Mesfin in K.B. 48: 473 (1993); U.K.W.F. ed. 2: 217 (1994). Type: Burundi, Gitega, *Muller* 50 (BR!, holo.)

Perennial herb 0.3–1.2 m high or long; stems several to many from thickened rootstock, much branched, often straggling, reddish or pale purple, ± glabrous. Leaves ovate in outline, pinnatifid, up to 6 cm long, up to 4.5 cm wide, with 3–7 coriaceous linear segments to 4 × 0.2 cm, margins scabrid in upper part, apex callose-acute to acuminate; petiole to 2 cm long. Capitula in lax terminal corymbose cymes; involucre 5–9 mm long; outer phyllaries 6–12, green with pale margins, glabrous or pubescent; inner phyllaries yellow-orange, 6–8. Ray florets yellow, neuter, 8, rays 3.7–11(–13) × 1.2–2.5 mm, tube to 2.3 mm long; disc florets yellow, tubular, 3.4–5.5 mm long. Achenes black, narrowly obovoid, 3–5 mm long; aristae yellow, 2(–4), 1.5–3 mm long, antrorsely barbed or rarely smooth.

UGANDA. Karamoja District: Debasien, May 1939, *A.S.Thomas* 2938! & Mt Moroto, Imagit peak, Sep. 1956, *J.Wilson* 262! & Mt Moroto, June 1963, *Tweedie* 2652!

KENYA. Northern Frontier District: Mt Nyiru, Aug. 1971, *Archer* 707!; Nakuru District: Menengai Crater, Aug. 1967, *Mwangangi* 198!; Masai District: Oldebesi Lemoko, Apr. 1961, *Glover et al.* 801!

TANZANIA. Mbulu District: Nou Forest Reserve, June 1969, *Carmichael* 1651!; Lushoto District: Sunga–Manolo road, May 1953, *Drummond & Hemsley* 2796! & 3 km W of Mlalo, June 1953, *Drummond & Hemsley* 2954!

DISTR. **U** 1; **K** 1–6; **T** 2, 3; Ethiopia, Rwanda, Burundi

HAB. Grassland on rocky hillsides, secondary bushland, forest clearings, thicket edges, often on rocky slopes; 1800–2900 m

USES. Attractive to bees (*Bytebier*)

CONSERVATION NOTES. Least concern (LC)

SYN. *B. kirkii* (Oliv. & Hiern) Sherff var. *flagellata* Sherff in B.J.B.B. 13, 4: 287 (1935) & in Field Mus. Nat. Hist., Bot. Ser. 16: 562, fig. 145/g (1937) & in Bot. Leafl. 5: 15 (1951); Blundell, Wild Fl. E. Afr.: pl. 355 (1987)

B. kirkii (Oliv. & Hiern) Sherff var. *ciliato-vaginata* Cufod., Nuov. Giuorn. Bot. Ital. 50: 110 (1943) & E.P.A.: 1138 (1967). Type: Ethiopia, Mega, *Corradi* 1899 (FT!, holo.)

B. kirkii sensu T.T.C.L.: 148 (1949), *non* (Oliv. & Hiern) Sherff

18. **Bidens fischeri** (*O.Hoffm.*) *Sherff* in Bot. Gaz. 76: 158 (1923) & in Field Mus. Nat. Hist., Bot. Ser. 16: 553, fig. 142/a–h (1937); Mesfin in K.B. 48: 475 (1993). Type: Tanzania, *Fischer* 354 (B†, holo.); neotype: Rungwe District: Poroto Mts, Ngozi, *Richards* 6498 (K!, neo.)

Trailing or straggling herb to 2 m long; stem rooting at lower nodes, leafy throughout. Leaves deltoid in outline, bipinnatisect, 4–13 cm long, 2.5–8 cm wide, pinnae ovate to lanceolate, pinnules deeply lobed, apices acute or acuminate, with appressed white hairs above; petiole to 55 mm long. Capitula terminal, solitary or in pairs, radiate; peduncle to 9 cm long; involucre 6–13 mm long; outer phyllaries 8, green, pilose, inner phyllaries 8, brownish orange in fruit, glabrous; paleae to 12 mm long, glabrous. Ray florets yellow, neuter, 8–13, rays 30–35 × 5–7.5 mm; disc florets yellow, 5 mm long. Achenes black or dark brown, oblong, compressed, outer 4–5 mm long, inner 7–9 mm long, glabrous; aristae 2, yellow, 2–3.5 mm long, retrorsely barbed.

TANZANIA. Iringa District: Udzungwa Mountain National Park, Luhombero Mt, Oct. 2000, *Luke et al.* 6997! & Livalonge Tea Estate, Nyalawa R., Aug. 1971, *Perdue & Kibuwa* 11261!; Rungwe District: Ngozi, Poroto Mts, Oct. 1956, *Richards* 6498! (type)
DISTR. **T** 7; not known elsewhere
HAB. Moist forest and its clearings; 1800–2350 m
USES. None recorded on specimens from our area
CONSERVATION NOTES. Known from only four specimens; at least vulnerable (VU-D2)

SYN. *Coreopsis fischeri* O.Hoffm. in P.O.A. C: 414 (1895)

19. **Bidens lineariloba** *Oliv.* in T.L.S. 29: 99, fig. 60 (1872); Oliv. & Hiern, F.T.A. 3: 394 (1877); Sherff in Field Mus. Nat. Hist., Bot. Ser. 16: 587, fig. 143/g–o (1937); Mesfin in K.B. 48: 476 (1993). Type: Tanzania, Kahama/Tabora District: Mininga [Mininja], *Grant* 187 (K!, holo.)

Annual herb 60–150 cm high, erect, much branched; stem usually solitary, sometimes with reddish marks, striate. Leaves irregularly bipinnatipartite, 4–16 cm long, 2–9 cm wide, segments linear or lanceolate, 0.5–3 mm wide, apices apiculate, scabridulous; petiole absent or up to 35 mm long. Capitula in lax terminal cymes; peduncle to 8 cm long; involucre 7–15 mm long; outer phyllaries 8(–13), green, sparsely hairy, inner phyllaries yellow to orange with scarious margins, glabrous or nearly so; paleae 6–10 mm long, brown. Ray florets yellow or golden yellow, neuter, 6–8, rays 20–30 × 6–12 mm; disc florets yellow or brown, 3.5–5 mm long. Achenes ± dimorphic, the outer obovoid, 4–6.5 mm long, with 2–3 mm long aristae; the inner linear, 13–17 mm long, with two 5–6 mm long barbed aristae; all aristae retrorsely barbed.

UGANDA. Acholi District: Amiel, Nov. 1937, *Tothill* 2656!; Karamoja District: Mt Moroto base, Oct. 1952, *Verdcourt* 783!; Teso District: Omunyal Swamp, Sep. 1954, *Lind* 351!
KENYA. West Pokot District: Kongelai S, Sep. 1969, *Tweedie* 3694!
TANZANIA. Masai District: Loliondo, Ngosaro Sambu, July 1956, *Williams Sangai* 709!; Kondoa District: Kondoa township, May 1983, *Kisena* 78!; Iringa District: Nyang'oro, Apr. 1983, *Magogo* 2385!
DISTR. **U** 1, 3; **K** 2; **T** 1–5, 7; Congo (Kinshasa), Burundi
HAB. Old cultivations, roadsides, woodland, wooded grassland, bushland; 500–1800 m
USES. Minor medicinal against strokes (*Koritschoner*); used as a vegetable (*Hartley*); 'elephants eat it and become fierce' (*Dyson-Hudson*)
CONSERVATION NOTES. Least concern (LC)

SYN. *B. lineariloba* Oliv. var. *deminuta* Sherff in Amer. J. Bot. 42, 6: 562 (1955). Type: Tanzania, Shinyanga, *Bax* 205 (K!, holo.)

20. **Bidens odora** (*Sherff*) *T.G.J.Rayner* in Phytologia 72: 101 (1992); Mesfin in K.B. 48: 477 (1993). Type: Tanzania, Njombe District: Mdapo, *Semsei* 1655 (EA!, holo., K!, iso.)

Shrub 1.2 m high; stems erect, glabrous. Leaves bipinnatisect, 2–7 cm long, 1–4 cm wide, segments filiform to linear, 0.4–1 mm wide, apex acute, glabrescent; petiole to 25 mm long. Capitula in lax terminal cymes; peduncle 1–3 cm long; involucre 5–7 mm long; outer phyllaries 8, glabrous or nearly so, inner phyllaries 7–8; paleae pale brown, 5–6 mm long. Ray florets yellow, 7–8, rays 12–15 mm long; disc florets yellow, 5–5.5 mm long. Achenes grey or black, oblong-obovoid, 4–4.5 mm long, outer surface hispid, inner glabrous, with corky wings; aristae 2, yellow, 2–2.5 mm long, densely antrorsely barbed.

TANZANIA. Njombe District: Mdapo, Mar. 1954, *Semsei* 1655! (type)
DISTR. **T** 7; known from only the type
HAB. Grassland; ± 2200 m
USES. None recorded on specimens from our area

CONSERVATION NOTES. At least vulnerable (VU-D2)

SYN. *B. odora* Willd. & Schlecht., Suppl.: 56 (1814), *nom. nud.*
  Coreopsis odora Sherff in K.B. 11: 445 (1956)

21. **Bidens elliottii** (*S.Moore*) *Sherff* in Bot. Gaz. 59: 309 (1915) & in Field Mus. Nat. Hist., Bot. Ser. 16: 576, fig. 156/i–p (1937); Maquet in Fl. Rwanda 3: 634, fig. 195/2 (1985); Lisowski, Asterac. Fl. Afr. Centr. 1: 154, fig. 36 (1991); Mesfin in K.B. 48: 477 (1993). Type: Uganda, Toro District: Kivata, *Scott Elliot* 7724 (BM!, holo., K!, iso.)

Perennial herb, woody, erect or ascending, 0.9–4.5 m high; stem much branched near apex, striate, sparsely hairy. Leaves rhombic in outline, bi- to tri-pinnatisect, 8–17 cm long, 4–15 cm wide, segments linear to narrowly lanceolate, sparsely hairy on veins above, glandular-punctate; pseudopetiole to 45 mm long, winged. Capitula in terminal lax corymbose cymes; peduncle to 4 cm long; involucre 8–10 mm long; outer phyllaries 8–13, 9–15 mm long, sparsely hairy, inner phyllaries 8–13, orange-striped, pilose; paleae yellowish, orange-striped, 4–7 mm long. Ray florets yellow, pistillate and fertile, 7–13, rays 15–22 × 3–9 mm, tube 2.2 mm long; disc florets yellow, 4.9–7.3 mm long. Achenes black, oblong to obovoid, graded monomorphic, 3.6–8 mm long, margins narrowly winged; aristae 2(–3), 1.6–3 mm long, antrorsely barbed.

UGANDA. Ruwenzori Mts, Jan. 1967, *Magogo* 25! & idem, Kichuchu, July 1951, *Osmaston* 3866!; Kigezi District: Mt Muhavura, June 1939, *Purseglove* 787!
DISTR. **U** 2; Congo (Kinshasa), Rwanda
HAB. Grassland, woodland, heath, bamboo; 1800–3150 m
USES. None recorded on specimens from our area
CONSERVATION NOTES. Least concern (LC) – a fairly large number of collections from a common habitat, though no collections in Uganda since 1967 !

SYN. *Coreopsis elliotii* S.Moore in J.L.S. 35: 346 (1902); Z.A.E.: 381 (1911)
  C. mildbraedii Muschl. in Z.A.E.: 381 (1911). Type: Congo (Kinshasa), Ruwenzori, Butagu Valley, *Mildbraed* 2539 (B†, holo.)
  Bidens amoena Sherff in Bot. Gaz. 76: 144 (1923). Type: Uganda, Ruwenzori Mts, *Doggett* s.n. (K!, holo.)
  B. kigeziensis Sherff in Amer. J. Bot. 38, 1: 67 (1951). Type: Uganda, Kigezi District: Muhavura–Mgahinga saddle, *Purseglove* 2189 (EA!, holo., K!, iso.)

22. **Bidens elgonensis** (*Sherff*) *Agnew*, U.K.W.F.: 466 (1974); Mesfin in K.B. 48: 479 (1993); U.K.W.F. ed. 2: 217 (1994). Type: Uganda, Mt Elgon, *Dummer* 3304 (K!, holo.)

Perennial herb or shrub 0.4–1.2 m high; stems erect or leaning on other vegetation, branched near apex, thin and brittle, glabrous. Leaves obovoid in outline, pedately tripartite, 2–5 cm long, 1.5–4 cm wide, the segments lobed or serrate, 2–7(–12) mm wide, apex acute to apiculate, hispid especially on midrib and margins, black-punctulate; petiole absent or up to 5 mm long. Capitula terminal and solitary or few in lax cymes; peduncle 2–9 cm long; involucre 8–10 mm long; outer phyllaries 6–11, pilose, inner phyllaries ± 8, green with yellow margins, sparsely pilose; paleae light brown with purple apex, 6–7 mm long. Ray florets yellow or golden yellow, neuter, 8–13, rays 8–13 × 3–5 mm; disc florets yellow, ± 5 mm long. Achenes black, oblong-ellipsoid, 4–6 mm long, glabrous; pappus absent.

UGANDA. Karamoja District: Mt Moroto, Feb. 1959, *J.Wilson* 684!; Mbale District: Mt Elgon, Gabaralome, Dec. 1938, *A.S.Thomas* 2689! & Mt Elgon, Sasa Trail, Jan. 1997, *Wesche* 858!
KENYA. Trans Nzoia District: Elgon, Endebess Bluff, Aug. 1997, *Wesche* 1680! & Cherangani, Kaibibich, Nov. 1966, *Tweedie* 3381! & Embobut forest, Jan. 1971, *Tweedie* 3902!
DISTR. **U** 1, 3; **K** 3; endemic to Mt Moroto, Mt Elgon and the Cherangani Mts

HAB. Upland grassland and in *Erica* bushland, low scrub (regenerating) and *Hagenia-Hypericum* zone; 2500–3600 m

USES. None recorded on specimens from our area

CONSERVATION NOTES. No specific threats to these habitats apart from occasional fires; wide altitudinal range; least concern (LC)

SYN. *Coreopsis elgonensis* Sherff in Bot., Gaz. 80: 374 (1925); F.P.S. 3: 20 (1956); A.V.P.: 218 (1957)
    C. morotonensis Sherff in Amer. J. Bot. 34: 157 (1947). Type: Uganda, Karamoja District: Mt Moroto, *Dale* U261 (EA!, holo., K!, photo.)
    Bidens morotonensis (Sherff) Agnew, U.K.W.F.: 466 (1974)
    B. elgonensis (Sherff) Agnew subsp. *morotonensis* (Sherff) T.G.J.Rayner in Phytologia 72: 101 (1992)
    B. elgonensis (Sherff) Agnew subsp. *cheranganiensis* T.G.J.Rayner in Phytologia 72: 101 (1992). Type: Kenya, Cherangani Hills, *Tweedie* 4192 (EA!, holo.)

23. **Bidens nobiloides** *Sherff* in Bot. Leafl. 7: 20 (1952); Mesfin in K.B. 48: 480 (1993). Type: Kenya, Northern Frontier District: Mt Kulal, *Bally* 5616 (EA, holo., C, K!, iso.)

Annual or ?perennial herb 0.9–1.2 m high; stem branched in upper part, red-brown, sparsely hairy. Leaves deltoid in outline, bipinnatisect, 3–10 cm long, 2–4 cm wide, segments linear-lanceolate, 1–3 mm wide, margins revolute with ± 2 rows of hairs or glabrous, densely gland-dotted; petiole to 5 mm long, winged. Capitula terminal in lax cymes; peduncle to 6.5 cm long; involucre 5–9 mm long; outer phyllaries 8–10, sparsely hairy or glabrous, inner phyllaries 8–9, orange with yellow-green margins, pilose; paleae 5 mm long, striate. Ray florets yellow or orange-yellow, neuter, 8, rays 9–15 × 4–6 mm; disc florets yellow, 3–3.5 mm long. Achenes black, obovoid, 2–4.2 mm long, glabrous or pubescent, margins narrowly winged; pappus absent or with 2 antrorsely barbed aristae 1.5–2.5 mm long.

KENYA. Northern Frontier District: Mt Kulal North, Sep. 1944, *Joy Bally* 3914! & Marsabit, June 1959, *Joy Adamson* 9! & Ndoto Mts, Sirwan, Jan. 1959, *Newbould* 3378!

DISTR. **K** 1; endemic to Mt Marsabit, Mt Kulal, Ndoto Range and Mathews Range

HAB. Forest margins or clearing; (?1200–)1950–2400 m

USES. None recorded on specimens from our area

CONSERVATION NOTES. A fairly wide distribution in a common habitat; probably least concern (LC)

SYN. *Coreopsis scopulorum* Sherff in Bot. Gaz. 88: 302 (1929) & in Field Mus. Nat. Hist., Bot. Ser. 11: 362 (1936). Type: Kenya, Northern Frontier District: Mt Warges, *Heller* s.n. (US, holo.)
    Bidens kigeziensis Sherff var. *subsessilis* Sherff in Amer. J. Bot. 42, 6: 561 (1955). Type: Kenya, Northern Frontier District: Mt Kulal or Marsabit, *Martin* 202 (K!, holo., BR!, iso.)
    B. scopulorum (Sherff) T.G.J.Rayner in Phytologia 72: 99 (1992)

24. **Bidens negriana** (*Sherff*) *Cufod.*, E.P.A.: 1139 (1967); Mesfin in Symb. Bot. Upsal. 24: 56, fig. 30 (1984) & in K.B. 48: 481 (1993). Type: Ethiopia, Shewa, Lake Zwai, *Negri* 915b (FT!, holo.)

Annual herb 0.2–1.8 m high; stems often striped reddish below, simple or branched, sparsely pubescent or glabrous. Leaves ovate to deltoid in outline, bipinnatisect, 3–8 cm long, 2–4 cm wide, segments linear or narrowly lanceolate, 1–4.5 mm wide, hispid-pubescent mainly on main vein and margins; petiole to 20 mm long. Capitula terminal, in lax cymes; peduncle to 6 cm long; involucre 3–5 mm long; outer phyllaries 5–10, pubescent and punctate, inner phyllaries 6–8, orange-brown, sparsely hairy; paleae yellow-brown, 3–5 mm long. Ray florets yellow, neuter, 5–8, rays 4–12 × 3.5–5 mm, striate, tube 0.8–1.3 mm long; disc florets yellow, 2.5–4 mm long. Achenes black, 1.8–3.6 mm long, hairy or glabrous, margins often with corky wings; aristae 2, antrorsely barbed, (1–)2–4 mm long or sometimes absent.

UGANDA. West Nile District: Mt Otzi, Oct. 1959, *Scott* in *EA* 11775!; Acholi District: Mt Rom, no date, *Eggeling* 2383!; Mbale District: Buligenyi, Aug. 1932, *A.S.Thomas* 381!
DISTR. U 1, 3; Sudan, Ethiopia
HAB. Rocky outcrops, short grassland; 900–2200 m
USES. None recorded on specimens from our area
CONSERVATION NOTES. Least concern (LC)

SYN. *Coreopsis negriana* Sherff in Bot. Gaz. 90: 397 (1930) & in Field Mus. Nat. Hist., Bot. Ser. 11: 366 (1936)
    Bidens setigeroides Sherff in Bot. Gaz. 92: 310 (1931) & in Field Mus. Nat. Hist., Bot. Ser. 16: 630, fig. 189/a–i (1937). Type: Uganda, Mbale District: Bukedi, Buwalasi [Wallasi], *Snowden* 411 (K!, holo., BM!, iso.)
    B. setigeroides Sherff var. *munita* Sherff in Amer. J. Bot. 38, 1: 68 (1951). Type: Uganda, Acholi District: Chua, Agoro, *A.S.Thomas* 4385 (EA!, holo.)
    B. navicularia Sherff in Amer. J. Bot. 38, 1: 68 (1951). Type: Uganda, Acholi District: Kilak, Gulu, *A.S.Thomas* 4051 (K!, holo.)
    Coreopsis microglossa Sherff in Amer. J. Bot. 42, 6: 564 (1955). Type: Uganda, West Nile District: Oraba, *Hazel* 699 (K!, holo., BR!, iso.)
    Bidens natator I.Friis & Vollesen in K.B. 37: 468 (1982). Type: Sudan, Imatong Mts, between Gilo & Mt Konoro, *Friis & Vollesen* 322 (C!, holo., K!, KHF, iso.)

25. **Bidens grantii** (*Oliv.*) *Sherff* in Bot. Gaz. 59: 309 (1915) & in Field Mus. Nat. Hist., Bot. Ser. 16: 539 (1937); F.P.U.: 172, fig. 112 (1962); Maquet in Fl. Rwanda 3: 634, fig. 195/1 (1985); Blundell, Wild Fl. E. Afr.: pl. 353 (1987); Lisowski, Asterac. Fl. Afr. Centr. 1: 141 (1991); Mesfin in K.B. 48: 486 (1993); U.K.W.F. ed. 2: 218 (1994). Type: Tanzania, Bukoba District: Karagwe, *Grant* 448 (K!, holo., BR!, iso.)

Annual herb 0.5–2.5 m high; stem purplish, branched above, sparsely pilose to glabrous. Leaves narrowly deltoid in outline, bipinnatifid to pinnately 3–5-partite, 3.5–10.5 cm long, 2.5–7.5 cm wide, segments rhomboidally ovate or lanceolate, dentate, scabrid-pubescent above, pubescent beneath, glandular-punctate. Capitula terminal, in lax cymes; peduncle to 5.5 cm long; involucre 5–7 mm long; outer phyllaries 8–10, hispid, inner phyllaries 8, orange, sparsely pubescent or glabrous; paleae to 8 mm long. Ray florets yellow or orange-yellow, neuter, 8–10, rays 10–25 × 6–7 mm; disc florets yellow, 3.5–4 mm long. Achenes black, narrowly ellipsoid, 4–7 mm long, hairy in upper part; aristae 2, 1–2.2 mm long, antrorsely barbed.

UGANDA. Ankole District: Kyibego, Apr. 1970, *Katende & Lye* 130!; Kigezi District: Kambuga, Apr. 1941, *A.S. Thomas* 3806!; Masaka District: near Luunga in Jubiya Forest, May 1969, *Lye & Morrison* 2968!
KENYA. Nakuru District: Molo, 1930, *?R.W. Mettam* 255!
TANZANIA. Bukoba District: Mukigando, undated, *Ford* 765!; Ngara District: Rusomo Falls, Mar. 1960, *Tanner* 4768!; Mbulu District: Mt Hanang, Nov. 1932, *Geilinger* 3470!
DISTR. U 2, 4; K 3; T 1, 2; Congo (Kinshasa), Rwanda, Burundi
HAB. Grasslands, shallow soil over rock, weed of cultivation and ruderal sites; 1100–2100 m
USES. None recorded on specimens from our area
CONSERVATION NOTES. Least concern (LC)

SYN. *Coreopsis grantii* Oliv. in Trans. Linn. Soc. 29: 98, fig. 65 (1873); Oliv. & Hiern, F.T.A. 3: 388 (1877)
    Bidens kivuensis Sherff in Bot. Gaz. 96: 145 (1934). Type: Congo (Kinshasa), Mulungu, *Lebrun* 5467 (BR!, holo.)
    B. grantii (Oliv.) Sherff var. *dawei* Sherff in Bot. Gaz. 96: 145 (1934). Type: Uganda, Masaka District: Buddu, *Dawe* 243 (K!, holo.)
    B. grantii (Oliv.) Sherff var. *scaettae* Sherff in Bot. Gaz. 96: 145 (1934). Type: Rwanda, Nyabihus, *Scaetta* 2286 (BR!, holo.)
    B. kivuensis Sherff var. *armata* Sherff in Bot. Gaz. 96: 145 (1934). Type: Congo (Kinshasa), Nyagezi basin and surrounding mountains, *Boutakoff* 57 (BR!, holo.)
    B. steppia (Steetz) Sherff var. *humbertii* Sherff in Bot. Gaz. 96: 145 (1934). Type: Congo (Kinshasa), Beiga Mts, *Humbert* 7593 (BM!, holo., P!, iso.)
    Coreopsis quarrei Sherff in Bot. Gaz. 96: 145 (1934). Type: Congo (Kinshasa), Katuba–Lubumbashi [Elisabethville], *Quarré* 419 p.p. (BR!, holo.)

C. giorgii Sherff in Bot. Gaz. 96: 145 (1934). Type: Congo (Kinshasa), Katompe, *De Giorgi* 170 (BR!, holo.)

Bidens straminoides Sherff in Amer. J. Bot. 22: 706 (1935). Type: Rwanda, Mt Bohanga, *Scaetta* 2272 (BR!, holo.)

B. snowdenii Sherff in Field Mus. Nat. Hist., Bot. Ser. 17: 586 (1939). Type: Uganda, Kigezi District: Mabungo, *Snowden* 1631 (K!, holo.)

NOTE. Said to be abundant in Ankole and Kigezi.

26. **Bidens steppia** (*Steetz*) *Sherff* in Bot. Gaz. 76: 82 (1923) & in Field Mus. Nat. Hist., Bot. Ser. 16: 542, fig. 137/j–r (1937); Wild in Kirkia 6: 19 excl. syn. p.p. (1967); Mesfin in K.B. 48: 487 (1993). Type: Mozambique, Rios de Sena, *Peters* 58 (B†, holo.)

Annual herb 7–200 cm high; stem often reddish, branching towards apex, angular, pubescent. Leaves deltoid in outline, incised to bi- or tri-pinnatisect or pinnatifid, 2.5–35 cm long, 2–26 cm wide, segments linear to narrowly ovate, variously lobed or cleft, glandular-punctate and pilose with shiny silky hairs to glabrous; petiole absent or up to 90 mm long. Capitula terminal and solitary or in lax cymes; peduncle to 12 cm long; involucre 7–10 mm long; outer phyllaries 8–16, up to 30 mm long in fruit, pilose, inner phyllaries ± 8, yellow with red striations, pilose; paleae 5–10 mm long. Ray florets yellow or orange, neuter or pistillate, 8–13, rays 25–35 × 5–12 mm; disc florets yellow or orange-yellow, 2.8–4 mm long. Achenes black, narrowly obovoid, 6–10 mm long, bristly; aristae 2, 1–2.5 mm long, antrorsely barbed, rarely without aristae or aristae nude.

TANZANIA. Musoma District: Zanaki, June 1959, *Tanner* 4388!; Mpanda District: Kabungu, July 1948, *Semsei* 77!; Songea District: Msena near Litenga Hills, May 1972, *Shabani* 838!
DISTR. **U** (see note); **T** 1, 4, 5, 7, 8; Cameroon, Central African Republic, Congo (Kinshasa), Angola, Burundi, Zambia, Malawi, Mozambique, Zimbabwe
HAB. Grassland, floodplain vegetation, bushed grassland, woodland, weed of cultivation; 750–1950 m
USES. Minor medicinal against chest pain (*Semsei*) or stomach pain (*Tanner*)
CONSERVATION NOTES. Least concern (LC)

SYN. *Coreopsis steppia* Steetz in Peters, Reise Mossamb. Bot. 2: 496 (1863); Oliv. & Hiern, F.T.A. 3: 388 (1877)
  Bidens steppia (Steetz) Sherff var. *leptocarpa* Sherff in Bot. Gaz. 90: 392 (1930). Type: Tanzania, Rungwe District: Tukuyu [Langenburg], *Stolz* 729 (B†, holo., G!, lecto., BM!, K!, LE!, LU!, M!, S!, WAG!, Z!, iso., chosen by Rayner 1992)
  B. steppia (Steetz) Sherff var. *elskensii* Sherff in B.J.B.B. 13: 286 (1935). Type: Rwanda, Kitega, *Elskens* 257 (BR!, holo.)
  Coreopsis multiflora Sherff in B.J.B.B. 13: 292 (1935) & in Field Mus. Nat. Hist., Bot. Ser. 11, 6: 383 (1936). Type: Congo (Kinshasa), Shaba, Kansenia, *Thoreau* 211 (BR!, holo.)
  C. vulgaris Sherff in B.J.B.B. 13: 291 (1935). Type: Congo (Kinshasa), Shaba, Mukishi, *Becquet* 62 (BR!, holo.)
  C. injucunda Sherff in B.J.B.B. 13: 293 (1935) & in Field Mus. Nat. Hist., Bot. Ser. 11, 6: 393 (1936). Type: Congo (Kinshasa), Kipilia–Lubumbashi [Elisabethville], *Quarré* 1662 (BR!, holo.)
  Bidens uhligii Sherff in B.J.B.B. 13: 286 (1935)) & in Field Mus. Nat. Hist., Bot. Ser. 16: 542 (1937). Type: Tanzania, Mwanza District: Ukerewe Is., *Uhlig* V 46 (K!, holo.)
  B. steppia (Steetz) Sherff var. *inarmata* Sherff in Amer. J. Bot. 34: 152 (1947). Type: Tanzania, Rungwe District: Tukuyu [Neu Langenburg], *Koslui* 7951 (EA!, holo.)
  B. steppia (Steetz) Sherff var. *kalamboensis* Sherff in Amer. J. Bot. 42, 6: 562 (1955). Type: Zambia, Kalambo Falls, *Richards* 1342 (K!, holo.)
  B. steppia (Steetz) Sherff var. *garusonis* Sherff in Ann. Mag. Nat. Hist. 10, 2: 42 (1957). Type: Mozambique, Garuso, *Gilliland* 1847 (BM!, holo.)

NOTE. Mesfin (1993) cites several specimens for Uganda; these are now missing at Kew, though one of them (*Jarrett* 479) turned out to be another species, *B. grantii*.

27. **Bidens oligoflora** (*Klatt*) *Wild* in Kirkia 6: 20 (1967); Lisowski, Asterac. Fl. Afr. Centr. 1: 143, t. 33 (1991); Mesfin in K.B. 48: 487 (1993). Type: Angola, Malange, *Buchner* 32 (B†, holo., GH, lecto., chosen by Wild)

Annual herb to 1 m high; stem erect, pilose to glabrous. Leaves pinnatipartite or bi-pinnatipartite, 3–20 cm long, 2–6 cm wide, segments ovate to linear (narrower in upper leaves), sparsely pilose on both surfaces; petiole to 7 cm long. Capitula terminal, in lax corymbose cymes; peduncle to 13 cm long; involucre 8–10 mm long; outer phyllaries 8–10, pilose, inner phyllaries brown, ± 8, with yellow margins, pilose; paleae 8–10 mm long, glabrous. Ray florets golden yellow to orange, neuter, 8, rays 10–30 × 2–12 mm; disc florets yellow. Achenes black or dark brown, broadly obovoid or orbicular, 6–8 mm long, scabrid; aristae 2, to 2 mm long, antrorsely barbed, rarely absent.

TANZANIA. Mpanda District: Mwese Hill, May 1975, *Kahurananga, Kibuwa & Mungai* 2609!
DISTR. **T** 4; Cameroon, Congo (Kinshasa), Angola, Zambia, Zimbabwe
HAB. Bushed grassland; ± 1828 m
USES. None recorded on specimens from our area
CONSERVATION NOTES. Least concern (LC)

SYN. *Coreopsis oligoflora* Klatt in Leopoldina 25: 107 (1889); Sherff in Field Mus. Nat. Hist., Bot. Ser. 11, 6: 386 (1936)
　　C. mattfeldii Sherff in Bot. Gaz. 76: 839 (1923) & in Field Mus. Nat. Hist., Bot. Ser. 11, 6: 392 (1936). Type: Zambia, Lake Tanganyika, Fwamba, *Carson* 75 (B†, holo., K!, lecto., chosen by Mesfin)
　　C. oligoflora Klatt var. *robusta* Sherff in Bot. Gaz. 90: 386 (1930) & in Field Mus. Nat. Hist., Bot. Ser. 11, 6: 386, 388 (1936). Type: Congo (Kinshasa), Bolobo, *Buttner* 408 (B†, holo.)
　　Bidens onisciformis Sherff in Bot. Gaz. 96: 144 (1934) & in Field Mus. Nat. Hist., Bot. Ser. 16: 559 (1937). Type: Congo (Kinshasa), W Katana [Kivu], *van der Houdt* 211 (BR!, holo.)
　　B. isokoensis Sherff in Amer. J. Bot. 42, 6: 565 (1955). Type: Zambia, Isoko Valley, *Richards* 1108 (K!, holo., SRGH!, iso.)
　　B. drummondii Wild in Kirkia 6: 17 (1967). Type: Zambia, 11 km W of Chizera, *Drummond & Rutherford-Smith* 7204 (SRGH!, holo., M!, iso.)

28. **Bidens acuticaulis** *Sherff* in Bot. Gaz. 59: 301 (1915) & in Field Mus. Nat. Hist., Bot. Ser. 16: 347, fig. 104/j–r (1937); Mesfin in K.B. 48: 492 (1993). Type: Angola, Kassuango-Kuiri, *Gossweiler* 4052 (BM!, holo.)

Annual herb 0.2–1 m high, erect; stem simple or branched, reddish or speckled with reddish, glabrous or pilose. Leaves pinnate to bipinnatifid, 2–13 cm long and to 6 cm wide, segments 3–5, narrowly linear to filiform, to 50 by 2 mm, minutely ciliate and sometimes sparsely hispidulous; (pseudo)petiole to 50 mm long, narrowly winged. Capitula solitary and terminal; peduncle to 15 cm long; involucre 4–8 mm long, lengthening to 13 mm in fruit; outer phyllaries 4–9, green to reddish, often reflexing, glabrous or ciliate, inner phyllaries ± 8, pale brown with dark red or purple margins or red all over, glabrous or pubescent. Ray florets yellow, the upper part paler, neuter, 5–8, rays 6–7 × 1–1.5 mm, striate; disc florets yellow, campanulate, the lobes glabrous or puberulous. Achenes black or dark brown, narrowly ellipsoid with beak, appressed hairy; beak up to 2.5 cm or up to 24 cm long, with 2–3 aristae 1–2.5 mm long, retrorsely barbed.

var. **acuticaulis**

Achene beak to 2.5 cm long, golden when ripe; aristae retrorsely barbed.

TANZANIA. Rungwe District: Kyimbila, Mulinda, July 1912, *Stolz* 1442!; Songea District: 9 km S of Lumecha Bridge, May 1956, *Milne-Redhead & Taylor* 9886! & 140 km E of Songea, June 1956, *Milne-Redhead & Taylor* 10582!
DISTR. **T** 7, 8; Congo (Kinshasa), Angola, Zambia, Malawi, Mozambique

HAB. *Brachystegia* woodland; 850–1050 m
USES. None recorded on specimens from our area
CONSERVATION NOTES. Least concern (LC)

SYN. *B. ciliata* De Wild. in F.R. 13: 203 (1914), *non B. ciliata* Fisch. & Mey. (1839), *nom. illegit.*
Type: Congo (Kinshasa), Shaba, Lubumbashi [Elisabethville], *Bequaert* 302 (BR!, holo.)
B. paupercula Sherff in Bot. Gaz. 76: 158, fig. 12/a–g (1923) & in Field Mus. Nat. Hist., Bot.
Ser. 11, 6: 380, fig. 94 (1936); Lisowski, Asterac. Fl. Afr. Centr. 1: 133, fig. 31 (1991).
Type: Tanzania, Rungwe District: Kyimbila, Mulinda, *Stolz* 1442 (B†, holo., M!, lecto., C!,
G!, K!, L!, S!, Z!, isolecto., chosen by Rayner, 1992)

var. **filirostris** (*P.Taylor*) *T.G.J.Rayner* in Phytologia 72: 101 (1992). Type: Tanzania, Songea
District: 140 km E of Songea, *Milne-Redhead & Taylor* 10547 (K!, lecto., B!, BR!, K!, S!,
SRGH!, isolecto.)

Achene beak to 24 cm long, pendulous, golden brown, twisted when immature, straight in
mature fruit; aristae retrorsely barbed only at apex.

TANZANIA. Mpanda District: Sabaga, June 1975, *Kahurananga et al.* 2739! & 48 km on
Mpanda–Uvinza road, May 1997, *Bidgood et al.* 4025!; Songea District: 140 km E of Songea,
June 1956, Milne-Redhead & Taylor 10547!
DISTR. **T** 4, 8; Zambia
HAB. *Brachystegia* woodland; 900–1500 m
USES. None recorded on specimens from our area
CONSERVATION NOTES. Three Tanzanian specimens; data deficient (DD)

SYN. *B. paupercula* Sherff var. *filirostris* P.Taylor in Hook. Ic. Pl.: t. 3580 (1962)

29. **Bidens diversa** *Sherff* in Bot. Gaz. 76: 159 (1923) & in Field Mus. Nat. Hist., Bot.
Ser. 16: 329, fig. 75/a,b,d–i (1937); Mesfin in K.B. 48: 493 (1993). Type: Angola,
Mounyino, *Antunes* 315 (B†, holo.); neotype: Angola, Huila, Lubango, *Borges* 167
(LISC, neo., M, P, PRE, SRGH, iso., chosen by Rayner 1992)

Annual herb, 30–50(–100) cm high; branches slender, glabrous. Leaves pinnate, to
10 cm long, with a filiform rachis and 3–5 filiform segments to 5 cm long, glabrous
or sparsely setulose; petiole to 3 cm long. Capitula solitary and terminal; peduncle to
12 cm long; involucre 4–5 mm long; outer phyllaries 8, inner phyllaries reddish
brown, glabrous or ciliate. Ray florets yellow, neuter, 8, rays 15–20 × 3–4 mm; disc
florets brownish-yellow or orange. Achenes black, narrowly ellipsoid, striate-sulcate
and bristly, graded monomorphic, 6–14 mm long; aristae 2, ± 1 mm long, retrorsely
barbed at apex.

TANZANIA. Ufipa District: 6 km on Tatanda–Sumbawanga road, Apr. 1997, *Bidgood et al.* 3485!
DISTR. **T** 4; Angola, Zambia, Malawi, Mozambique
HAB. Woodland on rocky outcrop; gregarious in the FZ area; ± 1700 m
USES. None recorded on specimens from our area
CONSERVATION NOTES. Least concern (LC)

SYN. *B. filiformis* Sherff in Field Mus. Nat. Hist., Bot. Ser. 17: 600 (1939); Wild in Kirkia 6: 21
(1967). Type: Zambia, Lake Chila, *Burtt* 6269 (F, holo., BM, BR, K!, iso.)
B. diversa Sherff subsp. *filiformis* (Sherff) T.G.Rayner in Phytologia 75, 2: 156 (1993).

30. **Bidens schimperi** *Walp.*, Repert. 6: 168 (1846); Oliv. & Hiern, F.T.A. 3: 393
(1877); Sherff in Bot.Gaz. 81: fig. 3 (1926) & in Field Mus. Nat. Hist., Bot. Ser. 16:
554, fig. 142/i–g, 143, 144 (1937); F.P.S. 3: 11 (1956); Wild in Kirkia 6: 23 (1967);
U.K.W.F.: 466 (1974); Cribb & Leedal, Mountain Flow. S. Tanz.: 151, fig. 40b (1982);
Mesfin in Symb. Bot. Upsal. 24: 110, fig. 56 (1984) & in K.B. 48: 497 (1993); U.K.W.F.
ed. 2: 217, t. 89 (1994). Type: Ethiopia, Tigray, Djeladjerrane, *Schimper* 1429 (P!,
holo., FT!, G!, K!, LE!, M!, MO!, S!, W!, iso.)

Annual herb 0.07–1.5 m high; stem 4-angled, simple to much branched, often with red or purple markings, sparsely to densely pubescent. Leaves broadly ovate in outline, tripartite with lobed parts, rarely simple or bi- to tri-pinnatisect, 2.5–20(–30) cm long, 1.5–10(–15) cm wide, leaf segments ovate to narrowly ovate, the margins lobed, entire, incised-dentate to crenate-serrate, pubescent, especially above, rarely almost glabrous, densely glandular-punctate; petiole 10–55(–120, fide EA) mm long. Capitula erect or nodding, in lax terminal corymbose cymes; peduncle 2–18 cm long; outer phyllaries 8, linear, 2.5–10 mm long, sparsely pubescent, inner phyllaries 8, 3.5–11 mm long, pubescent or glabrous, light brown to red with a wide yellow margin. Ray florets yellow or occasionally orange, neuter, 6–8, rays 6–20 mm long, tube ± 1.5 mm long; disc florets dull yellow to orange, ± 4 mm long. Achenes dark brown to black, 3–10 mm long (outer) to 5–16 mm long (inner), striate, strigose, compressed or flat, sometimes margins callose-thickened; aristae 2, 0.2–4 mm long, retrorsely barbed.

KENYA. Nairobi National Park, Athi Gate, Jan. 1962, *Verdcourt* 3261!; Masai District: Olodungoro, July 1961, Glover et al. 2102!; Kwale District: Shimba Hills, past Pengo hill, July 1968, *Gillett* 18677!
TANZANIA. Musoma District: Seronera R., 1.5 km to Banagi, Mar. 1961, *Greenway* 9936!; Kondoa District: Kolo Hills, June 1973, *Ruffo* 719!; Mufindi District: Ruaha R. source, between Mafinga and Mbeya, Apr. 1983, *Macha* 250!
DISTR. **K** 4, 6, 7; **T** 1–8; Southern Egypt, Sudan, Ethiopia, Djibouti, Somalia, Angola, Zambia, Malawi, Mozambique, Zimbabwe, Botswana, Namibia, South Africa
HAB. May be abundant in disturbed, overgrazed or burnt soil, and may form stands in wheat fields; also in grassland, bushed or wooded grassland (especially on black cotton soils), woodland, evergreen bushland and secondary thicket; 400–2600 m
USES. Minor medicinal against chest pain and as 'childrens remedy' (*Koritschoner*)
CONSERVATION NOTES. Least concern (LC)
VARIABILITY. Leaf dissection is very variable, as is the length of the ray florets and the inner achenes.

SYN. *Coreopsis exaristata* O.Hoffm. in E.J. 20: 414 (1895), *non Bidens exaristata* DC. Type: Tanzania, Lushoto District: Usambara, *Holst* 102 (B†, holo.)
    Bidens punctata Sherff in Bot. Gaz. 59: 302 (1915). Type: Malawi, Tumbi, *Johnston* 343 (K!, holo.)
    B. *microcarpa* Sherff in Bot. Gaz. 76: 84 (1923) & in Field Mus. Nat. Hist., Bot. Ser. 16: 567, fig. 149/a–h (1937). Type as for *C. exaristata* O.Hoffm.
    B. *acutiloba* Sherff in Bot. Gaz. 76: 147 (1923). Type: Tanzania, Kilimanjaro, Kwa Kinabo, *Volkens* 384 (B†, holo., BM!, lecto., chosen by Mesfin 1984, G!, K!, iso.)
    B. *schimperi* Walp. var. *leptocera* Sherff in Bot. Gaz. 81: 52 (1926) & in Field Mus. Nat. Hist., Bot. Ser. 16: 554 (1937). Type: Tanzania, Kilosa, *Swynnerton* 845a (BM!, holo.)
    B. *schimperi* Walp. var. *punctata* Sherff in Bot. Gaz. 85: 17 (1928). Type: Malawi, Thumbi Hill, Kabata, *Johnston* 341 (K!, holo.)
    B. *schimperi* Walp. var. *brachycera* Sherff in Field Mus. Nat. Hist., Bot. Ser. 17: 585 (1939). Type: Tanzania, near Shinyanga, *Bax* 206 (K!, holo.)
    B. *schimperi* Walp. var. *leiocera* Sherff in Field Mus. Nat. Hist., Bot. Ser. 17: 586 (1939). Type: Zambia, Mazabuka, *Martin* s.n. (K!, holo.)
    B. *schimperi* Walp. var. *greenwayi* Sherff in Amer. J. Bot. 42: 563 (1955). Type: Tanzania, Moshi District: Engare Nairobi, *Greenway* 6876 (K!, holo.)
    B. *grantii* (Oliv.) Sherff. var. *stapfioides* Sherff in Ann. Mag. Nat. Hist. 10, 2: 445 (1957). Type: Tanzania, Tabora, *Smith* 1134 (EA!, holo.)

31. **Bidens biternata** (*Lour.*) *Merr. & Sherff* in Bot. Gaz. 88: 293 (1929); Sherff in Field Mus. Nat. Hist., Bot. Ser. 16: 388, fig. 99/a, c–m (1937); F.P.S. 3: 12 (1956); C.D. Adams in F.W.T.A. ed. 2, 2: 234 (1963); Wild in Kirkia 6: 26 (1967); E.P.A.: 1135 (1967); U.K.W.F.: 466 (1974); Mesfin in Symb. Bot. Upsal. 24: 115, fig. 58 (1984); Maquet in Fl. Rwanda 3: 633 (1985); Lisowski, Asterac. Fl. Afr. Centr. 1: 135 (1991); D.J.N. Hind in Fl. Masc. 109: 195, t. 66 (1993); Mesfin in K.B. 48: 498 (1993); U.K.W.F. ed. 2: 217, t. 89 (1994). Type: none cited in protologue, no Loureiro specimens of this taxon extant; neotype: China, Kwangtung, Honam Is., *Merrill* 10122 (UC, neo., chosen by Mesfin)

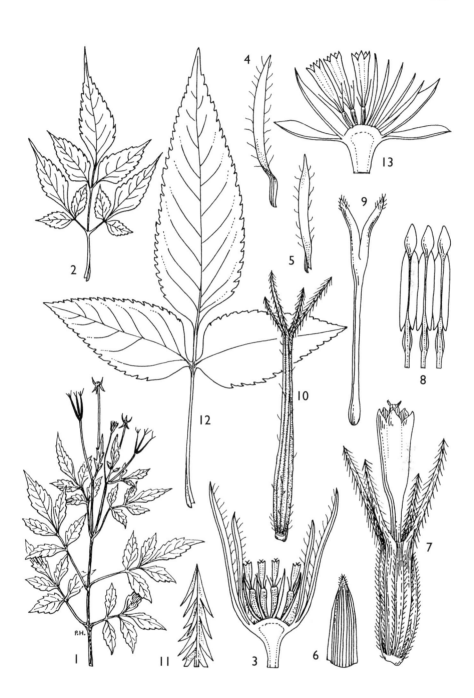

FIG. 172. *BIDENS BITERNATA* — **1**, habit, × ²/₃; **2**, lower leaf, × ²/₃; **3**, capitulum section, × 6; **4–5**, outer phyllaries, × 6; **6**, palea, × 6; **7**, floret, × 12; **8**, stamens, × 30; **9**, style, × 18; **10**, achene, × 6; **11**, pappus arista tip, much enlarged. *BIDENS PILOSA* — **12**, lower leaf, × ²/₃; **13**, capitulum section, × 6. 1 from *Boivin* 155 & *de l'Isle* 125, 2 from *Mesfin* 8157, 3–11 from *Boivin* 155, 12 from *Cadet* 339, 13 from *Cadet* 2219. Drawn by Pat Halliday, from Flore des Mascareignes.

Annual herb, erect, 0.1–2 m high; stem reddish tinged, 4-angled, simple or branched, pubescent or glabrous. Leaves opposite or rarely alternate near the apex, broadly ovate in outline, pinnately (3–)5–9-lobed or almost pinnatisect, 3–20 cm long, 2.5–12 cm wide, leaf segments ovate to lanceolate, lobed or bilobed at base, margins crenate-serrate to almost lobulate, apices acute; petiole to 60 mm long. Capitula in terminal lax paniculate cymes; involucre cup-shaped; outer phyllaries 4–8(–15), linear to oblanceolate, 4–6(–12) mm long, to 20 mm long in fruit, pubescent, inner phyllaries 8–11, yellowish green, with paler margins, 3.5–7 mm long, pubescent; paleae 4.5–6 mm long. Ray florets yellow, rarely orange, neuter or with staminodes, (1–)2–5, tube 0.8–1.3 mm long, pubescent, rays 3–6 × 1.3–2.5 mm, striate; disc florets yellow to orange-yellow, 3–3.7 mm long, glabrous. Achenes black, 4–8-ribbed, linear-tetragonal, 6–16(–25) mm long, strigose in upper half; aristae (2–)3–4(–5), yellow-brown, 2–4 mm long, retrorsely barbed. Fig. 172 (page 805).

UGANDA. Karamoja District: between Kakamari and Kotido, 1930, *Liebenberg* 324!; Bugisu District: Budadiri, Jan. 1932, *Chandler* 448!
KENYA. Northern Frontier District: Dandu, May 1952, *Gillett* 13151!; Nairobi District: Njiro Farm, Jan. 1952, *Bogdan* 3387!; Masai District: Lemek area, Apr. 1961, *Glover et al.* 824!
TANZANIA. Masai District: Lemuta–Nasera Rock, Apr. 1987, *Chuwa* 2600!; Kilosa District: Mikumi National Park, Mkata R., May 1977, *ole Sayalel* 1277!; Iringa District: Mtera, May 1971, *Mhoro* 1143!
DISTR. U 1, 3, 4; **K** 1–7; **T** 1–8; widespread in tropical and subtropical Africa, Asia and Australia
HAB. Forest margins, weed of cultivation (especially in maize), ruderal sites, woodland, evergreen bushland, river-banks, grassland; 50–2500 m
USES. None recorded on specimens from our area
CONSERVATION NOTES. Least concern (LC)

SYN. *Coreopsis biternata* Lour., Fl. Cochinch. ed. 1: 508 (1790) & ed. 2: 622 (1793)
    Bidens abyssinica Walp., Repert. 6: 167 (1846). Type: Ethiopia, Tigray, near Adwa, *Schimper* 337 (P!, lecto., chosen by Mesfin, BM!, BR!, FT!, G!, M!, MO!, S!, UPS1, iso.)
    B. kotschyi Sch.Bip. in Walp., Repert. 6: 168 (1846). Type: Sudan, Nubia, Kordofan, Mt Arash-Cool, *Kotschy* 79 (P!, holo., G!, K!, L!, M!, MO!, S!, UPS!, iso.)
    B. abyssinica Walp. var. *glabrata* Vatke in Linnaea 39: 500 (1875). Type: Tigray, Scholoda, *Schimper* 285 (B†, holo., Z!, lecto., chosen by Rayner 1992)
    B. quadriseta Oliv. & Hiern in F.T.A. 3: 393 (1877). Type: Ethiopia, Tigray, Dschadscha, *Schimper* 2181 (B†, holo., G, lecto., chosen by Mesfin, MO!, P!, S!, Z!, iso.)
    B. bipinnata auct. mult., *non* L.; e.g. Oliv. & Hiern, F.T.A. 3: 393 (1877), p.p.; Lanza in Zavattari, Miss. Biol. Borana 4: 262 (1939); F.P.S. 3: 11 (1956); E.P.A.: 1135 (1967); Lisowski, Asterac. Fl. Afr. Centr. 1: 133 (1991)
    B. pilosa L. var. *quadriseta* (Oliv. & Hiern) Engl. in Abh. Konigl. Akad. Wiss.: 437 (1892)
    B. pilosa L. var. *glabrata* (Vatke) Engl. in Abh. Konigl. Akad. Wiss.: 437 (1892)
    B. abyssinica Walp. var. *incisifolia* Chiov. in Ann. Ist. Bot. Rome 8: 186 (1904). Type: Ethiopia, Gonder, near Gageros, *Schimper* 105 (B†, holo., Z!, lecto., chosen by Mesfin, G!, iso.)
    B. pilosa L. var. *abyssinica* Fiori in Nuovo Giorn. Bot. Ital. 20: 390 (1913). Type: Eritrea, Hamasen, Ghinda, *Fiori* 1832 (FT, holo.)
    B. chinensis (L.) De Wild. var. *abyssinicus* (Walp.) O.E.Schultz in E.J. Suppl. 50: 180 (1914)
    B. cylindrica Sherff in Bot. Gaz. 81: 28, fig. 91/g–l (1926). Type: Zambia, Borume, Zambezi, *Menyhart* 1110 (WU!, holo.)
    B. biternata (Lour.) Merr. & Sherff var. *glabrata* (Vatke) Sherff forma *abyssinica* (Walp.) Sherff in Bot. Gaz. 90: 389 (1930)
    B. biternata (Lour.) Merr. & Sherff var. *glabrata* (Vatke) Sherff forma *lasiocarpa* (O.E.Schultz) Sherff in Bot. Leafl. 9: 10 (1954)

NOTE. Seeds stick to the coats of animals and to clothes.

32. **Bidens pilosa** *L.*, Sp. Pl. 1, 2: 832 (1753); Oliv. & Hiern, F.T.A. 3: 392 (1877); Sherff in Field Mus. Nat. Hist., Bot. Ser. 16: 412, fig. 99/b, 102/1–b, e–j (1937); U.O.P.Z.: 146 (1949); F.P.S. 3: 11 (1956); F.P.U.: 174, fig. 113 (1962); Wild in Kirkia 6: 24, excl. syn. (1967); E.P.A.: 1139 excl. syn. (1967); U.K.W.F.: 466 (1974); Mesfin in Symb. Bot. Upsal. 24: 122, fig. 61, fig. 58 (1984); Maquet in Fl. Rwanda 3: 634, fig.

195/3 (1985); Blundell, Wild Fl. E. Afr.: pl. 90 (1987); Lisowski, Asterac. Fl. Afr. Centr. 1: 136, fig. 32 (1991); D.J.N. Hind in Fl. Masc. 109: 195 (1993); Mesfin in K.B. 48: 500 (1993); U.K.W.F. ed. 2: 217, t. 89 (1994). Type: without locality or collector (LINN 975.8, lecto., designated by D'Arcy in Ann. Miss. Bot. Gard. 62: 1178 (1975))

Annual herb, erect, 0.1–1.5 m high; stem much branched or sometimes hardly branched, pale green or sometimes reddish, 4-angled, sparsely pubescent to glabrous. Leaves pinnately 3–5-lobed or occasionally simple in upper or lowermost part of stem, up to 15(–20) cm long, segments ovate to ovate-lanceolate, margins serrate or crenate-serrate, sparsely pubescent to glabrous, terminal segment 5–10 cm long, 2.5–5 cm wide, lateral segments 2–5 cm long, 1–2 cm wide and asymmetric, often lobed or bilobed; petiole to 70 mm long. Capitula terminal and solitary or few together; peduncle to 16 cm long; outer phyllaries 7–10, 3–4 mm long, reflexed at anthesis, inner phyllaries 5–8, yellow-green to pale brown, 3–4.5 mm long with yellow scarious margins, glabrous but for the apex; receptacle club-shaped in fruit; paleae light brown, striate, 3–5 mm long. Ray florets white or cream, sometimes absent, neuter or with pistillodes or staminodes, 4–8, rays (when fully developed) 7–15 × 3–4.5 mm, tube 0.5–0.8 mm long; disc florets yellow or yellow-orange, ± 4 mm long, pubescent at base. Achenes black, 4–6-ribbed, linear-tetragonal, 4–12 mm long, strigose or verrucose; aristae 2–3(–4), 2–4 mm long, retrorsely barbed. Fig. 172/12–13 (page 805).

Uganda. West Nile District: Metu Rest Camp, Sep. 1953, *Chancellor* 303!; Kigezi District: Kachwekano Farm, July 1949, *Purseglove* 2980!; Mengo District: Mabira Forest, Nov. 1938, *Loveridge* 15!
Kenya. Northern Frontier District: Mt Kulal, Jan. 1977, *Masheti & Gagah* H316!; South Nyeri District: Mwea Research Station, June 1976, *Kahurananga & Kibui* 2799!; Masai District: Chyulu Hills, Dec. 1991, *Luke* 2988!
Tanzania. Pare District: South Pare Mts, Chome Shengena Forest, Sep. 1999, *Mlangwa & Masanyika* 537!; Mpanda District: Kabungu, July 1946, *Semsei* 64!; Mbeya District: Mbeya Peak Forest Reserve, Nov. 1958, *Gaetan Myembe* 100!; Zanzibar Is., 1873, *Hildebrandt* 1123!
Distr. **U** 1, 2, 4; **K** 1–6; **T** 1–8, **Z**, **P**; pan-tropical and -sub-tropical, also extending into some temperate areas
Hab. A common weed of cultivation and ruderal sites, often on alluvial sites, also in forest margins and grassland, in swamps and along streams; may be locally common; 0–60 and 750–2500 m
Uses. Boiled leaves are used as a vegetable (reported by six collectors). Minor medicinal for infected eyes and cuts (*Tanner*) and for joint pains and stomach in children (*Hakizayo*)
Conservation notes. Least concern (LC)

Syn. *B. sundaica* Blume in Bijdr. Nat. Wetensch. 1: 913 (1826). Type: Indonesia, Java, *Blume* in herb. L 900, 146.70 (L, holo.)
    B. sundaica Blume var. *minor* Blume in Bijdr. Nat. Wetensch. 1: 914 (1826). Type: Indonesia, Java, *Blume* in herb. L 900, 146.72 (L, holo.)
    B. pilosa L. var. *minor* (Blume) Sherff in Bot. Gaz. 80: 387 (1925) & in Field Mus. Nat. Hist., Bot. Ser. 16: 421, fig. 102/c–d, k–r (1937)
    B. biternata (Lour.) Merr. & Sherff var. *glabrata* auct., *non* (Vatke) Sherff; Lanza in Zavattari, Miss. Biol. Borana 4: 263 (1939)

Note. Often known as 'blackjack'. One of the commonest East African weeds, spreading by the achenes which adhere to clothes and skin, to animal coats and bird feathers.

SPECIES OF UNCLEAR AFFINITY

Coreopsis arenicola S.Moore in J.L.S. 37: 170 (1905). Type: Uganda, Masaka District: Misozi [Musozi], *Bagshawe* 12.
Described by Moore as a low shrub, with leaves 1.5–2.5 cm long with narrow segments to 1 mm wide; rays 11 × 3 mm; achenes 6–7 mm long, with aristae 1 mm long. Possibly *B. negriana* but the description is rather vague, and the specimen has not been seen.

127. COSMOS

Cav., Ic. 1: 9 (1791); Sherff in Field Mus. Nat. Hist., Bot. Ser. 8(6): 401–447 (1932); D.J.N. Hind in Fl. Masc. 109: 198 (1993)

Annual or perennial herbs, sometimes shrubs. Leaves opposite, pinnatisect or simple. Capitula terminal and solitary or in few-headed terminal or axillary cymes, radiate; involucre often fleshy at base, the phyllaries connate at base, 2-seriate with the outer leafy and the inner scarious; paleae present. Ray florets neuter; disc florets many, anthers with ovate or cordate apical appendage; style branches truncate, with caudate appendage. Achenes fusiform, and typically attenuate into a beak; pappus absent or of retrorsely barbed aristae.

About 30 species from the Americas; some species cultivated.

1. Leaves with segments linear; achenes and their beak not
   exserted from capitulum . . . . 1. *C. bipinnatus*
   Leaves with segments lanceolate or elliptic-ovate, rarely
   linear; achenes and their beak exserted at maturity 2
2. Ray florets sulphur yellow or rich orange, the ray 18–30 mm
   long; inner phyllaries (much) shorter than the outer . . . . 2. *C. sulphureus*
   Ray florets pale purple, pink or white, the ray 10–18 mm
   long; inner phyllaries about equal to the outer . . . . 3. *C. caudatus*

1. **Cosmos bipinnatus** *Cav.*, Ic. 1: 10, t. 14 (1791); D.J.N. Hind in Fl. Masc. 109: 199 (1993). Type: from a plant from Mexico cultivated at Madrid (MA, ?holo)

Annual herb, erect, 0.6–1.3 m high. Leaves pinnatisect, 6–11 cm long, 3–5 cm wide, the segments linear, 0.4–1.2 mm wide. Capitula solitary or in few-headed cymes; peduncle 1.5–20 cm long. Ray florets 8, pink, white or red; ray 25–40 mm long; disc florets many, yellow, 6–7 mm long. Achenes 7–16 mm long, the beak 1–6 mm long; pappus of two aristae, 1–1.5 mm long.

KENYA. Baringo District: Maji Moto, Oct. 1978, *Vincent* 213
TANZANIA. Lushoto District: Lushoto, July 1971, *Mbailwa* 84!
DISTR. **K** 3–5, 7; **T** 3, 7; originally from tropical America, now a common tropical ornamental which becomes sub-naturalised
HAB. Introduced as ornamental, possibly naturalized in disturbed sites
USES. None recorded on specimens from our area
CONSERVATION NOTES. Least concern (LC)

SYN. *Coreopsis formosa* Bonato, Pisaura autom. Coreopsis formosa: 22, t. 2 (1793). Type: the illustration in Bonato (iconotype)
  Bidens formosa (Bonato) Sch.Bip. in Seemann, Bot. Voy. Herald: 307 (1856); Wild in Kirkia 6: 32 (1967); Lisowski, Asterac. Fl. Afr. Centr. 1: 172 (1991)

2. **Cosmos sulphureus** *Cav.*, Ic. 1: 56 (1791); F.P.S. 3: 1 (1956); D.J.N. Hind in Fl. Masc. 109: 199, t. 68 (1993). Type: from a plant cultivated at Madrid from Mexican seed (MA, ?holo)

Annual herb 0.3–1.2 m high; stem 4-gonous, pubescent to glabrous. Leaves ovate in outline, deeply dissected, 5–15 cm long and 4–8 cm wide, the segments 2–5 mm wide, glabrous; petiole absent or to 3 cm long. Capitula solitary and terminal; peduncle 2–22 cm long; involucre with phyllaries 5–12 mm long; paleae 9–10 mm long. Ray florets 7–10, rich orange or sulphur yellow, 18–30 mm long; disc florets orange-yellow, 7–9 mm long. Achenes blackish, 16–28 mm long, the beak 6–9 mm long; pappus of 2(–3) retrorsely barbed aristae 4–7 mm long. Fig. 173 (page 809).

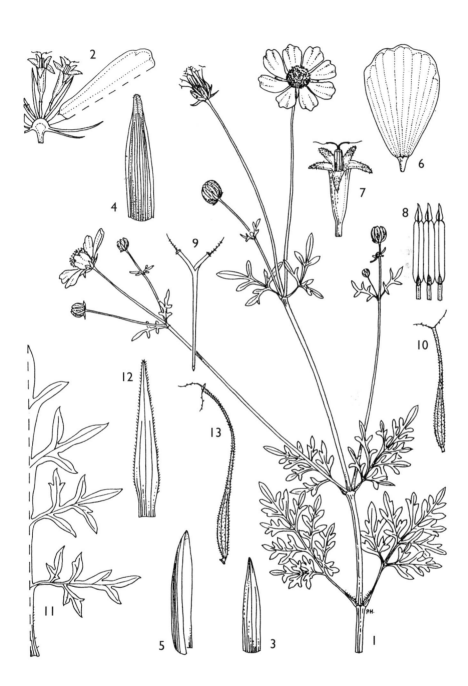

FIG. 173. *COSMOS SULPHUREUS* — **1**, habit, × ²/₃; **2**, capitulum section, × 2; **3–4**, outer and inner phyllaries, × 4; **5**, palea, × 4; **6**, ray floret, × 2; **7**, disc floret, ovary removed, × 4; **8**, stamens, × 6; **9**, disc floret style, × 6; **10**, achene, × 2. *COSMOS CAUDATUS* — **11**, part of lower leaf, × ²/₃; **12**, phyllary, × 4; **13**, achene, × 2. 1–10 from *Osman* in MAU 20871, 11–13 from *Rochecouste* 428. Drawn by Pat Halliday, from Flore des Mascareignes.

UGANDA. Mengo District: Ziku Forest, Dec. 2001, *Mbatudde* 103! & 107!
KENYA. North Kavirondo District: 4 km on Webuye–Kitale road, Oct. 1981, *Gilbert & Mesfin*
     6580!; Teita District: Taveta, July 1957, *Meinertzhagen* 3 (fide EA)
TANZANIA. Uzaramo District: Dar es Salaam University campus, Apr. 1986, *Kisena* 257!; Songea
     District: 1.5 km E of Songea, Apr. 1956, *Milne-Redhead & Taylor* 9690! & Mapera village, May
     1991, *Ruffo & Kisena* 3237!
DISTR. U 4; K 3, 5, 7; T 6–8; originally from Mexico but a widespread weed in tropical Africa
HAB. Roadsides, waste ground; 0–1650 m
USES. None recorded on specimens from our area
CONSERVATION NOTES. Least concern (LC)

SYN. *Bidens sulphureus* (Cav.) Sch.Bip. in Seemann, Bot. Voy. Herald: 308 (1856); Wild in Kirkia
     6: 32 (1967); Lisowski, Asterac. Fl. Afr. Centr. 1: 171 (1991), as *sulphurea*

3. **Cosmos caudatus** *Kunth* in H.B.K., Nov. Gen. & Sp. ed. fol. 4: 188 (1818);
D.J.N. Hind in Fl. Masc. 109: 199 (1993). Type: Cuba, Havana, *Humboldt & Bonpland*
s.n. (P, holo.)

Annual herb 0.6–2.5 m high; stems pilose or glabrous. Leaves ovate in outline,
much dissected, 5–15 cm long, the segments 3–10 mm wide, with scabrid veins and
margins. Capitula solitary and terminal; peducle 10–30 cm long; involucre 6–15 mm
long. Ray florets 8, ray pale purple, pink or white, 10–18 mm long; disc florets yellow,
8–10 mm long. Achenes 12–35 mm long, with a beak 10–15 mm long; pappus absent
or of (2–)3 aristae 2–5 mm long. Fig. 173/11–12 (page 809).

UGANDA. Mengo District: Old Entebbe, Dec. 1956, *Harker* 608!
DISTR. U 4; originally from tropical America, now a pantropical weed
HAB. Ruderal site; ± 1150 m
USES. None recorded on specimens from our area
CONSERVATION NOTES. Least concern (LC)

SYN. *Bidens caudata* (Kunth) Sch.Bip. in Seemann, Bot. Voy. Herald: 308 (1856)

128. CHRYSANTHELLUM

Rich. in Pers., Syn. Pl. 2: 471 (1807); B.L.Turner in Phytologia 64: 410–444 (1988)

Annual or rarely perennial herbs. Leaves alternate or all radical. Capitula
terminal or axillary to upper leaves, solitary, heterogamous, radiate; involucre 1–2-
seriate; paleae flat, persistent. Ray florets pistillate; disc florets hermaphrodite;
anthers sagittate at base; style bifid, the branches densely papillose. Achenes oblong,
with thick pale margins those of disc florets flattened; pappus absent or reduced to
a small disc.

Ten pantropical species.

Chrysanthellum indicum *DC.*, Prodr. 5: 631 (1836). Type: India, Gojpur and
Sukanaghur, *Wallich Herb.* 3231 (K!, holo.)

Annual herb 3–45 cm high, glabrous; stems prostrate to ascending or erect, green
and sometimes with red markings. Leaves radical as well as alternate (the cauline
ones), slightly fleshy, glaucous beneath, ovate in outline, 2–3-pinnatifid, 1–5 cm long,
1–4 cm wide, the segments narrowly lanceolate to linear, apiculate; petiole 10–30 mm
long, clasping the stem at base. Capitula solitary or in few-headed cymes, but
seemingly forming leafy panicles; peduncle to 6 cm long; involucre shortly
cylindrical, 3–4 mm long; phyllaries with scarious margins and turning orange in
fruit; paleae to 3 mm long. Ray florets yellow or orange-yellow, ray 1–2.5 mm long,

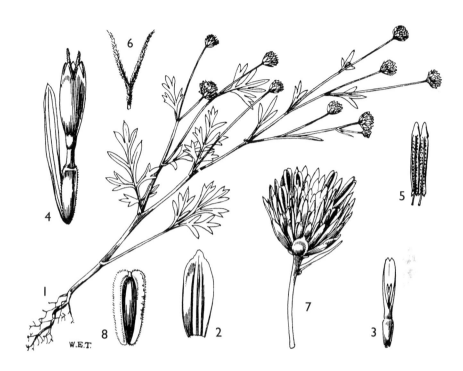

FIG. 174. *CHRYSANTHELLUM INDICUM* — **1**, habit; **2**, phyllary; **3**, ray floret; **4**, disc floret with palea; **5**, stamens; **6**, style arms; **7**, fruiting head; **8**, achene. Drawn by W.E. Trevithick, from the Flora of West Tropical Africa.

tube 0.5–0.7 mm long; disc florets (8–)13–34, yellow or orange, 0.8–1.8 mm long. Achenes oblong, 2–6 mm long, those of disc florets compressed, puberulous, with a narrow wing; pappus absent. Fig. 174.

subsp. **afroamericanum** *B.L. Turner* in Phytologia 51, 4: 291 (1982); Maquet in Fl. Rwanda 3: 640, fig. 198/2 (1985); Lisowski, Asterac. Fl. Afr. Centr. 1: 173, t. 38 (1991). Type: Argentina, *O'Donnel & Rodrigues* 501 (A, holo., F, UC, iso.)

UGANDA. Acholi District: Choli, Oct. 1967, *Buzigye* 9!; Bunyoro District: Masindi, Dec. 1932, *Hazel* 275!; Mengo District: Butuntumula–Kakoge road km 8, Apr. 1956, *Langdale-Brown* 2037!
KENYA. Baringo District: 6 km N of Kampi ya Samaki, June 1977, *Gilbert* 4749!; Meru District: Kieiga Forest to Ruiri School, June 1974, *Faden & Faden* 74/875!; Central Kavirondo District: Paponditi [Pop Onditi], Dec. 1968, *Kokwaro* 1637!
TANZANIA. Musoma District: Mugango, June 1959, *Tanner* 4288!; Ulanga District: Mahenge, Mbangala, Feb. 1932, *Schlieben* 1776!; Songea District: 12 km E of Songea by Nonganonga stream, Dec. 1955, *Milne-Redhead & Taylor* 7763!
DISTR. **U** 1–4; **K** 3–5; **T** 1, 4, 6, 8; pantropical weed, widespread in tropical Africa
HAB. Weed of cultivation and ruderal sites, in grassland, on black cotton soil; 750–1900 m
USES. None recorded on specimens from our area
CONSERVATION NOTES. Least concern (LC)

SYN. *C. procumbens* Pers., Syn. Pl. 2: 471 (1807), *nom. illegit.*; Oliv. & Hiern, F.T.A. 3: 395 (1877)
C. americanum auctt. e.g. F.P.S. 3: 17 (1956); F.P.U.: 174 (1962); C.D. Adams in F.W.T.A. ed. 2, 2: 234, t. 250 (1963); Wild in Kirkia 6: 34 (1967); U.K.W.F. ed. 2: 217 (1994), *non* (L.) Vatke

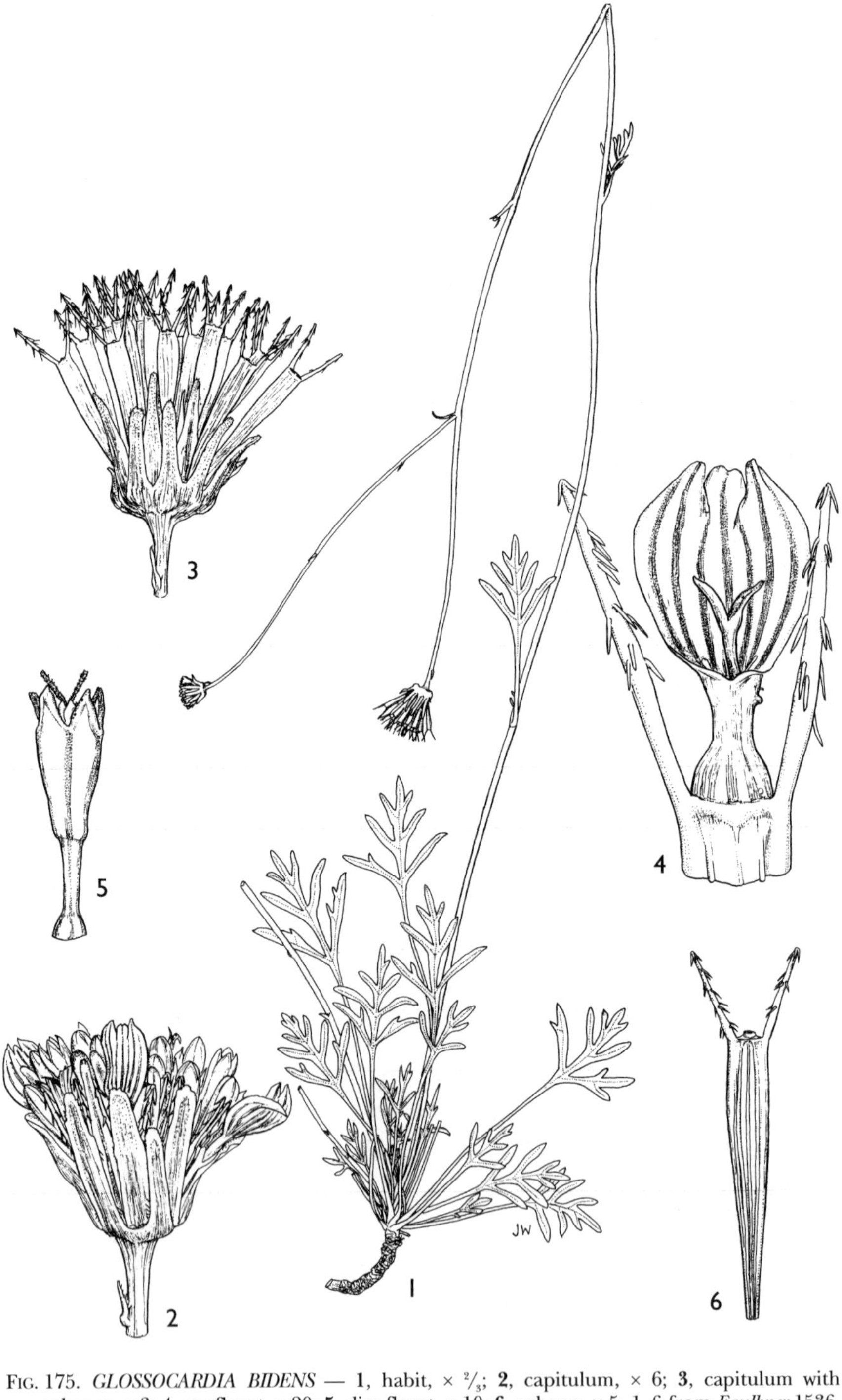

FIG. 175. *GLOSSOCARDIA BIDENS* — **1**, habit, × ²⁄₃; **2**, capitulum, × 6; **3**, capitulum with achenes, × 3; 4, ray floret, × 20; **5**, disc floret, × 10; **6**, achene, × 5. 1–6 from *Faulkner* 1536. Drawn by Juliet Williamson.

129. **GLOSSOCARDIA**

Cass. in Bull. Soc. Philom. (1817): 138 (1817) & in Dict. Sc. Nat. 19: 62 (1821)

Annual or perennial herbs. Leaves alternate or sometimes opposite or rosulate, pinnatisect or simple. Capitula solitary or few together, radiate; involucre 1–4-seriate. Ray florets female or neuter, yellow-white or purple; disc florets often functionally male, often 4-lobed. Achenes compressed; pappus of 2 aristae or a crown.

Twelve species in S Asia, Australia and the Pacific; one species possibly introduced to Africa, though C.Jeffrey (pers. comm.) considers this to be a native, circum-maritime species.

Glossocardia bidens (*Retz.*) *Veldk.* in Blumea 35 (2): 468 (1991). Type: India, Bengala, *König* s.n. (LD, holo.)

Perennial herb, with few mostly decumbent branches to 50 cm long, often almost leafless. Leaves mainly basal, alternate, also tufted at nodes under lateral branches, the lowermost sometimes simple, but usually deltoid in outline, pinnatifid, 1–4 cm long, 0.5–3 cm wide, with few narrow lobes, subglabrous. Capitula few, on long peduncles, radiate; phyllaries ovate, grading into paleae. Ray florets 5–12, yellow with purple veins, female, to 5 mm long; disc florets 7–12, yellow, 2.5–3.5 mm long; anthers black. Achenes dark brown to black, ellipsoid, 5–8(–13) mm long, glabrous or with barbs along the ribs; with 2 aristae to 5 mm long, with retrorse barbs. Fig 175 (p. 812).

TANZANIA. Tanga District: Tongoni, Nov. 1954, *Faulkner* 1536!
DISTR. **T** 3; possibly an introduction; Nepal, India and Sri Lanka to Korea, Japan and the Pacific
HAB. Coastal wooded grassland; ± 0 m
USES. None recorded on the single specimen from our area
CONSERVATION NOTES. Least concern (LC)

SYN. *Zinnia bidens* Retz., Obs. Bot. 5: 28 (1788)
    Glossogyne bidens (Retz.) Alston in Trimen, Handb. Fl. Ceylon 3: 41 (1895)

130. **AMBROSIA**

L., Sp. Pl.: 987 (1753)

Annual or perennial herbs, less often shrubs, aromatic. Leaves alternate or sometimes opposite. Capitula unisexual, monoecious. Male capitula small, in racemes or spikes; involucre cupuliform, of connate phyllaries; corolla tubular; anthers ± free, with linear apical appendage; ovary abortive. Female capitula solitary, axillary to upper leaves but below male heads, sessile, 1-flowered; involucre enveloping the achene, becoming hardened, with 4–6 small spines; corolla absent; style branches long and linear, exserted from involucre. Achenes ovoid, smooth, enveloped in hardened involucre; pappus absent.

40 species from the Americas; several have become pantropical weeds.

Ambrosia maritima *L.*, Sp. Pl.: 987 (1753); Oliv. & Hiern, F.T.A. 3: 370 (1877); C.D. Adams in F.W.T.A. ed. 2, 2: 268 (1963); Wild in Kirkia 6: 2 (1967); Lisowski, Asterac. Fl. Afr. Centr. 1: 110 (1991); U.K.W.F. ed. 2: 215 (1994). Type: Mediterranean, 'habitat in Hetruriae, Cappadociae maritimis arenosis', Herb. Clifford: 443, Ambrosia 1 (BM, lecto., designated by Alavi in Jafri & El-Gadi, Fl. Libya 107: 120 (1983))

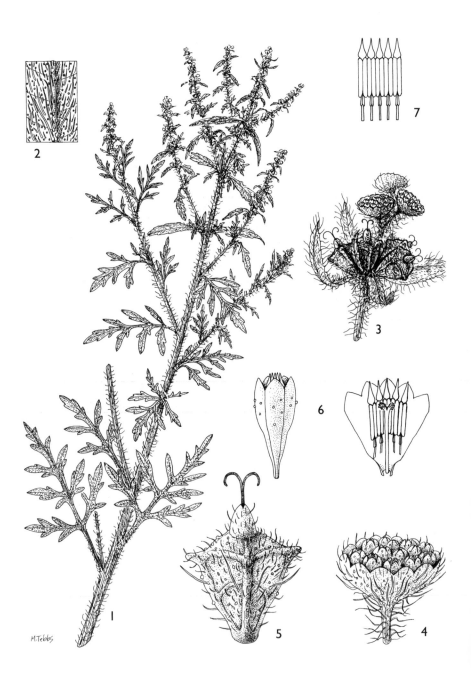

FIG. 176. *AMBROSIA MARITIMA* — **1**, habit, × ²⁄₃; **2**, detail of lower surface of leaf, × 9; **3**, ♀ and ♂ capitula, × 4; **4**, ♂ capitulum, × 10; **5**, ♀ floret, × 14; **6**, ♂ floret and floret section, × 14; **7**, stamens, × 14. All from *Bautista et al.* 3457. Drawn by Margaret Tebbs, from Flora de Bahia.

Annual herb 0.4–1.2 m high, aromatic; stems pilose. Leaves alternate, ovate in outline, deeply lobed, the lobes again lobed, 3–13 cm long, 1.5–7 cm wide, the lobes obtuse with crenate margins, pubescent on both surfaces but more densely so beneath; petiole to 5 cm long. Male capitula in racemes to 13 cm long, sessile or nearly so; involucre to 5 mm in diameter; florets 15–20, white, campanulate, 2–2.5 mm long. Female florets a few together below the base of the male inflorescence, involucre 5–6 mm long with 4–5 spines 2 mm long. Achenes dark grey, ovoid, 1.2 mm long, smooth; pappus absent. Fig. 176 (p. 814).

UGANDA. Bunyoro, Kitoba, Apr. 1942, *Purseglove* 1218!; Mengo District: Kampala, Port Bell, Dec. 1934, *Chandler* 1192!; Masaka District: Lake Nabugabo, Bale, May 1972, *Lye* 6992!
KENYA. South Kavirondo District: Kindu–Homa Point road, Sep. 1933, *Napier* 5355!
TANZANIA. Mwanza District: Ngwanganda, May 1953, *Tanner* 1531!; Dodoma District: Ikowa Reservoir, June 1974, *Mhoro & Backeus* 1953!; Rufiji District: Mafia I., Kanga, Dec. 1977, *Wingfield* 4491a!
DISTR. **U** 1, 2, 4; **K** 5; **T** 1, 5–7; a widespread weed
HAB. Shores of the sea, lakes, dams, rivers, also in cultivations; 0–1200 m
USES. Minor medicine against snakebite (*Tanner*)
CONSERVATION NOTES. Least concern (LC)

SYN. *A. senegalensis* DC., Prodr. 5: 525 (1836). Types: Senegal, *Perrotet* s.n. (P!, syn.) & *Bacle* s.n. (G-DC, syn.)

NOTE. The first record from East Africa is from Entebbe, Uganda, in 1923.

131. **PARTHENIUM**

L., Sp. Pl.: 988 (1753)

Herbs or shrubs. Leaves alternate. Capitula in terminal panicles, radiate, heterogamous; involucre with phyllaries in 2–3 series; receptacle paleate. Female florets with minute rays; disc florets infundibuliform; anthers with apical appendage, obtuse at base; style of ray florets bifid, of disc florets entire and obtuse. Achenes of ray florets compressed, keeled, enveloped by phyllary and two paleae; pappus of 2 recurved awns enclosed in the paleae.

Twenty species from tropical and subtropical America; one pantropical weed.

Parthenium hysterophorus *L.*, Sp. Pl.: 988 (1753); Wild in Kirkia 6: 8 (1967); D.J.N. Hind in Fl. Masc. 109: 214, t. 74 (1993). Type: Jamaica (LINN 1115/1, lecto., chosen by Stuessy in Ann. Miss. Bot. Gard. 62: 1094 (1975))

Annual herb to 60 cm high; stems puberulous or scabridulous. Leaves elliptic to ovate, bipinnately lobed, 6–30 cm long, 3–10 cm wide, lobes 1.5–4 mm wide and obtuse, puberulous on both surfaces. Capitula hemispherical, in lax panicles; involucre 2–3 mm long, the phyllaries pubescent. Ray florets 5, whitish, almost round, 0.5–1 mm; disc florets yellowish, narrowly funnel-shaped, ± 2.7 mm long. Achenes black, obovoid, ± 2 mm long, with 2 recurved pappus awns. Fig. 177 (page 816).

KENYA. Kiambu District: Ruiru, Dec. 1975, *Cordingley* in EA 15978! & June 1985, *Terry* 3529! & Kiambu, Kiamara Estate, July 1979, *Mathara* in EA 16418!
DISTR. **K** 4; from Caribbean America, now a pantropical weed
HAB. Weed of coffee plantations; ± 1600 m
USES. None recorded on specimens from our area
CONSERVATION NOTES. Least concern (LC)

NOTE. The first record from East Africa is from Ruiru, Kenya, in 1973; resistant to most herbicides.

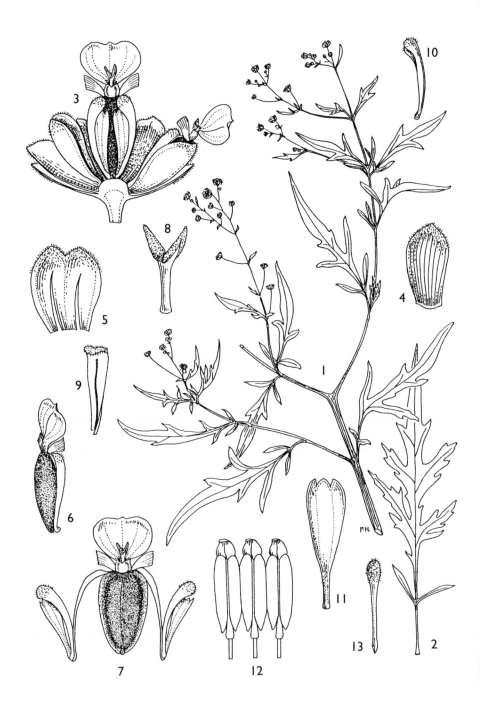

FIG. 177. *PARTHENIUM HYSTEROPHORUS* — **1**, habit, × ²/₃; **2**, basal leaf, × ²/₃; **3**, capitulum section, × 12; **4–5**, outer and inner phyllaries, × 12; **6**, ray floret complex, side view, × 12; **7**, ray floret complex, × 12; **8**, ray floret style, × 36; **9**, palea, × 12; **10**, palea side view, × 12; **11**, disc floret corolla, × 24; **12**, stamens, × 36; **13**, disc floret style, × 36. 1–13 from *Cadet* 2522. Drawn by Pat Halliday, from Flore des Mascareignes.

132. **XANTHIUM**

L., Sp. Pl.: 987 (1753)

Annual herbs. Leaves alternate. Capitula unisexual, monoecious, axillary or terminal, discoid. Male capitula globose, many-flowered; involucre with few phyllaries; receptacle paleate; corolla subcylindric, 5-dentate; anthers with minute apical appendage; style undivided. Female capitula beneath the male, 2-flowered; involucre connate, 2-beaked at apex and covered with hooked bristles; corolla absent; style bifid. Achenes enclosed in the hardened involucre; pappus absent.

Either 2 or 13 species (the taxonomy is disputed); global weeds.

Xanthium strumarium *L.*, Sp. Pl.: 987 (1753); Oliv. & Hiern, F.T.A. 3: 371 (1877); Wild in Kirkia 6: 3 (1967); D.J.N. Hind in Fl. Masc. 109: 218, t. 75 (1993). Type: no locality given (LINN 1113/1, lecto., chosen by Rechinger in Fl. Iranica 164: 39 (1989))

Coarse annual herb to 1.5 m high; stems scabridulous, often reddish. Leaves broadly deltate or broadly ovate, simple or 3–5-lobed, 5–12 cm long, 4–18 cm wide, base cordate as well as broadly cuneate, margins coarsely crenate, apex acute or obtuse, scabrid on both surfaces; 3-veined from base; petiole to 12 cm long. Male capitula terminal on short axillary branches; involucre ± 3 mm long; corolla ± 2.5 mm long. Female capitula to 4 mm long, with connate involucre. Fruit ellipsoid, 1.5–2.3 cm long, covered in hooked spines 3–4 mm long, apex acute. Fig. 178 (page 818).

Kenya. North Kavirondo District: Kakamega town, July 1968, *Tweedie* 3550!; Nairobi, Mathare village, Sep. 1971, *Mwangangi & Kasyoki* 1791!; Teita District: Tsavo East National Park, Sala–Galdessa, June 1995, *Luke et al.* 4360!
Tanzania. Masai District: Ngorongoro, Kakesio, May 1989, *Chuwa* 2775!; Dodoma District: Dodoma town, May 1978, *Sigara* in *Ruffo* 1134! & *Ruffo* 1304!
Distr. **K** 4, 5; **T** 2, 5; probably originally from North America; now a global weed
Hab. Ruderal sites; 20–1650 m
Uses. None recorded on specimens from our area
Conservation notes. Least concern (LC)

Syn. *X. pungens* Wallr. in Beitr. Bot. 1: 231 (1844); Widder in Phyton 12: 182 (1967); U.K.W.F. ed. 2: 215 (1994). Type: Germany, no further details found

Note. The first record from East Africa is from Kakamega, Kenya, in 1962.
    The related *X. spinosum* has not (yet) been found in our area; it is a widespread tropical weed and differs from *X. strumarium* in the 3-partite yellow axillary spines on the stem.

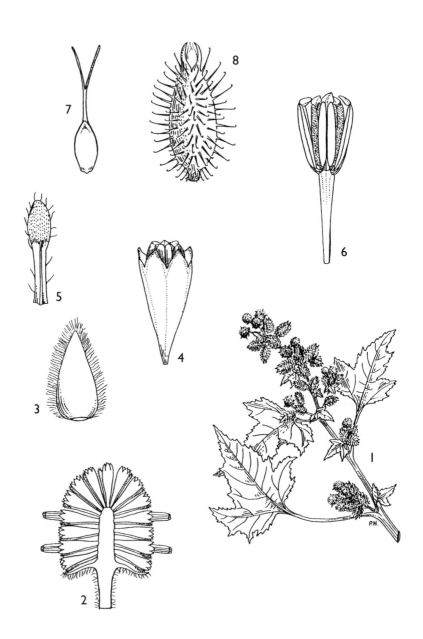

Fig. 178. *XANTHIUM STRUMARIUM* — **1**, habit, × ²/₃; **2**, ♂ capitulum section, × 6; **3**, phyllary, × 12; **4**, ♂ floret, × 16; **5**, palea, × 16; **6**, stamens, × 16; **7**, ♀ floret, × 6; **8**, phyllary of ♀ capitulum, × 2. 1–8 from *Guého* in MAU 18696. Drawn by Pat Halliday, from Flore des Mascareignes.

EUPATORIEAE*

Cass. in J. Phys. Chim. Hist. Nat. Arts 88: 150–169, 189–204 (1819); King &
Robinson, Genera of the Eupatorieae (1987)

Heliantheae Cass. supersubtribus *Eupatoriodinae* (Cass.) C.Jeffrey in Bot. Zhurn. 87,
11: 11 (2002)

Annual herbs, or more usually shrubs, sub-shrubs, rarely trees (not in East Africa),
climbing, procumbent or erect. Leaves simple, usually opposite, often glandular-
punctate and often with three veins from the base. Inflorescence terminal or
sometimes axillary, usually a corymbose panicle. Capitula homogamous, discoid;
phyllaries in (1–)2–many series; receptacle usually naked. Florets hermaphrodite;
corolla regular, 5-lobed, lobes short and often glandular or pubescent; anthers with
flat apical appendages; style-arms long, obtuse, long-exserted, often forming most
conspicuous part of floret. Achenes black when mature; carpopodium usually
evident; pappus of awns, bristles, hairs, gland-tipped pegs, a corona or absent.

About 2400 species in 170 genera, primarily from the New World. A number of the taxa
occurring in East Africa are pantropic weeds.

1. Pappus of 3–5 viscid-tipped knobs . . . .   133. **Adenostemma**
   Pappus of bristles, hairs or scales, or occasionally
   absent 2
2. Pappus of ± 5 awned scales, or awnless, reduced, or
   occasionally absent . . . .   135. **Ageratum** (p. 827)
   Pappus of hairs or bristles 3
3. Climbing shrubs or sub-shrubs; phyllaries and florets
   4; phyllaries often with swollen bases . . . .   137. **Mikania** (p. 833)
   Erect shrubs or subshrubs; phyllaries more than 4,
   without swollen bases 4
4. Phyllaries and florets numerous; phyllaries distinctly
   ribbed; pappus setae fewer than 15 . . . .   136. **Ageratina** (p. 831)
   Phyllaries 5–7; florets 4–5; phyllaries not ribbed;
   pappus dense, setae numbering more than 20 . .   134. **Stomatanthes** (p. 825)

133. ADENOSTEMMA

J.R.Forst. & G.Forst., Char. Gen. Pl.: 89 (1776); R.M. King & H. Rob., Genera of the
Eupatorieae: 58 (1987); Bremer, Anderberg, Karis & Lundberg in Asteraceae Clad.
& Class.: 648 (1994)

Perennial herbs, poorly branched. Leaves opposite. Inflorescence a lax terminal
cyme, with foliaceous bracts. Involucre 1–2-seriate, spreading and then reflexed with
age, phyllaries basally connate; receptacle flat becoming convex with age, deeply
alveolate. Florets hermaphrodite, corolla white, funnel-shaped or almost cylindrical,
tube sparsely glandular-hairy, lobes short, triangular with distal collar of dense long
hairs which mat the corollas together; anther-bases obtuse, apical appendages
obtuse, broader than long; style yellow, arms swollen and flattened, glabrous or with
few short black hairs. Mature achenes oblong, slightly curved, distinctly 3-angled with
distinct apical nectary. Carpopodium distinct, asymmetrical; pappus of pegs, connate
about the nectary, tipped with dark brown, viscid, glandular knobs.

* by S.A.L. Smith, c/o the Herbarium, Royal Botanic Gardens, Kew, Richmond, Surrey
TW9 3AE, U.K.

About 24 species in Central America, West Indies, South America, Africa, Asia and the Pacific Islands.

Note: The structure of the capitula is ideal for the efficient dispersal of mature achenes. The hairs on the outer surface of the corolla become matted together so that, at maturity, the whole mass lifts away together, aided by the outward movement of the receptacle. This exposes the pappus pegs with their sticky tips to contact with animals and thus they are dispersed. Because of this, the genus is widespread and characterised by considerable variation which has in turn led to much confusion in its taxonomy. The group is in desperate need of revision, the last comprehensive work being that of de Candolle (1836). Since that time, a number of conflicting views have been published in the literature. In F.T.A., Oliver placed *A. mauritianum*, *A. caffrum* and *A. perrotteti* under the pantropic *A. viscosum* of J.R. & G. Forster. F.W.T.A. ed. 1 on the other hand treated *A. mauritianum*, *A. perrotteti* and *A. viscosum* as being the three species represented in the area, with *A. caffrum* appearing in the synonymy of *A. viscosum* Forst. This view was revised for F.W.T.A. ed. 2 where *A. caffrum* was given as an accepted species, with *A. viscosum* Forst. in synonymy. Lisowski in Asteraceae d'Afrique Centrale also recognised 3 species, namely *A. mauritianum*, *A. perrotteti* and *A. caffrum*. He considered that *A. viscosum* had been misused by authors for *A. perrottetii* and *A. caffrum*, but declined to treat *A. viscosum* sensu J.R. Forst. & G. Forst. As a result of my survey, I have chosen an alternative explanation. After inspection of *A. viscosum* J.R. Forst. & G. Forst. from other paleotropical regions, I have relegated *A. perrottetii* to the synonymy of *A. viscosum*, a view supported by King & Robinson (1987). *A. mauritianum* is easily distinguished by its glabrous achenes (other African taxa have tuberculate achenes), and is retained. *A. caffrum* differs from *A. viscosum* by size and number of capitula and by number of corolla lobes and anthers, and by sessile, lanceolate leaves. It is endemic to Africa, whereas *A. viscosum* has pantropic distibution.

1. Mature achenes without tubercles .... 3. *A. mauritianum*
    Mature achenes tuberculate 2
2. Leaves petiolate to 3.5 cm, blade ovate to rhomboid; capitula generally 5 mm diameter or less, few to numerous; corolla less than 1.5 mm long with 3–4(–5) lobes .... 1. *A. viscosum*
    Leaves sessile to shortly petiolate (to 1.6 cm), blade lanceolate to ovate-lanceolate (in East Africa); capitula generally more than 5 mm diameter, few; corolla 2 mm or longer with 5–6 lobes .... 2. *A. caffrum*

1. **Adenostemma viscosum** *J.R.Forst. & G.Forst.*, Char. Gen. Pl.: 90, t. 45 (1776); DC., Prodr. 5: 112 (1836); Oliv. & Hiern, F.T.A. 3: 299 (1877); P.O.A. C.: 406 (1895); D.J.N. Hind in Fl. Mascar. 109: 231 (1993); Retief & Herman, Pl. Northern Provs of S.A.: 290 (1997). Type: Society Is., Tahiti, *Forster* s.n. (K!, ?holo.)

Erect to semi-procumbent herb to 1.3 m; stem fleshy, glabrous or sparsely pubescent, sometimes rooting from lower nodes. Leaves petiolate, petiole 0.7–4 cm, blade ovate to rhomboid, 2–15(–22) cm long, 1–7(–12.5) cm wide, base attenuate to cuneate, sometimes forming narrow wings along petiole, margins crenulate to serrate, apex obtuse to acute, glabrous or sub-glabrous, 3-veined from base. Inflorescence bracts 0.6–5 cm long at nodes of inflorescence branches. Capitula few to numerous, ± 3–5 mm long, 4–5 mm diameter when mature; stalks of individual capitula 1.2–2.5 cm, glandular-pubescent; phyllaries 2–3 mm long, 0.5–0.75 mm wide, flat, apex obtuse to acute, margins entire or ciliate, basally pubescent. Florets ± 30–42; corolla white, 1–1.5 mm long, basally wider, then narrowing into tube before expanding again into campanulate limb with 3–4(–5) (variable within capitulum) short, broadly triangular lobes; anthers free, 3–4(–5), ± 0.2 mm long; style ± 2 mm, bifid for $\frac{1}{3}$ of its length, arms swollen. Achenes 2.2–3 mm long, immature or aborted achenes purplish, generally smooth with small glands, becoming brown, tuberculate and glandular with maturity; pappus of 3 pegs, less than 1 mm long.

UGANDA. Bunyoro District: Bujenje, Budongo Forest, 24th Oct. 1971, *Synnott* 690!; Mbale District: Bufumbo, Bugisu [Bugishu], Nov. 1932, *Chandler* 999!; Masaka District: Bugala I., Sese Is., 6 Oct. 1958, *Symes* 462!

KENYA. Northern Frontier: Mt Nyiru, on track from Tuun on W side over Mt to South Horr, 29th Oct. 1978, *Gilbert, Gachathi & Gatheri* 5210!; Trans-Nzoia District: Kitale, 30th Oct. 1953, *Bogdan* 3806!; Kwale District: Shimba Hills, Risley's Ridge, 21st Mar. 1991, *Luke & Robertson* 2755!

TANZANIA. Musoma District: Klein's Camp, 23rd May 1962, *Greenway* 10658!; Moshi District: above Machame Central Girl's School, 22nd Feb. 1955, *Huxley* 135!; Mpanda District: Kasoje, Kungwe Mt, 17th July 1959, *Newbould & Harley* 4442!; Zanzibar I., Jozani Forest, 26 Nov. 1960, *Faulkner* 2733!

DISTR. **U** 1–4; **K** 1, 3, 4, 6, 7; **T** 1–8; **Z**; Senegal, Guinea Bissau, Sierra Leone, Liberia, Ivory Coast, Ghana, Nigeria, Cameroon, Bioko, São Tomé, Annobon, Gabon, Central African Republic, Congo (Kinshasa), Burundi, Sudan, Ethiopia, Angola, Zambia, Malawi, Zimbabwe, Mascarenes, South Africa; India, Sri Lanka, Australia, Pacific Is., Indonesia

HAB. Wet ground, often near streams, and in shade, sometimes actually growing in water; 1–2000 m

USES. None recorded on specimens from our area

CONSERVATION NOTES. Least concern (LC)

SYN. *A. perrottetii* DC., Prodr. 5: 110 (1836); Brenan in Mem. N.Y. Bot. Gard. 8: 463 (1954); F.P.S. 3: 8 (1956); C.D. Adams in F.W.T.A. ed. 2, 2: 286 (1963); E.P.A.: 1079 (1966); H.M. Burkill, Useful Pl. W. Trop. Afr. ed. 2, 1: 442 (1985); Maquet in Fl. Rwanda 3: 568, fig. 173/1 (1985); Lisowski, Aster. Afr. Centr. 2: 469 (1991); U.K.W.F. ed. 2: 203 (1994). Type: Cape Verde Is., *Perrottet* 27 (G-DC, holo., K!, microfiche)

NOTE. The dimorphic nature of the achenes in this species could be the source of some confusion. This can be clearly observed on many of the specimens held at K, but it does seem to be correlated with age of the capitulum. Perhaps the glands seen on the immature, smooth, deeply pigmented achenes are responsible for secreting the tuberculate integument seen on mature achenes. If mature achenes are present they remain a reliable character for separating this taxon from *A. mauritianum*, but the number of corolla lobes and anthers can be used for confirmation. Some sheets at K have been annotated with the name *A. lavenia* (L.) Kuntze. This species was regarded by King & Robinson (1987) as being endemic to Sri Lanka. Further revisionary work on the Old World species is required before it can be decided whether or not it is distinguisable from *A. viscosum.*

2. **Adenostemma caffrum** *DC.*, Prodr. 5: 112 (1836); Harv. in Fl. Cap.: 58 (1865); F.P.S. 3: 8 (1956); C.D. Adams in F.W.T.A. ed. 2, 2: 286 (1963); E.P.A.: 1078 (1966); H.M. Burkill, Useful Pl. W. Trop. Afr. ed. 2, 1: 442 (1985); Maquet in Fl. Rwanda 3: 568, fig. 173/2 (1985); Lisowski, Aster. Afr. Centr. 2: 471 (1991); U.K.W.F. ed. 2: 203, t. 82 (1994). Type: South Africa [Kafferland], *Drege* 5081 (G-DC, holo., K!, microfiche)

Vigorous, erect or semi-procumbent herb to 1.5 m, rooting at nodes on wet ground; stem ± fleshy, large stems woody and hollow, sometimes ribbed, glabrous or sparsely pubescent, sometimes with glandular hairs. Leaves sessile or shortly petiolate, blade triangular, broadly ovate, ovate to lanceolate, 4–21 cm long, 0.3–8.5 cm wide, margins sub-entire to serrate, crenulate or sharply dentate, sometimes minutely revolute, apex obtuse to acuminate, glabrous or pubescent. Inflorescence a loose terminal cyme, bracts ± amplexicaul, to 4.4 cm long; capitula few to several, 6–10 mm diameter when mature; stalks of individual capitula 0.4–7.2 cm, pubescent; phyllaries 4–6.5 mm long, 1–1.5 mm wide, flat, basally pubescent, apex obtuse to acute. Florets ± 50; corolla 3–4 mm long, white, 5–6 lobed; anthers free or weakly fused, 5(–6), ± 1 mm long; style conspicuous, 4–6.5 mm long, bifid for ± $^{2}/_{3}$ of its length, arms swollen and flattened, glabrous or with a few black hairs. Mature achenes brown, 2–3.5 mm long, tuberculate, glands present on tubercules (and sometimes between); pappus of (2–)3–4(–5) pegs.

Leaves broadly lanceolate to ovate lanceolate, 2.5–6.5 times
    longer than wide . . . . var. *asperum*
Leaves linear-lanceolate, 10–15 times longer than wide . . . . var. *angustifolia*

var. **asperum** Brenan in Mem. N.Y. Bot. Gard. 8: 462 (1954). Type: Malawi, Kota-kota District, Nchisi Mt, *Brass* 16979 (K, holo., lost?, NY, para.)

Leaves sessile or shortly petiolate (to 1.6 cm), blade lanceolate to ovate-lanceolate, 4–12.5 cm long, 1.3–4.8(–6.2) cm wide, margins sub-entire to serrate or crenulate, sometimes minutely revolute, apex obtuse.

UGANDA. Kigezi District: Kanungu, 7 June 1952, *Lind* 58!; Toro District: Nyakasura, on hill near crater, 7 Jan. 1936, *Hancock* 113/36!; Masaka District: NW side of lake Nabugabo, 9 Oct. 1953 *Drummond & Hemsley* 4681!
KENYA. Trans-Nzoia District: Kitale, Prison dam, Oct. 1969, *Tweedie* 3722! & Mt Elgon, 22 Mar. 1931, *E.J. & C. Lugard* 574!; Uasin Gishu District: Kipkarren, no date, *Brodhurst-Hill* 151!
TANZANIA. Ufipa District: Tatanda mission, near Zambian frontiers on Mbala–Sumbawanga road, 9 June 1980, *Hooper, Townsend & Mwasumbi* 1918!; Iringa District: Mufindi, W side of Lake Ngwazi, 16 Feb. 1986, *Bidgood & Lovett* 6!; Songea District: 2 km E of Songea by Nonganonga stream, 27 Apr. 1956, *Milne-Redhead & Taylor* 9836!
DISTR. **U** 2, 4; **K** 3, 5; **T** 1, 4, 7, 8; Guinea, Ghana, Nigeria, Cameroon, Central African Republic, Congo (Kinshasa), Rwanda, Burundi, Sudan, Ethiopia, Angola, Zambia, Malawi, Zimbabwe, Botswana, South Africa
HAB. Wet ground and swamps, especially in disturbed sites; 1050–2300 m
USES. Minor medicinal for chest diseases (*Brodhurst-Hill*)
CONSERVATION NOTES. Least concern (LC)

SYN. *A. schimperi* A.Rich., Tent. Fl. Abyss. I: 382 (1848). Type: Ethiopia, near Adoua, *Quartin Dillon* s.n. & *Schimper* 112 (?P, syn., BM!, isosyn.)
      [*A. viscosum* sensu Oliv. & Hiern in F.T.A. 3: 299 (1877) pro parte, *non* J.R. Forst. & G. Forst.]

VARIATION. The number of pegs forming the pappus appears to be far more variable as a character than previously recognised. The predominant number is 3 but up to 5 have been recorded. This variability can be seen within a capitulum. The presence of tubercles and/or glands on the achenes is considered a reliable character for identifying species of *Adenostemma*, but there is a collection from Iringa district, Black Camp Bay, Lake Ngwazi (*Renvoize & Car* 1970) which on first inspection appears to be close to *A. caffrum* var. *asperum* but has completely glabrous achenes.

NOTE. Brenan described var. *asperum* from Malawi on the basis of stronger denser indumentum of leaves and stem, and lanceolate to ovate-lanceolatae leaves with shallower marginal serrations than the typical var. described from South Africa. Material at Kew from South Africa, Malawi and Tropical East Africa shows a huge range of variation in indumentum, and I can find no correlation between indumentum and leaf shape or serration. However var. *asperum* is upheld here on the basis of leaf shape and margins type alone. The leaves of East African material differs from the typical variety in that they tend to be lanceolate or ovate-lanceolate, not triangular to triangular-ovate, and the margins are entire to serrate or crenulate and not sharply dentate. There are a couple of Tanzanian specimens with leaves approaching the typical variety in shape, but not margins type (Ufipa District: Malonje Plateau, near Molo village, *Richards* 15855; Iringa District: 30 km from Mafinga [Sao Hill] on Mbeya Road, *Bidgood, Mwasumbi & Vollesen* 806). Generally it appears that var. *caffrum* is a South African taxon, creeping up into Angola, Zambia & Malawi, and these two Tanzanian specimens may represent var. *caffrum* at its most northern limit. Alternatively, they may be intermediates. Likewise, var. *asperum* is a more northern taxon, being well represented in West, central and tropical S Africa. There are a few specimens from South Africa which resemble var. *asperum* in leaf shape, or margins type, and these may represent the southernmost limit of the distribution, or may be intermediates.

var. **longifolium** (*Chiov.*) *S.A.L.Smith* **comb. nov.** Type: Kenya, Naivasha District: S of Lake Naivasha, *Mearns* 687 (US!, holo., BM!, iso.)

Semi-procumbent to erect herb to 90 cm; stem fleshy, glabrous, often tinged red. Leaves sessile, membranous, linear-lanceolate, 4–21 cm long, 0.3–1.7 cm wide, margins sub-entire to minutely to coarsely serrate, apex acuminate, glabrous. Inflorescence bracts minute, 5 mm or less; capitula 2–8, stalks of individual capitula (1.4–)3.3–7.2 cm, sparsely to densely pubescent; phyllaries 4–5 mm long, 1–1.5 mm wide, sometimes tinged red. Corolla 4 mm long; style-arms sparsely covered with short black hairs. Achenes tuberculate or occasionally glandular without apparent tubercles.

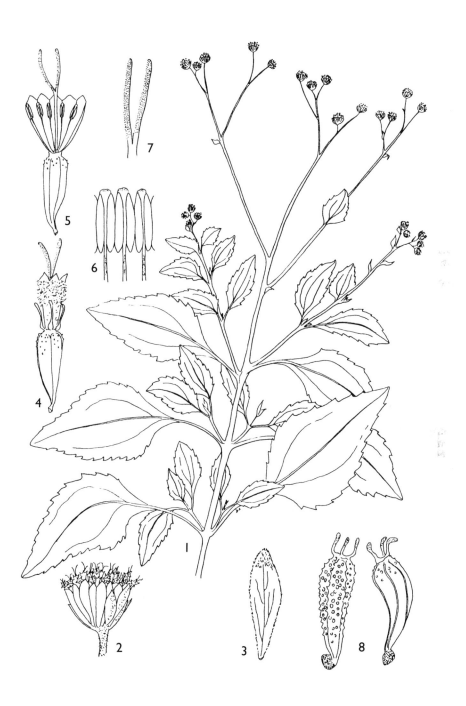

FIG. 179. *ADENOSTEMMA MAURITIANUM* — **1**, habit, × ¹/₂; **2**, capitulum, × 4; **3**, phyllary, × 6;
4, floret, × 6; **5**, opened floret, × 6; **6**, stamens, × 12; **7**, style branches, × 12; **8**, achenes, × 12.
1–8 from *Commerson* s.n. Drawn by Pat Halliday, from Flore des Mascareignes.

UGANDA. Teso district: Bugondo, Lake Kioga, Dec. 1931, *Chandler* 313! & July 1934, *Synge* 754!; Masaka District: Lwera, 32 km on Masaka–Kampala road, 11 Feb. 1971, *Kabuye* 338!
KENYA. Naivasha District: Lake Naivasha, June 1932, *Soland* 2105! & S side of Lake Naivasha, 12-16 July 1909, *Mearns* 683! & 687!
TANZANIA. Buha District: Kaberi swamp, 10 Aug. 1950, *Bullock* 3126!
DISTR. U 3, 4; **K** 3; **T** 4; Sierra Leone, Guinea, ?Botswana
HAB. Lake sides or swamp land, sometimes in water up to 3 feet deep; 1100–1900 m
USES. None recorded on specimens from our area
CONSERVATION NOTES. Least concern (LC)

SYN. *A. lavenia* (L.) Kuntze var. *longifolium* Chiov., Racc. Bot. Miss. Consol. Kenya: 62 (1935)

NOTE. There is little material of this subspecies at Kew. It may turn out to be related to the *A. angustifolia* Arn. described from India which also has narrow lanceolate leaves, but has the smaller heads of *A. viscosum*. Again, this taxon is poorly represented at Kew.

3. **Adenostemma mauritianum** *DC.*, Prodr. 5: 110 (1836); Bojer, Hort. Maurit. 177 (1837); F.P.S. 3: 8 (1956); E.P.A.: 1079 (1966); C.D. Adams in F.W.T.A. ed. 2, 2: 286 (1968); Maquet in Fl. Rwanda 3: 568, fig. 173/3 (1985); Lisowski, Aster. Afr. Centr. 2: 467 (1991); D.J.N. Hind in Fl. Mascar. 109: 229, t. 80 (1993); U.K.W.F. ed. 2: 203 (1994). Type: Mauritius, *Sieber* 119 & 364 (G-DC, syn.; K!, microfiche)

Herb to 1.2 m; stem fleshy, ascending, rooting at base, sometimes striate, glabrous or sparsely pubescent, more so towards the apex. Leaves petiolate, blade broadly triangular to ovate, 3.2–16 cm long, 1.7–9(–11) cm wide, base attenuate to form narrow wings along petiole, margins serrate, dentate or crenulate, apex acute, lamina membranous, darker above than beneath, glabrous or sub-glabrous with few scabrid hairs; petiole 0.5–7.2 cm. Inflorescence branching ± dichotomously, bracts 0.5–5 cm long, sessile; capitula 4–10 mm diameter when mature; stalks of individual capitula 1.2–5 cm, densely pubescent; phyllaries green, flat, narrowly elliptic to oblong, 4–5 mm long, 1–1.5 mm wide, sparsely pubescent, margins entire or ciliate, apex obtuse to acute. Florets 21–28; corolla white, 2–3 mm long, tube sparsely glandular-stipitate, gradually widening into limb, lobes (4–)5; anthers free, 5, ± 1 mm long; style 3.5–4 mm long, bifid for more than half its length, arms swollen and flattened, glabrous. Mature achenes 2.5–4 mm long, smooth, glabrous or sometimes with minute stipitate glands, never tuberculate; pappus of 3–4 pegs, 1–1.2 mm long. Fig. 179 (page 824).

UGANDA. Karamoja District: Mt Debasien, Mareyo, 1936, *Eggeling* 2768!; Toro District: Msandama [Musandama] Hill, 16 Dec. 1925, *Maitland*!; & Ruwenzori Mts, near Minimba Camp, 21 Jan. 1962, *Loveridge* 366!
KENYA. Trans-Nzoia District: NE Elgon, Aug. 1951, *Tweedie* 921!; North Kavirondo District: Kakamega Forest station, 22 Dec. 1967, *Perdue & Kibuwa* 9415!; Embu District: Castle Forest Station, 12 Dec. 1966, *Gillett* 18046!
TANZANIA. Mbulu/Masai District: Ngorongoro conservation area, Oldeani [Oldean] Mt, 10 Oct. 1988, *Chuwa* 2645!; Morogoro District: Uluguru, 20th July 1958, *Gilli* 564!; Tanga District: Muheza, East Usambara Mts, Lutindi Forest Reserve, Nilo peak area, 11 May 1987, *Iversen, Persson & Pettersson* 87474!
DISTR. U 1–3; **K** 3–7; **T** 2–4, 6, 7; Sierra Leone, Ivory Coast, Ghana, Nigeria, Cameroon, Bioko, Congo (Kinshasa), Rwanda, Burundi, Zambia, Malawi, Zimbabwe, South Africa, Madagascar, Mascarenes
HAB. In forest, forest edges, stream banks, bamboo zone; 1300–2750 m
USES. None recorded on specimens from our area
CONSERVATION NOTES. Least concern (LC)

SYN. *Lavenia erecta* Sieber, Fl. Maurit. II, no. 119 & 364 in sched.
    [*Adenostemma viscosum* sensu Oliv. & Hiern, F.T.A. 3: 299 (1877) pro parte; Humbert, Fl. Madag. 189: 200, fig. 39/1–4 (1960), *non* J.R. Forst. & G. Forst.]

134. STOMATANTHES

R.M.King & H.Rob. in Phytologia 19: 430 (1970) & the Genera of the Eupatorieae: 69 (1987); Bremer et al. in Aster. Clad. & Class.: 664 (1994)

Perennial herbs, shrubs or subshrubs. Leaves alternate, opposite or ternate. Inflorescence paniculate or corymbose; involucral bracts 4–12 in few series; receptacle glabrous. Florets with corolla funnel-shaped or nearly tubular, lobes triangular; anther filaments basally thick, apical anther appendages ovate; style branches linear to filiform or with clavate apices. Achenes 5–8-ribbed, often setuliferous with distinct carpopodium; pappus of many scabrid bristles.

Fifteen species, found mainly in Brazil and Uruguay, with three species occurring in tropical Africa.

Stomatanthes africanus (*Oliv. & Hiern*) *R.M.King & H.Rob.* in Phytologia 19: 430 (1970); H.M. Burkill, Useful Pl. W. Trop. Afr. ed. 2, 1: 492 (1985); R.M. King & H. Rob., the Genera of the Eupatorieae: 71, t. 6 (1987); Lisowski, Asterac. Afr. Centr. 2: 452, fig. 93 (1991); U.K.W.F., ed. 2: 204, t. 83 pro parte (1994). Type: Sudan-Congo (Kinshasa) border, Niamniam Land, Gumba, *Schweinfurth* 2897 (K!, lecto., selected here, US, isolecto.)

Perennial sub-shrub 20–120 cm high, with 1-several stems arising from a xylopodium 1–3.5 cm in diameter, the stems erect, rigid, woody, somewhat branched, colour from straw-colour through reddish brown to black, short-pubescent, more densely so on apical part. Leaves alternate to sub-opposite, sessile to shortly petiolate, petiole to 6 mm, blade lanceolate to ovate, 0.8–6.2(–8) cm long, 0.2–2.2(–4.5) cm wide, margins often slightly thickened, entire to serrate, sometimes irregularly lobulate, pubescent, especially on veins, sometimes glabrous or pubescence confined to leaf margins, glandular-punctate. Inflorescence of dense corymbs; bracts 3.5–10 mm long, linear to linear-lanceolate and foliaceous; capitula 8–14 mm long, 3–6 mm wide, stalks of individual capitula 2–10 mm, apparently extending in fruit up to 18 mm; phyllaries 5–7, 1–2-seriate, if outer series present 3–5 mm long, 1–1.5 mm wide, somewhat shorter than inner series, 4–9 mm long, 1–3.5 mm wide, green, sometimes with reddish or purplish tips, lanceolate to oblong-ovate, pubescent, ± glandular-punctate towards apices, with scarious, fimbriate margins, apex obtuse; receptacle flat. Florets 4–5, corolla 5–7 mm long, whitish, narrowly campanulate at the apex, 5-lobed, outer surface ± covered with hairs and/or small glands; anthers brown, basally obtuse with ovate apical appendages, anther-collars thickened; style-arms linear, very long, often contorted, shortly appendaged. Achenes brown, 3–4 mm long, ± 6-ribbed, setuliferous or with some evidence of setae and often with glands, straw-coloured carpopodium evident; pappus uniseriate, straw-coloured, 5–8 mm long, dense. Fig. 180 (page 826).

UGANDA. Kigezi District: Shumba Hill, N Rukiga, Aug. 1949, *Purseglove* 3084!; W Nile District: Lindu, 18 Mar. 1945, *Greenway & Eggeling* 7223! & Logiri, Apr. 1939, *Hazel* 725!
KENYA. North Kavirondo District: Kakamega, Jan. 1944, *Carroll* H5!; Trans-Nzoia District: S Cherangani, 18 Feb. 1958, *Symes* 282! & Kitale, Apr. 1964, *Tweedie* 2788!
TANZANIA. Bukoba District: Ndama, Oct. 1931, *Haarer* 2220!; Mpanda District: Mahali Mts, 24 Aug. 1958, *Jefford & Newbould* 1771!; Njombe District: 27 km from Itulo sheep farm HQ on road to Njombe, 13 Nov. 1966, *Gillett* 17806!
DISTR. U 1–4; K 3, 5; T 1, 4, 5, 7, 8; Guinea, Sierra Leone, Ivory Coast, Ghana, Nigeria, Cameroon, Gabon, Congo (Brazzaville), Congo (Kinshasa), Rwanda, Burundi, Sudan, Ethiopia, Angola, Zambia, Malawi, Mozambique, Zimbabwe, South Africa
HAB. Pyrophyte, often found on regularly burned highland grassland, a weed of cultivation, also in woodland; grows in dense clumps; (350–)1050–2450 m
USES. Used for making brooms and brushes (*Corall*)
CONSERVATION NOTES. Least concern (LC)

Fig. 180. *STOMATANTHES AFRICANUS* — **1**, habit, × ²/₃; **2**, lower stems and xylopodium, × ²/₃; **3**, detail of leaf under-surface, × 1.5; **4**, capitulum, × 4; **5**, floret, × 5; **6**, floret with glands, × 6; **7**, floret with glands and hairs, × 6; **8**, achene, × 5. 1 & 3–4 & 6 from *Richards* 20574, 2 from *Harder, Gereau & Kayombo*, 5 & 8 from *Jefford & Newbould* 1771, 7 from *Lea* 11. Drawn by Juliet Williamson.

SYN. *Eupatorium africanum* Oliv. & Hiern, F.T.A. 3: 301 (1877); F.P.S. 3: 29 (1956); C.D. Adams in F.W.T.A. ed. 2, 2: 285 (1963); E.P.A.: 1080 (1966); Maquet in Fl. Rwanda : 572, fig. 174/2 (1985)

    Vernonia humilis C.H.Wright in K.B. 1897: 260 (1897). Types: Malawi, Mt Mlanje, *McClounie* 30 (K!, photo of syn.), Tanzania, Njombe District: Lower plateau north of Lake Nyasa, *Thomson* s.n. (K!, photo of syn.)

    V. malosana Baker in K.B. 1898: 148 (1898). Type: Malawi, Malosa and Zomba Mts, *Whyte* s.n. (K!, holo.)

NOTE. As is common in pyrophytic plants, a high degree of variation in leaf characters was observed in *S. africanus*. The following specimen showed atypical leaf characters: *Bally* 7356! from Kenya, Subukia, has leaves that are up to 9 cm long with irregularly spaced, long-protruding lobes that give the plant a superficial resemblance to *S. zambiensis* R.M. King & H. Rob. (K.B. 30: 465 (1975)). However, Bally's collection differs from this taxon in that the leaves are not in whorls of 3, phyllaries are wider and more obtuse, achenes are not glabrous, and the inflorescence is a dense corymb as opposed to laxly cymose. Therefore I prefer to retain the specimen within *S. africanus*.

    The altitude figure in brackets refers to the Kirk syntype.

135. AGERATUM

L., Sp. Pl.: 839 (1753); M.F. Johnson in Ann. Missouri Bot. Gard. 58(1): 6 (1971); R.M. King & H. Rob., Genera of the Eupatorieae: 142 (1987); V.S. Sharma in F.R. 98: 557 (1987); Bremer et al. in Asterac., Clad. & Class.: 649 (1994)

Annuals or short-lived sub-shrubby perennials. Leaves opposite or sub-opposite apically, glandular-punctate beneath. Inflorescence of loose to dense terminal corymbs. Involucre campanulate; phyllaries bi-seriate, equal; receptacle broadly conical, epaleate. Corollas tubular to funnel-shaped, tube externally glandular-stipitate, 5-lobed; anthers short with ovate apical appendages; style-arms clavate, papillate. Achenes black, oblong, 5-ribbed, with short setae on the ribs; carpopodium large, white, asymmetrical; pappus of ± 5 awned scales, or reduced, awnless and occasionally absent.

About 44 species, mostly in Central and South America, and some in the West Indies with some species (e.g. *A. conyzoides* L. and *A. houstonianum* Mill.) widely cultivated and weedy.

1. Leaf base obtuse or broadly cuneate, never cordate; phyllaries glabrous to sparsely pilose, eglandular; style-arms exserted for less than 1.6 mm . . . . 1. *A. conyzoides*

Leaf base cordate to truncate; phyllaries hirsute and stipitate-glandular; style-arms exserted for more than 1.8 mm . . . . 2. *A. houstonianum*

    1. **Ageratum conyzoides** *L.*, Sp. Pl.: 839 (1753); Bojer, Hort. Maurit.: 177 (1837); Harv. in Fl. Cap.: 57 (1865); Oliv. & Hiern, F.T.A. 3: 300 (1877); P.O.A. C.: 406 (1895); U.O.P.Z.: 110 (1949); F.P.S. 3: 8 (1956); Humbert, Fl. Madag. 189: 201, fig. 39/10–13 (1960); F.P.U.: 182, fig. 122 (1962); C.D. Adams in F.W.T.A. ed. 2, 2: 287 (1963); E.P.A. 1079 (1966); M.F. Johnson in Ann. Missouri Bot. Gard. 58: 26 (1971); Maquet in Fl. Rwanda 3: 570, fig. 174/1 (1985); H.M. Burkill, Useful Pl. W. Trop. Afr. ed. 2, 1: 443 (1985); E.K. Banda & B. Morris, Common Weeds Malawi: 10, fig. 11 (1986); Lisowski, Asterac. Fl. Afr. Centr.: 473, fig. 95 (1991); D.J.N. Hind in Fl. Mascar. 109: 234, t. 81/1–7 (1993); U.K.W.F. ed. 2: 204, t. 80 (1994). Type: Herb. Clifford: 396, *Ageratum* 1 (BM, lecto., designated by Grierson in Dassanayake & Fosberg, Rev. Handb. Fl. Ceylon 1: 141 (1980))

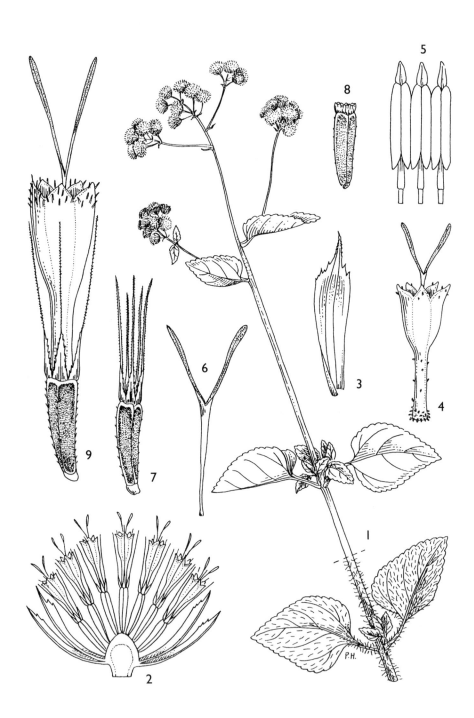

FIG. 181. *AGERATUM CONYZOIDES* — **1**, habit, × ²/₃; **2**, capitulum section, × 8; **3**, phyllary, × 12; **4**, floret, ovary removed, × 18; **5**, stamens, × 24; **6**, style, × 18; **7–8**, achenes, × 12. *AGERATUM HOUSTONIANUM* — **9**, floret, × 18. 1–7 from *McWhirter* 100, 8 from *McWhirter* 102, 9 from *Rouillard* 344. Drawn by Pat Halliday, from Flore des Mascareignes.

Annual herb to 1 m tall, malodorous; stem erect to occasionally decumbent, terete, sparsely pilose and whitish pubescent, especially at nodes, to subglabrous. Leaves opposite to occasionally sub-opposite above, with pubescent petiole 0.3–3 cm, blade ovate to rhomboid, 2.2–8.5 cm long, 1.2–6 cm wide, base obtuse, margins crenate to serrate, apex acute or obtuse, upper surface subglabrous to sparsely pilose, lower surface moderately pilose. Inflorescence much branched, open, of 1–several bunches of ± 7–40 heads borne in open to dense cymose clusters on pubescent and pilose bracteolate peduncles; capitula 4–5 mm in diameter; stalks of individual capitula 2–12 mm, extending in fruit to 37 mm; phyllaries bi-seriate, green, ovate to oblong-lanceolate, usually 2-ribbed, outer series 3.5–4.5 mm long, 0.9–1 mm wide, inner slightly narrower, margins basally entire, erose to lacerate at apex, rarely entire, apex abruptly contracted to point, occasionally short-acuminate, often purple, glabrous to subglabrous to very sparsely pilose. Florets 40–57, corolla tubular to sub-urceolate, 2–2.2 mm long, tube white, limb externally glabrous to sub-glabrous, lobes acute, blue, white, mauve, or purple; anthers basally rounded; style-arms blue, white, mauve or purple, exserted for 0.8–1.2 mm, papillate. Achenes 1.2–1.8 mm long, glabrous between ribs; pappus of 5 triangular scales 1.9–2 mm long, acuminate into a slender awn (see note), margins scabrid. Fig. 181/1–8 (page 828).

UGANDA. Karamoja District: Moroto Township, Oct. 1956, *Wilson* 283!; Kigezi District: Kachwekano Farm, July 1949, *Purseglove* 3010!; Busoga District: 5 km W of Busesa, 10 Oct. 1962, *Lewis* 6028!
KENYA. Northern Frontier: Moyale, 8 July 1952, *Gillett* 13528!; Kericho District: Arroket, North Sotik, July 1961, *Kerfoot* 3107!; Masai District: R. Romosha, Kilgoris, 8 Sep. 1961, *Glover et al.* 2593!
TANZANIA. Pangani District: Boza, 3 July 1956, *Tanner* 2975!; Mpanda District: Utahya, 9 Aug. 1958, *Newbould & Jefford* 1607!; Lindi District: Nachingwea, 16 June 1953, *Anderson* 911!; Zanzibar I., Kisimbani, 27 May 1961, *Faulkner* 2841!
DISTR. U 1–4; K 1–7; T 1–8; Z; P; pantropical
HAB. Weedy escape in cultivation and post-cultivation, often a pioneer on cleared land such as logging tracks, also in grassland and *Acacia* woodland; 0-2150 m
USES. Used by Maasai as adornment (*Glover*)
CONSERVATION NOTES. Least concern (LC)

NOTE. This species shows an incredible range of variation in response to edaphic factors which has led to the description of many infraspecific taxa and the proliferation of superfluous names. However, in East Africa, the key distinguishing features of *A. conyzoides* are sufficiently reliable to allow the species to be treated as one widespread taxon. Johnson (1971) formed a new combination, *A. conyzoides* subsp. *latifolium* (Cav.) M.F.Johnson (synonym *A. latifolium* Cav.), to include plants where the pappus is truncate and awnless (c.f. *A. houstonianum* var. *muticescens* B.L.Rob.), arguing that the taxon did not merit specific status. His decision to relegate to subspecies level was made on the basis that ssp. *latifolium* is diploid, 2n = 20, and ssp. *conyzoides* is tetraploid 2n = 40, therefore hindering production of viable offspring. A survey of the material held at Kew shows collections of *A. conyzoides* in the FTEA area to be of the typical subspecies. There is one possible exception, *Mgaza* 373 (Tanzania, Lushoto District, Dochi Agri Nursery) which has phyllary characters and size of capitula close to *A. conyzoides*, but leaf shape and style-arm length close to *A. houstonianum*. It completely lacks a pappus.
     The earliest effective typification appears to be by Grierson's who indicated Clifford material as type. As discussed by Barrie & al. (Taxon 41: 510. 1992), this material is identifiable as *Eclipta prostrata* (L.) L.; there are no grounds for rejecting the typification. Reveal (in Jarvis & al., Regnum Veg. 127: 15. 1993; a work conceived as a possible Names in Current Use (NCU) List) indicated a Hermann illustration (Parad. Bat.: ic. 161. 1698), which does agree with current usage of the name, as lectotype. However, as the NCU proposals were not adopted, this later typification does not have priority. It appears that, to avoid disruption to the generic names *Eclipta* and *Ageratum* a conservation proposal will be necessary (although the application of these names may be temporarily protected by Art. 57.1). (Charlie Jarvis & Mark Spencer of the Linnaean Project at the British Museum (Natural History) kindly supplied this note and the correct typification – as they have done for all Linnean typifications throughout this volume).

2. **Ageratum houstonianum** *Mill.*, Gard. Dict. ed. 8, no. 2 (1768); B.L. Rob. in Proc. Amer. Acad. Arts 51: 532 (1916) & in Contr. Gray Herb. 68: 6 (1923); C.D. Adams in F.W.T.A. ed. 2, 2: 287 (1968); M.F. Johnson in Ann. Missouri Bot. Gard. 58(1): 21 (1971); H.M. Burkill, Useful Pl. W. Trop. Afr. ed. 2, 1: 445 (1985); Lisowski, Asterac. Afr. Centr.: 477, fig. 96 (1991); D.J.N. Hind in Fl. Mascar. 109: 234, t. 81/9 (1993); U.K.W.F., ed. 2: 204 (1994). Type: Mexico: Vera Cruz, *Houston* s.n. (BM!, holo.).

Usually annual or short-lived perennial, erect to decumbent, simple or sparingly branched herb or sub-shrub to 1 m; stem terete, whitish pubescent and with some long, simple hairs. Leaves opposite to occasionally sub-opposite above, with pubescent petiole 0.4–3.2 cm, blade triangular to ovate, 1.3–5.8 cm long, 1–5.3 cm wide, base cordate to abruptly truncate, margins crenate to serrate, apex acute or obtuse, dark green, moderately hirsute above, a little paler, glabrous to sparsely hirsute beneath. Inflorescence terminal, ± 7–60 heads borne in tight cymose clusters on pubescent and hirsute bracteolate common stalks; capitula 4–7 mm diameter, stalks of individual capitula 0–6 mm extending to 12 mm in fruit; phyllaries green, lanceolate, usually 2-ribbed, outer series 3.5–4.5 mm long, 0.8–0.9 mm wide, inner slightly narrower, margins entire, apex acuminate, often purple, hirsute and glandular-stipitate especially basally. Florets 70–103, corolla-funnel shaped, 2.5–3 mm long, tube white, limb blue, mauve, or white, lobes acute, sparsely to moderately short-pubescent; anthers basally subcordate; style-arms blue, mauve or white, clavate, exserted for ± 2.5 mm, papillate. Achenes 1.2–2 mm long, sub-glabrous, with occasional setae between ribs; pappus 0–3 mm long, of 5 free triangular scales, lacerate, acuminate into a slender awn with scabrid margins, or pappus awnless, scales coroniform, truncate, lacerate, or occasionally pappus absent. Fig. 181/9 (page 828).

Pappus 1.5–3 mm long, of scales always or mostly with apical
  awns . . . .   var. *houstonianum*
Pappus <0.5 mm long, always awnless, much reduced,
  coroniform, or apparently absent . . . .   var. *muticescens*

var. **houstonianum**; R.M. King & H. Rob., Genera of the Eupatorieae: t. 43, p. 143 (1987)

Pappus 1.5–3 mm long, of lacerate scales, acuminate into a slender awn with scabrid margins.

KENYA. Nairobi District: Nairobi R., 22 Aug. 1963, *Thaipu* 136! & 27 Sep. 1951, *Kirika* 251! & 2 Aug. 1978, *Kinyua* 5!
TANZANIA. Lushoto District: Vugiri, 10th June 1963, *Archbold* 185!; Tanga District: E Usambara Mts, Muheza, 25–26 Oct 1986, *Borhidi, Iversen & Steiner* 86159!; Rungwe District: Rungwe Mission, Jan. 1954, *Semsei* 1553!
DISTR. **K** 4; **T** 2, 3, 7; Cameroon, Congo (Kinshasa), Sudan, Ethiopia, Zambia, Malawi, Swaziland, South Africa, Madagascar, Mascarenes; originally from Central America, and widespread as a garden escape
HAB. Weedy escape on stream-banks and moist waste ground; 0-1700 m
USES. None recorded on specimens from our area
CONSERVATION NOTES. Least concern (LC)

NOTE. This species is indigenous to central and southern Mexico, southwards to Central America, but has been introduced into the tropics and subtropics and is a common escape from cultivation.
    The specimens at K all show blue or purplish corolla pigmentation. However, label information states that other colour morphs have been observed. The different colour morphs have previously been assigned infraspecific ranks mainly to enable detection of garden escapes, but these names are now considered superfluous and as such have been relegated to synonymy (see Johnson, 1971, and note under var. *muticescens*).

var. **muticescens** *B.L.Rob.*, Proc. Amer. Acad. Arts 51: 532 (1916). Type: Mexico, Huasteca Province, Wartenberg, near Tantoyuca, *Ervendberg* 100 (MO, holo.)

Identical to typical variety except that pappus is less than 0.5 mm, awnless, truncate, coroniform, or occasionally absent.

KENYA. Meru District: Meru, or Arusha District: Mount Meru, May 1951, *Hancock* 9! (see note)
TANZANIA. Arusha District: Duluti, 27 Sep. 1957, *Carmichael* 656! & Boma road, 23 July 1970, *Mshana* 94!; Lushoto District: Magamba golf course, 2 Sep. 1962, *Semsei* 3518!
DISTR. **K** 4?; **T** 2, 3; Cameroon, Congo (Kinshasa), Ethiopia, Angola, Malawi, South Africa; originally from Central America and elsewhere as a garden escape.
HAB. Weedy escape, roadsides, riverbanks; to 1160 m (altitude based on information from two specimens).
USES. None recorded on specimens from our area
CONSERVATION NOTES. Least concern (LC)

NOTE. The label on the Hancock specimen is ambiguous as it has the **K** 4 locality printed on it, and the **T** 2 locality hand-written on at a later date.
　This variety has not previously been recorded from Africa. It is readily distinguishable by the obvious lack of pappus awns and therefore is retained at varietal level. In 1923 Robinson described the variations seen in cultivated *Ageratum houstonianum* and gave them the rank of forma in order to aid tracing of escaped material. Two colour forms of *A. houstonianum* var. *muticescens* were recognised; the typical blue form, forma *isochroum*, and a whitish form, forma *vesicolor*. Robinson described specimens grown in the Missouri Botanic Gardens which, for the most part, had short and muticous scales but show some variability on one plant or even in the same head, certain florets, especially the central ones and those of the terminal heads tending to have awned scales in the manner of the typical form. This was also observed in a specimen at K from Malawi (Blantyre district, Michiru Hills, 12 July 1988, *Banda & Tawakali* 3320). It is quite probable that hybridization occurs, resulting in intermediate forms. The material from the FTEA region appears to be true, with a high degree of consistency in the pappus morph.

136. AGERATINA

Spach, Hist. Nat. Veg. Phan. 10: 286. 1841; R.M. King & H. Rob., Genera of the Eupatorieae: 42 (1987)

Perennial herbs or shrubs. Leaves opposite. Inflorescence corymbose; involucres campanulate; phyllaries in 2–3 series, sub-equal, often costate; receptacle slightly convex, epaleate. Corollas with slender tube and campanulate limb. Apical anther appendages large, ovate-oblong; style base enlarged, arms linear, papillate. Achenes black, prismatic or fusiform, 5-ribbed; carpopodium large, symmetrical; pappus of usually deciduous barbellate bristles.

About 248 species native to the New World, two of which (*A. adenophora* and *A. riparia*) have become pernicious weeds.

Ageratina adenophora (*Spreng.*) *R.M.King & H.Rob.* in Phytologia 19: 211 (1970); U.K.W.F. ed. 2: 204 (1994); K.T.S.L.: 553, fig. (1994). Type: Mexico, *Humboldt & Bonpland* s.n. (P, holo.)

Shrub to 2 m; stems erect, somewhat woody, branched, with short glandular hairs, becoming more densely pubescent towards the apex, with an annular scar at nodes. Leaves opposite, petiole to 2.5 cm long, large leaves sometimes with additional small leaves in pseudowhorl, blade rhomboid, 1–6.5 cm long, 0.5–4.5 cm wide, margins serrate, apex acuminate, shortly scabrid-hairy, mainly 3-veined from the base. Inflorescence much branched, somewhat leafy, paniculate, with numerous capitula; common stalks long, pubescent; capitula 4–7 mm in diameter; phyllaries bi-seriate, lanceolate, 3.5–5 mm long, 0.5–1 mm wide, 2-ribbed, margins scarious and with longer hairs, apex acute, glandular-hairy. Florets ± 72; corolla 3.5 mm long, white to cream with filiform tube and campanulate limb, tube prominently 5-veined, lobes triangular, ± 0.2 mm long, thickened, sparsely hairy externally; anthers ± 1 mm long,

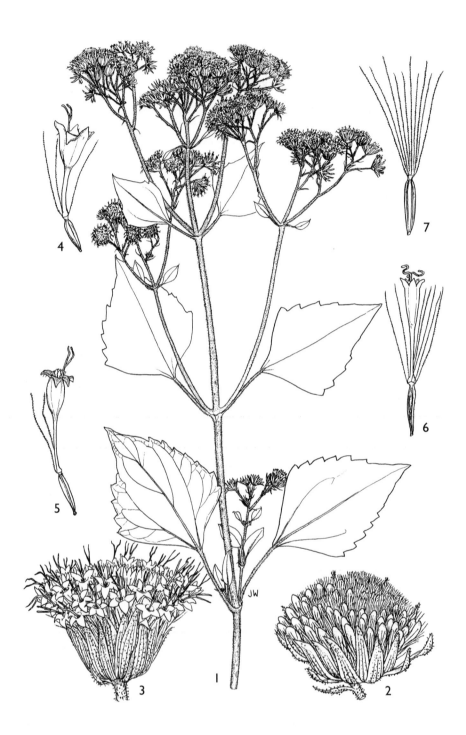

FIG. 182. *AGERATINA ADENOPHORA* — **1**, habit, × ²/₃; **2**, capitulum in bud, × 6; **3**, capitulum, × 6; **4–6**, florets, × 8; **7**, achene, × 8. 1 & 6–7 from *Gillett* 18523, 2–4 from *Gilbert & Mesfin* 6779, 5 from *Smith, Beentje & Muasya* 184. Drawn by Juliet Williamson.

basally saggitate, apically with ovate appendages; style 3.5–4.5 mm long, bifid for 0.5–1.5 mm, style-arms exserted for ± 1.5 mm, papillate. Achenes 1 mm long, 5-ribbed, glabrous, with distinct carpopodium; pappus of ± 10 barbellate, deciduous setae, ± 3 mm long, basally connate, slightly broader towards apex. Fig. 182 (p. 832).

UGANDA. Elgon, Sasa trail at Park border, Sep. 1997, *Lye & Pocs* 23135!
KENYA. Trans Nzoia District: Eldoret–Kitale, halfway from Moi's Bridge to junction with Bungoma road, 8 Oct. 1981, *Gilbert & Mesfin* 6528!; Elgon, near Elephants cave, Aug. 1963, *Tweedie* 2679!; Nakuru District: Turi near Molo, 21 Jan. 1968, *Gillett* 18523!
DISTR. U 3; K 3; Angola, Zambia, South Africa; a fairly widespread weed, although still quite restricted in Africa
HAB. Montane forest and forest margins, lower bamboo zone, riverine and swamp forest; 1950–2600 m
USES. None recorded on specimens from our area
CONSERVATION NOTES. Least concern (LC)

SYN. *Eupatorium glandulosum* Kunth, Nov. Gen. Sp. ed. fol. 4: 96 (1818), *nom. illegit.*; type as for *Ageratina adenophora*; non *E. glandulosum* Michx. (1802)
   E. adenophorum Spreng. in Syst. Veg. Ed. 16, 3: 420 (1826); Blundell, Wild Fl. E. Afr.: 166, fig. 93 (1987)

137. **MIKANIA**

Willd., Sp. Pl. ed. 4, 3: 1742 (1803), *nom. cons.*; B.L. Rob. in Contr. Gray Herb. 104: 55 (1934); R.M. King & H. Rob., Genera of the Eupatorieae: 419 (1987); Bremer, Anderberg, Karis & Lundberg in Asterac. Clad. & Class.: 647 (1994)

Climbing shrubs or sub-shrubs. Leaves opposite, 3–5(–7)-veined, glandular-punctate beneath. Inflorescence cymose to corymbose, much branched. Capitula borne on individual stalks, subtended by a sub-involucral bract; phyllaries 4, in 1 series, often swollen at the base; receptacle flat, glabrous. Florets 4; corolla funnelform or tubular opening into a campanulate 5-lobed limb, glabrous, lobes usually glandular; anther appendages large, ovate to triangular; style-arms long, variously twisted and contorted, papillose, distally tapering. Achenes 5-ribbed, glandular; carpopodium usually evident; pappus of barbellate hairs, often connate at base.

About 430 species, mainly in the New World. All Old World species (excepting *M. carteri* Baker) were grouped under the name *M. cordata* (Burm.f.) B.L.Rob. This taxon was recognised by Robinson as being related to the North American *M. scandens*, differing on account of its more open inflorescence, larger heads and minor characters of the phyllaries and pappus. Holmes retained this name to refer only to Asian-East Indian *M. cordata*.

1. Corolla lobes long (⅓ or more of the corolla); carpopodium rarely evident; common stalk of inflorescence usually trichotomously branched; stalks of individual capitula 2–9 mm long . . . .   1. *M. microptera*
   Corolla lobes shorter; carpopodium evident; common stalk of inflorescence irregularly branched; stalks of individual capitula less than 6 mm long 2
2. Leaves distinctly sagittate, ± 3 times longer than wide, achenes 1–2 mm long . . . .   2. *M. sagittifera*
   Leaves ovate with cordate to auriculate base, less than 3 times longer than wide; achenes 2 mm or more long . .   3. *M. chenopodiifolia*

1. **Mikania microptera** *DC.*, Prodr. 5: 196 (1836); W.C. Holmes in E.J. 103: 237, fig. 12 & 13 (map) (1982); Lisowski, Aster. Afr. Centr. 2: 463, fig. 94 (1991). Type: Brazil, Bahia, *Blanchet* 1710 (G-DC, holo., K!, MO, iso.)

Shrub to 5 m; stems reddish-brown, poorly branched, usually 6-angled, sometimes minutely winged, glabrous to pubescent around nodes and along ridges. Leaves opposite, with petiole 1.5–9 cm, blade ovate, paler green beneath, 3–11.5 cm long, 2.7–11.3 cm wide, base cordate, margins dentate to crenate to minutely toothed, apex acuminate, lamina thin, glabrous to sub-glabrous, glandular-punctate beneath, usually with 3 distinct and 2 less distinct veins originating from base. Inflorescence a fairly lax leafy corymb borne on a pubescent, axillary peduncle, 4–12 cm long, usually conspicuously trichotomously branched; capitula 6–9 mm long, 1.5–2 mm diameter; individual stalks of capitula 2–9 mm, pubescent; phyllaries pale green, linear-lanceolate, boat-shaped, sometimes swollen at base, 4–7 mm long, inner 2 wider than outer 2, glabrous to pubescent, more so towards apex, 3-veined, outer 2 consistently more hairy and sometimes glandular, chaffy. Corolla 3–4 mm long, white, glandular, especially at lobes, lobes accounting for at least $^3/_4$ of limb; anther appendages 1–2 times longer than wide, protruding beyond corolla lobes; style-arms exserted for ± 2 mm. Achenes fawn to black, 3–4 mm long, with scattered glands; carpopodium not evident or obscurely so; pappus 2.5–4 mm long, barbellate, connate at base.

UGANDA. Kabarole District: Kibale Forest, Kanyawara, 19 Sep. 1997, *Eilu* 431!; Mengo District: Entebbe, Sep. 1922, *Maitland* 120!
TANZANIA. Bukoba District: Minziro Forest Reserve, SE of Minziro village, July 2001, *Festo, Bayona & Wilbard* 1612!; Buha District: Kakombe valley, 24 Feb. 1964, *Pirozynski* 442!; Kigoma District: Gombe National Park bottom of Mkenke valley at Lake Tanganyika shore, 27 Feb. 1996, *Gereau, Mbago & Kayombo* 5805!
DISTR. U 2, 4; T 1, 4; Sierra Leone, Liberia, Ivory Coast, Ghana, Benin, Nigeria, Cameroon, Gabon, Central African Republic, Congo (Kinshasa), Angola; South America
HAB. Rain-forest, riverine forest and -thicket, swamp forest; 200–1800 m
USES. None recorded on specimens from our area
CONSERVATION NOTES. Least concern (LC)

SYN. *M. scandens* Willd. var. *microptera* (DC.) Baker in Mart., Fl. Bras. 6(2): 250 (1876)
    [?*M. scandens* sensu Oliv. & Hiern, F.T.A. 3: 301 (1877) pro parte, *non* Willd., Sp. Pl. ed. 4, 3: 1743 (1803)]

NOTE. As discussed by Holmes, *M. microptera* shows the only case of continental disjunction within the genus. There is no evidence that it has been introduced to either South America or Africa. It is retained as a separate species due to the less dense inflorescence, noticeably longer stalks of individual capitula and corolla lobes, and the lower altitudes at which it is found.

2. **Mikania sagittifera** *B.L.Rob.* in Contr. Gray Herb. 104: 68 (1934); W.C. Holmes in E.J. 103: 242, fig. 15 (1982); Maquet in Fl. Rwanda 3: 573, fig. 174/4 (1985); Lisowski, Aster. Afr. Centr. 2: 466 (1991). Type: Botswana, Kabulabula, on the Chobe R., July 1930, *van Son* 28729 (GH, holo., BM!, K!, iso.)

Shrub to 2 m; stem little branched, straw-coloured to reddish brown, terete to sub-angled, glabrous to pubescent, more densely so at nodes, with simple, short, multicellular hairs. Leaves opposite, with petiole 0.6–3.3 cm, blade distinctly sagittate, 1.5–6.3(–9.5) cm long, 0.4–3.2 cm wide, margins minutely toothed to serrate, apex long-acuminate, upper surface glabrous to very sparsely pubescent, lower surface glandular-punctate, glabrous to pubescent usually with 3 distinct and 2 less distinct veins originating from the base. Inflorescence a dense, leafy corymb on an axillary peduncle 3.5–9.3 cm long, branched, clusters of capitula in panicles; capitula 6–8 mm long, 1.5–2 mm diameter; stalks of individual capitula up to 6 mm long, becoming pendulous after fruit dispersal; phyllaries equal, light green, sometimes with lilac tinge, chaffy, lanceolate, boat-shaped, 3-veined, 4–7 mm long, 1–1.5 mm wide, apical margins fimbriate, glabrous to sparsely hairy at base. Corolla 3.5–4.5 mm long, dull white, with scattered glands, lobes acute, less than 1 mm long, margins thickened; anthers purplish, with white, lanceolate appendages, protruding beyond corolla lobes; style-arms exserted for ± 2–3 mm. Achenes brown-black, 1–2 mm long, glandular, carpopodium evident, cream to straw-coloured; pappus 3–5 mm long, barbellate, connate at base, ± spreading.

Uganda. Kigezi District: Lake Mutanda, Mushongero, 1 Feb. 1939, *Loveridge* 461!; Masaka District: Sango Bay, Masaka, 16 Aug. 1951, *Norman* 33! & Lake Nabugabo, Aug. 1935, *Chandler* 1365!
Kenya. Trans Nzoia District: Maboonde, Kitale, June 1966, *Tweedie* 3293!
Tanzania. Karagwe District: Lake Mujanju (Rwakanjunju), 11 July 1975, *Kiss* in *EA* 15912!; Bukoba District: Ishozi, Aug. 1931, *Haarer* 2107!; Mbeya District: Pungaluma Hills, 4 June 1990, *Kayombo* 1045!
Distr. **U** ?1, 2, 4; **K** 3; **T** 1, 4, 7, Cameroon, Congo (Kinshasa), Rwanda, Burundi, Angola, Zambia, Malawi, Zimbabwe, Botswana, Namibia, South Africa
Hab. Swampy forest and papyrus or sedge swamps; 900–1850 m
Uses. None recorded on specimens from our area
Conservation notes. Least concern (LC)

Syn. [*M. scandens* sensu Oliv. & Hiern in F.T.A. 3: 301 (1877) pro parte, *non* Willd.]
    M. scandens Willd. f. *angustifolia* O.Hoffm. in Warburg, Kunene-Sambesi Exped.: 405 (1903). Type: Angola, Lazingua, margins of Longa River, *Baum* 679 (BR, K!, iso.)
    M. angustifolia (O.Hoffm.) R.E.Fr., Wiss. Ergebn. Schwed. Rhod.-Kongo Exp. 1: 328 (1916), *non* Kunth
    [*M. cordata* sensu Lind & Tallantire, F.P.U.: 182 fig. pro parte quoad ic. 121, & U.K.W.F. ed. 2: 204 (1994) pro parte, *non* (Burm. f.) B.L. Rob.]

Note. Three specimens from outside the FTEA area, including the *van Son* isotype material, showed leaves at least 2 cm longer than any other of the specimens examined. They are all from the southern-most extent of the distribution. One of these (*Munro* 14) was also from the margins of the Chobe River in Botswana. The other was collected by *Kolberg & Kubirske* 195, on the banks of the Okavango River, Namibia.

3. **Mikania chenopodiifolia** *Willd.*, Sp. Pl. 3: 1745 (1803); W.C. Holmes in E.J. 103: 221, fig. 5 & 6 (map) (1982); Maquet in Fl. Rwanda 3: 573 (1985); Lisowski, Aster. Afr. Centr. 2: 460 (1991). Type: Sierra Leone, *Thunberg* s.n. (B-W)

Shrub to 5–9 m; stem terete to sub-angled, somewhat branched, sometimes sub-winged, brown to reddish brown, glabrous to pubescent, sometimes hairy at nodes. Leaves with petiole 1.1–7.5 cm, sparsely hairy to pubescent, blade ovate, cordate, 2.2–10 cm long, 1.1–6.7 cm wide, base auriculate, sub-sagittate to hastate, margins dentate to laxly serrate, ± revolute, apex acuminate to long-acuminate, glabrous or sparsely hairy above, glabrous or sparsely hairy to sometimes pubescent and glandular-punctate beneath, usually with 3 distinct and 2 less distinct raised veins originating from the base. Inflorescence a dense, leafy corymb on an axillary peduncle, 2.5–16 cm long, branched, clusters of capitula arranged in panicles; capitula 6–8.5 mm long, 2–3 mm in diameter; stalks of individual capitula 1.5–6 mm; phyllaries green, sometimes with purplish tinge, equal, narrowly lanceolate, boat-shaped, 4.5–8 mm long, sometimes appearing swollen at the base, apical margins shortly fimbriate, glabrous to pubescent, sometimes glandular. Corolla 3.5–5 mm long, white, glandular, especially at lobes, lobes short, triangular, margins thickened; anthers with lanceolate appendages, 1–2 $\frac{1}{2}$ times longer than wide, protruding beyond corolla lobes; style exserted for ± 1–2.6 mm. Achenes fawn to brown or black, 2–3 mm, glandular; small carpopodium evident, cream to straw-coloured; pappus 3–6 mm, minutely barbellate, connate at base. Fig. 183 (page 836).

Uganda. Mengo District, Entebbe, Oct. 1922, *Maitland* 176!; Masaka District: Bugala, Toa, 2 Mar. 1933, *Thomas* 928!; Toro District: Ruwenzori, Bujuku [Bujuhu] valley, 1950, *Osmaston* 3648!
Kenya. Kwale District: Kivukoni entrance to Shimba Hills National Reserve, 23 Nov. 1971, *Bally & Smith* B14358!; Teita District: Tsavo West, 17 Oct. 1962, *Lewis* 6034!; Kwale District: Gongoni forest, 3 June 1990, *Luke & Robertson* 2387!

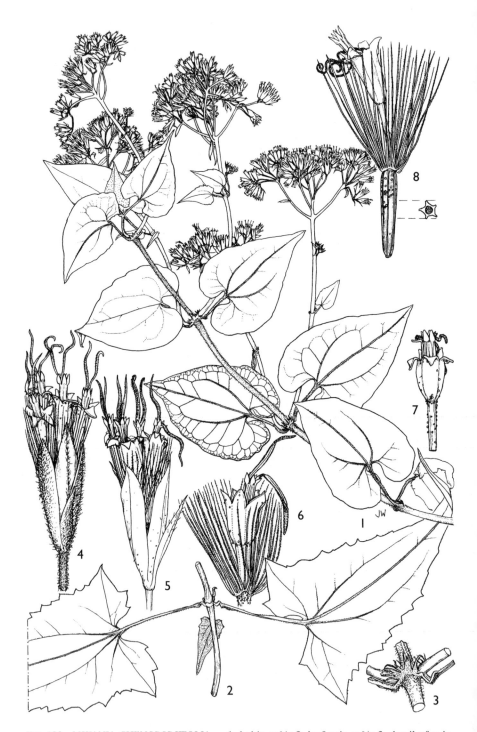

FIG. 183. *MIKANIA CHENOPODIIFOLIA* — **1**, habit, × ²/₃; **2**, leaf pair, × ²/₃; **3**, detail of axis stalk & petioles, × 2; **4–5**, capitulum, × 6; **6–7**, floret, × 8; **8**, achene, × 8. 1 & 3 & 5–6 from *Faulkner* 978, 2 & 8 from *Greenway & Kanuri* 12234, 4 & 7 from *Newbould* 2177. Drawn by Juliet Williamson.

TANZANIA. Mbulu District: Endanok ravine between Babati and Galappo, 26 Aug. 1951, *Welch* 105!; Morogoro District, Uluguru Mts, Bunduki, 17 June 1970, *Batty* 1076!; Kilosa District: Mamboya, May 1930, *Haarer* 1926!; Zanzibar I.: Mwora swamp, 15 July 1997, *Faulkner* 2641!

DISTR. **U** 2–4; **K** 1/4, 3–7; **T** 1–8; **Z**; **P**; Senegal, Gambia, Guinea Bissau, Sierra Leone, Liberia, Ivory Coast, Ghana, Togo, Nigeria, Cameroon, Equatorial Guinea, Bioko, Principe, São Tomé, Annobon, Gabon, Central African Republic, Congo (Kinshasa), Rwanda, Burundi, Sudan, Ethiopia, Angola, Zambia, Malawi, Mozambique, Zimbabwe, South Africa, Madagascar, Comoro Is.

HAB. Widespread, in forest, swamp forest, on river-banks, and bamboo edges; ± 0–3000 m

USES. Leaves eaten by gorilla (*Schaller*)

CONSERVATION NOTES. Least concern (LC)

SYN. *M. capensis* DC., Prodr. 5: 198 (1836); Harv. in Fl. Cap.: 59 (1865); W.C. Holmes in E.J. 130: 214, fig. 1 & 2 (map) (1982); Maquet in Fl. Rwanda 3: 573, fig. 174/3 (1985); Lisowski, Aster. Afr. Centr. 2: 458 (1991). Type: South Africa, [in sylvis Coloniae distr. orient.], *Burchell* 3674 (G-DC, lecto., K!, isolecto.), lectotypified by W.C. Holmes in E.J. 130: 214 (1982)

[*M. scandens* sensu Oliv. & Hiern, F.T.A. 3: 301 (1877); T.T.C.L.: 156 (1949); Humbert, Fl. Madag. 189: 202, fig. 39/5–9 (1960), *non* Willd.]

[*M. cordata* sensu F.P.S. 3: 43 (1956); C.D. Adams in F.W.T.A. ed. 2, 2: 286 (1963); E.P.A. 1080 (1966); ?U.K.W.F. ed. 2, 204 (1994) pro parte, *non* (Burm. f.) B.L. Rob.]

NOTE. Holmes (1982) retained 2 separate taxa here, namely *M. chenopodiifolia* and *M. capensis* which he separates on the size of the capitula and the size of the sub-involucral bract. However, the material studied at K and BM shows a continuous range of variation in both of these characters, so I feel unable to support this view and am reducing *M. capensis* DC. to synonymy here. *M. chenopodiifolia* is widespread in Africa and very variable and although some extremes are seen, intermediates are numerous. Three specimens which have larger capitula characters are: Uganda, Toro District, Ruwenzori, Bujuhu valley, 1950, *Osmaston* 3648 & Ruwenzori Exp., 1893–1894, *Scott Elliot* 9807; Tanzania, Iringa District, Rungwe-Wald, 16 Sep. 1932, *Geilinger* 2368.

Since the publication of Compositae part 1 in 2000 and part 2 in 2002, certain errors have come to light, and several collectors have provided specimens which have extended the ranges of species. I take this opportunity to correct and annotate; I would like to thank Charles Jeffrey and Rafael Govaerts for pointing out some of the following mistakes and/or omissions:

Part 1

page 75: **Taraxacum sp. agg**. also found in Uganda, **U** 2, 4 (Lye, pers. comm.). Kigezi District: Shoko Hill 8 km N of Kabale, Aug. 2001, *Lye & Namaganda* 25178!

page 185: **Vernonia pteropoda** also occurs in **T** 7

page 198: Robinson puts **V. conferta** as a synonym of **V. doniana** DC., treating it under the genus **Brenandendron** H.Rob. 1999, as *Brenandendron donianum* (DC.) H.Rob. This genus is, by the way, only 8 days earlier than its almost-homonym *Brenaniodendron* J.Léonard (Leguminosae). We prefer to maintain **Vernonia** sensu lato for this Flora.

Vernonia luhomeroensis Q. Luke & Beentje has been published from **T** 7; see Kew Bull. 58, 4: 977–980

page 297: Key lead 2a, for phyllaries 5–23 cm long, please read mm!

Range extensions:

Festo has collected **Bothriocline longipes** in **T** 1;

Freidberg has collected a second specimen of **Vernonia sp. E** on Mt Rungwe at 2200 m altitude;

Luke has collected **Lactuca inermis** in **K** 7; **Sonchus schweinfurthii** in **K** 7; **Gutenbergia cordifolia** var. **pulchra** in **T** 7, at 1220 m; **Bothriocline amplifolia** in **T** 7; **Bothriocline longipes** in **K** 7 and **T** 7; **Struchium sparganophora** in **T** 8; **Vernonia zanzibarensis** in **T** 7; **Vernonia thomsoniana** in **T** 7; **Vernonia poskeana** in **K** 3 and **T** 5; **Vernonia nestor** in **T** 7;

Mhoro has collected **Ethulia bicostata** subsp. **meruensis** in **T** 6, at 750 m; **Vernonia popeana** in **T** 6;

Mlangwa has collected **Vernonia cinerea** var. **cinerea** in **T** 3;

Mungai et al. collected **Vernonia anthelmintica** in **K** 7.

Part 2

page 319: **Ammobium alatum** R.Br. (Inuleae) has been cultivated in Nairobi – July 1953, *Verdcourt* 999! It has winged stems and conspicuous white phyllaries, like *Helichrysum.*

page 320: **Blumea bovei** has been put in **Doellia bovei** (DC.) Anderb.; similarly, **Blumea caffra** has been changed to **Doellia cafra** (DC.) Anderb. See Willdenowia 25: 21–23 (1995), where Anderberg states these two taxa are different from *Blumea balsamifera* in style branches, achene epidermis and anthers; and from all other *Blumea* species in the longitudinal red resiniferous ducts on the achenes. **Blumea axillaris** is not mentioned, but Anderberg feels that most *Blumea* are probably nearer *Duhaldea* than to *Blumea* sensu stricto.

page 322: the synonym *Inula bakeriana* is a valid substitute for the synonym *Bojeria vestita.*

page 344: **Anisopappus chinensis** subsp. **buchwaldii** should really be subsp. **africanus** (Hook.f.) Ortiz & Paiva; and var. **buchwaldii** should be var. **macrocephalus** (Humb.) S.Ortiz et al., as autonym rules do not apply here.

page 399: **Pseudognaphalium luteo-album** should read **luteoalbum.**

page 416: the correct authorship for the synonym *Helichrysum fruticosum* is Vatke, not (Forssk.) Vatke, and the name is not illegitimate; the authorship citation for *H. cymosum* subsp. *fruticosum* should be (Vatke) Hedb.

page 418: the varietal name *Gnaphalium globosum* var. *rhodochlamys* is not illegitimate, merely incorrect.

page 423: the correct authorship for the synonym *Helichrysum undatum* is Less., not (J.F.Gmel.) Less.

page 491: in the key to **Conyza**, the rays of **C. bonariensis** are often very difficult to see – and if the rays are overlooked, and the leaves are not lobed, specimens of this taxon would key out to *C. gouanii* (key lead 9).

page 503: **Conyza pedunculata** is an illegitimate name, as there is an earlier *Conyza pedunculata* Mill. (1768). **Conyza boranensis** (S.Moore) Cufod. in B.J.B.B. 36, Suppl.: 1087 (1966), a synonym I overlooked, is probably the best name for this taxon.

page 508: the synonym taxon *Conyza floribunda* Kunth was published in 1818 on p. 57 of the earlier folio edition, vol. 4.

page 517: *Chrysanthemum segetum* should be **Glebionis segetum** (L.) Fourr.; *Chrysanthemum frutescens* should be **Argyranthemum frutescens** (L.) Sch.Bip., and *Chrysanthemum coronarium* should be **Glebionis coronarium** Spach.

page 517: *Mlangwa & Msangi* 1592 from Shengena Forest Reserve is a cultivated specimen, **Argyranthemum gracile** Sch. Bip.

Range extensions:

Festo collected **Laggera brevipes** in **T** 1, and **Gamochaeta purpurea** in **T** 1; *Mlangwa* collected **Helichrysum nitens** in **T** 2;
Muasya et al. collected **Inula paniculata** in **K** 4;
Mwangoka collected **Helichrysum odoratissimum** in **T** 6.

INDEX TO COMPOSITAE

841

Anthemis cotula L., 520
  var. *atromarginata* Vatke, 520
Anthemis cotula auct., 520
Anthemis tigrensis *A.Rich.*, 518
Antunesia O.Hoffm., 156
Arctoteae, 285
Arctotideae *Cass.*, 285, 1, 4
Arctotis lanata Thunb., 287, 288
Arctotis rueppellii (Sch.Bip.) O.Hoffm., 290
Arctotis scaposa (Harv.) O.Hoffm., 287
Arctotis venusta T.Norl., 285
Argyranthemum frutescens (L.) Sch.Bip., 839
Argyranthemum gracile Sch.Bip., 839
Arnica hirsuta Forssk., 13
Arnica piloselloides L., 13
Artemisia *L.*, 528
Artemisia afra *Willd.*, 529
  var. *friesiorum* Chiov., 529
Artemisia arborescens L., 529
Artemisia maderaspatana L., 468
Artemisiopsis *S.Moore*, 393
Artemisiopsis linearis S.Moore, 393
Artemisiopsis villosa (*O.Hoffm.*) *Schweik.*, 393
Ascaricidia adoensis (Walp.) Steetz, 246
Aspilia *Thouars*, 746
Aspilia abyssinica (Sch.Bip.) Vatke, 748
Aspilia africana (*Pers.*) *Adams*, 750, 753
  subsp. *magnifica* (Chiov.) Wild, 750
Aspilia asperifolia O.Hoffm., 749
Aspilia aspilioides (Baker) S.Moore, 753
Aspilia brachyphylla S.Moore, 749
Aspilia brachystephana O.Hoffm., 748
Aspilia chrysops S.Moore, 753
Aspilia ciliata (*Schumach.*) *Wild*, 748, 753
Aspilia congoensis S.Moore, 750, 753
Aspilia dewevrei O.Hoffm., 748
Aspilia fischeri O.Hoffm., 753
Aspilia gillettii Wild, 753
Aspilia gondensis O.Hoffm., 749
Aspilia helianthoides (Schumach. & Thonn.)
  Oliv. & Hiern
  subsp. *ciliata* (Schumach.) Adams, 748
  subsp. *prieuriana* (DC.) Adams, 748
Aspilia holstii Engl., 753
Aspilia involucrata O.Hoffm., 749
Aspilia kotschyi (*Hochst.*) *Oliv.*, 747
  var. **alba** *Berhaut*, 747
  var. kotschyi, 747
Aspilia latifolia Oliv. & Hiern, 750
Aspilia macrorrhiza *Chiov.*, 750
Aspilia mendoncae Wild, 749
Aspilia mildbraedii Muschler, 753
Aspilia monocephala Bak, 753
Aspilia mossambicensis (*Oliv.*) *Wild*, 751, 749
Aspilia multiflora Oliv. & Hiern, 753
Aspilia natalensis (Sond.) Wild, 753
Aspilia natalensis auct., 753
Aspilia pluriseta *Schweinf.*, 749
  subsp. *gondensis* (O.Hoffm.) Wild, 748
Aspilia polycephala S.Moore, 748
Aspilia ritellii Chiov., 753
Aspilia schimperi (A.Rich.) Oliv. & Hiern, 748
Aspilia subpandurata O.Hoffm., 754

Aspilia tanganyikensis Lawalrée, 753
Aspilia vernayi Brenan, 753
Aspilia vulgaris N.E.Br., 749
Aspilia wedeliaeformis Vatke, 753
Aspilia zombensis Baker, 742
  var. *longifolia* S.Moore, 742
Astephania africana Oliv., 347
Aster *L.*, 468
Aster milanjiensis S.Moore, 470
Aster muricatus Thunb., 474
Aster tansaniensis *W.Lippert*, 470
Astereae, 1
Astereae *Cass.*, 457
Asteroideae, 1, 2
Athrixia *Ker-Gawl.*, 454
Athrixia diffusa Baker, 301
Athrixia felicioides Hiern, 349
Athrixia rosmarinifolia (*Walp.*) *Oliv. & Hiern*, 456
  var. foliosa (S.Moore) Kroner, 456
  var. **rosmarinifolia**, 456
Athrixia subsimplex *Brenan*, 456
Athrixia stenophylla Baker, 273
Athroisma *DC.*, 714
Athroisma boranense Cuf., 718
Athroisma gracile (*Oliv.*) *Mattf.*, 716, 718
  subsp. **gracile**, 716
  subsp. **psylloides** (*Oliv.*) *T.Eriksson*, 716
Athroisma gracile × boranense, 718
Athroisma haareri (Dandy) Mattf., 718
Athroisma hastifolium *Mattf.*, 715
Athroisma inevitabile *T.Eriksson*, 718
Athroisma psyllioides (Oliv.) Mattf., 718
Athroisma pusillum *T.Eriksson*, 715
Athroisma stuhlmannii *O.Hoffm.*, 718, 719
Austrosynotis *C.Jeffrey*, 611
Austrosynotis rectirama (*Baker*) *C.Jeffrey*, 611

Baccharis dioscoridis L., 363
Baccharis ovalis Pers., 366
Baccharis resiniflua DC., 516
Baccharis senegalensis Pers., 180
Baccharoides Moench, 159
Baccharoides adoensis (Walp.) H.Rob., 246
Baccharoides calvoana (Hook.f.) Isawumi, 238
  subsp. *calvoana*
    var. *hymenolepis* (A.Rich.) Isawumi, 241
Baccharoides lasiopus (O.Hoffm.) H.Rob., 244
Baccharoides pumila (Kotschy & Peyr.) Isawumi, 249
Baccharoides tayloriana Isawumi, 313
Barkhausia adenothrix A.Rich., 71
Barkhausia carbonaria (Sch.Bip.) A.Rich., 72
Barkhausia schimperi A.Rich., 72
Barkhausia schultzii A.Rich., 73
Barnadesieae, 1, 2, 3
Barnadesiinae, 3
Barnadesioideae, 1
Berkheya Ehrh., 290
Berkheya antunesii O.Hoffm., 300
Berkheya bipinnatifida (*Harv.*) *Roessler*, 292
  subsp. bipinnatifida, 293
  subsp. **echinopsoides** (*Baker*) *Roessler*, 292

var. **cordata** *G.V.Pope*, 293
   var. **echinopsoides** (*Baker*) *G.V.Pope*, 293
Berkheya echinacea (*Harv.*) *Burtt Davy*, 291
   subsp. echinacea, 292
   subsp. **polyacantha** (*Baker*) *Roessler*, 291
Berkheya echinopsoides Baker, 293
Berkheya gracilis O.Hoffm., 301
Berkheya johnstoniana Britten, 296
Berkheya parvifolia Baker, 291
Berkheya polyacantha Baker, 291
Berkheya spekeana *Oliv.*, 294
Berkheya subteretifolia Thell., 294
Berkheya zeyheri *Oliv. & Hiern*, 293
   subsp. rehmannii (Thell.) Roessler, 294
   subsp. zeyheri, 294
Berkheyopsis O.Hoffm., 296
Berkheyopsis diffusa O.Hoffm., 300
Bidens *L.*, 778
Bidens abyssinica Sch.Bip., 806
   var. *glabrata* Vatke, 806
   var. *incisifolia* Chiov., 806
Bidens acuticaulis *Sherff*, 802
   var. **acuticaulis**, 802
   var. **filirostris** (*P.Taylor*) *T.G.J.Rayner*, 803
Bidens acutiloba Sherff, 804
Bidens amoena Sherff, 798
Bidens angustata Sherff, 794
Bidens articulata Sherff, 794
Bidens baumii (*O.Hoffm.*) *Sherff*, 788
Bidens bequaertii De Wild., 789
Bidens bipinnata L., 806
Bidens bipinnata auctt. mult., 806
Bidens biternata (*Lour.*) *Merr. & Sherff*, 804
   var. *glabrata* (Vatke) Sherff, 807
     forma *abyssinica* (Sch.Bip.) Sherff, 806
     forma *lasiocarpa* (O.E.Schultz)Sherff, 806
   var. *glabrata* auct., 807
Bidens bruceae Sherff, 786
   var. *pubescentior* Sherff, 786
   var. *swynnertonii* Sherff, 786
Bidens buchneri (*Klatt*) *Sherff*, 785
Bidens caudata (Kunth) Sch.Bip., 810
Bidens chandleri Sherff, 792
Bidens chinensis (L.) De Wild.
   var. *abyssinicus* (Sch.Bip.) O.E.Schultz, 806
Bidens ciliata B.Wild., 803
Bidens ciliata Fisch. & Mey., 803
Bidens cinerea *Sherff*, 794
   var. *tricuspidata* Sherff, 795
Bidens cinereoides Sherff, 795
Bidens coriacea (O.Hoffm.) Sherff, 786
Bidens crataegifolia (O.Hoffm.) Sherff, 785
   var. *burttii* Sherff, 785
Bidens crocea *O.Hoffm.*, 789
Bidens cuspidata Sherff, 785
Bidens cylindrica Sherff, 806
Bidens dielsii Sherff, 793
   var. *incisior* Sherff, 793
   var. *intermedia* Sherff, 787
   var. *medusoides* Sherff, 793
Bidens diversa *Sherff*, 803
   subsp. *filiformis* (Sherff) T.G.J.Rayner, 803

Bidens dolosa Sherff, 788
Bidens drummondii Wild, 802
Bidens elgonensis (*Sherff*) *Agnew*, 798
   subsp. *cheranganiensis* T.G.J.Rayner, 799
   subsp. *morotonensis* (Sherff) T.G.J.Rayner, 799
Bidens elliottii (*S.Moore*) *Sherff*, 798
Bidens exilis (Sherff) Lisowski, 787
Bidens filiformis Sherff, 803
Bidens fischeri (*O.Hoffm.*) *Sherff*, 796
Bidens flagellata (*Sherff*) *Mesfin*, 796
Bidens formosa (Bonato) Sch.Bip., 808
Bidens gardullensis Cufod., 793
Bidens gracilior (O.Hoffm.) Sherff, 795
   var. *ukewensis* Sherff, 795
Bidens grantii (*Oliv.*) *Sherff*, 800, 801
   var. *dawei* Sherff, 800
   var. *scaettae* Sherff, 800
   var. *stapfioides* Sherff, 804
Bidens hildebrandtii *O.Hoffm.*, 782
   var. *boranensis* Lanza, 783
Bidens hoffmannii Sherff, 794
   var. *angustata* Sherff, 794
Bidens holstii (*O.Hoffm.*) *Sherff*, 787
   var. *rupestris* Sherff, 787
Bidens incumbens Sherff, 783
   var. *muthicola* Sherff, 783
Bidens insignis Sherff, 785
Bidens isokoensis Sherff, 802
Bidens jacksonii (S.Moore) Sherff, 774
Bidens kasaiensis Lisowski, 789
Bidens kigeziensis Sherff, 798
   var. *subsessilis* Sherff, 799
Bidens kilimandscharica (*O.Hoffm.*) *Sherff*, 783
   var. *oxymera* Sherff, 785
   var. *retrorsa* Sherff, 785
Bidens kirkii (Oliv. & Hiern) Sherff, 796
   var. *ciliato-vaginata* Cufod., 796
   var. *flagellata* Sherff, 796
Bidens kirkii auct., 796
Bidens kivuensis Sherff, 800
   var. *armata* Sherff, 800
Bidens kotschyi Sch.Bip., 806
Bidens leptoglossa (Sherff) Lisowski, 785
Bidens lindblomii Sherff, 783
Bidens linearifolia Sherff, 792
Bidens lineariloba *Oliv.*, 797
   var. *deminuta* Sherff, 797
Bidens lineata Sherff, 795
   var. *tenuipes* Sherff, 795
Bidens lynesii Sherff, 788
Bidens magnifolia *Sherff*, 788
   var. *versuta* Sherff, 788
Bidens meruensis Sherff, 785
Bidens microcarpa Sherff, 804
Bidens morotonensis (Sherff) Agnew, 799
Bidens napierae Sherff, 787
Bidens natator Friis & Vollesen, 800
Bidens navicularia Sherff, 800
Bidens negriana (Sherff) Cufod., 799, 807
Bidens neumannii Sherff, 793

Helichrysum stenopterum *DC.*, 409, 410, 412
Helichrysum steudelii A.Rich., 441
Helichrysum stuhlmannii *O.Hoffm.*, 441
   var. **aberdaricum** F.R.Fr., 448
   var. *ducisaprutii* (Chiov.) Moeser, 441
   var. **keniense** F.R.Fr., 448
   var. *latifolium* De Wild., 442
   var. *rigidum* Moeser, 441
Helichrysum sulphureofuscum *Baker*, 433
Helichrysum taxon A, 450
Helichrysum taylorii S.Moore, 431
Helichrysum theresae Lisowski, 426
Helichrysum tillandsiifolium *O.Hoffm.*, 412, 452
Helichrysum tithonioides Wild., 449
Helichrysum traversii *Chiov.*, 413
Helichrysum uhligii *Moeser*, 452
Helichrysum undatum Less., 423, 839
Helichrysum velatum Moeser, 422
Helichrysum verbascifolium S.Moore, 426
Helichrysum vernonioides Wild, 426
Helichrysum volkensii O.Hoffm., 444, 446
Helichrysum wittei *Hutch. & B.L.Burtt*, 448, 452
Helichrysum wollastonii S.Moore, 446
Helichrysum xanthosphaerum Baker, 431
Helichrysum zairense Lisowski, 435
Helichrysum zombense Moeser, 426
Herderia nyiroensis Buscal. & Muschl., 123
Herderia somalensis O.Hoffm., 123
Herderia stellulifera Benth., 212
Hieraciodes oliverianum Kuntze, 70
Hieracium capense L., 66
Hilliardiella H.Rob., 159
Hilliardiella aristata (DC.) H.Rob., 205
Hilliardiella calyculata (S.Moore) H.Rob., 208
Hilliardiella oligocephala (DC.) H.Rob., 207
Hilliardiella smithiana (Less.) H.Rob., 207
Hippia integrifolia L.f., 459
Hirpicium *Cass.*, 296
Hirpicium angustifolium (*O.Hoffm.*) *Roessler*, 297
Hirpicium antunesii (*O.Hoffm.*) *Roessler*, 300, 301
Hirpicium beguinotii (Lanza) Cufod., 300
Hirpicium diffusum (*O.Hoffm.*) *Roessler*, 298, 301
Hirpicium gracile (*O.Hoffm.*) *Roessler*, 300
Hoehnelia Schweinf., 111
Hoehnelia vernonioides Schweinf., 119
Homalocline schimperi (A.Rich.) Schweinf., 92
Hymenatherum tenuilobum DC, 706
Hypericophyllum *Steetz*, 719
Hypericophyllum angolense (*O.Hoffm.*) *N.E.Br*, 719
Hypericophyllum compositarum *Steetz*, 720
Hypericophyllum elatum (*O.Hoffm.*) *N.E.Br.*, 720
Hypericophyllum scabridum N.E.Br., 720
Hypochaeris *L.*, 73
Hypochaeris glabra *L.*, 73
Hypochaeris radicata L., 73
Hystrichophora *Mattf.*, 284

Hystrichophora macrophylla *Mattf.*, 284

Inula *L.*, 322
Inula acervata S.Moore, 325
Inula bakeriana O.Hoffm., 323, 838
Inula bequaertii De Wild., 325
Inula decipiens E.A.Bruce, 325
Inula eminii (*O.Hoffm.*) *O.Hoffm.*, 328
Inula glomerata *Oliv. & Hiern*, 324, 325
Inula indica L., 334
Inula leptoclada Webb., 336
Inula macrophylla Kar. & Kir, 325
Inula macrophylla (A.Rich.) Sch.Bip., 325
Inula mannii (*Hook.f.*) *Oliv. & Hiern*, 326, 328
Inula paniculata (*Klatt*) *Burtt Davy*, 325, 324, 839
Inula rungwensis *Beentje*, 328
Inula shirensis *Oliv.*, 323
Inula stolzii *Mattf.*, 323, 328
Inula stuhlmannii O.Hoffm., 329
Inula subscaposa *S.Moore*, 324
Inulaster macrophyllus A.Rich., 325
Inuleae *Cass.*, 1, 2, 3, 315
Iphiona rotundifolia Oliv. & Hiern, 332
Iphionopsis *Anderb.*, 329
Iphionopsis rotundifolia (*Oliv. & Hiern*) *Anderb.*, 329

Jaumea angolensis O.Hoffm., 720
Jaumea compositarum (Steetz) Benth., 720
Jaumea elata O.Hoffm., 722
Jaumea johnstonii Baker, 720
Jeffreya Cabrera, 479
Jeffreya *Wild*, 457
Jeffreya decurrens (L.) Cabrera, 393
Jeffreya palustris (*O.Hoffm.*) *Wild.*, 477
Jeffreya petitiana (*Lisowski*) *Beentje*, 477

Kleinia *Mill.*, 680
Kleinia abyssinica (*A.Rich.*) *A.Berger*, 694
   var. **abyssinica**, 696
   var. **hildebrandtii** (*Vatke*) *C.Jeffrey*, 696
Kleinia amaniensis (*Engl.*) *A.Berger*, 688
Kleinia barbertonica (Klatt) Burtt Davy, 648
Kleinia barbertonicus auct., 648
Kleinia breviflora C.Jeffrey, 683
Kleinia coccinea (Oliv. & Hiern) A.Berger, 694
Kleinia gregorii (*S.Moore*) *C.Jeffrey*, 683
Kleinia implexa (*Bally*) *C.Jeffrey*, 685
Kleinia kleinioides (Sch.Bip.) M.R.F.Tayl., 683, 684
Kleinia kleinioides auctt., 684
Kleinia leptophylla *C.Jeffrey*, 690
Kleinia longiflora DC., 683, 685
Kleinia longiflora auctt., 685
Kleinia mweroensis (*Baker*) *C.Jeffrey*, 689
Kleinia negrii Cufod., 683
Kleinia odora (*Forssk.*) *DC.*, 684
Kleinia oligodonta C.Jeffrey, 687
Kleinia patriciae *C.Jeffrey*, 690
Kleinia pendula (*Forssk.*) *DC.*, 692
Kleinia petraea (*R.E.Fr.*) *C.Jeffrey*, 688
Kleinia picticaulis (*Bally*) *C.Jeffrey*, 689

New names validated in this part

Adenostemma caffrum *DC.* var. **longifolium** (*Chiov.*) *S.A.L.Smith* **comb. nov.**
Crassocephalum rubens (*Jacq.*) *S.Moore* var. **sarcobasis** (*DC.*) *C.Jeffrey & Beentje* **comb. nov.**
Melanthera pungens *Oliv. & Hiern* var. **albinervia** (*O.Hoffm.*) *Beentje* **comb. et stat. nov.**

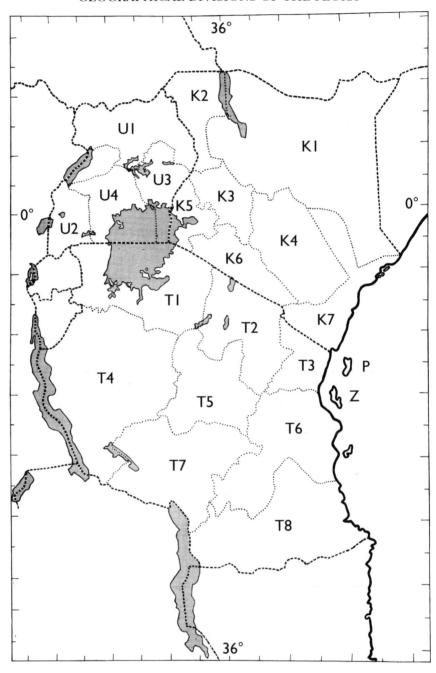

PLANTS PEOPLE
POSSIBILITIES

First published in 2005 by
Royal Botanic Gardens, Kew
Richmond, Surrey, TW9 3AB, UK
www.kew.org

ISBN 1 84246 106 0

Design by Media Resources, typesetting and page layout by Margaret Newman,
Information Services Department,
Royal Botanic Gardens, Kew.

For information or to purchase all Kew titles please visit
www.kewbooks.com or email publishing@kew.org

LIST OF ABBREVIATIONS

A.V.P. = O. Hedberg, Afroalpine Vascular Plants; **B.J.B.B.** = Bulletin du Jardin Botanique de l'Etat, Bruxelles; Bulletin du Jardin Botanique Nationale de Belgique; **B.S.B.B.** = Bulletin de la Société Royale de Botanique de Belgique; **C.F.A.** = Conspectus Florae Angolensis; **E.J.** = A. Engler, Botanische Jahrbücher für Systematik, Pflanzengeschichte und Pflanzengeographie; **E.M.** = A. Engler, Monographieen Afrikanischer Pflanzen-Familien und Gattungen; **E.P.** = A. Engler, Das Pflanzenreich; **E.P.A.** = G. Cufodontis, Enumeratio Plantarum Aethiopiae Spermatophyta; in B.J.B.B. 23, Suppl. (1953) et seq.; **E. & P. Pf.** = A. Engler & K. Prantl, Die Natürlichen Pflanzenfamilien; **F.A.C.** = Flore d'Afrique Centrale (*formerly* F.C.B.); **F.C.B.** = Flore du Congo Belge et du Ruanda-Urundi; Flore du Congo, du Rwanda et du Burundi; **F.E.E.** = Flora of Ethiopia & Eritrea; **F.D.-O.A.** = A. Peter, Flora von Deutsch-Ostafrika; **F.F.N.R.** = F. White, Forest Flora of Northern Rhodesia; **F.P.N.A.** = W. Robyns, Flore des Spermatophytes du Parc National Albert; **F.P.S.** = F.W. Andrews, Flowering Plants of the Anglo-Egyptian Sudan *or* Flowering Plants of the Sudan; **F.P.U.** = E. Lind & A. Tallantire, Some Common Flowering Plants of Uganda; **F.R.** = F. Fedde, Repertorium Speciorum Novarum Regni Vegetabilis; **F.S.A.** = Flora of Southern Africa; **F.T.A.** = Flora of Tropical Africa; **F.W.T.A.** = Flora of West Tropical Africa; **F.Z.** = Flora Zambesiaca; **G.F.P.** = J. Hutchinson, The Genera of Flowering Plants; **G.P.** = G. Bentham & J.D. Hooker, Genera Plantarum; **G.T.** = D.M. Napper, Grasses of Tanganyika; **I.G.U.** = K.W. Harker & D.M. Napper, An Illustrated Guide to the Grasses of Uganda; **I.T.U.** = W.J. Eggeling, Indigenous Trees of the Uganda Protectorate; **J.B.** = Journal of Botany; **J.L.S.** = Journal of the Linnean Society of London, Botany; **K.B.** = Kew Bulletin, *or* Bulletin of Miscellaneous Information, Kew; **K.T.S.** = I. Dale & P.J. Greenway, Kenya Trees and Shrubs; **K.T.S.L.** = H.J. Beentje, Kenya Trees, Shrubs and Lianas; **L.T.A.** = E.G. Baker, Leguminosae of Tropical Africa; **N.B.G.B.** = Notizblatt des Botanischen Gartens und Museums zu Berlin-Dahlem; **P.O.A.** = A. Engler, Die Pflanzenwelt Ost-Afrikas und der Nachbargebiete; **R.K.G.** = A.V. Bogdan, A Revised List of Kenya Grasses; **T.S.K.** = E. Battiscombe, Trees and Shrubs of Kenya Colony; **T.T.C.L.** = J.P.M. Brenan, Check-lists of the Forest Trees and Shrubs of the British Empire no. 5, part II, Tanganyika Territory; **U.K.W.F.** = A.D.Q. Agnew (or for ed. 2, A.D.Q. Agnew & S. Agnew), Upland Kenya Wild Flowers; **U.O.P.Z.** = R.O. Williams, Useful and Ornamental Plants in Zanzibar and Pemba; **V.E.** = A. Engler & O. Drude, Die Vegetation der Erde, IX, Pflanzenwelt Afrikas; **W.F.K.** = A.J. Jex-Blake, Some Wild Flowers of Kenya; **Z.A.E.** = Wissenschaftliche Ergebnisse der Deutschen Zentral-Afrika-Expedition 1907–1908, 2 (Botanik).

FAMILIES OF VASCULAR PLANTS REPRESENTED IN
THE FLORA OF TROPICAL EAST AFRICA

The family system used in the Flora has diverged in some respects from that now in use at Kew and the herbaria in East Africa. The accepted family name of a synonym or alternative is indicated by the word "see". Included family names are referred to the one used in the Flora by "in" if in accordance with the current system, and "as" if not. Where two families are included in one fascicle the subsidiary family is referred to the main family by "with".

PUBLISHED PARTS

Foreword and preface
*Glossary
Index of Collecting Localities

Acanthaceae
  Part 1
*Actiniopteridaceae
*Adiantaceae
Aizoaceae
Alangiaceae
Alismataceae
*Alliaceae
*Aloaceae
*Amaranthaceae
*Amaryllidaceae
*Anacardiaceae
*Ancistrocladaceae
Anisophyllaceae — as Rhizophoraceae
Annonaceae
*Anthericaceae
Apiaceae — see Umbelliferae
Apocynaceae
  *Part 1
*Aponogetonaceae
Aquifoliaceae
*Araceae
Araliaceae
Arecaceae — see Palmae
*Aristolochiaceae
Asparagaceae
*Asphodelaceae
Aspleniaceae
Asteraceae — see Compositae
Avicenniaceae — as Verbenaceae
*Azollaceae

*Balanitaceae
*Balanophoraceae

*Balsaminaceae
Basellaceae
Begoniaceae
Berberidaceae
Bignoniaceae
Bischofiaceae — in Euphorbiaceae
Bixaceae
Blechnaceae
*Bombacaceae
*Boraginaceae
Brassicaceae — see Cruciferae
Brexiaceae
Buddlejaceae — as Loganiaceae
*Burmanniaceae
*Burseraceae
Butomaceae
Buxaceae

Cabombaceae
Cactaceae
Caesalpiniaceae — in Leguminosae
*Callitrichaceae
Campanulaceae
Canellaceae
Cannabaceae
Cannaceae — with Musaceae
Capparaceae
Caprifoliaceae
Caricaceae
Caryophyllaceae
*Casuarinaceae
Cecropiaceae — with Moraceae
*Celastraceae
*Ceratophyllaceae
Chenopodiaceae
Chrysobalanaceae — as Rosaceae
Clusiaceae — see Guttiferae
Cobaeaceae — with Bignoniaceae
Cochlospermaceae

Papaveraceae
Papilionaceae — in Leguminosae
*Parkeriaceae
Passifloraceae
Pedaliaceae
Periplocaceae — see Apocynaceae (Part 2)
Phytolaccaceae
*Piperaceae
Pittosporaceae
Plantaginaceae
Plumbaginaceae
Poaceae — see Gramineae
Podocarpaceae
Podostemaceae
Polemoniaceae — see Cobaeaceae
Polygalaceae
Polygonaceae
*Polypodiaceae
Pontederiaceae
*Portulacaceae
Potamogetonaceae
Primulaceae
*Proteaceae
*Psilotaceae
*Ptaeroxylaceae
*Pteridaceae

*Rafflesiaceae
Ranunculaceae
Resedaceae
Restionaceae
Rhamnaceae
Rhizophoraceae
Rosaceae
Rubiaceae
  Part 1
  *Part 2
  *Part 3
*Ruppiaceae
*Rutaceae

*Salicaceae
Salvadoraceae
*Salviniaceae
Santalaceae
*Sapindaceae
Sapotaceae
*Schizaeaceae
Scrophulariaceae

Scytopetalaceae
Selaginellaceae
Selaginaceae — in Scrophulariaceae
*Simaroubaceae
*Smilacaceae
Sonneratiaceae
Sphenocleaceae
Strychnaceae — in Loganiaceae
*Surianaceae
Sterculiaceae

Taccaceae
Tamaricaceae
Tecophilaeaceae
Ternstroemiaceae — in Theaceae
Tetragoniaceae — in Aizoaceae
Theaceae
Thelypteridaceae
Thismiaceae — in Burmanniaceae
Thymelaeaceae
*Tiliaceae
Trapaceae
Tribulaceae — in Zygophyllaceae
*Triuridaceae
Turneraceae
Typhaceae

Uapacaceae — in Euphorbiaceae
Ulmaceae
*Umbelliferae
*Urticaceae

Vacciniaceae — in Ericaceae
Valerianaceae
Velloziaceae
*Verbenaceae
*Violaceae
*Viscaceae
*Vitaceae
*Vittariaceae

*Woodsiaceae

*Xyridaceae

*Zannichelliaceae
*Zingiberaceae
*Zosteraceae
*Zygophyllaceae

FORTHCOMING PARTS

Acanthaceae
Part 2
Apocynaceae
Part 2

Asclepiadaceae — see Apocynaceae
Commelinaceae
Cyperaceae
Solanaceae

Editorial adviser, National Museums of Kenya: Quentin Luke
Adviser on Linnaean types: C. Jarvis

Parts of this Flora, unless otherwise indicated, are obtainable from:
Royal Botanic Gardens, Kew, Richmond, Surrey TW9 3AB, England. www.kew.org or www.kewbooks.com

* only available through CRC Press at:
UK and Rest of World (except North and South America):
CRS Press/ITPS,
Cheriton House, North Way, Andover, Hants SP10 5BE.
e: uk.tandf@thomsonpublishingservices. co.uk

North and South America:
CRC Press,
2000NW Corporate Blvd, Boco Raton, FL 33431-9868,
USA.
e: orders@crcpress.com

Information on current prices can be found at www.kewbooks.com or www.tandf.co.uk/books/